T0289886

Fungal Infections: An Issue of Infectious Disease Clinics

Fungal Infections: An Issue of Infectious Disease Clinics

Editor: Sanda Warne

Cataloging-in-Publication Data

Fungal infections : an issue of infectious disease clinics / edited by Sanda Warne.
p. cm.
Includes bibliographical references and index.
ISBN 978-1-63927-673-8
1. Mycoses. 2. Mycoses--Diagnosis. 3. Mycoses--Treatment.
4. Communicable diseases. 5. Infection. I. Warne, Sanda.
RC117 .F863 2023
616.969--dc23

American Medical Publishers,
41 Flatbush Avenue,
1st Floor, New York,
NY 11217, USA

ISBN 978-1-63927-673-8 (Hardback)

Contents

Preface

The world is advancing at a fast pace like never before. Therefore, the need is to keep up with the latest developments. This book was an idea that came to fruition when the specialists in the area realized the need to coordinate together and document essential themes in the subject. That's when I was requested to be the editor. Editing this book has been an honour as it brings together diverse authors researching on different streams of the field. The book collates essential materials contributed by veterans in the area which can be utilized by students and researchers alike.

Fungal infection, also known as mycosis, is a disease that occurs due to fungi. On the basis of the body part affected by fungi, fungal infections are divided into systematic, superficial and subcutaneous. Symptoms of fungal infections range widely, generally a rash can appear in case of a superficial infection. There can be changes and lumps beneath the skin. Meningitis or symptoms similar to pneumonia can occur during a systematic or deeper infection. Fungal infections can happen when a spore is breathed in or comes in interaction with the skin, or when it enters the body via an injection, wound or cut. Treatment usually involves antifungal medicines, which can be an injection, oral medication or a cream, based on the type and extent of infection. In certain cases, treatment requires surgery to cut out infected tissue. This book is a compilation of chapters that discuss the most vital aspects of fungal diseases. It presents researches and studies performed by experts across the globe. Those in search of information to further their knowledge will be greatly assisted by this book.

Each chapter is a sole-standing publication that reflects each author´s interpretation. Thus, the book displays a multi-facetted picture of our current understanding of application, resources and aspects of the field. I would like to thank the contributors of this book and my family for their endless support.

Editor

Estimated Burden of Fungal Infections in Namibia

Cara M. Dunaiski [1,*] **and David W. Denning** [2]

[1] Department of Health and Applied Sciences, Namibia University of Science and Technology, 13 Jackson Kaujeua Street, Windhoek 9000, Namibia

[2] National Aspergillosis Centre, Wythenshawe Hospital and the University of Manchester, Manchester M23 9LT, UK

* Correspondence: cdunaiski@nust.na

Abstract: Namibia is a sub-Saharan country with one of the highest HIV infection rates in the world. Although care and support services are available that cater for opportunistic infections related to HIV, the main focus is narrow and predominantly aimed at tuberculosis. We aimed to estimate the burden of serious fungal infections in Namibia, currently unknown, based on the size of the population at risk and available epidemiological data. Data were obtained from the World Health Organization (WHO), Joint United Nations Programme on HIV/AIDS (UNAIDS), and published reports. When no data existed, risk populations were used to estimate the frequencies of fungal infections, using the previously described methodology. The population of Namibia in 2011 was estimated at 2,459,000 and 37% were children. Among approximately 516,390 adult women, recurrent vulvovaginal candidiasis (≥4 episodes /year) is estimated to occur in 37,390 (3003/100,000 females). Using a low international average rate of 5/100,000, we estimated 125 cases of candidemia, and 19 patients with intra-abdominal candidiasis. Among survivors of pulmonary tuberculosis (TB) in Namibia 2017, 112 new cases of chronic pulmonary aspergillosis (CPA) are likely, a prevalence of 354 post-TB and a total prevalence estimate of 453 CPA patients in all. Asthma affects 11.2% of adults, 178,483 people, and so allergic bronchopulmonary aspergillosis (ABPA) and severe asthma with fungal sensitization (SAFS) were estimated in approximately 179/100,000 and 237/100,000 people, respectively. Invasive aspergillosis (IA) is estimated to affect 15 patients following leukaemia therapy, and an estimated 0.13% patients admitted to hospital with chronic obstructive pulmonary disease (COPD) (259) and 4% of HIV-related deaths (108) — a total of 383 people. The total HIV-infected population is estimated at 200,000, with 32,371 not on antiretroviral therapy (ART). Among HIV-infected patients, 543 cases of cryptococcal meningitis and 836 cases of *Pneumocystis* pneumonia are estimated each year. Tinea capitis infections were estimated at 53,784 cases, and mucormycosis at five cases. Data were missing for fungal keratitis and skin neglected fungal tropical diseases such as mycetoma. The present study indicates that approximately 5% of the Namibian population is affected by fungal infections. This study is not an epidemiological study—it illustrates estimates based on assumptions derived from similar studies. The estimates are incomplete and need further epidemiological and diagnostic studies to corroborate, amend them, and improve the diagnosis and management of these diseases.

Keywords: Namibia; HIV/AIDS; fungal infections; opportunistic infections; pulmonary infections

---

1. Introduction

Namibia is a middle-income country in southern Africa, with a population of approximately 2.5 million inhabitants [1]. It consists of five geographical areas: the Central Plateau, the Namib Desert, the Great Escarpment, the Bushveld, and the Kalahari Desert. Together, these areas make Namibia one of the most arid landscapes south of the Sahara. Because of its location between the Namib and Kalahari deserts, the country has the least rainfall in sub-Saharan Africa.

The climate in Namibia is arid, semi-arid, and subtropical (in the northernmost regions), with a large temperature range (between 5 and 20 °C). Fog can occur along the temperate desert coast. The hottest months of the year are January and February. Average daytime temperatures range from 9 to 30 °C. During the winter months, May to September, temperatures can fluctuate from between −6 and 10 °C at night to 20 °C in the day. Winter days are generally clear, cloudless, and sunny. Frost can occur over large areas of the country during the coldest months. Overall, Namibia experiences rainfall in the summer, with limited showers beginning in October and continuing until April [2].

The Namibian health system has a public health service run by the Ministry of Health and Social Services (MoHSS) and a relatively well-established private health sector [3]. Windhoek, the capital city of Namibia, is the main referral centre with generally good health services. The Windhoek State Hospital, illustrated in Figure 1, is Namibia's main referral hospital. However, these services are deficient in the rural and remote populations of the country. All public pathology is managed by the Namibian Institute of Pathology—approximately 30% of the private healthcare facilities, is also managed by this institution [3].

Figure 1. The Windhoek Central Hospital.

Mycotic infections pose an increasing threat to public health for several reasons. Opportunistic infections such as *Pneumocystis* pneumonia (PCP), candidiasis, cryptococcosis and aspergillosis are becoming increasingly problematic as the number of people with weakened immune systems increases, particularly people with HIV/AIDS or cancer [4–8]. In 2017, 8% of the Namibian population was HIV positive [9], which is one of the highest infection rates in the world [10,11]. In addition to HIV/AIDS, Namibia also has one of the highest tuberculosis infection rates in the world, with 63.5% of tuberculosis cases being HIV positive [10]. However, there is a light at the end of the tunnel. According to a recent study comparing Namibia to nine other African countries and the United States of America, Namibia has the highest percentage of HIV-positive patients aware of their status and on antiretroviral (ARV) treatment. In addition, Namibia also has the highest viral suppression rate. Therefore, despite the high HIV numbers, Namibia is on its way to epidemic control [12].

The most noteworthy mycotic infections are opportunistic in HIV/AIDS patients. The latest health facility census states that the availability of health services varies throughout Namibia, providing some form of care and support services for HIV patients. These care and support services cater for opportunistic infections related to HIV. However, the main focus is narrow and is predominantly aimed at tuberculosis. According to the census, only 30% of care and support services can treat topical fungal infections, including all hospitals and 50% of sick bays but only 9% of clinical care and support service facilities have at least two medicines to treat cryptococcosis. Furthermore, systemic intravenous treatment for specific fungal infections are available in only two of every ten facilities in Namibia [3].

The Namibia Standard Treatment Guidelines (2011) [13] proscribe the management of a variety of opportunity infections including PCP and bacterial pneumonia. However, there is limited use of laboratory diagnosis, particularly antimicrobial susceptibility testing, so the guidelines rely on empirical treatment. The only fungal tests performed are the Cryptococcal Antigenaemia (CrAg) Test and the Cryptococcal Antigen Lateral Flow Assay [14], and no *Pneumocystis* or *Aspergillus* diagnostics are done. This lack of diagnostic testing following clinical examination was confirmed by a study on the compliance of these guidelines by Nakwatumbah et al 2017 [14]. Life-saving drugs such as flucytosine [15] are not available through most African health care providers, including in Namibia. Thus, most countries depend on fluconazole induction monotherapy to treat cryptococcal meningitis, even though it has been proven to be less effective than amphotericin B and flucytosine [15,16]. Opportunistic infections complicating HIV continues to be the sole leading cause of infirmity and mortality amid HIV/AIDS patients in settings where resources are insufficient [6]. Opportunistic infections give rise to an assortment of detrimental circumstances, ranging from premature death to reducing the quality of life of HIV-infected persons, speeding up the rate of progression to full blown AIDS, reducing patients' response to antiretroviral treatment especially when HIV-positive patients are co-infected with tuberculosis (partly through drug interactions and pill burden overload) and many social issues including increased stigma, limited ability to work and high medical care costs [17].

Because HIV and/or AIDS has such a major impact in Namibia, the government of Namibia has prioritized the prevention and treatment of HIV/AIDS to curtail its morbidity and mortality. However, in spite of national endeavours, premature deaths from HIV/AIDS endure and echo deficiencies in the health care system in Namibia [11]. In 2017, there were 2700 adult and child deaths related to HIV [9], with no indication whether the causative agent may be a fungal pathogen. Therefore, here, we estimate the prevalence of and incidence of serious fungal infections in Namibia.

2. Materials and Methods

A full literature search was conducted to ascertain all epidemiological papers publicising fungal infection rates in Namibia. Medline, the PubMed website, Google Scholar, and the African Journals Online (AJOL) databases were utilized in the literature search to identify all published papers reporting fungal infection rates from Namibia. Worldwide epidemiological articles of fungal infections were reviewed for citations of publications from Namibia and Africa. The search terms used were: "fungal infections", "Namibia", "fungi and Namibia", "fungal infections and Namibia," "mycosis and Namibia", "candidemia and Namibia," "candidiasis and Namibia", "aspergillosis and Namibia," "cryptococcosis and Namibia," "histoplasmosis and Namibia," and "fungal keratitis and Namibia," to identify specific disease conditions. We used specific populations at risk to estimate national incidence or prevalence where no relevant studies were available. National or local records were favoured but, where unobtainable, data from other sources were utilised.

Prevalence data for conditions which may become complicated by serious fungal infections were obtained from national surveys, registries, or published estimates. Namibian population estimates and AIDS-related deaths were obtained from the WHO Health statistics [18], the Namibia 2011 Population and Housing Census Indicators [19], and UNAIDS country factsheets [9]. National tuberculosis data (2016) were obtained from the World Health Organization [1]. The national prevalence for acute myeloid leukaemia was obtained from the National Cancer Registry [20]. The prevalence of asthma was assumed to be 11.3% based on a study by Hamatui et al. (2017) [21]. The prevalence of AIDS patients presenting with cryptococcal meningitis was assumed to be 3.3% based on a study by Sawadago et al. [22]. We assumed that 11% with CD4 counts <200 cells/µl are at risk of developing PCP over 2 years, as not all patients will present to hospital in one year, based on a study by Mgori and Bash [11]. Invasive aspergillosis (IA) was anticipated to complicate several immunocompromising and critical care conditions. IA was assumed to complicate 10% of cases of acute myeloid leukaemia (AML) [23]. IA is assumed to complicate 2.6% of lung cancer, based on a study by Yan in 2009 [24]. In addition, IA was assumed to complicate 1.3% of severe acute exacerbations of chronic obstructive pulmonary disorder (COPD) requiring hospitalization [25,26]. Lastly, we assumed that 4% of AIDS-related deaths were caused by IA, based on a study by Antinori [27]. Mucormycosis was assumed to affect 2 per million of the population based on data from Europe [28,29].

Chronic pulmonary aspergillosis was assumed to complicate tuberculosis with and without cavity lesions in 6.5 and 0.2% of cases, respectively [30]. We have also assumed that pulmonary tuberculosis is the underlying diagnosis in 80% of all CPA cases [31]. Allergic bronchopulmonary aspergillosis (ABPA) was assumed to occur in 2.5% of adult asthmatics [31], based on a study by Hamatui [21]. Although ABPA also occurs in cystic fibrosis, no estimate of the prevalence of this disease in Namibia was attainable. Severe asthma with fungal sensitization (SAFS) was estimated at 33% of the most severe asthmatics [32].

The burden of candidemia was assumed to occur in intensive care units (ICUs) and non-ICU inpatient settings at a ratio of 7:3 [33] and, therefore, we determined the number of ICU beds based on data from the Namibia Health Facilities Census [3]. In the absence of local data, we assumed that post-surgical *Candida* peritonitis occurs in 50% of the one-third of candidemia patients in ICU, based on a French study [34]. Oral candidiasis was assumed to affect 90% of untreated HIV patients, based on a study in Tanzania [35]. Oesophageal candidiasis was assumed to affect 0.5% of HIV patients on ARV treatment [36,37]. As elsewhere, vulvovaginal candidiasis recurrences are not reported in Namibia, we assumed that recurrent vulvovaginal candidiasis had a prevalence of 9% among adult females, based on a study by Denning et al. [38]. We could not find data on fungal keratitis, tinea capitis or histoplasmosis. However, using data from South Africa, we derived an estimation of tinea capitis infection from the overall prevalence of 15% at a conservative estimate of 6% of Namibian children under the age of 15 years [39].

3. Results

The population of Namibia in 2015 was approximately 2.5 million, of whom approximately 69% are female [18]. Approximately 37% are younger than 15 years of age and 7% are older than 60 years. Among all females, 56% are women between the ages of 15 and 54 years of age [3,40]. Namibia is classified by the World Bank as an upper middle-income (UMI) country with a gross domestic product (GDP) of USD 4620 per capita 2016 [41]. The total estimated population with HIV infection was 200,000 in 2017, with 167,629 (83%) on ART. In 2017, there were an estimated 190,000 adults (≥15 years) living with HIV, of which, 30% of men, 5% of women and 24% of children were ART naïve. UNAIDS estimated that there were 2700 AIDS-related deaths in 2017 [9].

Table 1 shows the estimates of the total burden of serious fungal infections and the number of infections classified according to the major at-risk groups as well as the rate per 100,000 inhabitants.

In total, we estimated the occurrence of 112,870 cases of serious fungal infections in Namibia each year (Table 1). We estimate that there are approximately 543 cases of HIV-related cryptococcal meningitis in Namibia. Fungal pneumonia, especially PCP, is commonly associated with AIDS epidemic [4]. A study conducted in Namibia in 2015, found 11.3% with PCP infections among HIV-positive patients [11]. Assuming a rate of 10% among those with a CD4+ T-lymphocyte count <200 × 106/L, we anticipate 836 new cases annually (33.6/100,000). This may be an underestimation of the total cases as many patients go undiagnosed due to a lack of diagnostic testing. Even microscopy for *Pneumocystis* (which is occasionally performed in Namibia) underestimates disease by 25% compared with PCR, based on a comparative study performed in Windhoek [4].

Namibia's population was approximately 2.5 million people in 2015 [1], with approximately 80% of women below the age of 60 [3]. Across the world, approximately 5–9% of women report four episodes or more of *Candida* vulvovaginitis (VVC) per year [42], which we have assumed arbitrarily is 6% [38]. In Namibia, this equals circa 37,390 patients aged 15–50 affected annually (3003/100,000 females) from recurrent VVC (≥4 episodes/annum).

Table 1. Estimated burden of serious fungal infections in Namibia.

Infection	Number of Infections Per Underlying Disorder Per Year						
	No Underlying Disease	HIV/AIDS	Respiratory	Cancer	ICU	Total Burden	Rate/100K *
Cryptococcal meningitis		543				543	21.8
Pneumocystis pneumonia		836				836	33.6
Invasive aspergillosis		108	1	15	259	383	15.4
Chronic pulmonary aspergillosis			453			453	18.2
Allergic bronchopulmonary aspergillosis (ABPA)			4462			4462	179
Severe asthma with fungal sensitisation (SAFS)			5892			5892	237
Candidemia				87	37	125	5.00
Candida peritonitis					19	19	0.75
Oral candidiasis		6660				6660	267
Oesophageal candidiasis		2318				2318	93.1
Recurrent *Candida* vaginitis (≥4x/year)	37,390					37,390	3,003
Mucormycosis				5		5	0.20
Tinea capitis	53,784					53,784	2160
Total serious fungal infection burden						112,870	

*Rate per 100,000 population except recurrent *Candida* vaginitis when rate per 100,000 females

Except for corticosteroids, most of sub-Saharan Africa lacks high-intensity cancer treatments, transplantation and immune modulation therapy, possibly reducing the incidence of invasive aspergillosis (IA). We have, however, assumed that the rates of infection are similar in acute myeloid leukaemia as elsewhere (10%) and that an equal number of cases of IA occur in those with all other haematological disorders [23]. This estimate provides an estimate of approximately 383 cases (15.4/100,000), which includes HIV-associated and COPD cases, as well as other corticosteroid treatment risk [23].

Chronic pulmonary aspergillosis may be mistaken for and follow confirmed pulmonary TB with a frequently delayed diagnosis. Misdiagnosis is also possible, since it has similar clinical features to TB relapse/reinfection or multi-drug resistant pulmonary TB [43,44]. It was estimated in 2013 that there are 67,788 Pulmonary TB survivors, of which 42% are HIV-positive patients [41], using assumptions as described in a recent study in Uganda [30] (i.e., 6.5% CPA rate in those with residual cavities following TB and 0.2% in those without residual cavities). The incidence of new cases of CPA was estimated at 112 (4.50/100,000) cases [44]. With a less than 5% annual mortality rate [41], the CPA prevalence after TB is estimated at 354 (14.20/100,000) cases. The overall estimate for CPA was 453 (18.2/100,000).

Fungal allergy augments asthma, particularly in adults. The prevalence of asthma in adults in Namibia is 3.39% based on the World Health Survey [45], and 11.2% based on a small study conducted in Namibia by Hamatui and Beyon in 2017 [21]. The prevalence of allergic bronchopulmonary aspergillosis (ABPA) in adult asthmatics is taken to be 2.5%, partly based on data from South Africa [46]. We estimated a total of 4462 cases in Namibia in those with underlying asthma, based on 11.2% asthma prevalence. Severe asthma with fungal sensitization (SAFS) prevalence is approximately 0.33% of the worst 10% adult asthmatics, thus 3% of asthma patients have SAFS [32,47]. We estimate approximately 5890 cases of SAFS in Namibia (237/100,000). Resembling ABPA, SAFS is also treatable with oral itraconazole with great improvement in the quality of life of individuals who respond [32].

Mucormycosis was estimated to have a burden of five cases. There have been no cases of histoplasmosis or tinea capitis reported in Namibia. Using a conservative estimate that 6% of Namibian children would be affected by tinea capitis, we estimated 53,784 such cases. However, disseminated histoplasmosis does occur in Africa, having been documented initially in 1957 [48]. More recently, cases have been recognized in HIV patients [49]. *Emergomyces* infection has also been found to be the most common dimorphic fungus causing human fungal infection in South Africa [50]. However, no cases have been reported in Namibia. Fungal keratitis is accountable for human blindness, except when diagnosed and treated in a timely fashion. There was one case report on a foreign national travelling to Namibia who contracted *Fusarium* keratitis [51]. However, there are no reports from Namibia regarding fungal keratitis.

4. Discussion

Namibia is a country in south-western Africa that covers an area of 824,000 square kilometres. With a population of 2.5 million, it is one of the least densely populated countries in the world [2]. There are 13 Regional Health Directorates overseeing service delivery in 34 health districts in Namibia. Public health services are provided through 30 public district hospitals, 44 health centres, and 269 clinics. Because of the vastness of the country, the sparse distribution of the population, and the lack of access to permanent health facilities in some communities, outreach (mobile clinic) services are available. Three intermediate hospitals and the national referral hospital provide support to the district hospitals [2].

Namibia faces a number of important health care challenges, most notably HIV and associated infections including tuberculosis. Understanding the magnitude of respective problems is critical for prioritizing limited resources, and national data to help quantify the burden of fungal disease in Namibia remains scarce. This study serves to provide an estimate of the fungal disease burden in Namibia. Our burden estimate of 112,870 serious fungal disease cases indicates significant morbidity and probably mortality in Namibia.

The most frequent serious fungal diseases are recurrent vulvovaginal candidiasis and oesophageal candidiasis. While recurrent vulvovaginal candidiasis (RVVC) is not mortal, it is a significantly more severe clinical form than VVC. This is because of the recurrences of symptoms, by definition, four or more episodes per year, and for its fractiousness. RVVC affects approximately 30% of women globally at some point in their lifetime. It has an annual prevalence ranging from 4–10% [52]. Here, we estimate a rate of 37,390. Among the Southern African Development Community (SADC) countries, Namibia's RVVC rate is the fourth highest among estimated burdens [53–56]. Djomand et al. described the prevalence of *Candida* vaginitis in newly infected HIV patients in Namibia [57]. This estimate, however, does not include women on hormone replacement therapy, and is an annual prevalence estimate, not a lifetime experience estimate [58]. Recurrent VVC is an disagreeable condition for women, which may have psychological and economic consequences [52,58–60]. Fluconazole resistance among *Candida spp.* isolated from women with VVC have been identified by recent studies which has a large impact on health and well-being of affected women [61]. These data, however, do not give a true reflection of the burden of RVVC, as it only elucidated new cases. This, therefore, illustrates a dire need for the improvement of disease surveillance.

Oesophageal candidiasis is common with HIV and may supplement oral candidiasis. It is, however, often distinct and more debilitating. Recent global estimates found ~1,300,000 cases from HIV-positive cases only, of which 20% had CD4 counts <200 and 5% of those were on ARVs [62]. The prevalence of oral and oesophageal candidiasis ranged from 14% in pregnant women with HIV to 67% of Senegalese in patients with AIDS, in a review of the opportunistic infections related to HIV in sub-Saharan Africa [6]. In ART-naive patients in low-to-middle income countries, the summary risk was highest (>5%) in oral candidiasis [63]. Oral candidiasis was found to be the most common opportunistic infection (OI) in Nigeria [64]. The most frequent OI before the initiation of highly active antiretroviral therapy (HAART) was oral candidiasis in Uganda [17]. Oral candidiasis (OC) is the most common opportunistic fungal infection among immunocompromised individuals [65]. In this study, we estimated 2318 cases of oesophageal candidiasis and 6660 cases of oral candidiasis. These high numbers are correlated with the high HIV prevalence in Namibia [9].

Cryptococcal meningitis is the leading cause of meningitis is southern Africa [15,22,66–70] and a leading cause of death among people living with HIV. We estimated the occurrence of 543 cases of cryptococcal meningitis per year in Namibia. This is within the range estimated for Namibia in by Rajasingham et al, who estimated the occurrences of 501–1000 cases of HIV-associated cryptococcal meningitis annually [68]. Sawadago et al. estimated CrAg prevalence of ≤3% among HIV-infected patients with CD4+ <100 cells/μL [22]. They also found that 11.3% of HIV-related deaths were caused by PCP [11], which is comparable to that of South Africa, which had an overall prevalence of 3.4% [69]. We can consequently accept that based on the assumption that untreated cryptococcal meningitis is consistently terminal, many deaths occurred before individuals present to the hospital. Therefore, establishing CrAg positivity in at-risk patients on a national basis will be helpful in mapping areas with a high HIV-related cryptococcal disease burden [69]. Fortunately, CrAg screening and preventative treatment among HIV-infected patients in Namibia are in accordance with WHO recommendations, and have a demonstrated improved survival [68]. Despite this, only 9% of health care facilities in Namibia have at least two medicines to treat cryptococcal disease [13], leaving a lot of room for improvement.

PCP signalled the inception of the global HIV pandemic—prior to 1995, it was projected that two-thirds of HIV-infected persons would ultimately develop PCP [71]. Because PCP can be found in a variety of patients with cellular immune defects, the true incidence of PCP is, however, difficult to determine [72]. We estimated 836 cases of PCP per year. The number of cases proven by microscopy or by molecular testing are expected to be substantially less than our estimate, given the limitations in the availability and in the sensitivity of microscopy and molecular diagnostics [73,74]. Undeniably, this diagnostic deficiency compels clinicians to treat patients based on risk factor analysis, clinical and chest radiography evidence [4].

The prevalence of CPA is estimated at 18.2/100,000—a high rate even in Africa. The Republic of South Africa's proportion of CPA has the highest estimate, at 175.8/100,000 [55], compared to rates for Kenya, Egypt, Algeria, Senegal, Nigeria, Malawi, Mozambique, and Tanzania following tuberculosis, which was estimated at 32/100,000, 13.8/100,000, 2.2/100,000, 19/100,000, 78/100,000, 24/100,000, 16.4/100,00, and 69.9/100,000, respectively [54,56,75–79]. The high number in Namibia reflects the high burden of tuberculosis [41,80,81] and clearly requires validation. *Aspergillus spp.* were the dominant fungi isolated from the bare patches on the Giribes plains in Namibia [82]. However, there are no published data on aspergillosis in Namibia, and most clinical isolates are referred to South Africa. Therefore, in areas with a high prevalence of tuberculosis, i.e., Namibia, criteria for the diagnosis of CPA should be established, as clinical evidence is very similar to pulmonary tuberculosis. Serological testing for *Aspergillus* IgG and/or precipitins, and sputum culture should be included in the identification of *A. fumigatus* [31]. TB is only one of the underlying diseases supplementary with CPA [26]. However, as there are very few Namibians older than 60 years, the bulk of the CPA cases will be imputable to TB. Relapsed or non-responding TB is a common clinical conundrum facing clinicians and, without Aspergillus IgG serology, CPA is not a diagnosable disease in Namibia currently [83]. Thus, it is essential to bridge this diagnostic gap to improve the incidence rate of CPA in Namibia.

Fungal asthma, which is comparatively common, can be attributed to ABPA and SAFS. Asthma prevalence in adults has been described at 3.39% [45], and more recently at 6% in adult Africans [84]. A higher prevalence of asthma, specifically 11.3%, and other respiratory symptoms was observed in a study that measured particulate pollution concentration in Windhoek, Namibia [21]. Because aspergillosis is not tested for and fungal sensitization studies have not been conducted in Namibia, our estimate of the burden for ABPA and SAFS is rudimentary at best. Nonetheless, with estimated rates of 179/100,000 and 236/100,000, respectively, fungal asthma, amenable to antifungal therapy as well as usual treatments for asthma [13], needs to be addressed in Namibia as it is may lead to more severe diseases and complications such as bronchiectasis and CPA [32].

We estimated 125 episodes of candidemia each year. Due to the developments in intensive care units (ICUs), cancer care and organ transplantation, many diseases are no longer lethal. However, hospitalization (i.e., ICUs) has triggered an upsurge in numerous opportunistic infections, such as candidiasis [5,33,34]. Therefore, candidemia cases were based on the estimate of ICU beds (4) in each district hospital (30) [2] and the frequency of acute myeloid leukaemia patients in Namibia. Data are required to clarify the true incidence of candidemia in Namibia, especially in the wake of multi-drug resistant yeasts, such as *Candida auris* [85].

Several important fungal infections are omitted, including histoplasmosis, fungal sinus disease, all skin and nail infections, fungal keratitis and some rare invasive infections, even though all are treatable [13]. These fungal infections are often misdiagnosed, and diagnostic delays are detrimental [50]. These and other fungal diseases are not notifiable in Namibia [13], so true incidences are unclear. It is interesting to note that several of these serious fungal infections have been reported in neighbouring sub-Saharan countries. Fungal eye infections are more frequent in HIV patients [86,87]; a study in Tanzania showed that 50% of the referred patients with an ophthalmic complaint had microbial keratitis [88]. In Uganda, keratitis is responsible for 25% of children with impaired vision [89]. Emergomycosis, histoplasmosis, and sporotrichosis are AIDS-related systemic mycoses which are endemic to South Africa and, have on occasion, been the cause of outbreaks [50,90,91]. Dermatophyte infections, especially tinea capitis, are also common among children all over Africa, particularly in areas of poor socioeconomic standing, urbanization (i.e., the increase in informal settlements), and poor sanitary conditions [92–96]. Tinea infections are a public health problem due to their contagious nature.

As mentioned before, Namibia is considered a semi-arid to arid country, with some climatic diversity. Arid soil systems are the harshest terrestrial environments on Earth. The Namib Desert, as the oldest hyper-arid desert on Earth, is thought to have exhibited hyper-arid conditions for the last 5 million years. Water shortage creates an environment where specific organisms able to survive on low water resources can survive. Extended dry periods can result in high cellular mortality due to desiccation and oxidative damage. A study by Frossard et al. (2015) demonstrated the responsiveness of shallow subsurface soil desert bacterial and fungal communities to water availability changes. According to recent climate change models, precipitation intensities, as well as the duration of dry periods between wetting events, are projected to increase in the country [97].

The Namib Desert, despite its hyper-arid nature, experiences coastal fogging on its extensive shoreline. Fog can reach 50 km inland. The fog supplies high levels of biodiversity and productivity—in some regions, it is the only source of water. Fog water can provide vital support to microbial communities, driving the majority of plant litter decomposition in water-limited systems. Since fog forms through activation of ambient ground-level aerosol particles, sources for these aerosols should play a key role in determining the microbial composition of fog. The ocean surface can be a dominant source of aerosols to the coastal environment, and ocean bacteria are present in aerosols, non-strata clouds, and adventive coastal fog. If coastal waters were polluted, then coastal aerosols will contain bacteria associated with that pollution—these may include pathogens. Fog has the potential to transport and maintain pathogenic viability. Evans et al. (2019) were able to isolate 10 fungal isolates from Namib fog, and an additional five isolates from clear air samples, which were closely related to extremophilic fungi, including Ascomycota, Basidiomycota, Chytridiomycota, Glomeromycota, and Zygomycota. These groups contained pathogenic fungal species, including suspected plant pathogens and those causing respiratory infections in immunocompromised people [98].

Several fungal isolates have been isolated from various forms of food contaminants in Namibia. Several mycotoxins were found in sorghum malts processed for brewing Namibian traditional beverages, some of which were above the EU allowable limit [99]. In an attempt determine the presence of fungal pathogens in Marama beans, an endemic perennial wild tuberous Fabacea that is valued by the indigenous people of the semi-arid land of the Kalahari for its nutritional and medicinal properties, Uzabakiriho et al. (2013) identified two fungal isolates, namely *Alternaria tenuissima* and *Phoma* spp., which are both part of the Ascomycota family. *A. tenuissima*, can infect cereal grains and be a source of food contamination [100]. A toxic isolate of *F. chlamydosporum* was obtained from millet from households of patients suffering from the haemorrhagic disease Onyalai in Namibia [101]. Nawases et al. 2018 conducted a study on the large open markets in Windhoek that provide staple foods that are accessible to the poor people of the city. There were 17 different fungal isolates from the food samples, including known aflatoxigenic forms of *Aspergillus* species [102].

5. Conclusions

The true burden of fungal infections is unknown in Namibia because of limited research studies and a lack of systematic diagnosis and data collection. More epidemiological studies and health impact studies are necessary to accurately measure the burden and impact of fungal diseases in different patient groups and clinical settings.

Increasing funding for combatting diseases in poor countries has been instrumental in reducing the burden of disease and improving the overall economy of many poor countries. However, funding is normally geared towards HIV/AIDS and tuberculosis and dwarfs that of any other infectious diseases including tuberculosis [103].

Socio-economic factors need to be considered in Namibia, one of the most unequal countries in the world [104]. Poor living conditions marked by poor sanitation, housing, limited water supply, and low economic power are just some of the predisposing factors. Urbanization and increasing urban poverty characterize much of Southern Africa—this results in poor urban health. People living in informal settlements are at great risk to diseases spread through air or contact as a result of living conditions and overcrowding both in school and living environments [93,105]. Patient- and caregiver-related factors such as delays in seeking health care, access to critical care, dialysis and ventilator support, and poor monitoring of patients were identified as the main modifiable factors identified in the Oshakati Intermediate Hospital, which contributed to HIV and/or AIDS inpatient mortality [106]. All of these factors can be attributed to some form of poverty or resource-limited situation. The National TB budget in 2017 was 56 million USD [1]. The increase in National TB and Leprosy Plan (NTLP) financial resources resulted in an increase in staff, staff training and a swift surge in treatment strategies. However, the NLTP is exceedingly reliant on external funding and, therefore, advancement can swiftly be lost when external funding starts diminishing without other funding for TB control [107]. Chinsembu described this phenomenon well, "Despite this impressive progress, Chinsembu cautioned that Namibia's ART programme is like a candle in the wind as it battles to glimmer against the inevitable possibility of dying from another form of AIDS - 'Acquired Income Deficiency Syndrome'. There are concerns that the country's free public sector ART programme is not sustainable due its heavy reliance on donor funds" [107].

Novel efforts are vital to spearhead local strategies to assess the actual burden of CPA, cryptococcal meningitis, PCP, fungal keratitis, mucormycosis, dermatophytosis, and other unreported fungal infections. The present level of knowledge is not sufficient and there are inadequate infrastructures to diagnose and treat these serious problems. According to the Namibia Health Facility Census (2009), only 30% of facilities that offer care and support services for HIV/AIDS, provide treatment of opportunistic infections, including topical fungal infections [3]. This is not acceptable. It is crucial that national programs build toward amplifying the ability to identify and accordingly manage HIV-related illnesses and opportunistic infections.

6. Limitations

This study has several limitations. The identified manuscripts published on fungal infections in Namibia lack the necessary figures while others are not updated enough to be incorporated in our model. Also, the estimates on the statistics of other patients with immunosuppression in Namibia are not precise. The National Cancer Registry's latest version was published on data collected between 2010 and 2014—it has not been updated since. Organ transplantation is not routinely performed in Namibia—patients are referred to outside the country for these services. To our knowledge, there is no official registry of all patients who have had an organ transplant living in Namibia. Again, a centralized registry on all surgeries performed in the country is missing. We believe that the lack of data on these groups of patients may have moderately underestimated the number of some invasive fungal conditions such as candidemia. Finally, due to the nature of the modelling we conducted, and based on the lack of robust data from Namibia, our estimates probably have wide confidence limits. Despite these shortcomings, we feel these very first country-wide assessments and conjectures provide an acceptable basis to raise awareness of the unaddressed burden of fungal disease, and point at the need for improved registries, the effective use of available diagnostic tools and the implementation of new diagnostics tools, as well as strengthening of the availability of and access to first-line antifungals in Namibia.

Author Contributions: Conceptualization, Methodology and Formal Analysis—D.W.D. and C.M.D.; Writing—Original Draft Preparation, C.M.D.; Writing—Review and Editing, C.M.D. and D.W.D. All authors read and approved the final manuscript.

References

1. World Health Organization Namibia Tuberculosis Profile. Available online: https://extranet.who.int/sree/Reports?op=Replet&name=/WHO_HQ_Reports/G2/PROD/EXT/TBCountryProfile&ISO2=NA&outtype=html (accessed on 8 October 2018).
2. Ministry of Health and Social Services (MoHSS). *ICF International Namibia Demographic and Health Survey 2013*; MoHSS: Windhoek, Namibia; ICF International: Rockville, MD, USA, 2014.
3. Ministry of Health and Social Services (MoHSS). [Namibia] and ICF Namibia Health Facility Census 2009. Available online: https://www.unicef.org/namibia/Namibia_Health_Facilities_Census_2009_key_find_HIV_AIDS_TB_STI.pdf (accessed on 17 September 2018).

4. Nowaseb, V.; Gaeb, E.; Fraczek, M.G.; Richardson, M.D.; Denning, D.W. Frequency of Pneumocystis jirovecii in sputum from HIV and TB patients in Namibia. *J. Infect. Dev. Ctries.* **2014**, *8*, 349–357. [CrossRef] [PubMed]
5. Yapar, N. Therapeutics and Clinical Risk Management Dovepress Epidemiology and risk factors for invasive candidiasis. *Ther. Clin. Risk Manag.* **2014**, *2014*, 95–105. [CrossRef] [PubMed]
6. Holmes, C.B.; Losina, E.; Walensky, R.P.; Yazdanpanah, Y.; Freedberg, K.A. Review of Human Immunodeficiency Virus Type 1—Related Opportunistic Infections in Sub-Saharan Africa. *Clin. Infect. Dis.* **2003**, *36*, 652–662. [CrossRef] [PubMed]
7. Rubaihayo, J.; Tumwesigye, N.M.; Konde-Lule, J. Trends in prevalence of selected opportunistic infections associated with HIV/AIDS in Uganda. *BMC Infect. Dis.* **2015**, *15*, 1. [CrossRef] [PubMed]
8. Dunaiski, C.M.; Janssen, L.; Erzinger, H.; Pieper, M.; Damaschek, S.; Schildgen, O.; Schildgen, V. Inter-Specimen Imbalance of Mitochondrial Gene Copy Numbers Predicts Clustering of Pneumocystis jirovecii Isolates in Distinct Subgroups. *J. Fungi* **2018**, *4*, 84. [CrossRef] [PubMed]
9. UNAIDS Country factsheets Namibia 2017 HIV and AIDS. Available online: https://aidsinfo.unaids.org/%0D (accessed on 5 September 2018).
10. Chinsembu, K.C.; Hedimbi, M. An ethnobotanical survey of plants used to manage HIV/AIDS opportunistic infections in Katima Mulilo, Caprivi region, Namibia. *J. Ethnobiol. Ethnomed.* **2010**, *6*, 1–9. [CrossRef] [PubMed]
11. Mgori, N.K.; Mash, R. HIV and/or AIDS-related deaths and modifiable risk factors: A descriptive study of medical admissions at Oshakati Intermediate Hospital in Northern Namibia. *Afr. J. Prim. Health Care Fam. Med.* **2015**, *7*, 1–7. [CrossRef] [PubMed]
12. *Joint United Nations Program on HIV/AIDS. 90-90-90: An ambitious treatment target to help end the AIDS epidemic*; UNAIDS: Geneva, Switzerland, 2014.
13. Republic of Namibia, Ministry of Health and Social Services. Namibia Standard Treatment Guidelines. Available online: http://www.healthnet.org.na/documents.html (accessed on 2 March 2019).
14. Nakwatumbah, S.; Kibuule, D.; Godman, B.; Haakuria, V.; Kalemeera, F.; Baker, A.; Mubita, M.; Mwangana, M. Compliance to guidelines for the prescribing of antibiotics in acute infections at Namibia's national referral hospital: A pilot study and the implications. *Expert Rev. Anti-Infect. Ther.* **2017**, *15*, 713–721. [CrossRef] [PubMed]
15. Molloy, S.F.; Kanyama, C.; Heyderman, R.S.; Loyse, A.; Kouanfack, C.; Chanda, D.; Mfinanga, S.; Temfack, E.; Lakhi, S.; Lesikari, S.; et al. Antifungal Combinations for Treatment of Cryptococcal Meningitis in Africa. *N. Engl. J. Med.* **2018**, *378*, 1004–1017. [CrossRef]
16. Ministry of Health and Social Services (MoHSS). *NEMLIST—Namibia Essential Medicines List*; Ministry of Health and Social Services: Windhoek, Namibia, 2016.
17. Rubaihayo, J.; Tumwesigye, N.M.; Konde-Lule, J.; Wamani, H.; Nakku-Joloba, E.; Makumbi, F. Frequency and distribution patterns of opportunistic infections associated with HIV/AIDS in Uganda. *BMC Res. Notes* **2016**, *9*, 1–16. [CrossRef]
18. WHO. World Health Statistics SDGs. In *World Health Statistics 2016*; WHO: Geneva, Switzerland, 2016.
19. Kafidi, L. *Namibia Statistics Agency Namibia 2011, Population and Housing, Census Indicators*; FGI House: Windhoek, Namibia, 2013.
20. Namibia National Cancer Registry (NNCR)—Cancer Incidences in Namibia 2010–2014. Available online: https://afcrn.org/images/M_images/attachments/125/CancerinNamibia2010-2014.pdf (accessed on 3 December 2018).
21. Hamatui, N.; Beynon, C. Particulate Matter and Respiratory Symptoms among Adults Living in Windhoek, Namibia: A Cross Sectional Descriptive Study. *Int. J. Environ. Res. Public Health* **2017**, *14*, 110. [CrossRef] [PubMed]
22. Sawadogo, S.; Makumbi, B.; Purfield, A.; Ndjavera, C.; Mutandi, G.; Maher, A.; Kaindjee-Tjituka, F.; Kaplan, J.E.; Park, B.J.; Lowrance, D.W. Estimated prevalence of Cryptococcus antigenemia (CrAg) among HIV-infected adults with advanced immunosuppression in Namibia justifies routine screening and preemptive treatment. *PLoS ONE* **2016**, *11*, 1–9. [CrossRef] [PubMed]
23. Lortholary, O.; Gangneux, J.; Sitbon, K.; Lebeau, B.; De Monbrison, F.; Le Strat, Y.; Coignard, B.; Dromer, F. Epidemiological trends in invasive aspergillosis in France: The SAIF. *Clin. Infect. Dis.* **2011**, *17*, 1882–1889. [CrossRef] [PubMed]

24. Yan, X.; Li, M.; Jiang, M.; Zou, L.Q.; Luo, F.; Jiang, Y. Clinical characteristics of 45 patients with invasive pulmonary aspergillosis: Retrospective analysis of 1711 lung cancer cases. *Cancer* **2009**, *115*, 5018–5025. [CrossRef] [PubMed]
25. Guinea, J.; Torres-Narbona, M.; Gijón, P.; Muñoz, P.; Pozo, F.; Peláez, T.; De Miguel, J.; Bouza, E. Pulmonary aspergillosis in patients with chronic obstructive pulmonary disease: Incidence, risk factors, and outcome. *Clin. Microbiol. Infect.* **2009**, *16*, 1–8. [CrossRef] [PubMed]
26. Smith, N.L.; Denning, D.W. Underlying conditions in chronic pulmonary aspergillosis including simple aspergilloma. *Eur. Respir. J.* **2011**, *37*, 865–872. [CrossRef] [PubMed]
27. Antinori, S.; Nebuloni, M.; Magni, C.; Fasan, M.; Adorni, F.; Viola, A.; Corbellino, M.; Galli, M.; Vago, G.; Parravicini, C.; et al. Trends in the Postmortem Diagnosis of Opportunistic Invasive Fungal Infections in Patients With AIDS. *Am. J. Clin. Pathol.* **2009**, *132*, 221–227. [CrossRef] [PubMed]
28. Torres-narbona, M.; Mun, P.; Gadea, I. Impact of Zygomycosis on Microbiology Workload: A Survey Study in Spain. *J. Clin. Microbiol.* **2007**, *45*, 2051–2053. [CrossRef] [PubMed]
29. Bitar, D.; Van Cauteren, D.; Lanternier, F.; Dannaoui, E.; Che, D.; Dromer, F.; Desenclos, J.; Lortholary, O. Increasing Incidence of Zygomycosis (Mucormycosis), France, 1997–2006. *Emerg. Infect. Dis.* **2009**, *15*, 1395–1401. [CrossRef] [PubMed]
30. Page, I.D.; Byanyima, R.; Hosmane, S.; Onyachi, N.; Opira, C.; Richardson, M.; Sawyer, R.; Sharman, A.; Denning, D.W. Chronic pulmonary aspergillosis commonly complicates treated pulmonary tuberculosis with residual cavitation. *Eur. Respir. J.* **2019**, *53*, 1801184. [CrossRef] [PubMed]
31. Denning, D.W.; Pleuvry, A.; Cole, D.C. Global burden of chronic pulmonary aspergillosis as a sequel to pulmonary tuberculosis. *Bull. World Health Organ.* **2011**, *89*, 864–872. [CrossRef] [PubMed]
32. Denning, D.W.; Pashley, C.; Hartl, D.; Wardlaw, A.; Godet, C.; Giacco, S. Del Fungal allergy in asthma—State of the art and research needs. *Clin. Transl. Allergy* **2014**, *4*, 1–23. [CrossRef] [PubMed]
33. Arendrup, M.C. Epidemiology of invasive candidiasis. *Curr. Opin. Crit. Care* **2010**, *16*, 445–452. [CrossRef] [PubMed]
34. Montravers, P.; Mira, J.-P.; Gangneux, J.-P.; Leroy, O.; Lortholary, O. A multicentre study of antifungal strategies and outcome of Candida spp. peritonitis in intensive-care units. *Clin. Microbiol. Infect.* **2011**, *17*, 1061–1067. [CrossRef] [PubMed]
35. Matee, M.I.; Scheutz, F.; Moshy, J. Occurrence of oral lesions in relation to clinical and immunological status among HIV-infected adult Tanzanians. *Oral Dis.* **2000**, *6*, 106–111. [CrossRef] [PubMed]
36. Smith, E.; Orholm, M. Trends and patterns of opportunistic diseases in Danish AIDS patients 1980–1990. *Scand. J. Infect. Dis.* **1990**, *22*, 665–672. [CrossRef] [PubMed]
37. Buchacz, K.; Baker, R.K.; Palella, F.J., Jr.; Chmiel, J.S.; Lichtenstein, K.A.; Novak, R.M.; Wood, K.C.; Brooks, J.T.; HOPS Investigators. AIDS-defining opportunistic illnesses in US patients, 1994–2007: A cohort study. *Aids* **2010**, *24*, 1549–1559. [CrossRef] [PubMed]
38. Denning, D.W.; Kneale, M.; Sobel, J.D.; Rautemaa-Richardson, R. Global burden of recurrent vulvovaginal candidiasis a systematic review. *Lancet Infect. Dis.* **2018**, *18*, 339–347. [CrossRef]
39. Young, C.N. Scalp ringworm among Black children in South Africa and the occurrence of Trichophyton yaoundei. *S. Afr. Med. J.* **1976**, *50*, 705–707.
40. Namibia Statistics Agency. *Namibia 2011 Population and Housing Census Main Report*; Namibia Statistics Agency: Windhoek, Namibia, 2011.
41. Republic of Namibia, Ministry of Health and Social Services. Third Medium Term Strategic Plan for Tuberculosis and Leprosy 2017/18—2021/22. Available online: http://www.mhss.gov.na/documents/119527/563974/Strategic+plan+TBL+Booklet+new+with+cover.pdf/224cc849-0067-4634-82fc-78ac3f9464e6 (accessed on 8 October 2018).
42. Sobel, J.D. Vulvovaginal candidosis. *Lancet* **2007**, *369*, 1961–1971. [CrossRef]
43. Brown, G.D.; Denning, D.W.; Gow, N.A.R.; Levitz, S.M.; Netea, M.G.; White, T.C. Hidden killers: Human fungal infections. *Sci. Transl. Med.* **2012**, *4*, 165rv13. [CrossRef] [PubMed]
44. Kosmidis, C.; Denning, D.W. The clinical spectrum of pulmonary aspergillosis. *Thorax* **2015**, *70*, 270–277. [CrossRef] [PubMed]
45. To, T.; Stanojevic, S.; Moores, G.; Gershon, A.S.; Bateman, E.D.; Cruz, A.A.; Boulet, L.-P. Global asthma prevalence in adults: Findings from the cross-sectional world health survey. *BMC Public Health* **2012**, *12*, 204. [CrossRef] [PubMed]

46. Denning, D.W.; Pleuvry, A.; Cole, D.C. Global burden of allergic bronchopulmonary aspergillosis with asthma and its complication chronic pulmonary aspergillosis in adults. *Med. Mycol.* **2013**, *51*, 361–370. [CrossRef] [PubMed]
47. Denning, D.W.; O'Driscoll, B.R.; Hogaboam, C.M.; Bowyer, P.; Niven, R.M. The link between fungi and severe asthma: A summary of the evidence. *Eur. Respir. J.* **2006**, *27*, 615–626. [CrossRef] [PubMed]
48. Davies, P. A Fatal case of histoplasmosis contracted in Kenya. *East Afr. Med. J.* **1957**, *34*, 555–557. [PubMed]
49. Oladele, R.O.; Ayanlowo, O.O.; Richardson, M.D.; Denning, D.W. Histoplasmosis in Africa: An emerging or a neglected disease? *PLoS Negl. Trop. Dis.* **2018**, *12*, 1–17. [CrossRef] [PubMed]
50. Schwartz, I.S.; Kenyon, C.; Lehloenya, R.; Claasens, S.; Spengane, Z.; Prozesky, H.; Burton, R.; Parker, A.; Wasserman, S.; Meintjes, G.; et al. AIDS-Related Endemic Mycoses in Western Cape, South Africa, and Clinical Mimics: A Cross-Sectional Study of Adults With Advanced HIV and Recent-Onset, Widespread Skin Lesions. *Open Forum Infect. Dis.* **2017**, *4*, 1–7. [CrossRef] [PubMed]
51. Zaigraykina, N.; Tomkins, O.; Garzozi, H.J.; Potasman, I. Fusarium keratitis acquired during travel to Namibia. *J. Travel Med.* **2010**, *17*, 209–211. [CrossRef]
52. Zhu, Y.-X.; Li, T.; Fan, S.-R.; Liu, X.-P.; Liang, Y.-H.; Liu, P. Health-related quality of life as measured with the Short-Form 36 (SF-36) questionnaire in patients with recurrent vulvovaginal candidiasis. *Health Qual. Life Outcomes* **2016**, *14*, 15. [CrossRef] [PubMed]
53. Kalua, K.; Zimba, B.; Denning, D. Estimated Burden of Serious Fungal Infections in Malawi. *J. Fungi* **2018**, *4*, 61. [CrossRef] [PubMed]
54. Faini, D.; Maokola, W.; Furrer, H.; Hatz, C.; Battegay, M.; Tanner, M.; Denning, D.W.; Letang, E. Burden of serious fungal infections in Tanzania. *Mycoses* **2015**, *58*, 70–79. [CrossRef] [PubMed]
55. Schwartz, I.; Denning, D. The estimated burden of fungal diseases in South Africa. *J. Infect. Public Health* **2019**, *12*, 124. [CrossRef]
56. Sacarlal, J.; Denning, D. Estimated Burden of Serious Fungal Infections in Mozambique. *J. Fungi* **2018**, *4*, 75. [CrossRef] [PubMed]
57. Djomand, G.; Schlefer, M.; Gutreuter, S.; Tobias, S.; Patel, R.; Deluca, N.; Hood, J.; Sawadogo, S.; Chen, C.; Muadinohamba, A.; et al. Prevalence and Correlates of Genital Infections Among Newly Diagnosed Human Immunodeficiency Virus–Infected Adults Entering Human Immunodeficiency Virus Care in Windhoek, Namibia. *Sex. Transm. Dis.* **2016**, *43*, 698–705. [CrossRef] [PubMed]
58. Foxman, B.; Muraglia, R.; Dietz, J.-P.; Sobel, J.D.; Wagner, J. Prevalence of recurrent vulvovaginal candidiasis in 5 European countries and the United States: Results from an internet panel survey. *J. Low. Genit. Tract Dis.* **2013**, *17*, 340–345. [CrossRef] [PubMed]
59. De Bernardis, F.; Arancia, S.; Sandini, S.; Graziani, S.; Norelli, S. Studies of Immune Responses in Candida vaginitis. *Pathogens* **2015**, *4*, 697–707. [CrossRef] [PubMed]
60. Muzny, C.A.; Schwebke, J.R. Bio fi lms: An Underappreciated Mechanism of Treatment Failure and Recurrence in Vaginal Infections. *Clin. Infect. Dis.* **2015**, *61*, 601–606. [CrossRef]
61. Wang, F.J.; Zhang, D.; Liu, Z.H.; Wu, W.X.; Bai, H.H.; Dong, H.Y. Species Distribution and In Vitro Antifungal Susceptibility of Vulvovaginal Candida Isolates in China. *Chin. Med. J.* **2016**, *129*, 1161–1165. [CrossRef]
62. Bongomin, F.; Gago, S.; Oladele, R.; Denning, D. Global and Multi-National Prevalence of Fungal Diseases—Estimate Precision. *J. Fungi* **2017**, *3*, 57. [CrossRef]
63. Drouin, O.; Bartlett, G.; Nguyen, Q.; Low, A.; Gavriilidis, G.; Easterbrook, P.; Muhe, L. Incidence and Prevalence of Opportunistic and Other Infections and the Impact of Antiretroviral Therapy Among HIV-infected Children in Low- and Middle-income Countries: A Systematic Review and Meta-analysis. *Clin. Infect. Dis.* **2016**, *62*, 1586–1594.
64. Akinyemi, J.O.; Ogunbosi, B.O.; Fayemiwo, A.S.; Adesina, O.A.; Obaro, M.; Kuti, M.A.; Awolude, O.A.; Olaleye, D.O.; Adewole, I.F. Demographic and epidemiological characteristics of HIV opportunistic infections among older adults in Nigeria. *Afr. Health Sci.* **2017**, *17*, 315–321. [CrossRef] [PubMed]
65. Mushi, M.F.; Bader, O.; Taverne-Ghadwal, L.; Bii, C.; Groß, U.; Mshana, S.E. Oral candidiasis among African human immunodeficiency virus-infected individuals: 10 years of systematic review and meta-analysis from sub-Saharan Africa. *J. Oral Microbiol.* **2017**, *9*, 1317579. [CrossRef] [PubMed]
66. Jarvis, J.N.; Meintjes, G.; Williams, A.; Brown, Y.; Crede, T.; Harrison, T.S. Adult meningitis in a setting of high HIV and TB prevalence: Findings from 4961 suspected cases. *BMC Infect. Dis.* **2010**, *10*, 67. [CrossRef] [PubMed]

67. Veltman, J.A.; Bristow, C.C.; Klausner, J.D. Meningitis in HIV-positive patients in sub-Saharan Africa: A review. *J. Int. AIDS Soc.* **2014**, *17*, 1–10. [CrossRef]
68. Rajasingham, R.; Smith, R.M.; Park, B.J.; Jarvis, J.N.; Govender, N.P.; Chiller, T.M.; Denning, D.W.; Loyse, A.; Boulware, D.R. Global burden of disease of HIV-associated cryptococcal meningitis: An updated analysis. *Lancet Infect. Dis.* **2017**, *17*, 873–881. [CrossRef]
69. Coetzee, L.M.; Cassim, N.; Sriruttan, C.; Mhlanga, M.; Govender, N.P.; Glencross, D.K. Cryptococcal antigen positivity combined with the percentage of HIV-seropositive samples with CD4 counts <100 cells/μl identifies districts in South Africa with advanced burden of disease. *PLoS ONE* **2018**, *13*, 1–11. [CrossRef]
70. Parkes-Ratanshi, R.; Achan, B.; Kwizera, R.; Kambugu, A.; Meya, D.; Denning, D.W. Cryptococcal disease and the burden of other fungal diseases in Uganda; Where are the knowledge gaps and how can we fill them? *Mycoses* **2015**, *58*, 85–93. [CrossRef]
71. Wasserman, S.; Engel, M.E.; Griesel, R.; Mendelson, M. Burden of pneumocystis pneumonia in HIV-infected adults in sub-Saharan Africa: A systematic review and meta-analysis. *BMC Infect. Dis.* **2016**, *16*, 1–9. [CrossRef]
72. Denning, D.W.; Alanio, A. GAFFI Fact Sheet Pneumocystis pneumonia. Available online: https://www.gaffi.org/wp-content/uploads/Briefing-note-PCP-GAFFI-December-2017-V4.pdf (accessed on 12 November 2018).
73. Oladele, R.O.; Otu, A.A.; Richardson, M.D.; Denning, D.W. Diagnosis and Management of Pneumocystis Pneumonia in Resource-poor Settings. *J. Health Care Poor Underserved* **2018**, *29*, 107–158. [CrossRef]
74. Nondumiso, C.; Du Toit, M.; Wasserman, S.; Niehlsen, K. Outcomes of HIV-associated pneumocystis pneumonia at a South African referral hospital. *PLoS ONE* **2018**, *13*, 1–13.
75. Guto, J.A.; Bii, C.C.; Denning, D.W. Estimated burden of fungal infections in Kenya. *J. Infect. Dev. Ctries.* **2016**, *10*, 777–784. [CrossRef] [PubMed]
76. Zaki, S.M.; Denning, D.W. Serious fungal infections in Egypt. *Eur. J. Clin. Microbiol. Infect. Dis.* **2017**, *36*, 971–974. [CrossRef] [PubMed]
77. Chekiri-Talbi, M.; Denning, D.W. Burden of fungal infections in Algeria. *Eur. J. Clin. Microbiol. Infect. Dis.* **2017**, *36*, 999–1004. [CrossRef] [PubMed]
78. Badiane, A.S.; Ndiaye, D.; Denning, D.W. Burden of fungal infections in Senegal. *Mycoses* **2015**, *58*, 63–69. [CrossRef]
79. Oladele, R.O.; Denning, D.W. Burden of serious fungal infection in Nigeria. *West Afr. J. Med.* **2014**, *33*, 107–114. [PubMed]
80. Services, Ministry of Health and Social Services. Tuberculosis Infection Control Guidelines Ministry of Health and Social Services. Available online: https://www.challengetb.org/publications/tools/country/Namibia_Infection_Control_Guidelines.pdf (accessed on 3 March 2019).
81. Ricks, P.M.; Mavhunga, F.; Modi, S.; Indongo, R.; Zezai, A.; Lambert, L.A.; DeLuca, N.; Krashin, J.S.; Nakashima, A.K.; Holtz, T.H. Characteristics of multidrug-resistant tuberculosis in Namibia. *BMC Infect. Dis.* **2012**, *12*, 1. [CrossRef]
82. Schutte, A.L. An Overview of Aspergillus (Hyphomycetes) and Associated Teleomorphs in Southern Africa. *Bothalia* **1994**, *24*, 171–185.
83. Denning, D.W.; Cadranel, J.; Beigelman-aubry, C.; Ader, F.; Chakrabarti, A.; Blot, S.; Ullmann, A.J. Chronic pulmonary aspergillosis: Rationale and clinical guidelines for diagnosis and management. *Eur. Respir. J.* **2016**, *47*, 45–68. [CrossRef]
84. Kwizera, R.; Musaazi, J.; Meya, D.B.; Worodria, W.; Bwanga, F.; Kajumbula, H.; Fowler, S.J.; Kirenga, B.J.; Gore, R.; Denning, D.W. Burden of fungal asthma in Africa: A systematic review and meta-analysis. *PLoS ONE* **2019**, *14*, e0216568. [CrossRef]
85. Mhlanga, M.; Govender, N.P.; Thomas, J.; Matlapeng, P.; Mpembe, R.; Lowman, W.; Corcoran, C.; Magobo, R.E.; Govind, C.; Senekal, M. Candida auris in South Africa, 2012–2016. *Emerg. Infect. Dis.* **2018**, *24*, 2036–2040.
86. Leck, A.K.; Thomas, P.A.; Hagan, M.; Kaliamurthy, J.; Ackuaku, E.; John, M.; Newman, M.J.; Codjoe, F.S.; Opintan, J.A.; Kalavathy, C.M.; et al. Aetiology of suppurative corneal ulcers in Ghana and south India, and epidemiology of fungal keratitis. *Br. J. Ophthalmol.* **2002**, *2002*, 1211–1215. [CrossRef]
87. Burton, M.J.; Pithuwa, J.; Okello, E.; Afwamba, I.; Jecinta, J.; Oates, F.; Chevallier, C.; Hall, A.B. Europe PMC Funders Group Microbial Keratitis in East Africa: Why are the outcomes so poor? *Ophthalmic Epidemiol.* **2011**, *18*, 158–163. [CrossRef] [PubMed]

88. Poole, T.R.G.; Hunter, D.L.; Maliwa, E.M.K.; Ramsay, A.R.C. Aetiology of microbial keratitis in northern Tanzania. *Br. J. Ophthalmol.* **2002**, *86*, 941–942. [CrossRef]
89. Waddell, K.M. Childhood blindness and low vision in Uganda. *Eye* **1998**, *12*, 184–192. [CrossRef] [PubMed]
90. Govender, N.P.; Maphanga, T.G.; Zulu, T.G.; Patel, J.; Walaza, S.; Jacobs, C.; Ebonwu, J.I.; Ntuli, S.; Naicker, S.D.; Thomas, J. An Outbreak of Lymphocutaneous Sporotrichosis among Mine-Workers in South Africa. *PLoS Negl. Trop. Dis.* **2015**, *9*, 1–15. [CrossRef]
91. Maphanga, T.G.; Britz, E.; Zulu, T.G.; Mpembe, R.S.; Naicker, S.D.; Schwartz, I.S.; Govender, N.P. In Vitro Antifungal Susceptibility of Yeast and Mold Phases of Isolates of Dimorphic Fungal Pathogen Emergomyces africanus (Formerly Emmonsia sp.) from HIV-Infected South African Patients. *J. Clin. Microbiol.* **2017**, *55*, 1812–1820. [CrossRef]
92. Kallel, A.; Hdider, A.; Fakhfakh, N.; Belhadj, S.; Belhadj-Salah, N.; Bada, N.; Chouchen, A.; Ennigrou, S.; Kallel, K. Tinea capitis: Main mycosis child. Epidemiological study on 10years. *J. Mycol. Medicale* **2017**, *27*, 345–350. [CrossRef]
93. Chepchirchir, A.; Bii, C.; Ndinya-Achola, J.O. Dermatophyte Infections in Primary School Children in Kibera Slums of Nairobi. *East Afr. Med. J.* **2009**, *86*. [CrossRef]
94. Emele, F.E.; Oyeka, C.A. Tinea capitis among primary school children in Anambra state of Nigeria. *Mycoses* **2008**, *51*, 536–541. [CrossRef]
95. Nweze, E.I.; Eke, I.E. Dermatophytes and dermatophytosis in the eastern and southern parts of Africa. *Med. Mycol.* **2018**, *56*, 13–28. [CrossRef] [PubMed]
96. Ali, J.; Yifru, S.; Woldeamanuel, Y. Prevalence of tinea capitis and the causative agent among school children in Gondar, North West Ethiopia. *Ethiop. Med. J.* **2009**, *47*, 261–269. [PubMed]
97. Frossard, A.; Ramond, J.-B.; Seely, M.; Cowan, D.A. Water regime history drives responses of soil Namib Desert microbial communities to wetting events. *Sci. Rep.* **2015**, *5*, 12263. [CrossRef] [PubMed]
98. Evans, S.E.; Dueker, M.E.; Logan, J.R.; Weathers, K.C. The biology of fog: Results from coastal Maine and Namib Desert reveal common drivers of fog microbial composition. *Sci. Total Environ.* **2019**, *647*, 1547–1556. [CrossRef] [PubMed]
99. Nafuka, S.; Misihairabgwi, J.; Bock, R.; Ishola, A.; Sulyok, M.; Krska, R. Variation of Fungal Metabolites in Sorghum Malts Used to Prepare Namibian Traditional Fermented Beverages Omalodu and Otombo. *Toxins* **2019**, *11*, 165. [CrossRef] [PubMed]
100. Uzabakiriho, J.D.; Shikongo, L.; Chimwamurombe, P.M. Co-infection of Tylosema esculentum (Marama bean) seed pods by Alternaria tenuissima and a Phoma spp. *Afr. J. Biotechnol.* **2013**, *12*, 32–37. [CrossRef]
101. Kiehn, T.E.; Nelson, P.E.; Bernard, E.M.; Edwards, F.F.; Armstrong', D. Catheter-Associated Fungemia Caused by Fusarium chlamydosporum in a Patient with Lymphocytic Lymphoma. *J. Clin. Microbiol.* **1985**, *21*, 501–504.
102. Nawases, B.; Uzabakiriho, J.-D.; Chimwamurombe, P. Identification of Fungi Associated with Processed-Food Contamination at Open Markets of Windhoek, Namibia. *J. Pure Appl. Microbiol.* **2018**, *12*, 1489–1494. [CrossRef]
103. GAFFI Improving Outcomes for Patients with Fungal Infections across the World—A Road Map For the Next Decade. Available online: https://www.gaffi.org/wp-content/uploads/GAFFI_Road_Map_interactive-final0415.pdf (accessed on 2 December 2018).
104. English, J. Inequality and Poverty in Namibia: A Gaping Wealth Gap. Available online: https://borgenproject.org/inequality-and-poverty-in-namibia/ (accessed on 2 May 2019).
105. Nickanor, N.; Kazembe, L.N. Increasing Levels of Urban Malnutrition with Rapid Urbanization in Informal Settlements of Katutura, Windhoek: Neighbourhood Differentials and the Effect of Socio-Economic Disadvantage. *World Health Popul.* **2016**, *16*, 5–21. [CrossRef]
106. Gorkom, J. van TB Control in Namibia 2002–2011: Progress and Technical Assistance. *Open Infect. Dis. J.* **2013**, *7*, 23–29. [CrossRef]
107. Chinsembu, K.C. Model and experiences of initiating collaboration with traditional healers in validation of ethnomedicines for HIV/AIDS in Namibia. *J. Ethnobiol. Ethnomed.* **2009**, *5*, 30. [CrossRef] [PubMed]

2

Identification by MALDI-TOF MS of *Sporothrix brasiliensis* Isolated from a Subconjunctival Infiltrative Lesion in an Immunocompetent Patient

Aline M. F. Matos [1], Lucas M. Moreira [2], Bianca F. Barczewski [1], Lucas X. de Matos [1], Jordane B. V. de Oliveira [1], Maria Ines F. Pimentel [3], Rodrigo Almeida-Paes [2], Murilo G. Oliveira [4], Tatiana C. A. Pinto [5], Nelson Lima [6], Magnum de O. Matos [7], Louise G. de M. e Costa [8], Cledir Santos [9,*] and Manoel Marques Evangelista Oliveira [10,11,*]

1 Department of Ophthalmology, University Hospital of the Federal University of Juiz de Fora, Juiz de Fora 36038-330, Brazil; alinemotafreitas@yahoo.com.br (A.M.F.M.); biancabarczewski@hotmail.com (B.F.B.); lucas_xavier_jf@hotmail.com (L.X.d.M.); vargasjordane@yahoo.com.br (J.B.V.d.O.)

2 Laboratório de Micologia, Instituto Nacional de Infectologia Evandro Chagas, Fundação Oswaldo Cruz, Rio de Janeiro 21040-900, Brazil; l.machadomoreira@gmail.com (L.M.M.); rodrigo.paes@ini.fiocruz.br (R.A.-P.)

3 Laboratório de Pesquisa Clínica e Vigilância em Leishmanioses, Instituto Nacional de Infectologia Evandro Chagas, Fundação Oswaldo Cruz, Rio de Janeiro 21040-900, Brazil; maria.pimentel@ini.fiocruz.br

4 Department of Pharmacy, Federal University of Juiz de Fora, Juiz de Fora 36036-900, Brazil; murilogo@hotmail.com

5 Instituto de Microbiologia Paulo de Goes, Universidade Federal do Rio de Janeiro, Rio de Janeiro 21941-901, Brazil; tati.micro@gmail.com

6 CEB—Biological Engineering Centre, University of Minho, Campus de Gualtar, 4710-057 Braga, Portugal; nelson@ie.uminho.pt

7 Imaging Department of Instituto Oncológico, Hospital Nove de Julho, Juiz de Fora 36010-510, Brazil; magnummatos@yahoo.com.br

8 Department of Pathology, Faculty of Medicine of Federal University of Juiz de Fora, Juiz de Fora 36038-330, Brazil; louisepatologia@gmail.com

9 Department of Chemical Science and Natural Resources, BIOREN-UFRO, Universidad de La Frontera, 4811-230 Temuco, Chile

10 Laboratório de Pesquisa Clínica em Dermatozoonoses em Animais Domésticos, Instituto Nacional de Infectologia Evandro Chagas, Fundação Oswaldo Cruz, Rio de Janeiro 21040-900, Brazil

11 Laboratório de Taxonomia, Bioquímica e Bioprospecção de Fungos, Instituto Oswaldo Cruz, Fundação Oswaldo Cruz, Rio de Janeiro 21040-900, Brazil

* Correspondence: cledir.santos@ufrontera.cl (C.S.); manoel.marques@ini.fiocruz.br (M.M.E.O.)

Abstract: Sporotrichosis is a globally distributed subcutaneous fungal infection caused by dimorphic fungi belonging to the *Sporothrix* species complex that affects the skin of limbs predominantly, but not exclusively. A rare case of ocular sporotrichosis in an immunocompetent Brazilian patient from the countryside of Rio de Janeiro State is reported. A 68-year-old woman presented with a subconjunctival infiltrative lesion in the right eye with pre-auricular lymphadenopathy of onset 4 months ago that evolved to suppurative nodular lesions on the eyelids. Conjunctival secretion was evaluated by histopathological examination and inoculated on Sabouraud Dextrose Agar (SDA). Histopathology showed oval bodies within giant cells and other mononucleated histiocytes. Fungus grown on SDA was identified as *Sporothrix* sp. by morphological observations. The isolated strain was finally identified by Matrix-Assisted Laser Desorption/Ionization Time-of-Flight Mass Spectrometry (MALDI-TOF MS) associated with an in-house database enriched with reference *Sporothrix* complex spectra. The strain presented a MALDI spectrum with the ion peaks of the molecular mass profile of *S. brasiliensis*. The patient was adequately treated with amphotericin B subsequently replaced by itraconazole. Due to scars left by the suppurative process, the patient presented poor final visual acuity. The present work presents an overview of ocular sporotrichosis and discusses the diagnostic difficulty that can lead to visual sequelae in these cases.

Keywords: sporotrichosis; *Sporothrix brasiliensis*; ocular sporotrichosis; fungal identification; MALDI-TOF MS

---

1. Introduction

Sporotrichosis is a globally distributed subcutaneous fungal infection caused by the dimorphic fungi belonging to the *Sporothrix* species complex, which can affect animals and humans [1–3]. Particularly common in tropical and subtropical areas, this infection is caused by transcutaneous trauma involving animals, soil, plants, or organic matter contaminated with these fungal species through which the fungal conidia enter into the host [4,5]. Such infection may progress into chronic cutaneous and/or it may spread to internal organs [4,5].

The establishment of this mycosis in ocular adnexa is rare [1,6,7], and typically limited to the eyelids and eyebrows with accompanying regional lymphadenopathy following traumatic inoculation, or as fixed-cutaneous ulceration or conjunctivitis following exposure [6,8,9]. However, the number of reported ocular sporotrichosis cases has increased in recent years [6–8,10–13], especially in Rio de Janeiro State, Brazil, where sporotrichosis is endemic [14,15].

Fungal taxonomy has historically been based mainly on morphological traits. However, the use of fungal macro- and micromorphological traits is not enough to differentiate fungal interspecific similarities. Genotypic traits provide the greatest number of variable characters for fungal taxonomy. However, even with the high level of sensitivity and resolution associated with molecular methods, some problems still arise when molecular techniques are applied for the identification of fungal genetic close related species [16,17]. Nevertheless, DNA housekeeping genes sequencing remains the gold standard for fungal identification [18].

Each morphological and molecular technique presents limitations and advantages when applied to filamentous fungi identification. Both of them can be used as complementary methods for reliable fungal identification. More recently, proteomic profiles generated by Matrix-Assisted Laser Desorption/Ionization Time-of-Flight Mass Spectrometry (MALDI-TOF MS) have been used as an important tool for the identification of filamentous fungi isolated from different substrates [17–22].

For the use of MALDI-TOF MS in clinical settings as a reliable, fast, and cost-effective fungal identification system, an associated and accurate database is essential [22]. In this present work, a clinical case report of ocular sporotrichosis in a patient where the isolate was assessed by classical taxonomy and histopathology and, finally, identified by MALDI-TOF MS associated with an in-house database enriched with reference *Sporothrix* complex spectra, is presented. Moreover, a brief overview of previously reported cases is presented and discussed.

2. Materials and Methods

2.1. Case Report

A 68-year-old female patient, housewife, living in the city of Paty do Alferes in the South Region of Rio de Janeiro State, Brazil, with a history of systemic arterial hypertension and depression, presented with ocular inflammation 4 months ago, which evolved into a suppurative lesion in the right eye (RE), associated with ocular pain. She was examined by different specialists and prescribed with (1) oral antibiotics cephalosporins and beta-lactam (Cefalexin 500 mg each 6 h for 10 days, and amoxicillin/clavulanate 875/125 mg each 12 h for 7 days); (2) topical antibiotics (tobramycin eye drops 3 mg/mL 1 drop each 6 h for 7 days, and moxifloxacin eye drops 0.5% 1 drop each 6 h for 7 days) and, (3) antivirals (acyclovir 400 mg/day for 14 days). The response to the several treatments was null.

The patient was attended at the ophthalmology outpatient clinic of the University Hospital of the Federal University of Juiz de Fora (Juiz de Fora, Minas Gerais State, Brazil), where she underwent an ophthalmologic examination and reported a history of home contact with a cat that had skin lesions. The main diagnostic hypothesis of sporotrichosis was established, but other infiltrative and neoplastic causes needed to be ruled out. The patient was hospitalized to perform imaging exams (Nuclear Magnetic Resonance) of the ocular orbit and biopsy of the conjunctival lesions. The material from the biopsy was sent to pathological examination and conjunctival secretion was sent to microbiological culture.

2.2. *Etiological Agent and Taxonomy*

The specimen underwent routine examinations, which involved bacterioscopic and direct examination with potassium hydroxide, culture on Sabouraud Dextrose Agar 2%, and Mycosel Agar (both from Difco™, Sparks, NV, USA). Cultures were incubated at 25 °C and were observed over 4 weeks for fungal growth. Dimorphism was demonstrated by conversion to the yeast-like form on Brain Heart Infusion Agar (Difco™, Sparks, NV, USA) at 37 °C for 1 week. In addition, a slide culture was performed on Potato Dextrose Agar (Oxoid, Basingstoke, UK)), incubated at 25 °C for 7 days and mounted with Lactophenol Cotton Blue (Fluka, Buchs, Switzerland) for *Sporothrix* identification [1,23].

Identification at the species level was performed using MALDI-TOF MS as previously described by Oliveira and collaborators [16]. Briefly, 10^6 yeast cells were transferred from the culture plate at the yeast phase (c.a. 1 μg) to a 500 μL tube containing 20 μL of 25% formic acid in water (*v*/*v*). The supernatant of each sample (1 μL) was transferred to a paraffin film surface and 2 μL of the matrix solution α-cyano-4-hydroxycinnamic acid (CHCA, Fluka, Buchs, Switzerland) was added and gently mixed. Each suspension (1 μL) was spotted onto the MALDI-TOF MS stainless plate (FlexiMass™, Shimadzu Biotech, Manchester, UK) in triplicate to test reproducibility. Previous to the spectra acquisition, the sample was air-dried at room temperature.

Spectra acquisition was performed on an Axima LNR equipment (Kratos Analytical, Shimadzu, Manchester, UK) equipped with a nitrogen laser (337 nm). Mass range from 2000 to 20,000 Da was recorded using the MALDI-TOF MS linear mode with a delay of 104 ns and an acceleration voltage of +20 kV. For the CHCA matrix, final spectra were generated by summing 2 laser shots accumulated per profile and 100 profiles produced per sample, leading to 200 laser shots per summed spectrum.

The resulting peak lists were exported into the SARAMIS® software package (Spectral Archiving and Microbial Identification System; AnagnosTec, Postdam-Golm, Germany) for data analyses, and final microbial identification was achieved. Peak lists of individual samples were compared to the SARAMIS™ database, enriched with in-house spectra of reference strains of the *Sporothrix* species complex, generating a ranked list of matching spectra.

2.3. *In Vitro Susceptibility Testing*

The susceptibility testing of the clinical *Sporothrix* strain was determinate according to the Clinical and Laboratory Standards Institute (CLSI) through the M38-A2 broth microdilution protocol [24]. According to the M38-A2 protocol, a filamentous form of the *Sporothrix* strain was used. The inocula were prepared with fungal conidia in sterile saline solution after incubation of each strain on PDA for 7 days at 35 °C. The antifungal susceptibility tests were conducted according to the M38-A2 protocol of the CLSI, using RPMI 1640 medium buffered to pH 7.0 with 0.165 mol/L morpholinepropanesulfonic acid (Thermo Fisher Scientific®, Waltham, MA, USA), resulting in the final concentration of 0.4×10^4 to 5×10^4 CFU cells/mL, and incubation at 35 °C for 72 h.

Minimal inhibitory concentrations (MIC) were the lowest drug concentrations that produced either complete fungal growth inhibition for amphotericin B, itraconazole, and posaconazole. For ketoconazole, the MIC was the lowest concentration that produced a 50% reduction in growth, and for terbinafine it was the lowest concentration that produced at least an 80% reduction in growth. All the tests were performed in triplicate for the isolate. Tests were validated only if MIC values for these strains were within the range described in the M38-A2 reference document. MICs were determined by visual inspection after 72 h of incubation at 35 °C, as described (CLSI 2008). The reference strains *Aspergillus fumigatus* ATCC 204305 and *A. flavus* ATCC 204304 were included in each experiment for control.

2.4. Ethics Statement

This study was approved by the Research Ethics Committee of INI/Fiocruz (CAAE 28063114.2.0000.5262, approved on 24 April 2014). The patient was an adult and provided written informed consent. All data analyzed were anonymized.

3. Results

The patient presented with a subconjunctival infiltrative lesion forming an erythematous aspect throughout the ocular circumference of the right eye, occupying the upper and lower fornixes. She also presented with lower eyelid lesions, erythema, and thickening of the skin, ulceration in the inner corner of the eye, and ipsilateral pre-auricular lymphadenopathy. The left eye (LE) had no alterations (Figure 1A,B).

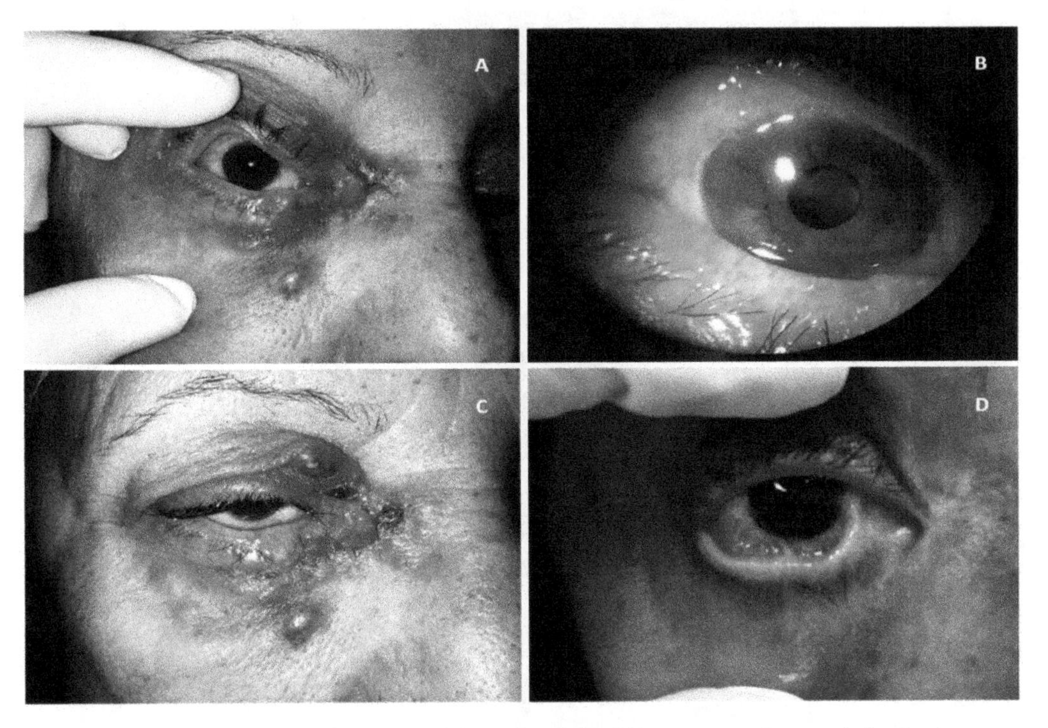

Figure 1. Lesions observed at the first clinical evaluation, (**A**) suppurative nodular lesions on eyelids and (**B**) subconjunctival infiltrative lesion occupying upper and lower fornices; (**C**) evolution of lesions in a short time interval, which increased in number and coalesced; (**D**) diffuse symblepharon after resolution of the acute phase, causing restriction of ocular abduction.

Nuclear Magnetic Resonance of the orbits revealed an expansive and infiltrative lesion involving the right eyelids and periorbital region extending to the ipsilateral lacrimal duct, with apparent involvement of the conjunctiva and lacrimal gland (Figure 2).

Bacterioscopy, direct examination with potassium hydroxide, and bacterial culture of the secretion collected from the conjunctival sac were all negative.

A biopsy of bulbar conjunctiva showed a granulomatous inflammatory process associated with suppurative areas containing small round to oval bodies within giant cells and other mononucleated histiocytes. Grocott methenamine silver and Periodic Acid-Schiff (PAS) (both from Agilent, Santa Clara, CA, USA) highlighted these rather conspicuous yeasts (Figure 3).

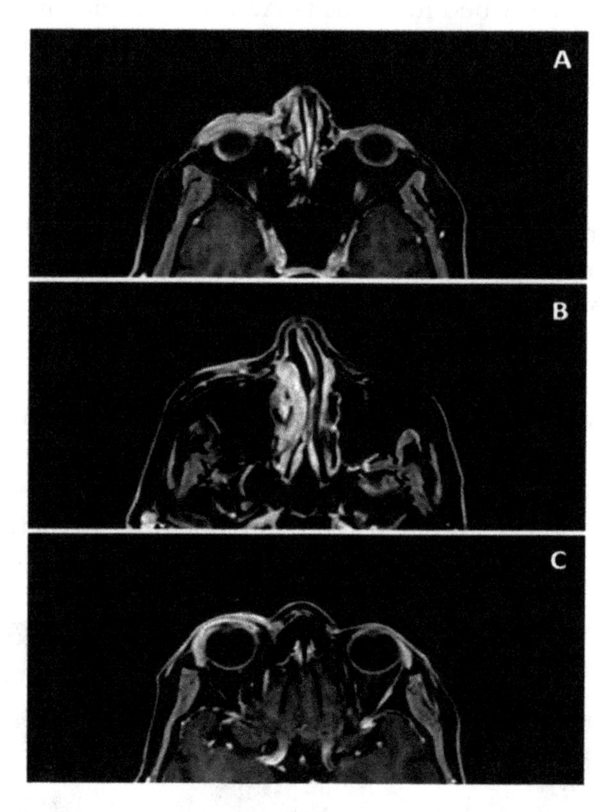

Figure 2. Nuclear Magnetic Resonance of orbits showing expansive lesion affecting: (**A**) conjunctiva and eyelids on the right side; (**B**) right inferior lacrimal system; (**C**) right lacrimal gland.

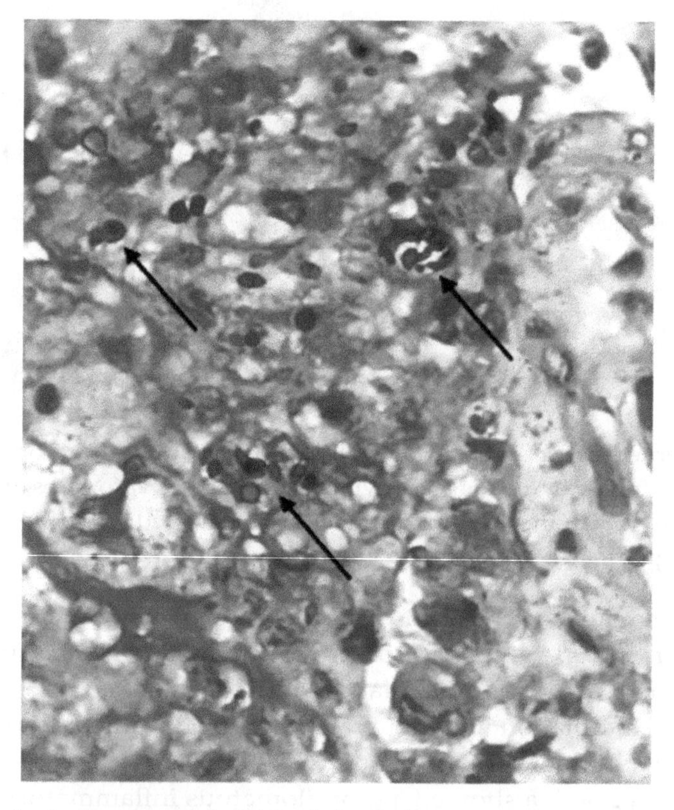

Figure 3. Histopathological stain with PAS of the bulbar conjunctiva (1000×)—arrows showing conspicuous round and oval yeasts within the granulomas.

Culture of the conjunctival secretion on Sabouraud Dextrose Agar showed fungal growth with raised moist colonies with a membranous aspect and a wrinkled or folded surface. Initially, the colonies were white to cream-colored, later turning brown to dark grey and black. Hyaline hyphae, septate and branched, with small conidiophores with claviform conidia implants in groups, were microscopically observed. It was then identified as *Sporothrix* sp. by traditional methodology through micro- and macromorphology, based on visual and light microscopy observations, respectively.

Subsequently, the isolate was assessed by MALDI-TOF MS and acquired spectra were compared with reference spectra stored in an accurate in-house MALDI spectra reference database containing spectral data for the *Sporothrix* species complexes. The obtained spectra presented ion peaks compatible with the molecular mass of *Sporohtrix brasiliensis* (Figure 4).

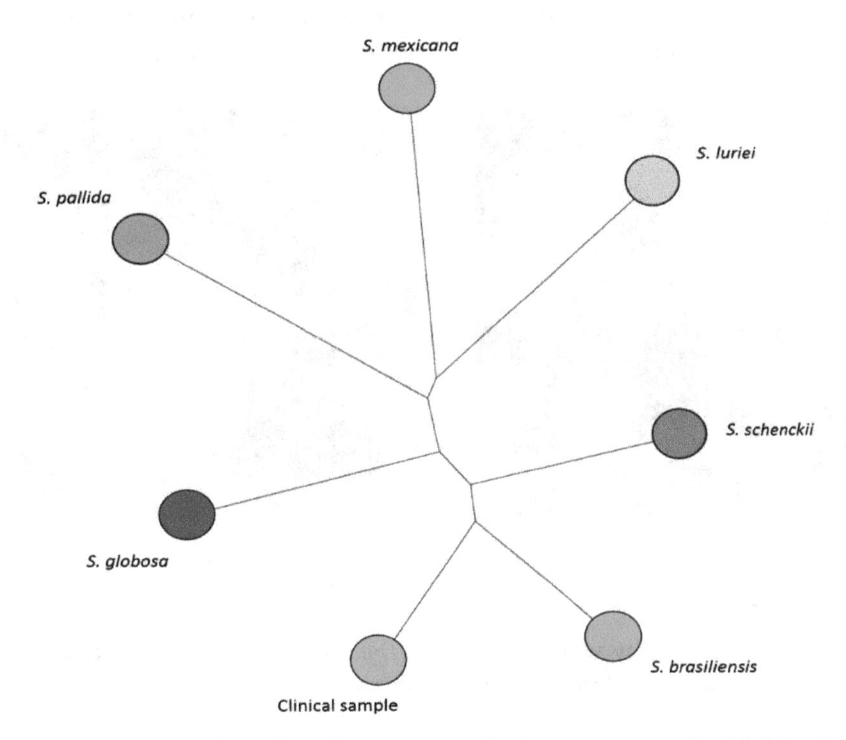

Figure 4. Neighbor-Joining tree based on Pearson correlation constructed with Matrix-Assisted Laser Desorption/Ionization Time-of-Flight Mass Spectrometry (MALDI-TOF MS) spectra of 7 *Sporothrix* isolates showing the 6 reference strains color-coded according to the *Sporothrix* species [*S. luriei* CBS 937.72 (Amber), *S. pallida* SPA8 (Teal), *S. mexicana* MUM 11.02 (Cyan), *S. schenckii* IPEC 27722 (Brown), *S. globosa* IPEC 27135 (Lavender) and *S. brasiliensis* CBS 120339/IPEC 16490 (Green)] and 1 clinical sample with color according to the *Sporothrix* species identified by MALDI-TOF MS.

While awaiting the results of the examinations, it was possible to observe the evolution of the skin lesions that became individualized, forming darkened nodules that suppurated (Figure 1C). After the diagnostic confirmation, in vitro susceptibility testing with amphotericin B, itraconazole, ketoconazole, posaconazole, and terbinafine were performed by the microdilution method [24] and revealed MICs of 1 μg/mL, 1 μg/mL, 0.25 μg/mL, 1 μg/mL and 0.125 μg/mL, respectively. According to the epidemiological cut-off values [25], this isolate was classified as a wild-type, thus justifying the initial antifungal treatment with amphotericin B 50 mg/day, subsequently replaced by itraconazole 200 mg/day.

The clinical condition evolved with noticeable improvement after the first week of treatment, with complete epithelialization of the ulcers after 1 month of medication. On the return visit, 3 months after the onset of the clinical condition, the patient presented with complete ptosis of the upper eyelid and symblepharon in the lower temporal region and upper and lower nasal regions, restricting ocular abduction (Figure 1D). The patient was treated with itraconazole 200 mg/day for a total period of

8 months. Treatment time was defined by the infectologist in charge of the case, who decided to keep the medication until the postoperative period of the symblepharon correction surgery.

The patient was then submitted to a symblepharon correction with a labial mucosa graft. After 8 months of symblepharoplasthy, the patient was evaluated for ptosis correction surgery that revealed severe ptosis with good levator muscle function in the affected eye. She was, therefore, submitted to reinsertion of the aponeurosis levator muscle of the upper eyelid.

Previously described surgeries have improved palpebral opening and ocular motility, but not all sequelae could be completely resolved with the surgery. A cicatricial epicantal fold was presented in the RE, with symblepharon recurrent in the nasal region, without diplopia in the primary position of gaze. Final corrected visual acuity was RE 20/80 (with rigid contact lenses) and LE 20/20. Low vision in the RE was probably due to residual astigmatism caused by scarring corneal irregularities and dry eye (Figure 5). Funduscopic examination revealed no abnormalities in both eyes.

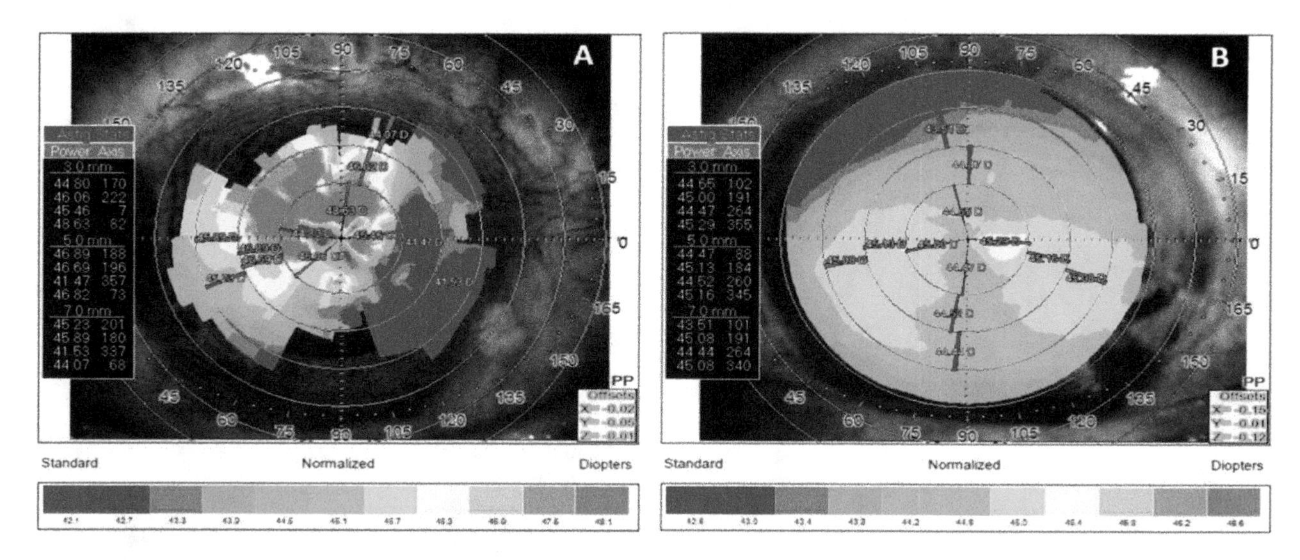

Figure 5. Comparative topography where (**A**) shows the right eye with irregular astigmatism caused by corneal scars and (**B**) shows the left eye with regular astigmatism.

4. Discussion

Sporotrichosis is a mycosis that predominantly affects the skin of the lower and upper limbs (hands, forearms, feet, and legs) and may also affect the face [26]. The eyes may be affected, with intraocular involvement, in the form of endophthalmitis, retinal granuloma, granulomatous necrotizing retinochoroiditis, and granulomatous uveitis, or the extraocular form, affecting tissues adjacent to the bulb and ocular adnexa [6,11,27,28].

Ramírez Soto found in a literature review of 16 publications related to endophthalmitis caused by sporotrichosis, 8 of which refer to exogenous endophthalmitis, an isolated form that occurs by direct ocular traumatic inoculation, and 10 cases of endogenous endophthalmitis, with hematogenous dissemination from other lesions [11]. In addition, the same author in a different study [6] observed 65 cases of patients with alterations only in the periocular appendages, from which the lymphocutaneous form was the most frequent, followed by the cutaneous-fixed form. However, other authors have described the fixed cutaneous form as the most prevalent, followed by lymphocutaneous [10,26].

The case herein described presented ocular involvement, with diffuse infiltration of the conjunctiva and corneal and extraocular alterations, with the involvement of the eyelids, lacrimal gland, and nasolacrimal duct, as evidenced by the imaging examination (Figure 2). The conjunctival and lacrimal sac involvement is even more rarely described [6] and the delay in the diagnosis of this patient may have contributed to the lesion extension. Other authors have also reported periods of 1–3 months until a definitive diagnosis is obtained [8,10]. Overall, the clinical presentation of the disease is variable [6] and can be misdiagnosed with other diseases such as hordeolum, lacrimal sac tumor [26], bacterial abscesses, and basal cell or sebaceous carcinoma [10] until a final diagnosis is obtained.

After the description of *Sporothrix schenckii* species complex by Marimon and collaborators [1], 150 cases of sporotrichosis with ocular involvement were published. Just 4.6% out of these cases had the isolation and fungal characterization at the species level (Table 1), being one case by *S. pallida* [29] and six other cases by *S. brasiliensis* [4,30–32]. However, in one of these studies [4], the fungus was not isolated from a clinical specimen recovered from the eyes, being disseminated disease presumed as a possible explanation to the diagnosis of ocular sporotrichosis associated to *S. brasiliensis*.

Table 1. Cases of sporotrichosis with ocular involvement were published after the description of *Sporothrix schenckii* species complex.

Authors	Cases (*n*)	Identification Method	Species	References
Negroni et al., 2007	1	Culture and Histopathology	*Sporothrix* spp.	[33]
Inokuma, et al., 2010	1	Culture, Histopathology and Serology	*Sporothrix* spp.	[34]
Iyengar et al., 2010	1	Culture	*Sporothrix* spp.	[8]
Kashima et al., 2010	1	Histopathology and Serology	*Sporothrix* spp.	[12]
Silva-Vergara et al., 2012	1	Culture and Molecular	*S. brasiliensis*	[30]
Morrison et al., 2013	1	Culture and Molecular	*S. pallida*	[22]
Ferreira et al., 2014	1	Culture	*Sporothrix* spp.	[35]
Freitas et al., 2014	4	Culture	*Sporothrix* spp.	[36]
Zhang et al., 2014	1	Culture	*Sporothrix* spp.	[37]
de Macedo et al., 2015	1	Culture and Molecular	*S. brasiliensis*	[31]
Fan et al., 2016	10	Culture	*Sporothrix* spp.	[10]
Ramírez Soto, 2016	21	Culture	*Sporothrix* spp.	[6]
Zhang et al., 2016	72	Culture and Serology	*Sporothrix* spp.	[26]
Biancardi et al., 2017	3	Culture and Molecular	*S. brasiliensis*	[4]
Yamagata et al., 2017	3	Culture	*Sporothrix* spp.	[15]
Ling et al., 2018	1	Culture and Histopathology	*Sporothrix* spp.	[38]
Lacerda Filho et al., 2019	1	Culture and Molecular	*S. brasiliensis*	[32]
Arinelli et al., 2019	26	Culture	*Sporothrix* spp.	[39]

MALDI-TOF MS has been proven capable to identify fungal species belonging to the different fungal genus, as showed in previous studies available in the literature [16–18,20–22].

Oliveira and collaborators [16] successfully identified species of *Sporothrix* and have established an accurate in-house MALDI spectral database, which is comprehensive in terms of *Sporothrix* MALDI spectra number and geographical distribution of strain origins. Moreover, all reference spectra available in the mentioned in-house database are validated by molecular biological analysis.

In the present study, the use of MALDI-TOF MS associated with an in-house database enriched with reference *Sporothrix* complex spectra, in addition to the histopathology and morphology observations, were performed to obtain a reliable fungal identification at the species level. The isolated species, in this case, was *Sporothrix brasiliensis*.

As shown in a previous study [16], the MALDI-TOF MS technique is as good as the partial sequence of genes (calmodulin, beta-tubulin, and chitin synthase, for example) for the identification of *S. brasiliensis*. The report of sporotrichosis by *S. brasiliensis* corroborates epidemiological studies that describe the current high circulation of this species in Rio de Janeiro State, especially involving adult women who are out of the labor market [5,23,40].

The *Sporothrix schenckii* complex is found worldwide. Clinical cases have commonly been reported in geographical regions with high humidity and temperatures, which presents optimal conditions for fungal growth. However, cases of sporotrichosis have also been reported in temperate climate regions, such as European countries [5,41] and Chile [42]. Individual species of *Sporothrix* have presented different geographical distribution. *Sporothrix globosa* and *S. schenckii* have been reported from several locations in both Eastern and Western Hemispheres [43]. *Sporothrix mexicana* has been recovered from the environment in Australia, Mexico, and Portugal, where it has been reported to cause occasional human infections [41,43].

In contrast, *S. brasiliensis* seems to be restricted to Latin America, and as of 2017, it has been described only in Brazil, where it is responsible for an ongoing epidemic in cats, other animals, and humans [2,3,23,44]. In a review of human sporotrichosis reported in Rio de Janeiro State, da Silva and collaborators [40] reported two cases in the municipality of Paty do Alferes in the period from 1997 to 2007. Data obtained for the period from 2008 to 2017, from the Management of Vector-Transmitted Diseases and Zoonoses Sector of the Health Secretariat of Rio de Janeiro State, demonstrated two additional cases from Paty do Alferes, which were reported in the year 2012. It is worth noting that the sporotrichosis endemic is slowly spreading through the municipalities of Rio de Janeiro State and other Brazilian states [44,45].

An important factor in understanding the spread of this epidemic is the social vulnerability of these patients and the lack of adequate information for the population. In the case reported in this article, the patient herself reported that it was common practice in her neighborhood to bury cats that died with signs of injury, which leads to soil contamination and disease spread. In addition, there is a lack of a referral center for the care of these cases, especially in the inner cities, which leads to delayed diagnosis, with a greater chance of disease transmission and greater sequelae for affected patients. In this context, it is here emphasized the absence of fast methodologies for diagnosis of sporotrichosis, with the MALDI-TOF MS a good tool in the diagnosis of sporotrichosis, when a reliable spectral database is available.

5. Conclusions

The *S. brasiliensis* strain isolated from the subconjunctival infiltrative lesion of the patient was successfully identified using a MALDI-TOF MS associated with an in-house database enriched with reference *Sporothrix* complex spectra. Stablishing an in-house database is a good way to overcome limitation associated with the use of MALDI-TOF MS in fungal identification. The patient was adequately treated with amphotericin B subsequently replaced by itraconazole. Due to scars left by the suppurative process, the patient presented poor final visual acuity.

Author Contributions: A.M.F.M., L.M.M., and M.M.E.O. conceived the study and performed the study design. A.M.F.M., B.F.B., L.X.d.M., and J.B.V.d.O., performed, attended examinations, and ophthalmologic treatment. L.M.M., M.G.O., and R.A.-P., performed the microbiological assays. T.C.A.P., N.L., C.S., and M.M.E.O., performed MALDI-TOF MS analysis. M.d.O.M. performed the analysis of the imaging exams. L.G.d.M.eC. performed the analysis of the histopathological exams. A.M.F.M., L.M.M., M.I.F.P., C.S., and M.M.E.O., wrote the paper. All authors have read and agreed to the published version of the manuscript.

Funding: Financial support for this work by Fundação de Amparo à Pesquisa do Rio de Janeiro [FAPERJ] (Grants: INST E-26/010.001784/2016; JCNE E-26/203.301/2017), by Conselho Nacional de Desenvolvimento Científico e Tecnológico [CNPq] (Grant Proc. 409227/2016-1). This study was financed in part by the Coordenação de Aperfeiçoamento de Pessoal de Nível Superior - Brasil (CAPES). MALDI-TOF MS analyses were partially developed using equipment funded by CONICYT/Chile through the project Fondequip EQM160054 2016. The Universidad de La Frontera (Temuco, Chile) partially funded this work through the Project DIUFRO PIA19-0001. Furthermore, this study was also supported by FCT under the scope of the strategic funding of UID/BIO/04469/2019 unit and BioTecNorte operation (NORTE-01-0145-FEDER-000004) funded by the European Regional Development Fund under the scope of Norte2020 - Programa Operacional Regional do Norte.

References

1. Marimon, R.; Cano, J.; Gené, J.; Sutton, D.A.; Kawasaki, M.; Guarro, J. Sporothrix brasiliensis, S. globosa, and S. mexicana, three new Sporothrix species of clinical interest. *J. Clin. Microbiol.* **2007**, *45*, 3198–3206. [CrossRef] [PubMed]
2. Zhang, Y.; Hagen, F.; Stielow, B.; Rodrigues, A.M.; Samerpitak, K.; Zhou, X.; Feng, P.; Yang, L.; Chen, M.; Deng, S.; et al. Phylogeography and evolutionary patterns in Sporothrix spanning more than 14,000 human and animal case reports. *Persoonia* **2015**, *35*, 1–20. [CrossRef] [PubMed]

3. Boechat, J.S.; Oliveira, M.M.E.; Almeida-Paes, R.; Gremião, I.D.F.; Machado, A.C.S.; Oliveira, R.V.C.; Figueiredo, A.B.F.; Rabello, V.B.S.; Silva, K.B.L.; Zancopé-Oliveira, R.M.; et al. Feline sporotrichosis: Associations between clinical-epidemiological profiles and phenotypic-genotypic characteristics of the etiological agents in the Rio de Janeiro epizootic area. *Mem. Inst. Oswaldo Cruz* **2018**, *113*, 185–196. [CrossRef] [PubMed]
4. Biancardi, A.L.; Freitas, D.F.; Valviesse, V.R.; Andrade, H.B.; de Oliveira, M.M.; do Valle, A.C.; Zancope-Oliveira, R.M.; Galhardo, M.C.; Curi, A.L. Multifocal choroiditis in disseminated sporotrichosis in patients with HIV/AIDS. *Retin. Cases Brief Rep.* **2017**, *11*, 67–70. [CrossRef] [PubMed]
5. Chakrabarti, A.; Bonifaz, A.; Gutierrez-Galhardo, M.C.; Mochizuki, T.; Li, S. Global epidemiology of sporotrichosis. *Med. Mycol.* **2015**, *53*, 3–14. [CrossRef] [PubMed]
6. Ramírez Soto, M.C. Sporotrichosis in the Ocular Adnexa: 21 Cases in an Endemic Area in Peru and Review of the Literature. *Am. J. Ophthalmol.* **2016**, *162*, 173–179.e3. [CrossRef]
7. Hampton, D.E.; Adesina, A.; Chodosh, J. Conjunctival sporotrichosis in the absence of antecedent trauma. *Cornea* **2002**, *21*, 831–833. [CrossRef]
8. Iyengar, S.S.; Khan, J.A.; Brusco, M.; FitzSimmons, C.J. Cutaneous Sporothrix schenckii of the human eyelid. *Ophthalmic Plast. Reconstr. Surg.* **2010**, *26*, 305–306. [CrossRef]
9. Schubach, A.; de Lima Barros, M.B.; Schubach, T.M.; Francesconi-do-Valle, A.C.; Gutierrez-Galhardo, M.C.; Sued, M.; de Matos Salgueiro, M.; Fialho-Monteiro, P.C.; Reis, R.S.; Marzochi, K.B.; et al. Primary conjunctival sporotrichosis: Two cases from a zoonotic epidemic in Rio de Janeiro, Brazil. *Cornea* **2005**, *24*, 491–493. [CrossRef]
10. Fan, B.; Wang, J.F.; Zheng, B.; Qi, X.Z.; Song, J.Y.; Li, G.Y. Clinical features of 10 cases of eyelid sporotrichosis in Jilin Province (Northeast China). *Can. J. Ophthalmol.* **2016**, *51*, 297–301. [CrossRef]
11. Ramírez Soto, M.C. Differences in clinical ocular outcomes between exogenous and endogenous endophthalmitis caused by *Sporothrix*: A systematic review of published literature. *Br. J. Ophthalmol.* **2018**, *102*, 977–982. [CrossRef] [PubMed]
12. Kashima, T.; Honma, R.; Kishi, S.; Hirato, J. Bulbar conjunctival sporotrichosis presenting as a salmon-pink tumor. *Cornea* **2010**, *29*, 573–576. [CrossRef] [PubMed]
13. Gremião, I.D.; Miranda, L.H.; Reis, E.G.; Rodrigues, A.M.; Pereira, S.A. Zoonotic Epidemic of Sporotrichosis: Cat to Human Transmission. *PLoS Pathog.* **2017**, *13*, e1006077. [CrossRef] [PubMed]
14. Freitas, D.F.; Valle, A.C.; da Silva, M.B.; Campos, D.P.; Lyra, M.R.; de Souza, R.V.; Veloso, V.G.; Zancopé-Oliveira, R.M.; Bastos, F.I.; Galhardo, M.C. Sporotrichosis: An emerging neglected opportunistic infection in HIV-infected patients in Rio de Janeiro, Brazil. *PLoS Negl. Trop. Dis.* **2014**, *8*, e3110. [CrossRef] [PubMed]
15. Yamagata, J.P.M.; Rudolph, F.B.; Nobre, M.C.L.; Nascimento, L.V.; Sampaio, F.M.S.; Arinelli, A.; Freitas, D.F. Ocular sporotrichosis: A frequently misdiagnosed cause of granulomatous conjunctivitis in epidemic areas. *Am. J. Ophthalmol. Case Rep.* **2017**, *8*, 35–38. [CrossRef] [PubMed]
16. Oliveira, M.M.; Santos, C.; Sampaio, P.; Romeo, O.; Almeida-Paes, R.; Pais, C.; Lima, N.; Zancopé-Oliveira, R.M. Development and optimization of a new MALDI-TOF protocol for identification of the *Sporothrix* species complex. *Res. Microbiol.* **2015**, *166*, 102–110. [CrossRef]
17. Santos, C.; Lima, N.; Sampaio, P.; Pais, C. Matrix-assisted laser desorption/ionization time-of-flight intact cell mass spectrometry to detect emerging pathogenic Candida species. *Diagn. Microbiol. Infect. Dis.* **2011**, *71*, 304–308. [CrossRef]
18. Lima, M.S.; Lucas, R.C.; Lima, N.; Polizeli, M.L.T.M.; Santos, C. Fungal Community Ecology Using MALDI-TOF MS Demands Curated Mass Spectral Databases. *Front. Microbiol.* **2019**, *315*, 1–4. [CrossRef]
19. Santos, C.; Paterson, R.R.; Venâncio, A.; Lima, N. Filamentous fungal characterizations by matrix-assisted laser desorption/ionization time-of-flight mass spectrometry. *J. Appl. Microbiol.* **2010**, *108*, 375–385. [CrossRef]
20. Pereira, L.; Dias, N.; Santos, C.; Lima, N. The use of MALDI-TOF ICMS as an alternative tool for Trichophyton rubrum identification and typing. *Enferm. Infecc. Microbiol. Clin.* **2014**, *32*, 11–17. [CrossRef]
21. Lima, N.; Santos, C. MALDI-TOF MS for identification of food spoilage filamentous fungi. *Curr. Opin. Food Sci.* **2017**, *13*, 26–30. [CrossRef]

22. Flórez-Muñoz, S.V.; Gómez-Velásquez, J.C.; Loaiza-Díaz, N.; Soares, C.; Santos, C.; Lima, N.; Mesa-Arango, A.C. ITS rDNA gene analysis versus MALDI-TOF MS for identification of Neoscytalidium dimidiatum isolated from onychomycosis and dermatomycosis cases in Medellin (Colombia). *Microorganisms* **2019**, *7*, 306. [CrossRef] [PubMed]
23. Oliveira, M.M.; Almeida-Paes, R.; Muniz, M.M.; Gutierrez-Galhardo, M.C.; Zancope-Oliveira, R.M. Phenotypic and molecular identification of Sporothrix isolates from an epidemic area of sporotrichosis in Brazil. *Mycopathologia* **2011**, *172*, 257–267. [CrossRef] [PubMed]
24. Clinical and Laboratory Standards Institute. *Reference Method for Broth Dilution Antifungal Susceptibility Testing of Filamentous Fungi*, 2nd ed.; Approved Standard M38-A2; Clinical and Laboratory Standards Institute: Wayne, PA, USA, 2008.
25. Espinel-Ingroff, A.; Abreu, D.P.B.; Almeida-Paes, R.; Brilhante, R.S.N.; Chakrabarti, A.; Chowdhary, A.; Hagen, F.; Córdoba, S.; Gonzalez, G.M.; Govender, N.P.; et al. Multicenter, International Study of MIC/MEC Distributions for Definition of Epidemiological Cutoff Values for *Sporothrix* Species Identified by Molecular Methods. *Antimicrob. Agents Chemother.* **2017**, *61*. [CrossRef]
26. Zhang, Y.; Wang, Y.; Cong, L.; Yang, H.; Cong, X. Eyelid sporotrichosis: Unique clinical findings in 72 patients. *Australas J. Dermatol.* **2016**, *57*, 44–47. [CrossRef]
27. Curi, A.L.; Félix, S.; Azevedo, K.M.; Estrela, R.; Villar, E.G.; Saraça, G. Retinal granuloma caused by *Sporothrix schenckii*. *Am. J. Ophthalmol.* **2003**, *136*, 205–207. [CrossRef]
28. Font, R.L.; Jakobiec, F.A. Granulomatous necrotizing retinochoroiditis caused by Sporotrichum schenkii. Report of a case including immunofluorescence and electron microscopical studies. *Arch. Ophthalmol.* **1976**, *94*, 1513–1519. [CrossRef]
29. Morrison, A.S.; Lockhart, S.R.; Bromley, J.G.; Kim, J.Y.; Burd, E.M. An environmental Sporothrix as a cause of corneal ulcer. *Med. Mycol. Case Rep.* **2013**, *2*, 88–90. [CrossRef]
30. Silva-Vergara, M.L.; de Camargo, Z.P.; Silva, P.F.; Abdalla, M.R.; Sgarbieri, R.N.; Rodrigues, A.M.; dos Santos, K.C.; Barata, C.H.; Ferreira-Paim, K. Disseminated Sporothrix brasiliensis infection with endocardial and ocular involvement in an HIV-infected patient. *Am. J. Trop. Med. Hyg.* **2012**, *86*, 477–480. [CrossRef]
31. de Macedo, P.M.; Sztajnbok, D.C.; Camargo, Z.P.; Rodrigues, A.M.; Lopes-Bezerra, L.M.; Bernardes-Engemann, A.R.; Orofino-Costa, R. Dacryocystitis due to Sporothrix brasiliensis: A case report of a successful clinical and serological outcome with low-dose potassium iodide treatment and oculoplastic surgery. *Br. J. Dermatol.* **2015**, *172*, 1116–1119. [CrossRef]
32. Lacerda Filho, A.M.; Cavalcante, C.M.; Da Silva, A.B.; Inácio, C.P.; de Lima-Neto, R.G.; de Andrade, M.C.L.; Magalhães, O.M.C.; Dos Santos, F.A.G.; Neves, R.P. High-Virulence Cat-Transmitted Ocular Sporotrichosis. *Mycopathologia* **2019**, *184*, 547–549. [CrossRef] [PubMed]
33. Negroni, R.; Maiolo, E.; Arechavala, A.; Santiso, G.; Bianchi, M.H. Clinical cases in Medical Mycology: Case no. 25. *Rev. Iberoam. Micol.* **2007**, *24*, 79–81. [CrossRef]
34. Inokuma, D.; Shibaki, A.; Shimizu, H. Two cases of cutaneous sporotrichosis in continental/microthermal climate zone: Global warming alert? *Clin. Exp. Dermatol.* **2010**, *35*, 668–669. [CrossRef] [PubMed]
35. Ferreira, C.P.; Nery, J.A.; de Almeida, A.C.; Ferreira, L.C.; Corte-Real, S.; Conceição-Silva, F. Parinaud's oculoglandular syndrome associated with Sporothrix schenckii. *IDCases* **2014**, *1*, 38–39. [CrossRef]
36. Freitas, D.F.; Lima, I.A.; Curi, C.L.; Jordão, L.; Zancopé-Oliveira, R.M.; Valle, A.C.; Galhardo, M.C.; Curi, A.L. Acute dacryocystitis: Another clinical manifestation of sporotrichosis. *Mem. Inst. Oswaldo Cruz* **2014**, *109*, 262–264. [CrossRef]
37. Zhang, Y.; Pyla, V. Palpebral sporotrichosis. *Int. J. Dermatol.* **2014**, *53*, e356–e357. [CrossRef]
38. Ling, J.L.L.; Koh, K.L.; Tai, E.; Sakinah, Z.; Nor Sharina, Y.; Hussein, A. A Case of Acute Granulomatous Conjunctivitis Caused by Cat-transmitted Sporothrix schenckii. *Cureus* **2018**, *10*, e3428. [CrossRef]
39. Arinelli, A.; Aleixo, A.L.Q.D.; Freitas, D.F.S.; do Valle, A.C.F.; Almeida-Paes, R.; Gutierrez-Galhardo, M.C.; Curi, A.L.L. Ocular Sporotrichosis: 26 Cases with Bulbar Involvement in a Hyperendemic Area of Zoonotic Transmission. *Ocul. Immunol. Inflamm.* **2019**, *14*, 1–8. [CrossRef]
40. Silva, M.B.; Costa, M.M.; Torres, C.C.; Galhardo, M.C.; Valle, A.C.; Magalhães, M.E.A.; Sabroza, P.C.; Oliveira, R.M. Urban sporotrichosis: A neglected epidemic in Rio de Janeiro, Brazil. *Cad. Saude Publica* **2012**, *28*, 1867–1880. [CrossRef]
41. Dias, N.M.; Oliveira, M.M.; Santos, C.; Zancope-Oliveira, R.M.; Lima, N. Sporotrichosis caused by *Sporothrix mexicana*, Portugal. *Emerg. Infect. Dis.* **2011**, *17*, 1975–1976. [CrossRef]

42. Rodrigues, A.M.; Cruz Choappa, R.; Fernandes, G.F.; de Hoog, G.S.; de Camargo, Z.P. Sporothrix chilensis sp. nov. (Ascomycota: Ophiostomatales), a soil-borne agent of human sporotrichosis with mild-pathogenic potential to mammals. *Fungal Biol.* **2016**, *120*, 246–264. [CrossRef] [PubMed]
43. Rodrigues, A.M.; de Hoog, S.; de Camargo, Z.P. Emergence of pathogenicity in the Sporothrix schenckii complex. *Med. Mycol.* **2013**, *51*, 405–412. [CrossRef] [PubMed]
44. Oliveira, M.M.; Maifrede, S.B.; Ribeiro, M.A.; Zancope-Oliveira, R.M. Molecular identification of *Sporothrix* species involved in the first familial outbreak of sporotrichosis in the state of Espírito Santo, southeastern Brazil. *Mem. Inst. Oswaldo Cruz* **2013**, *108*, 936–938. [CrossRef] [PubMed]
45. Rodrigues, A.M.; de Melo Teixeira, M.; de Hoog, G.S.; Schubach, T.M.; Pereira, S.A.; Fernandes, G.F.; Bezerra, L.M.; Felipe, M.S.; de Camargo, Z.P. Phylogenetic analysis reveals a high prevalence of Sporothrix brasiliensis in feline sporotrichosis outbreaks. *PLoS Negl. Trop. Dis.* **2013**, *7*, e2281. [CrossRef]

3

Golgi Reassembly and Stacking Protein (GRASP) Participates in Vesicle-Mediated RNA Export in *Cryptococcus neoformans*

Roberta Peres da Silva [1,2,†], Sharon de Toledo Martins [3,†], Juliana Rizzo [4], Flavia C. G. dos Reis [3], Luna S. Joffe [5], Marilene Vainstein [6], Livia Kmetzsch [6], Débora L. Oliveira [5], Rosana Puccia [1], Samuel Goldenberg [3], Marcio L. Rodrigues [3,4] and Lysangela R. Alves [3,*]

1 Departamento de Microbiologia, Imunologia e Parasitologia da Escola Paulista de Medicina-UNIFESP, São Paulo, SP 04023-062, Brazil; roberta.peresdasilva@nottingham.ac.uk (R.P.d.S.); ropuccia@gmail.com (R.P.)
2 School of Life Sciences, University of Nottingham, Nottingham NG7 2RD, UK
3 Instituto Carlos Chagas, Fundação Oswaldo Cruz, Fiocruz-PR, Curitiba, PR 81310-020, Brazil; sdt.martins@gmail.com (S.d.T.M.); flaviar23@gmail.com (F.C.G.d.R.); sgoldenb@fiocruz.br (S.G.); marciolrodrig@gmail.com (M.L.R.)
4 Instituto de Microbiologia Professor Paulo de Góes, Universidade Federal do Rio de Janeiro, Rio de Janeiro, RJ 21941-901, Brazil; juju.rizzo@gmail.com
5 Centro de Desenvolvimento Tecnológico em Saúde (CDTS), Fundação Oswaldo Cruz, Rio de Janeiro, RJ 21040-900, Brazil; lujoffe@gmail.com (L.S.J.); debora_leite@yahoo.com.br (D.L.O.)
6 Centro de Biotecnologia e Departamento de Biologia Molecular e Biotecnologia, Universidade Federal do Rio Grande do Sul, Porto Alegre, RS 91501-970, Brazil; mhv@cbiot.ufrgs.br (M.V.); liviak@cbiot.ufrgs.br (L.K.)
* Correspondence: lysangela.alves@fiocruz.br
† These authors contributed equally to this work.

Abstract: Golgi reassembly and stacking protein (GRASP) is required for polysaccharide secretion and virulence in *Cryptococcus neoformans*. In fungal species, extracellular vesicles (EVs) participate in the export of polysaccharides, proteins and RNA. In the present work, we investigated if EV-mediated RNA export is functionally connected with GRASP in *C. neoformans* using a *graspΔ* mutant. Since GRASP-mediated unconventional secretion involves autophagosome formation in yeast, we included the *atg7Δ* mutant with defective autophagic mechanisms in our analysis. All fungal strains exported EVs but deletion of *GRASP* or *ATG7* profoundly affected vesicular dimensions. The mRNA content of the *graspΔ* EVs differed substantially from that of the other two strains. The transcripts associated to the endoplasmic reticulum were highly abundant transcripts in *graspΔ* EVs. Among non-coding RNAs (ncRNAs), tRNA fragments were the most abundant in both mutant EVs but *graspΔ* EVs alone concentrated 22 exclusive sequences. In general, our results showed that the EV RNA content from *atg7Δ* and WT were more related than the RNA content of *graspΔ*, suggesting that GRASP, but not the autophagy regulator Atg7, is involved in the EV export of RNA. This is a previously unknown function for a key regulator of unconventional secretion in eukaryotic cells.

Keywords: *Cryptococus neoformans*; RNA; extracellular vesicles; GRASP; Atg7; unconventional secretory pathway

1. Introduction

Extracellular vesicle (EV) formation and release constitute a ubiquitous export mechanism of proteins, DNA and RNA [1,2]. EVs play key roles in processes of cell communication, homeostasis, immunopathogenesis and microbial virulence [1,2]. EV formation is a conserved mechanism in both prokaryotic and eukaryotic cells [3]. In fungi, EVs participate in the transport of macromolecules across the cell wall [4–6]. Fungal EVs transport a variety of macromolecules including proteins, lipids, glycans, pigments and, as more recently described, RNA [4,6–9].

EV biogenesis in fungi is still poorly understood. It has been hypothesized that EV biogenesis in eukaryotes is a complex process that is regulated at multiple levels [10,11]. EV formation is part of the unconventional secretion machinery in eukaryotes and general regulators of unconventional secretion have been identified. GRASP (Golgi reassembly and stacking protein) is a secretion regulator originally characterized in human cells as part of the Golgi cisternae stacking and ribbon formation [12,13]. During stress, GRASP is required for protein delivery to the plasma membrane or to the extracellular space by an unconventional pathway that involves autophagosome-like structures [14]. In mammalian cells, GRASP is also involved in the delivery of a mutant form of cystic fibrosis transmembrane conductance regulator to the plasma membrane in a Golgi-independent manner [15]. In *Drosophila melanogaster*, GRASP participates in the delivery of integrins from the ER directly to the plasma membrane, thus bypassing the Golgi [16]. In the amoeba *Dictyostelium discoideum*, a GRASP orthologue (GrpA) was necessary for acyl-coenzyme A-binding protein (AcbA) secretion during spore differentiation [17]. In the yeast species *Saccharomyces cerevisiae* and *Pichia pastoris*, another GRASP orthologous (Ghr1) was also required for starvation-induced secretion of AcbA [18,19].

In the yeast-like neuropathogen *Cryptococcus neoformans*, GRASP was required for polysaccharide export to the extracellular space. Polysaccharide secretion is fundamental for virulence in *C. neoformans* [20] and, in fact, a *graspΔ* mutant was hypovirulent in mice [20]. Polysaccharide export in *C. neoformans* is mediated by EVs but connections between GRASP functions and EV cargo remain uncharacterized.

Autophagy is a self-degradative process conserved in eukaryotes, presenting a housekeeping role by degrading dysfunctional components such as organelles and misfolded proteins [21]. The Atg7 is an autophagy regulator protein member of the ubiquitin-activating enzyme (E1) family involved in this process [22]. The Atg proteins have non-canonical roles in distinct cellular pathways. For example, *Toxoplasma gondii* Atg8 localizes to the apicoplast and is essential for organelle homeostasis and survival of the tachyzoite stage of the parasite [23]. Atg7 non-autophagic roles include cathepsin K secretion in bone osteoclasts [24], IFNγ-mediated antiviral activity against virus replication [25], adipogenesis in mice [26] and cell cycle regulation via p53 interaction and expression of p21 in mouse embryonic fibroblasts [27].

Autophagy regulators play key roles in cryptococcal physiology and, in fact, we have recently demonstrated that the putative autophagy regulator Atg7 affects both physiological and pathogenic mechanisms in *C. neoformans* [28]. In *D. discoideum*, GRASP-mediated unconventional secretion is mediated by autophagosomes, showing that there is a connection between these processes [18,29].

The role of unconventional secretion regulators in vesicular export of RNA is unknown but the functional connections between GRASP and Atg7 led us to evaluate whether these proteins affected extracellular RNA export in *C. neoformans*. Our results suggest that GRASP, but not Atg7, is a key regulator of vesicular export of RNA in *C. neoformans*.

2. Material and Methods

2.1. Fungal Strains and Growth Conditions

The *C. neoformans* strains used in this study included the parental isolate H99 and the mutant strains *atg7Δ* and *graspΔ*, which were generated in previous studies by our group [20,28]. Fungal cultures were maintained at 30 °C in Sabouraud dextrose plates (1% dextrose, 4% peptone).

Cells recovered from the stationary cultures were used to inoculate minimal medium composed of dextrose (15 mM), $MgSO_4$ (10 mM), KH_2PO_4 (29.4 mM), glycine (13 mM) and thiamine-HCl (3 μM) for further cultivation for three days at 30 °C, with shaking. All protocols adhered to the biosecurity demands of the Carlos Chagas Institute of Fiocruz (Curitiba, Brazil).

2.2. Extracellular Vesicle Isolation and Diameter Determination

EVs were isolated from fungal culture supernatants as previously described [4]. Briefly, cell-free culture supernatants were recovered by centrifugation at 4000× *g* for 15 min at 4 °C and the resulting supernatants were pelleted at 15,000× *g* for 30 min to remove small debris. The final supernatants were concentrated by a factor of 20 in an Amicon ultrafiltration system (100-kDa cutoff, Millipore, Burlington, VT, USA). Concentrated supernatants were centrifuged at 15,000× *g* for 30 min to ensure the removal of aggregates and the resulting supernatant was then ultracentrifuged at 100,000× *g* for 1 h to precipitate vesicles. Vesicle pellets were washed once in phosphate-buffered saline (PBS) and the final pellets were suspended in PBS. For analysis of EV dimensions, nanoparticle tracking analysis (NTA) was performed on a LM10 Nanoparticle Analysis System, coupled with a 488 nm laser and equipped with a sCMOS camera and a syringe pump (Malvern Panalytical, Malvern, UK). The data was acquired and analyzed using the NTA 3.0 Software (Malvern Panalytical). EVs from all samples were diluted 1:30 in filtered PBS (0.22 μM) and measured within the optimal dilution range previously described by Maas and colleagues (9×10^7–2.9×10^9 particles/mL) [30]. Polystyrene microspheres (100 nm) were used for equipment calibration. Samples were injected using a syringe pump speed of 50 and three videos of 60 s were captured per sample, with the camera level set to 15, gain set to 3 and viscosity set to water (0.954–0.955 cP). For data analysis, the gain was set to 10 and detection threshold was set to 5 for all samples. Levels of blur and max jump distance were automatically set. Particle detection values were normalized to the total number of cells in cultures from which each sample was obtained.

2.3. Small RNA Isolation

Small RNA (sRNA)-enriched fractions were isolated with the miRNeasy mini kit (Qiagen, Hilden, Germany) and then treated with the RNeasy MinElute Cleanup Kit (Qiagen), according to the manufacturer's protocol, to obtain small RNA-enriched fractions. The success of the sRNA extraction was assessed in representative EV preparations that were treated with 30 U DNase I (Qiagen) and characterized in an Agilent 2100 Bioanalyzer (Agilent Technologies, Santa Clara, CA, USA). To confirm that the RNA was confined within the EVs, vesicle samples were treated with 0.4 $\mu g\ \mu l^{-1}$ RNase (Promega, Madison, WY, USA) for 10 min at 37 °C before RNA extraction, as previously described [9].

2.4. RNA Sequencing

One hundred ng of purified sRNA were used for RNA-seq analysis from two independent biological replicates. The RNA-seq was performed in a SOLiD 3 plus platform using the RNA-Seq kit (Life Technologies, Carlsbad, CA, USA) according to the manufacturer's recommendations.

2.5. Cellular RNA Isolation and Quantitative PCR

Yeast cells were grown in minimal medium for 72 h, pelleted by 1 min centrifugation at 14.000× *g*, washed in PBS, suspended in the lysis buffer provided in miRCURY™ RNA Isolation Kit–Cell & Plant (Exiqon, Vedbaek, Danmark) and vortexed 5 times in acid washed glass beads (425–600 micron, Sigma-Aldrich, St. Louis, MO, USA). The lysate was centrifuged for 2 min at 14.000× *g* and the supernatants were collected for RNA isolation with the mirCURY™ kit, following the manufacturer's instructions. The RNAs were eluted in ultrapure water and treated with RQ1 RNase-Free Dnase (Promega) following the manufacturer's instructions. Reverse transcription reactions with the DNAse-treated RNAs were performed with a random primer and the ImProm-II™ Reverse Transcription System (Promega), following the manufacturer's instructions. Real time PCR reactions were performed using SYBR® Select Master Mix and run and analyzed using the LightCycler® 96 System (Roche, Basel, Switzerland). The primers corresponded to CNAG_03103 Cullin3 Forward GCCATACGGGAGATACAGAAC, Reverse GAGGTGTTGGACGATGAGAG, CNAG_07590 V_typeH Forward TCATGCTCAACGAAGTCAGG, Reverse GGAAGCAGTGGTTGTGAATG, CNAG_03337 hypothetic Forward CGGTCTTTATCGCTGCTGTAT, Reverse ATTGAAGAGTGGATGTCGTGG and CNAG_00483 Actin Forward CCACACTGTCCCCATTTACGA, Reverse CAGCAAGATCGATACGG AGGAT Each reaction was performed using 10 ng of cDNA. The experiment was performed in triplicates and the expression levels relative to actin were calculated according to Pfaffl's method using *t*-test for the statistical analysis [31].

2.6. In Silico Data Analysis

The sequencing data were analyzed using the version 9.1 of CLC Genomics Workbench©. The reads were trimmed on the basis of quality, with a threshold Phred score of 15. The reference genomes used for mapping were obtained from the NCBI database (*C. neoformans*-GCA_000149245.3). The alignment was performed as follows: additional 100-base upstream and downstream sequences; 10 minimum number of reads; 2 maximum number of mismatches; −2 nonspecific match limit and minimum fraction length of 0.9 for the genome mapping or 1.0 for the RNA mapping. The minimum reads similarity mapped on the reference genome was 80%. Only uniquely mapped reads were considered in the analysis. The libraries were normalized per million and the expression values for the transcripts were recorded in RPKM (reads per kilobase per million), we also analyzed the other expression values-TPM (transcripts per million) and CPM (counts per million).

2.7. Data Access

The data is deposited to the Sequence Read Archive (SRA) database of NCBI (Bethesda, MA, USA) under study accession number (SRA: SRX2793565 to 67).

3. Results

3.1. Lack of GRASP Results in Changes in the RNA Content of *Cryptococcus neoformans* Extracellular Vesicles

Our experimental model included wild type (WT) and two mutant strains of *C. neoformans*. WT cells corresponded to strain H99, a standard and widely investigated clinical isolate. Knockout mutant strains (KO) lacked expression of two regulators of cryptococcal pathogenicity, *GRASP* and *ATG7* [20,28].

We first asked whether the lack of either *GRASP* (*grasp*Δ) or *ATG7* (*atg7*Δ) expression would affect the EVs composition. The analysis of diameter distribution of wild type EVs by nanoparticle tracking analysis (Figure 1) revealed a major population of cryptococcal vesicles in the 50–250 nm range. Peaks of EVs corresponding to approximately 300, 410, 500 and 630 nm were also observed. Although the dimensions of cryptococcal EVs have been traditionally determined by dynamic light scattering and/or electron microscopy, the results obtained by nanoparticle tracking analysis were consistent with the previous literature [32]. Deletion of *GRASP* or *ATG7* produced a clear impact on the size distribution of cryptococcal EVs. In comparison to WT cells, peaks corresponding to sizes higher than 300 nm were no longer observed. A minor peak at 225 nm and major, sharp peaks at 100 and 140 nm were observed in EVs produced by both mutants. Complementation of mutant cells resulted in EV fractions enriched in the 100–300 nm range, but the minor peaks at 415, 500 and 600 nm observed in WT cells were still not detectable. Although deletion of *GRASP* or *ATG7* resulted in modified EV detection, no statistical differences were observed between the different samples. In summary, the nanoparticle tracking analysis revealed that deletion of *GRASP* and *ATG7* affected EV properties in *C. neoformans*. We then asked whether the differences in EV diameters correlated with the RNA content in *C. neoformans* EVs.

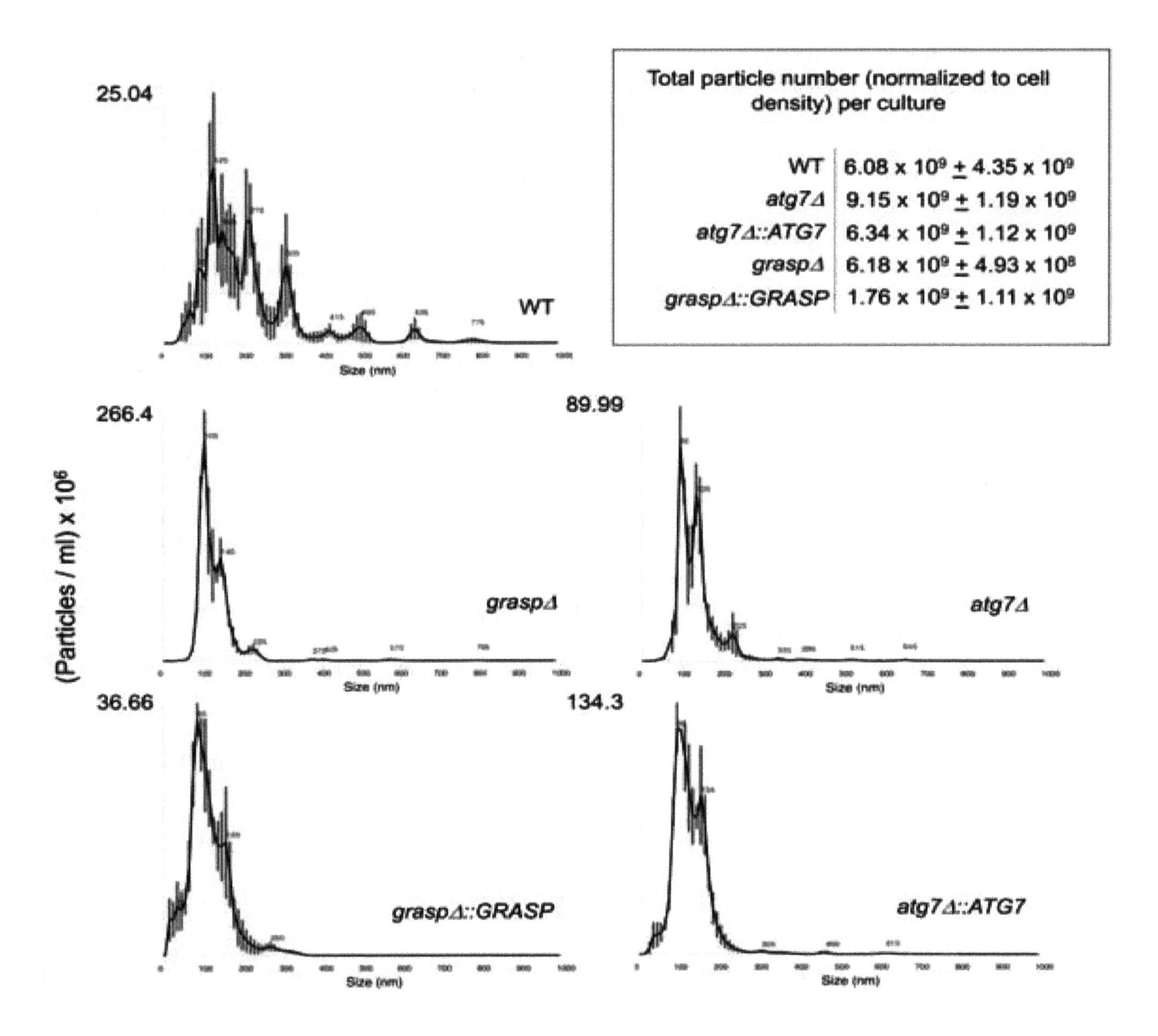

Figure 1. Nanoparticle tracking analysis of *Cryptococcus neoformans* extracellular vesicles (EVs) comparing wild type (WT), mutant (*grasp*Δ and *atg7*Δ) and complemented (*grasp*Δ::*GRASP* and *atg7*Δ::*ATG7*) cells. Results are representative of two independent biological replicates producing similar profiles. Particles were quantified in EV samples suspended in 150 mL phosphate-buffered saline (PBS). Particle detection values shown in the upper, right panel were normalized to the total number of cells in the cultures from which each sample was obtained.

Total RNA was isolated from fungal EVs and two independent biological replicates were subjected to RNA-seq (Figure S1). In order to compare the EV-RNA composition between the knockout (*atg7*Δ and *grasp*Δ) and the WT strains we first aligned the RNA-seq reads with the *C. neoformans* H99 genome (GCA_000149245.3) sequences. We used the raw data available for isolate H99 from our previous work [9] and compared them with the *atg7*Δ and *grasp*Δ EV RNA (Table 1). For all *C. neoformans* strains about 85% of the EV-RNA reads mapped to intronic regions, while less than 10% mapped to exons. A similar profile was observed for the *C. neoformans* WT strain (H99) in our previous work [9].

Analysis of EV-mRNAs showed that the correlation between WT and *atg7*Δ (r 0.71) sequences was greater than that for WT and *grasp*Δ (r 0.22) (Figure 2A,B). This result indicates that the mRNA content in WT EVs was closer to that of *atg7*Δ vesicles than to the content of *grasp*Δ EVs.

Table 1. RNA-seq mapping statistics. The values refer to the average of the replicates.

	C. neoformans					
	WT		*atg7Δ*		*graspΔ*	
	Uniquely Mapped	% of Total Mapped	Uniquely Mapped	% of Total Mapped	Uniquely Mapped	% of Total Mapped
Exon	5030	0.4	60,683	9.2	59,425	7.5
Exon-exon	10,664	0.6	1458	0.2	2350	0.3
Total exon	113,655	9.7	62,141	9.4	61,774	7.8
Total intron	1,003,971	90.3	568,003	84.9	758,109	86.9
Total gene	1,117,625	100	667,288	100.0	861,092	100.0

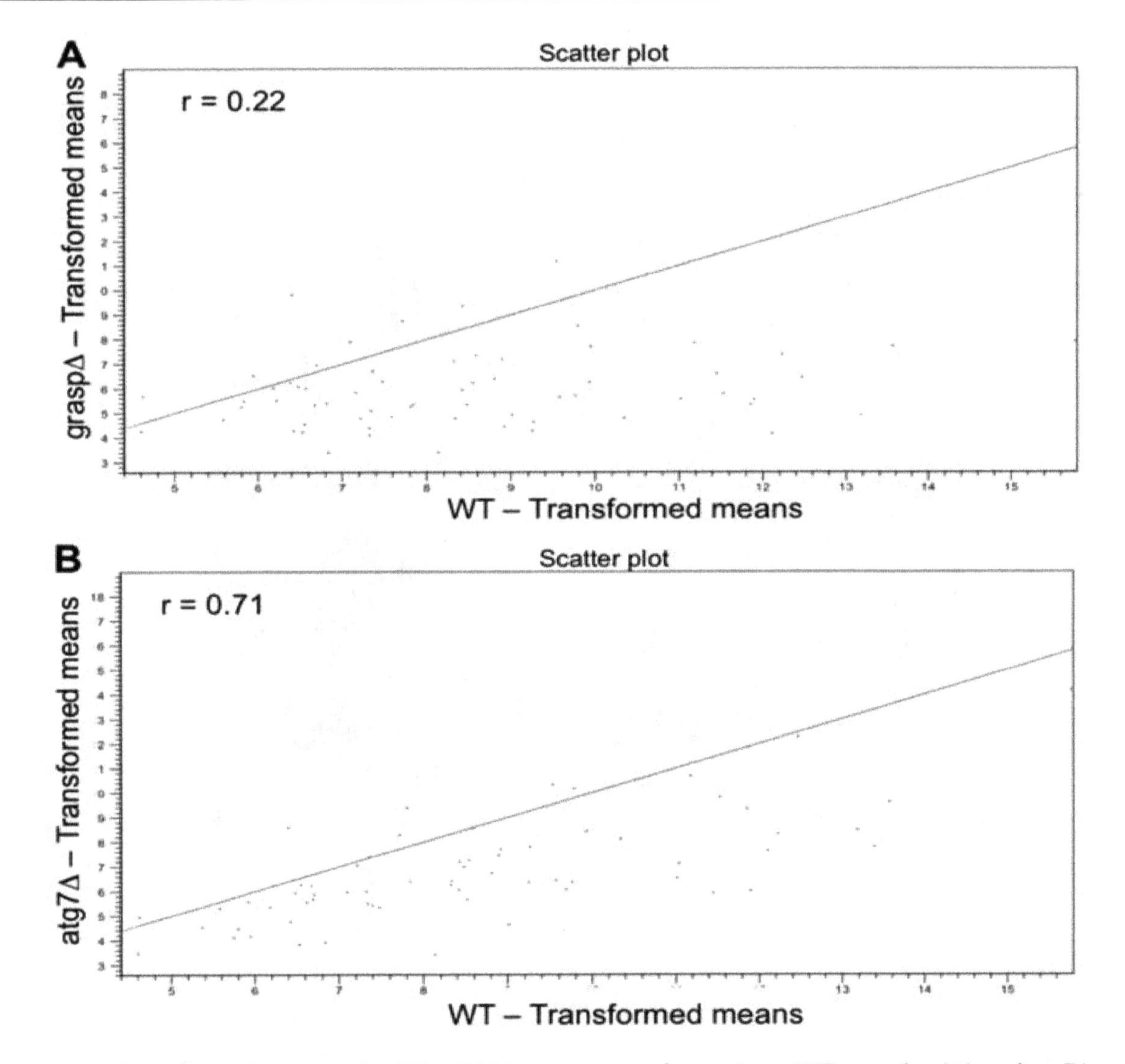

Figure 2. Correlation between the EV-mRNA sequences of *graspΔ* vs. WT samples (**A**) and *atg7Δ* vs. WT preparations (**B**). The transformed mean read values for WT EVs are in the X-axis, while those obtained from mutant vesicles are in the y-axis.

We next performed paired comparisons (WT versus *graspΔ* and WT versus *atg7Δ*) and applied the statistical negative binomial test [33] and the filters RPKM ≥ 50, log2 ≥ 2 and false discovery rate (FDR) ≤ 0.01. From the WT versus *graspΔ* analysis, 266 mRNAs were identified as enriched in the EVs from the *graspΔ* mutant (Table S1). From these transcripts, we observed enrichment in cellular components ($p \leq 0.03$) such as membrane and endoplasmic reticulum (Figure 3). For biological processes ($p \leq 0.03$), the enriched terms included organelle organization, cell cycle and gene expression (Figure 3). For the WT versus *atg7*Δ analysis, 74 mRNAs were found enriched in the *atg7*Δ compared to the WT strain (Table S2). The most abundant cellular components mRNAs ($n = 75$) in *atg7*Δ EVs were the nucleus and the mitochondrion (Figure 4). Biological processes were associated to transcription, transcription regulation and RNA processing (Figure 4). However, the score values for some terms did not meet the statistics criteria ($p \leq 0.03$).

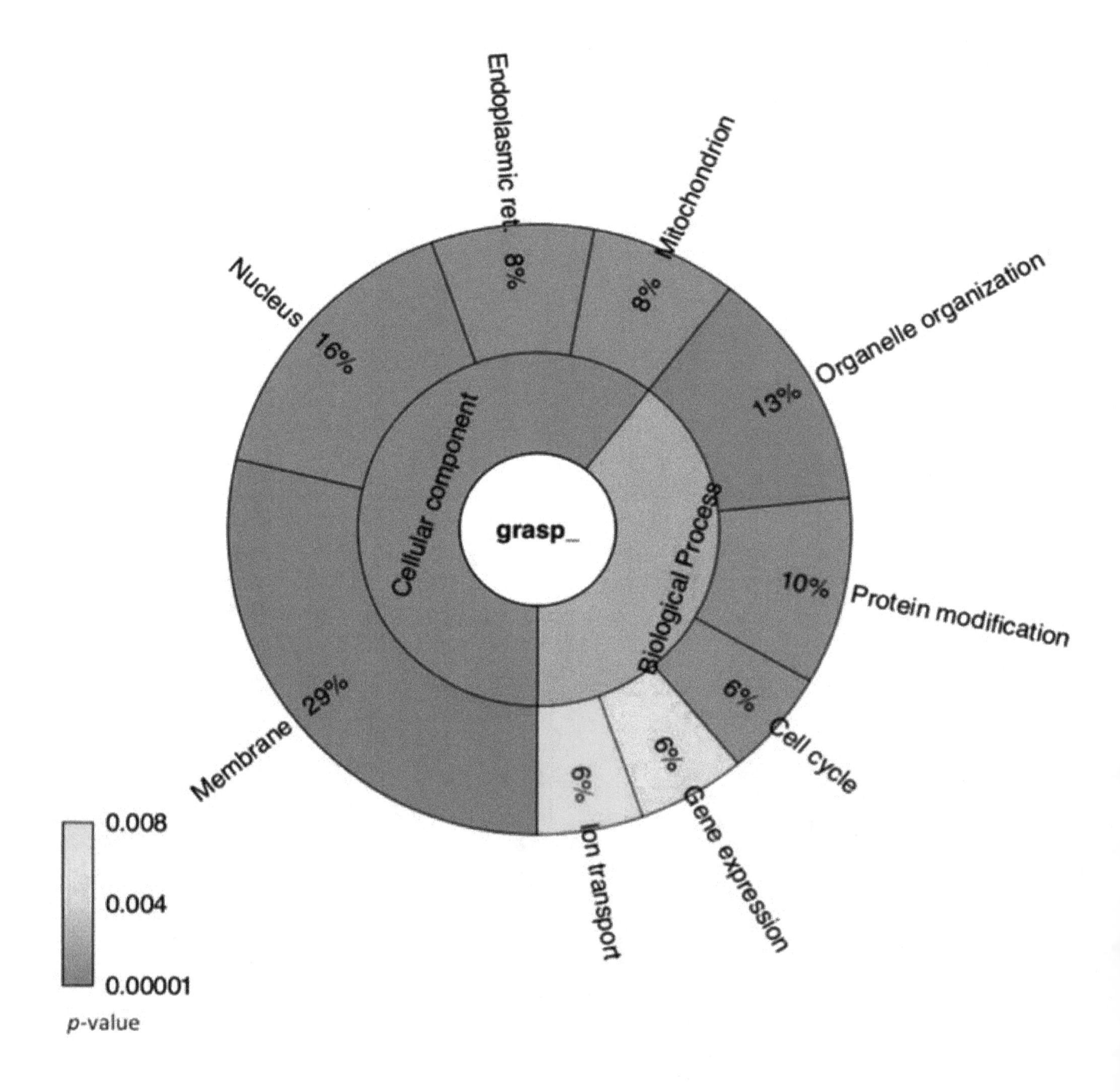

Figure 3. Krona chart representing the gene ontology of mRNA sequences enriched in EVs isolated from the *C. neoformans grasp*Δ mutant. The percentage refers to the relative enrichment for the Gene Ontology (GO) terms. The colors represent the *p*-value for each term plotted in the chart.

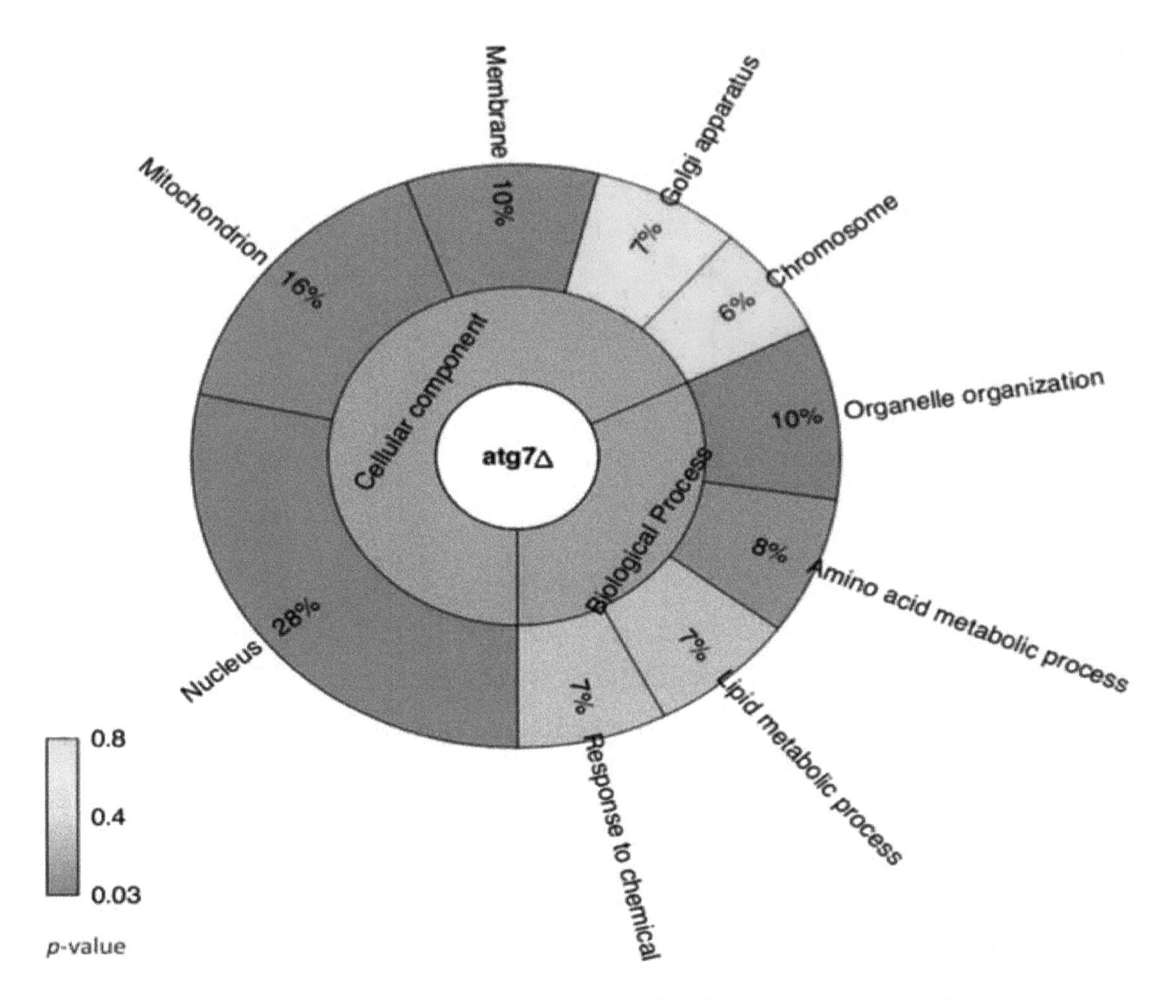

Figure 4. Krona chart representing the gene ontology of mRNA sequences enriched in EVs isolated from the *C. neoformans atg7*Δ mutant. The percentage refers to the relative enrichment for the GO terms. The colors represent the *p*-value for each term plotted in the chart.

Noteworthy, the second and third most abundant transcripts exclusively identified in *grasp*Δ EVs were those for the ER lumen protein retaining receptor and the regulator of vesicle transport through interaction with t-SNAREs 1 (Table S1). The former determines specificity of the luminal ER protein retention system and is required for normal vesicular traffic through the Golgi. The latter is involved in multiple transport pathways [34,35]. In addition, most of the transcripts were associated to organelles, such as the nucleus, the mitochondrion and the endoplasmic reticulum, suggesting that somehow the *GRASP* knockout resulted in altered population of transcripts composing the EVs. This enrichment profile was not observed in the *ATG7* knockout (Table S2), thus validating the differences observed in the *grasp*Δ mutant EVs.

As we observed this alteration in the EV-RNA composition for the *grasp*Δ mutant we asked if this difference was due to a general alteration in the cell transcriptome caused by the *GRASP* knockout. We then selected three of the most enriched transcripts found in the *grasp*Δ mutant EVs and assessed their expression value by qPCR in WT, mutant (*grasp*Δ and *atg7*Δ) and complemented (*grasp*Δ::*GRASP* and *atg7*Δ::*ATG7*) strains (Figure 5). The expression values of cullin 3, hypothetical protein CNAG_03337 and the V-type H transporting ATPase subunit C transcripts were similar in WT and *grasp*Δ mutant strains, despite the mRNA alteration in the EVs obtained from these two strains. The *atg7*Δ mutant showed the highest expression levels of these mRNAs when compared to the WT and *grasp*Δ strains (Figure 5). In addition, these transcripts had very low identification or were not detected in EVs from the *atg7*Δ strain (Table S2). Therefore, despite of the fact that *atg7*Δ mutant showed high expression levels, this variation did not correlate with the presence of these transcripts in EV fractions. Analysis of the complemented strains demonstrated a partial restoration of the wild-type phenotype in the *atg7Δ* system (Figure 5). Altogether, these results reinforce the notion that the *GRASP* deletion lead to a shift in the RNA composition of cryptococcal EVs.

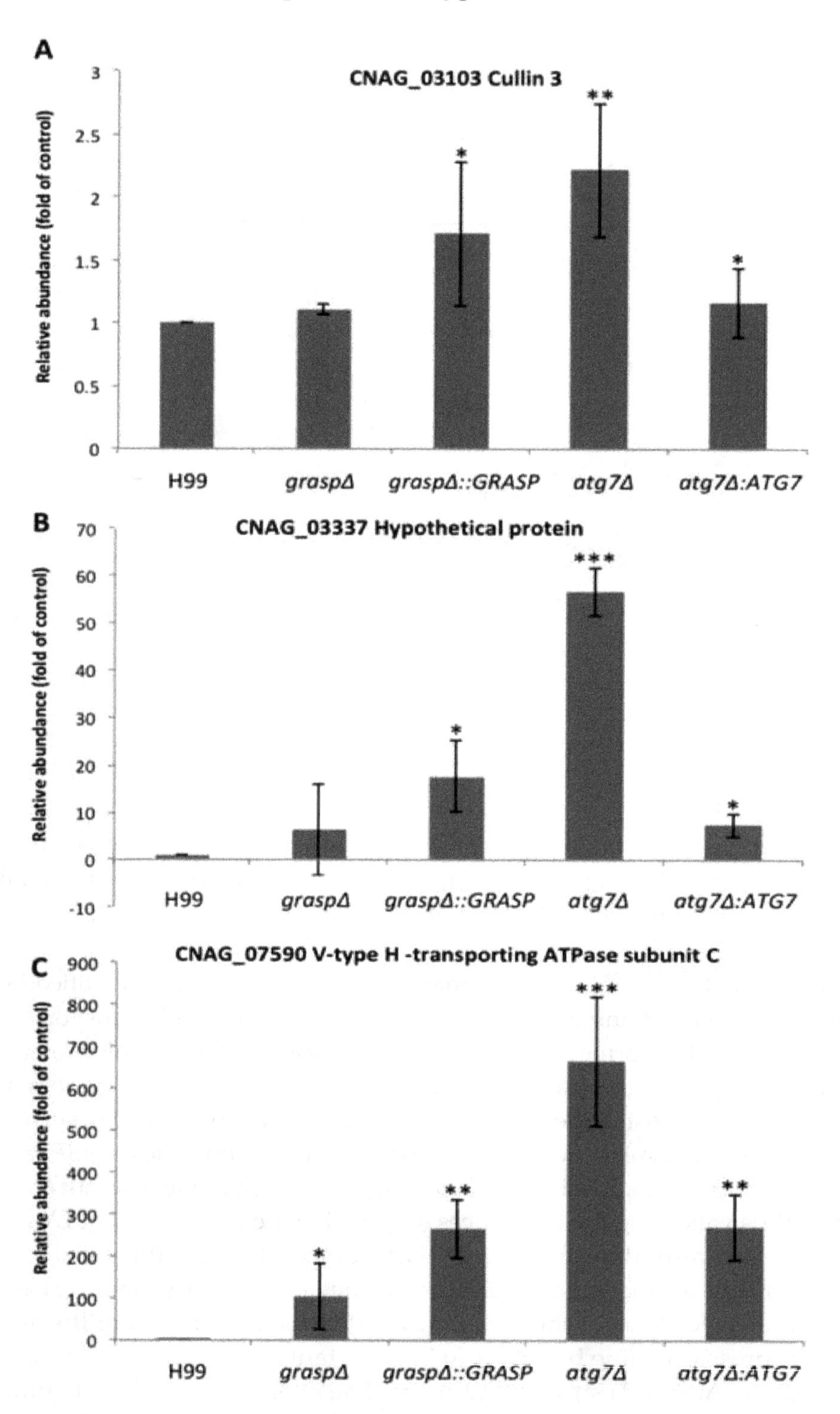

Figure 5. Analysis of the cellular transcription levels of three vesicular RNA sequences. Transcript levels for (**A**) Cullin 3; (**B**) hypothetical protein CNAG_03337 and (**C**) V-type H^+-transporting ATPase subunit C were normalized to the levels of actin transcripts. The X-axis corresponds to each strain analyzed (WT, *graspΔ*, *graspΔ::GRASP*, *atg7Δ* and *atg7Δ::ATG7*). The *y*-axis corresponds to the relative expression level of the mRNAs in the cell. Each bar represents the mean and standard error of triplicate samples. * $p < 0.05$; ** $p < 0.01$; *** $p < 0.001$.

3.2. Comparison of Cellular RNA Versus Extracellular Vesicle RNA Composition

The differences in the RNA composition of EVs produced by the *graspΔ* strain led us to question whether the mRNAs in the EVs correspond to those highly expressed in the cell, likely resulting from random incorporation into vesicular carriers [36]. To address this hypothesis, we compared the *C. neoformans* transcriptome (H99 strain) with the vesicular RNA sequences [9,37] (Table S3). After applying the differential gene expression analysis (DGE) we observed that, for several transcripts, there was an inversion between the expression patterns in the cell and the RNA abundance in the EVs (Figure 6 and Table 2). For example, one of the most enriched transcripts in the EVs presented low levels of expression in the cell (CNAG_06651 amidohydrolase). On the other hand, CNAG_03012 (encoding a quorum sensing-like molecule) had an RPKM value greater than 20,000 in the cell but showed low abundance in the EVs (average RPKM value of 36; Table 3 and Table S3). This observation indicates a lack of correlation between the most expressed cellular mRNAs and EV cargo, therefore reinforcing the supposition that RNA loading into WT or mutant EVs is not random.

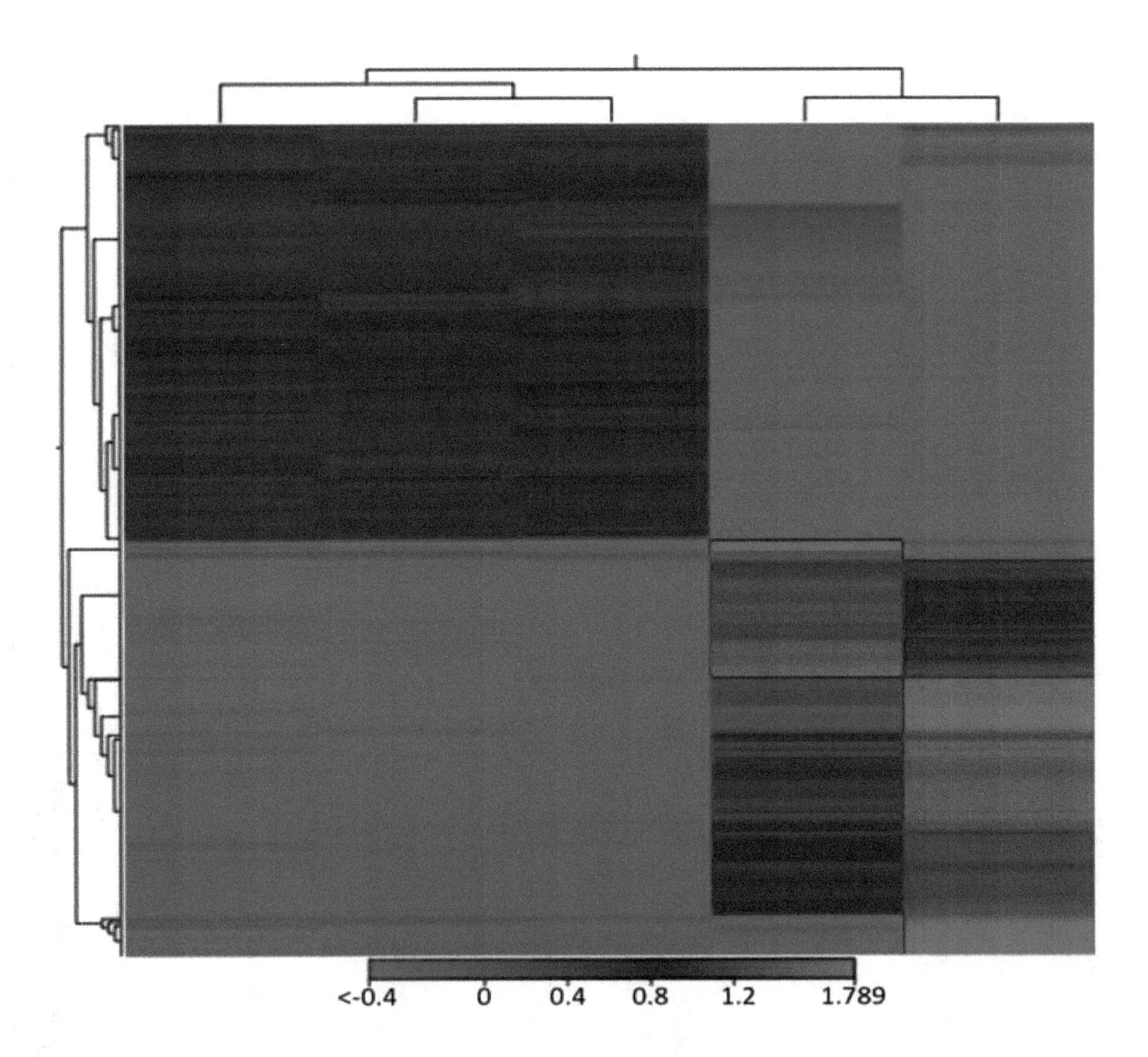

Figure 6. Heat map illustration of the comparison between cellular and EV RNAs. The expression levels are visualized using a gradient color scheme, where the red color is used for high expression levels and the blue color is used for low expression levels. Each line corresponds to a gene of the *C. neoformans* H99 strain.

Table 2. Comparison between cellular and vesicular RNA in *C. neoformans* (H99 strain). The top ten most expressed transcripts in the cell are shown in light blue. The most represented RNAs in the EVs are illustrated in light red.

Name	Product	EV vs. Cell-Log Fold Change	EV vs. Cell-FDR *p*-Value	SRR3199612 Cell 1 RPKM	SRR3199613 Cell 2-RPKM	SRR3199614Cell 3-RPKM	EV RNA 1-RPKM	EV RNA 2-RPKM
CNAG_03012	quorum sensing-like molecule	−5.06	0.00%	20,332.13	18,844.93	20,155.19	48.40	24.28
CNAG_06207	hypothetical protein	−6.93	0.00%	16,304.37	14,037.19	13,815.62	8.35	6.02
CNAG_04105	hypothetical protein	−2.31	2.29%	16,003.01	10,010.74	16,467.89	254.98	73.62
CNAG_03143	hypothetical protein	−2.09	3.87%	13,070.91	8338.53	12,401.56	231.90	78.39
CNAG_01735	hypothetical protein	−3.60	0.03%	9034.37	6373.02	7970.58	56.97	16.11
CNAG_06075	hypothetical protein	−2.98	0.39%	6021.45	5119.83	6051.19	66.75	13.02
CNAG_03007	hypothetical protein	−6.58	0.00%	5861.25	5356.39	4321.23	3.72	2.64
CNAG_06298	hypothetical protein	−7.02	0.00%	5319.83	5499.72	6635.36	2.65	3.96
CNAG_06101	ADP, ATP carrier protein	−3.04	0.11%	4475.82	5415.80	4535.39	44.21	32.06
CNAG_07466	U3 small nucleolar RNA-associated protein 7, U3 small nucleolar RNA-associated protein 7, variant 1, U3 small nucleolar RNA-associated protein 7, variant 2	10.05	0.00%	392.25	765.29	487.61	39,204.58	43,212.28
CNAG_01093	hypothetical protein	8.04	0.00%	45.01	41.14	33.25	831.32	480.72
CNAG_06651	amidohydrolase	12.70	0.00%	3.80	4.64	3.38	777.33	4354.12
CNAG_00311	3-hydroxyisobutyryl-CoA hydrolase	6.97	0.00%	62.61	78.16	54.48	650.07	373.07
CNAG_02129	hypothetical protein	2.12	3.76%	423.90	380.09	587.72	178.91	56.55
CNAG_05774	hypothetical protein, hypothetical protein, variant	4.48	0.00%	87.07	87.32	82.46	146.53	104.07
CNAG_05651	hypothetical protein	7.83	0.00%	5.28	8.41	7.53	138.73	51.57
CNAG_07515	hypothetical protein	4.79	0.00%	57.55	52.26	77.35	118.53	127.65
CNAG_04124	hypothetical protein	7.66	0.00%	6.60	6.26	5.59	113.90	21.83
CNAG_07028	26S proteasome regulatory subunit N11	4.03	0.00%	103.45	112.54	86.34	112.59	118.66

Table 3. Intron retention in EV RNAs.

ID	Data obtained from Gonzalez-Hilarion et al., 2016 [38]				RPKM	Unique Exon Reads	Unique Intron Reads	RPKM	Unique Exon Reads	Unique Intron Reads	RPKM	Unique Exon Reads	Unique Intron Reads	Product
	intron	Type	RPKM	Exons		WT			ΔAtg7			ΔGRASP		
CNAG_03602	Ic2-554 Ic2-555	in5UTR in5UTR	**6.66** **204.80**	5	**23.5**	4	228	**40.59**	3	87.5	**28.57**	1.5	81	U3 small nucleolar RNA-associated protein 5
CNAG_03645	Ic2-787 Ic2-788	inCDS inCDS	**8.01** **12.93**	8	**6.3**	1.5	73.5	**53.43**	5.5	9.5	**58.98**	6.5	14.5	NET1-associated nuclear protein 1 (U3 small nucleolar RNA-associated protein 17)
CNAG_04068	Ic2-3155	inCDS	**3.18**	4	**71.2**	3.5	10	**270.93**	6	11	**522.07**	12	4.5	large subunit ribosomal protein L28e
CNAG_07982	Ic4-247 Ic4-248	inCDS inCDS	**6.96** **17.47**	5	**1061.8**	60.5	100.5	**81.42**	5	3	**60.25**	3.5	7	hypothetical protein
CNAG_00930	Ic4-351 Ic4-349 Ic4-350	inCDS in5UTR in5UTR	**2.61** **1066.15** **81.59**	7	**50.1**	5.5	328	**81.47**	4	303.5	**106.30**	6	397	argininosuccinate synthase
CNAG_07884	Ic8-1359	inCDS	**10.84**	3	**7.4**	0.5	66	**18.73**	0.5	21.5	**85.27**	2.5	21	hypothetical protein
CNAG_07813	Ic12-778 Ic12-776	inCDS in5UTR	**8.46** **7.38**	5	**75.4**	6.5	231.5	**14.71**	0.5	141.5	**37.66**	1	208.5	hypothetical protein
CNAG_06167	Ic13-990	in5UTR	**26.36**	5	**131.1**	11	770	**103.14**	6.5	152.5	**146.05**	10.5	295	metal homeostatis protein bsd2
CNAG_01820	Ic3-1947	in3UTR	**28.14**	12	**342.4**	47	180.5	**236.94**	17	15.5	**344.85**	27.5	15	pyruvate kinase, pyruvate kinase, variant
CNAG_06033	Ic13-230	inCDS	**10.93**	7	**99.4**	15	22	**47.14**	3.5	5	**33.92**	3.5	63.5	pfkB family carbohydrate kinase superfamily
CNAG_03730	Ic2-1335	in5UTR	**35.87**	4	**41.7**	2	479.5	**102.24**	2	23.5	**0.00**	0	175.5	DNA-directed RNA polymerase II subunit RPB11
CNAG_06401	Ic14-772	in5UTR	**14.61**	11	**6.1**	0.5	23	**52.25**	4.5	162.5	**28.12**	2.5	27.5	hypothetical protein

3.3. Intronic Reads

We have previously observed that a great number of *C. neoformans* EV-RNA reads mapped to intronic regions of the genome [9], which is in agreement with our current findings with the knockout strains. To analyze intronic reads in mRNAs and exclude non-coding RNAs (ncRNAs), such as ribosomal RNA (rRNA) or transfer RNA (tRNAs), we used the presence of exons as a criterion to ensure ncRNAs were excluded (Table 3). We observed two types of patterns, including reads mapping to both exons and introns in variable proportions and those that mapped only to introns in the messenger RNAs. The intronic mapping shared by the EV RNAs from the WT and mutant strains of *C. neoformans* were associated to translation and also to transmembrane proteins (Table 3). From the 32 mRNAs with intronic reads found in the EV samples, 12 have previously been described as transcripts with intron retention [38]. It has already been reported that 59% of the genes from *C. neoformans* use alternative splicing (AS) that varies depending on the growth conditions. The intron retention (IR) is the prevalent AS mechanism in this fungus [38]. We also observed differences in abundance between the cell mRNAs compared to those in the EVs (Table 3). For example, the mRNA CNAG_07982 that codes for a hypothetical protein is 10 times more abundant in the EV than in the cell. A similar profile was observed for sequence CNAG_01820, which encodes a pyruvate kinase (Table 3). It has been speculated that the mRNAs that present IR are not the most expressed in cells based on a negative correlation between the highly expressed transcripts and the presence of IR [38]. However, our present data show that most of the reads that were considered as aligned in introns, are in fact the rRNAs 25S, 18S and 5.8S (data not shown). Nevertheless, we obtained highly abundant transcripts that are likely to be intron-retaining mRNAs (Table 3), suggesting that somehow these IR mRNAs might be directed to the EVs. The function of these transcripts needs to be further investigated.

3.4. Non-Coding RNAs

The EV-RNA sequences obtained in this work also mapped to ncRNAs. The most abundant molecules were the 25S, 18S and 5.8S rRNAs, accounting for more than 90% of the ncRNA and intronic reads (data not shown). As described for the mRNA analysis, we performed paired comparisons (WT versus *grasp*Δ and WT vs. *atg7*Δ) and applied the statistical negative binomial test [33] and the filters RPKM $\geq$ 50, log2 $\geq$ 2 and FDR $\leq$ 0.01. For the WT versus *grasp*Δ we observed 43 ncRNAs enriched in *grasp*Δ (Table S4). For WT versus *atg7*Δ 30 ncRNAs were enriched in the *atg7*Δ strain (Table S5). From these results, it was possible to observe that the tRNA-derived fragments (tRFs) were enriched in both knockouts (Table S4 and S5). tRFs have been identified in EVs from organisms in all kingdoms, including archaea, bacteria and eukaryotes, where they play different biological roles [39].

4. Discussion

Fungal extracellular vesicles might correspond to structures that randomly incorporate cytosolic molecules that are released extracellularly or in the cell wall [11]. Our current results, however, suggest that EV RNA cargo can be finely regulated. Our model consisted of an investigation of the role of *C. neoformans* proteins GRASP and Atg

7 in the vesicular export of RNA. Although these proteins are functionally connected in other systems [12,17,26,27], our findings suggest that GRASP, but not Atg7, has a fundamental role in addressing RNA to cryptococcal EVs. The Atg proteins, which are primarily linked to autophagy processes, have non-canonical roles in distinct cellular pathways. It seems clear, however, that despite the variety of functions played by Atg7 and the significant alterations that its gene deletion causes in *C. neoformans*, the RNA populations transported by EVs were not greatly affected by the *atg7* knockout in *C. neoformans*. The phenotypic characteristics of this mutant included more efficient melanization, larger cell size, autophagic bodies formation and virulence attenuation [28].

Remarkably, phenotypic traits including EV dimensions were only partially recovered in complemented strains. This observation is likely related to methodological particularities intrinsic to the genetic manipulation of *C. neoformans*. For instance, biolistic transformation usually results in large chromosomic alterations but most importantly, gene complementation results in random insertion of *ATG7*- or *GRASP*-containing cassettes in multiple chromosome loci. Under these conditions, many phenotypic traits can be unpredictably affected and complemented genes can have their expression altered. In the specific case of *GRASP*, complementation of the *graspD* strain used in this study resulted in *GRASP* overexpression [20,28], which might be related to the unique phenotypic properties of the complemented strain.

Sequencing analysis of vesicular RNA obtained from mutant strains suggested that important biological functions are associated with nucleic acid-containing fungal vesicles. For example, the tRF-3'end derived (or CCA) uses the canonical miRNA machinery to downregulate replication of protein A1 mRNA and other transcripts in B cell lymphoma [40]. Regulation of translation is also a potential process where tRFs participate. It was demonstrated that tRF derived from tRNA-Val in the archaebacteria *Haloferax volcanii* binds to the small ribosomal subunit, consequently repressing translation by preventing a peptidyl transferase activity [41]. tRFs are also associated to the regulation of cell viability, RNA turnover and RNA stability [42–45]. The roles of GRASP and Atg7 in these processes have not been established but the enrichment of specific classes of RNA in mutant EVs suggests the existence of robust connections between EV traffic and tRFs. In *Trypanosoma cruzi*, the causing agent of Chagas disease, tRF-containing EVs can be transferred to other parasites and/or to host cells to modulate gene expression or facilitate infection [46,47]. In EVs from dendritic and T cells there are different populations of tRFs indicating selective loading of these molecules into the vesicles [48]. Human semen EVs are enriched with tRFs that hypothetically act as translational repressors [49]. It is unknown whether fungal vesicles can be transferred to other cells and consequently regulate metabolism and gene expression but it is tempting to speculate this hypothesis based on the findings mentioned above.

The mRNA population from *graspΔ* EVs had low correlation with WT vesicles. In addition, ncRNA populations were also clearly distinct in EVs from WT and *graspΔ* cells, where snoRNA predominated in the WT and tRNA/tRFs in the KOs.

The distinct RNA cargo in the mutants analyzed in this study is in agreement with a key and general role of GRASP in unconventional secretion in *C. neoformans* and a minor participation of Atg7. Polysaccharides, which lack secretory tags, require GRASP for efficient secretion in *C. neoformans* [20]. Deletion of *ATG7*, however, did not affect polysaccharide export in this fungus [28]. Multivesicular body formation and consequent exosome release involve a number of cellular regulators whose functions directly affect EVs [50,51]. In fungi, a number of regulators affect biogenesis of exosome-like EVs, including the ESCRT machinery, flippases and GRASP [52]. It has been hypothesized that GRASP (Grh1) could participate in this process by acting as a chaperone and directly influencing the cargo of EVs [53]. This GRASP chaperone function could be linked to our current results since RNA cargo was deeply affected in the *graspΔ* mutant. Altogether, these results strongly indicate a novel function for the GRASP family in eukaryotes that could directly affect cell communication, gene expression and host-pathogen interactions.

Author Contributions: R.P.d.S.: obtained the EVs, isolated the RNA, performed the analysis; S.d.T.M.: isolated the EVs, performed NTA analysis and the qPCR validation; J.R.: isolated the EVs and interpreted NTA results; F.C.G.d.R. isolated the EVs and interpreted NTA results; L.S.J. performed the EVs characterization; M.V. produced the mutant strains; L.K. obtained the EVs; D.L.O. obtained the EVs; R.P. analyzed the data, discussed the results, wrote the manuscript; S.G. discussed the results, wrote the manuscript; M.L.R. analyzed the data, wrote the manuscript; L.R.A. performed the RNA-seq, analyzed the data, wrote the manuscript. All authors discussed the results, wrote and approved the final manuscript.

Acknowledgments: Rosana Puccia was supported by grants from the Brazilian agencies FAPESP, CNPq and CAPES. Marcio Lourenço Rodrigues was supported by grants from the Brazilian agencies FAPERJ and CNPq and by the Instituto Nacional de Ciência e Tecnologia de Inovação em Populações de Doenças Negligenciadas (INCT-IDPN). Samuel Goldenberg was supported by grants from the Brazilian agencies Fundação Araucária–PRONEX and CNPq.

References

1. Coakley, G.; Maizels, R.M.; Buck, A.H. Exosomes and other extracellular vesicles: The new communicators in parasite infections. *Trends Parasitol.* **2015**, *31*, 477–489. [CrossRef] [PubMed]
2. Tkach, M.; Théry, C. Communication by extracellular vesicles: Where we are and where we need to go. *Cell* **2016**, *164*, 1226–1232. [CrossRef] [PubMed]
3. Deatheragea, B.L.; Cooksona, B.T. Membrane vesicle release in bacteria, eukaryotes Eukaryotes and Archaea: A conserved yet underappreciated aspect of microbial life. *Infect. Immun.* **2012**, *80*, 1948–1957. [CrossRef] [PubMed]
4. Rodrigues, M.L.; Nimrichter, L.; Oliveira, D.L.; Frases, S.; Miranda, K.; Zaragoza, O.; Alvarez, M.; Nakouzi, A.; Feldmesser, M.; Casadevall, A. Vesicular polysaccharide export in *Cryptococcus neoformans* is a eukaryotic solution to the problem of fungal trans-cell wall transport. *Eukaryot. Cell* **2007**, *6*, 48–59. [CrossRef] [PubMed]
5. Albuquerque, P.C.; Nakayasu, E.S.; Rodrigues, M.L.; Frases, S.; Casadevall, A.; Zancope-Oliveira, R.M.; Almeida, I.C.; Nosanchuk, J.D. Vesicular transport in *Histoplasma capsulatum*: An effective mechanism for trans-cell wall transfer of proteins and lipids in ascomycetes. *Cell Microbiol.* **2008**, *10*, 1695–1710. [CrossRef] [PubMed]
6. Rodrigues, M.L.; Nakayasu, E.S.; Oliveira, D.L.; Nimrichter, L.; Nosanchuk, J.D.; Almeida, I.C.; Casadevall, A. Extracellular vesicles produced by *Cryptococcus neoformans* contain protein components associated with virulence. *Eukaryot. Cell* **2008**, *7*, 58–67. [CrossRef] [PubMed]
7. Eisenman, H.C.; Frases, S.; Nicola, A.M.; Rodrigues, M.L.; Casadevall, A. Vesicle-associated melanization in *Cryptococcus neoformans*. *Microbiology* **2009**, *155*, 3860–3867. [CrossRef] [PubMed]
8. Rizzo, J.; Oliveira, D.L.; Joffe, L.S.; Hu, G.; Gazos-Lopes, F.; Fonseca, F.L.; Almeida, I.C.; Frases, S.; Kronstad, J.W.; Rodrigues, M.L. Role of the Apt1 protein in polysaccharide secretion by *Cryptococcus neoformans*. *Eukaryot. Cell* **2014**, *13*, 715–726. [CrossRef] [PubMed]
9. Peres da Silva, R.; Puccia, R.; Rodrigues, M.L.; Oliveira, D.L.; Joffe, L.S.; César, G.V.; Nimrichter, L.; Goldenberg, S.; Alves, L.R. Extracellular vesicle-mediated export of fungal RNA. *Sci. Rep.* **2015**, *5*, 7763. [CrossRef] [PubMed]
10. Oliveira, D.L.; Nakayasu, E.S.; Joffe, L.S.; Guimarães, A.J.; Sobreira, T.J.; Nosanchuk, J.D.; Cordero, R.J.; Frases, S.; Casadevall, A.; Almeida, I.C.; et al. Biogenesis of extracellular vesicles in yeast: Many questions with few answers. *Commun. Integr. Biol.* **2010**, *3*, 533–535. [CrossRef] [PubMed]
11. Rodrigues, M.L.; Franzen, A.J.; Nimrichter, L.; Miranda, K. Vesicular mechanisms of traffic of fungal molecules to the extracellular space. *Curr. Opin. Microbiol.* **2013**, *16*, 414–420. [CrossRef] [PubMed]
12. Barr, F.A.; Puype, M.; Vandekerckhove, J.; Warren, G. GRASP65, a protein involved in the stacking of Golgi cisternae. *Cell* **1997**, *91*, 253–262. [CrossRef]
13. Shorter, J.; Watson, R.; Giannakou, M.E.; Clarke, M.; Warren, G.; Barr, F.A. GRASP55, a second mammalian GRASP protein involved in the stacking of Golgi cisternae in a cell-free system. *EMBO J.* **1999**, *18*, 4949–4960. [CrossRef] [PubMed]
14. Rabouille, C.; Malhotra, V.; Nickel, W. Diversity in unconventional protein secretion. *J. Cell Sci.* **2012**, *125*, 5251–5255. [CrossRef] [PubMed]
15. Gee, H.Y.; Noh, S.H.; Tang, B.L.; Kim, K.H.; Lee, M.G. Rescue of ΔF508-CFTR trafficking via a GRASP-dependent unconventional secretion pathway. *Cell* **2011**, *146*, 746–760. [CrossRef] [PubMed]
16. Grieve, A.G.; Rabouille, C. Extracellular cleavage of E-cadherin promotes epithelial cell extrusion. *J. Cell Sci.* **2014**, *127*, 3331–3346. [CrossRef] [PubMed]

17. Kinseth, M.A.; Anjard, C.; Fuller, D.; Guizzunti, G.; Loomis, W.F.; Malhotra, V. The Golgi-associated protein GRASP is required for unconventional protein secretion during development. *Cell* **2007**, *130*, 524–534. [CrossRef] [PubMed]
18. Duran, J.M.; Anjard, C.; Stefan, C.; Loomis, W.F.; Malhotra, V. Unconventional secretion of Acb1 is mediated by autophagosomes. *J. Cell Biol.* **2010**, *188*, 527–536. [CrossRef] [PubMed]
19. Manjithaya, R.; Anjard, C.; Loomis, W.F.; Subramani, S. Unconventional secretion of *Pichia pastoris* Acb1 is dependent on GRASP protein, peroxisomal functions and autophagosome formation. *J. Cell Biol.* **2010**, *188*, 537–546. [CrossRef] [PubMed]
20. Kmetzsch, L.; Joffe, L.S.; Staats, C.C.; de Oliveira, D.L.; Fonseca, F.L.; Cordero, R.J.; Casadevall, A.; Nimrichter, L.; Schrank, A.; Vainstein, M.H.; et al. Role for Golgi reassembly and stacking protein (GRASP) in polysaccharide secretion and fungal virulence. *Mol. Microbiol.* **2011**, *81*, 206–218. [CrossRef] [PubMed]
21. Glick, D.; Barth, S.; Macleod, K.F. Autophagy: Cellular and molecular mechanisms. *J. Pathol.* **2010**, *221*, 3–12. [CrossRef] [PubMed]
22. Nakatogawa, H.; Suzuki, K.; Kamada, Y.; Ohsumi, Y. Dynamics and diversity in autophagy mechanisms: Lessons from yeast. *Nat. Rev. Mol. Cell Biol.* **2009**, *10*, 458–467. [CrossRef] [PubMed]
23. Lévêque, M.F.; Berry, L.; Cipriano, M.J.; Nguyen, H.M.; Striepen, B.; Besteiro, S. Autophagy-related protein ATG8 has a noncanonical function for apicoplast inheritance in *Toxoplasma gondii*. *MBio* **2015**, *6*, e01446-15. [CrossRef] [PubMed]
24. DeSelm, C.J.; Miller, B.C.; Zou, W.; Beatty, W.L.; van Meel, E.; Takahata, Y.; Klumperman, J.; Tooze, S.A.; Teitelbaum, S.L.; Virgin, H.W. Autophagy proteins regulate the secretory component of osteoclastic bone resorption. *Dev. Cell* **2011**, *21*, 966–974. [CrossRef] [PubMed]
25. Dreux, M.; Chisari, F.V. Impact of the autophagy machinery on hepatitis Hepatitis C virus infection. *Viruses* **2011**, *3*, 1342–1357. [CrossRef] [PubMed]
26. Zhang, Y.; Goldman, S.; Baerga, R.; Zhao, Y.; Komatsu, M.; Jin, S. Adipose-specific deletion of autophagy-related gene 7 (atg7) in mice reveals a role in adipogenesis. *Proc. Natl. Acad. Sci. USA* **2009**, *106*, 19860–19865. [CrossRef] [PubMed]
27. Lee, I.H.; Kawai, Y.; Fergusson, M.M.; Rovira, I.I.; Bishop, A.J.; Motoyama, N.; Cao, L.; Finkel, T. Atg7 modulates p53 activity to regulate cell cycle and survival during metabolic stress. *Science* **2012**, *336*, 225–228. [CrossRef] [PubMed]
28. Oliveira, D.L.; Fonseca, F.L.; Zamith-Miranda, D.; Nimrichter, L.; Rodrigues, J.; Pereira, M.D.; Reuwsaat, J.C.; Schrank, A.; Staats, C.; Kmetzsch, L.; et al. The putative autophagy regulator Atg7 affects the physiology and pathogenic mechanisms of *Cryptococcus neoformans*. *Future Microbiol.* **2016**, *11*, 1405–1419. [CrossRef] [PubMed]
29. Bruns, C.; McCaffery, J.M.; Curwin, A.J.; Duran, J.M.; Malhotra, V. Biogenesis of a novel compartment for autophagosome-mediated unconventional protein secretion. *J. Cell Biol.* **2011**, *195*, 979–992. [CrossRef] [PubMed]
30. Maas, S.L.N.; De Vrij, J.; Van Der Vlist, E.J.; Geragousian, B.; Van Bloois, L.; Mastrobattista, E.; Schiffelers, R.M.; Wauben, M.H.M.; Broekman, M.L.D.; Nolte-'t Hoen, E.N. Possibilities and limitations of current technologies for quantification of biological extracellular vesicles and synthetic mimics. *J. Control. Release* **2015**, *200*, 87–96. [CrossRef] [PubMed]
31. Pfaffl, M.W. A new mathematical model for relative quantification in real-time RT-PCR. *Nucleic Acids Res.* **2001**, *29*, e45. [CrossRef] [PubMed]
32. Rodrigues, M.L.; Oliveira, D.L.; Vargas, G.; Girard-Dias, W.; Franzen, A.J.; Frasés, S.; Miranda, K.; Nimrichter, L. Analysis of yeast extracellular vesicles. *Methods Mol. Biol.* **2016**, *1459*, 175–190. [CrossRef] [PubMed]
33. Baggerly, K.A.; Deng, L.; Morris, J.S.; Aldaz, C.M. Differential expression in SAGE: Accounting for normal between-library variation. *Bioinformatics* **2003**, *19*, 1477–1483. [CrossRef] [PubMed]
34. Semenza, J.C.; Hardwick, K.G.; Dean, N.; Pelham, H.R. ERD2, a yeast gene required for the receptor-mediated retrieval of luminal ER proteins from the secretory pathway. *Cell* **1990**, *61*, 1349–1357. [CrossRef]
35. Von Mollard, G.; Stevens, T.H. The *Saccharomyces cerevisiae* v-SNARE Vti1p is required for multiple membrane transport pathways to the vacuole. *Mol. Biol. Cell* **1999**, *10*, 1719–1732. [CrossRef] [PubMed]
36. Abels, E.R.; Breakefield, X.O. Introduction to extracellular vesicles: Biogenesis, RNA cargo selection, content, release and uptake. *Cell. Mol. Neurobiol.* **2016**, *36*, 301–312. [CrossRef] [PubMed]

37. Li, C.; Lev, S.; Saiardi, A.; Desmarini, D.; Sorrell, T.C.; Djordjevic, J.T. Identification of a major IP5 kinase in *Cryptococcus neoformans* confirms that PP-IP5/IP7, not IP6, is essential for virulence. *Sci. Rep.* **2016**, *6*, 23927. [CrossRef] [PubMed]
38. Gonzalez-Hilarion, S.; Paulet, D.; Lee, K.T.; Hon, C.C.; Lechat, P.; Mogensen, E.; Moyrand, F.; Proux, C.; Barboux, R.; Bussotti, G.; et al. Intron retention-dependent gene regulation in *Cryptococcus neoformans*. *Sci. Rep.* **2016**, *6*, 32252. [CrossRef] [PubMed]
39. Keam, S.P.; Hutvagner, G. tRNA-Derived Fragments (tRFs): Emerging new roles for an ancient RNA in the regulation of gene expression. *Life* **2015**, *5*, 1638–1651. [CrossRef] [PubMed]
40. Maute, R.L.; Schneider, C.; Sumazin, P.; Holmes, A.; Califano, A.; Basso, K.; Dalla-Favera, R. tRNA-derived microRNA modulates proliferation and the DNA damage response and is down-regulated in B cell lymphoma. *Proc. Natl. Acad. Sci. USA* **2013**, *110*, 1404–1409. [CrossRef] [PubMed]
41. Gebetsberger, J.; Zywicki, M.; Künzi, A.; Polacek, N. tRNA-derived fragments target the ribosome and function as regulatory non-coding RNA in *Haloferax volcanii*. *Archaea* **2012**, *2012*, 260909. [CrossRef] [PubMed]
42. Lee, Y.S.; Shibata, Y.; Malhotra, A.; Dutta, A. A novel class of small RNAs: tRNA-derived RNA fragments (tRFs). *Genes Dev.* **2009**, *23*, 2639–2649. [CrossRef] [PubMed]
43. Haussecker, D.; Huang, Y.; Lau, A.; Parameswaran, P.; Fire, A.Z.; Kay, M.A. Human tRNA-derived small RNAs in the global regulation of RNA silencing. *RNA* **2010**, *16*, 673–695. [CrossRef] [PubMed]
44. Couvillion, M.T.; Bounova, G.; Purdom, E.; Speed, T.P.; Collins, K. A *Tetrahymena Piwi* bound to mature tRNA 3′ fragments activates the exonuclease Xrn2 for RNA processing in the nucleus. *Mol. Cell* **2012**, *48*, 509–520. [CrossRef] [PubMed]
45. Goodarzi, H.; Liu, X.; Nguyen, H.C.; Zhang, S.; Fish, L.; Tavazoie, S.F. Endogenous tRNA-derived fragments suppress breast cancer progression via YBX1 displacement. *Cell* **2015**, *161*, 790–802. [CrossRef] [PubMed]
46. Garcia-Silva, M.R.; Cabrera-Cabrera, F.; das Neves, R.F.; Souto-Padrón, T.; de Souza, W.; Cayota, A. Gene expression changes induced by *Trypanosoma cruzi* shed microvesicles in mammalian host cells: Relevance of tRNA-derived halves. *BioMed Res. Int.* **2014**, *2014*, 305239. [CrossRef] [PubMed]
47. Garcia-Silva, M.R.; das Neves, R.F.; Cabrera-Cabrera, F.; Sanguinetti, J.; Medeiros, L.C.; Robello, C.; Naya, H.; Fernandez-Calero, T.; Souto-Padron, T.; de Souza, W.; et al. Extracellular vesicles shed by *Trypanosoma cruzi* are linked to small RNA pathways, life cycle regulation and susceptibility to infection of mammalian cells. *Parasitol. Res.* **2014**, *113*, 285–304. [CrossRef] [PubMed]
48. Nolte-'t Hoen, E.N.; Buermans, H.P.; Waasdorp, M.; Stoorvogel, W.; Wauben, M.H.; 't Hoen, P.A. Deep sequencing of RNA from immune cell-derived vesicles uncovers the selective incorporation of small non-coding RNA biotypes with potential regulatory functions. *Nucleic Acids Res.* **2012**, *40*, 9272–9285. [CrossRef] [PubMed]
49. Vojtech, L.; Woo, S.; Hughes, S.; Levy, C.; Ballweber, L.; Sauteraud, R.P.; Strobl, J.; Westerberg, K.; Gottardo, R.; Tewari, M.; et al. Exosomes in human semen carry a distinctive repertoire of small non-coding RNAs with potential regulatory functions. *Nucleic Acids Res.* **2014**, *42*, 7290–7304. [CrossRef] [PubMed]
50. Huotari, J.; Helenius, A. Endosome maturation. *EMBO J.* **2011**, *30*, 3481–3500. [CrossRef] [PubMed]
51. Hanson, P.I.; Cashikar, A. Multivesicular body morphogenesis. *Annu. Rev. Cell Dev. Biol.* **2012**, *28*, 337–362. [CrossRef] [PubMed]
52. Oliveira, D.L.; Rizzo, J.; Joffe, L.S.; Godinho, R.M.; Rodrigues, M.L. Where do they come from and where do they go: Candidates for regulating extracellular vesicle formation in fungi. *Int. J. Mol. Sci.* **2013**, *14*, 9581–9603. [CrossRef] [PubMed]
53. Malhotra, V. Unconventional protein secretion: An evolving mechanism. *EMBO J.* **2013**, *32*, 1660–1664. [CrossRef] [PubMed]

4

Sporotrichosis in Immunocompromised Hosts

Flavio Queiroz-Telles [1,*], Renata Buccheri [2] and Gil Benard [3]

1 Department of Public Health, Federal University of Paraná, Curitiba 80060-000, Brazil
2 Emilio Ribas Institute of Infectious Diseases, São Paulo 05411-000, Brazil; renatabuccheri@gmail.com
3 Laboratory of Medical Mycology, Department of Dermatology, and Tropical Medicine Institute, University of São Paulo, Sao Paulo 05403-000, Brazil; bengil60@gmail.com
* Correspondence: queiroz.telles@uol.com.br

Abstract: Sporotrichosis is a global implantation or subcutaneous mycosis caused by several members of the genus *Sporothrix*, a thermo-dimorphic fungus. This disease may also depict an endemic profile, especially in tropical to subtropical zones around the world. Interestingly, sporotrichosis is an anthropozoonotic disease that may be transmitted to humans by plants or by animals, especially cats. It may be associated with rather isolated or clustered cases but also with outbreaks in different periods and geographic regions. Usually, sporotrichosis affects immunocompetent hosts, presenting a chronic to subacute evolution course. Less frequently, sporotrichosis may be acquired by inhalation, leading to disseminated clinical forms. Both modes of infection may occur in immunocompromised patients, especially associated with human immunodeficiency virus (HIV) infection, but also diabetes mellitus, chronic alcoholism, steroids, anti-TNF treatment, hematologic cancer and transplanted patients. Similar to other endemic mycoses caused by dimorphic fungi, sporotrichosis in immunocompromised hosts may be associated with rather more severe clinical courses, larger fungal burden and longer periods of systemic antifungal therapy. A prolonged outbreak of cat-transmitted sporotrichosis is in progress in Brazil and potentially crossing the border to neighboring countries. This huge outbreak involves thousands of human and cats, including immunocompromised subjects affected by HIV and FIV (feline immunodeficiency virus), respectively. We reviewed the main epidemiologic, clinical, diagnostic and therapeutic aspects of sporotrichosis in immunocompromised hosts.

Keywords: AIDS; IRIS; cat-transmitted sporotrichosis; immunocompromised hosts; mycoses of implantation; sporotrichosis; *Sporothrix brasiliensis*; *Sporothrix schenckii*; subcutaneous mycoses

---

1. Introduction

Sporotrichosis is a subacute to chronic fungal infection caused by several species of genus *Sporothrix*, a group of thermal dimorphic fungi. Although disease occurs worldwide, most cases are reported in tropical and subtropical zones from Latin America, Africa and Asia [1,2]. Usually, sporotrichosis is an implantation mycosis whose infectious propagules are inoculated from several environmental sources into skin, mucosal or osteoarticular sites [3]. Less frequently, infection may occur by inhalation, resulting in pulmonary disease [4], but in both modalities, immunocompetent and immunocompromised patients can be affected.

In endemic regions, this disease is mainly associated with plant transmission (sapronosis), the main etiologic agents being *S. schenckii* and *S. globosa* [5]. During the last three decades, a new, probable mutant species, *S. brasiliensis*, has emerged in the state of Rio de Janeiro, Brazil. This species is transmitted to humans by infected cats (zoonosis), causing the largest outbreak of sporotrichosis ever reported [6–8]. This epizootic outbreak continues to expand, affecting human and feline patients in several Brazilian regions, possibly reaching neighboring countries [9]. Feline sporotrichosis is unique among infections caused by endemic dimorphic fungi because it is directly transmitted in the yeast phase. The feline lesions typically harbor a high yeast-like fungal burden that can be acquired via cat scratches and bites, by non-traumatic ways, such as a cat's cough or sneezing, and direct contact between patients' integumental barriers and animal secretions [8].

Similar to other endemic mycoses, sporotrichosis in immunocompromised hosts is usually clinically remarkable. The increased clinical severity is related to a decrease of host immune and inflammatory responses, heavy fungal burden, extensive dissemination and higher mortality rates. In addition, in opportunistic sporotrichosis (OS), conventional serology may reveal false negative antibodies levels and long courses of systemic antifungal therapy are usually required.

This mycosis can affect anyone regardless of age, gender or comorbidities, mostly depending on exposure [1]. Human immunodeficiency virus (HIV)/AIDS changes the natural history of sporotrichosis and its opportunistic character depends on the immune status of the host. Comorbidities such as diabetes mellitus, chronic alcoholism, steroid treatment, hematologic cancer and organ transplantation have been sporadically described as risk factors for severe forms of the disease and case reports have focused on unusual manifestations in these scenarios. The aim of this review is to discuss the main epidemiological, clinical, diagnostic and therapeutic aspects of OS, with an emphasis on cat-transmitted sporotrichosis (CTS). In addition, in an attempt to better understand why certain comorbidities may predispose to OS, we performed a critical review of the data on the immune response in sporotrichosis.

2. Epidemiology and Clinical Manifestations

Sapronotic sporotrichosis is mainly related to several types of transcutaneous injuries, occurring in patients in contact with plant material or contaminated soil. Less frequently, animal associated trauma has been associated with *S. schenckii* and *S. globosa* and, to a lesser extent, with *S. pallida* clade (*S. mexicana, S. chilensis, S. luriei* and *S. pallida)* [5,10]. Zoonotic sporotrichosis is caused by *S. brasiliensis*, and although this expanding and uncontrolled outbreak is apparently limited to Brazil's borders, proven cases have been reported in Argentina and possibly Panama [9,11].

Sporotrichosis is a spectral disease, classified into two categories (cutaneous and extracutaneous sporotrichosis), which comprise four distinct clinical forms: lymphocutaneous (LC), fixed cutaneous (FC), disseminated cutaneous, and extra-cutaneous [12]. LC and FC forms are classical and most common clinical presentations [13,14]. Typically, but not exclusively, disseminated cutaneous and extra-cutaneous, considered severe forms, occur in hosts with depressed cellular immunity [15–19]. Indeed, findings from studies that represent the largest reported outbreaks of this mycosis in regions such as China, Japan, Peru and Brazil indicate the frequency of these severe forms ranges from 1.3% to 9% [13,14,20,21].

The disseminated cutaneous form is a rare variant of sporotrichosis characterized by multiple skin lesions at noncontiguous sites without extracutaneous involvement. It is important to emphasize that in some situations it is difficult to identify whether the clinical presentation is due to dissemination from a single lesion or to multiple inoculations [22]. The extracutaneous or disseminated forms are characterized by the involvement of organs and systems. Skin, eyes, lungs, liver, kidney, heart, central nervous system (CNS) and genitalia have already been described as affected sites. The osteoarticular form may occur by contiguity of the primary lesion or hematogenous spread dissemination from lungs [4,23–26].

Poorly explored until recently, the reemergence of this mycosis in different parts of the world led to a renewed interest in its study, mainly focusing on immunopathogenesis mechanisms and immune response against fungal molecular components; these topics have been reviewed recently [10,27,28]. Nevertheless, our knowledge on the immunopathogenesis of sporotrichosis is still fragmentary.

3. What We (Don´t) Know about the Immune Response in Human Sporotrichosis

In sporotrichosis, exposure does not necessarily result in overt disease since the proportion of those who will develop an illness is smaller than those who control the infection. Although the mechanisms underlying this observation are unknown, some studies suggest that different *Sporothrix* species may present different pathogenicity levels, which may lead to varying degrees in clinical manifestations, with some "susceptible" individuals developing the more benign FC form (which can eventually heal spontaneously), while others evolving to severe disseminated or extracutaneous forms.

Current available data on the immune response to *Sporothrix* spp. (or their components) is predominantly based on in vitro studies and in rodent experimental models, reprising the strategies used in the investigation of the immune reactivity of the other, better studied, endemic deep or subcutaneous mycoses; data gathered directly from human patients is scarce. While those studies differ widely in the species tested, the fungal phase used to infect or to obtain fungal components (conidia vs. yeast vs. germlings), the size and route of inoculation (mostly intraperitoneal and intravenous) and animal models employed (mouse strains, Wistar rats, golden hamster and, more recently, the great wax moth *Galleria mellonella*), the extent to which they contribute to the understanding of the immunopathogenesis of the human disease is still not clear since the pieces do not fit the puzzle. Only recently, mouse models mimicking the human disease (i.e., subcutaneous inoculation) have been explored [29–32].

As a first line of defense against pathogens, innate immunity is considered key to fungal control. Pioneering work by Kajiwara et al. showed that neutrophils and macrophages from mice with chronic granulomatous disease (CGD), who show defective NADPH oxidase complex function and fail to generate microbicidal reactive oxygen species (ROS), were unable to control the growth of *S. schenckii* yeast cells, and those animals developed a disseminated lethal disease upon subcutaneous inoculation, while wild-type counterparts were resistant to systemic infection and survived [32]. Translating those findings, however, to human context seems challenging: while Cunningham et al. observed phagocytosis and intracellular killing of *S. schenckii* by human polymorphonuclear cells in vitro, mediated by the H_2O_2–KI–myeloperoxidase system [33], Schafner et al. found that virulent *S. schenckii* was resistant to killing by neutrophils and H_2O_2 [34].

The controversial role of nitric oxide (NO) in sporotrichosis highlights the complexity of the host's immune response in this mycosis. Experimental data suggest a dual role for NO, supporting both its fungicidal activity against *S. schenckii* in vitro [35] and its association with T cell suppression and poorer outcome in murine models [36]. In patients' biopsies, expression of NO synthase-2 was higher in LC lesions, while FC injuries displayed more intense inflammation, tissue destruction and higher fungal burden [37,38].

Human macrophages were also shown to phagocytose and kill (probably through ROS release) *S. schenckii* conidia and yeast cells [39]. Some studies suggest that melanin expression would protect the isolates from macrophage phagocytosis and oxidative attack [35,40]. However, there are no studies analyzing the in situ expression of melanin by intralesional yeast cells in biopsies. Curiously, in the human monocytic cell line THP-1, engulfment of *S. schenckii* conidia preferentially occurs through mannose receptors while yeast cells internalization relies on complement receptors [39], suggesting the interplay of different receptors in fungi–host interaction.

In parallel to neutrophils and macrophages, it was shown that bone marrow-derived mouse dendritic cells (DC) also participate in the recognition process of fungal components and drive the cellular immune responses [41], regulating the magnitude and balance of Th-1 and Th-17 responses in vitro. The latter were associated with control of the fungal burden in an intraperitoneal infection mouse model [42,43]. Other immune cells, such as mast cells, can also amplify the acute response by releasing mediators (histamine and proinflammatory cytokines that attract neutrophils) that exacerbate the inflammatory process, but with deleterious effects to the host, rather than contributing to control of fungal burden [44,45].

Several studies addressed the recognition of *Sporothrix* spp. and their components by innate immunity receptors (pattern recognition receptors, PRR) and its influence in subsequent cellular immunity. Toll-like receptors (TLR) are conserved membrane-associated proteins that recognize a broad set of microbial components, such as *S. schenckii* lipid antigens, recognized via TLR4 [46], triggering diverse cell responses. TLR2 activation, for example, enhances in vitro phagocytosis of *S. schenckii* yeast cells by mouse macrophages and promotes the release an array of pro- (TNF-α, IL-1β, IL-12) and anti-inflammatory (IL-10) cytokines as well as effector/cytotoxic compounds (e.g., NO) [47,48]. Keratinocytes are also activated through TLR2 and TLR4 to release proinflammatory cytokines when challenged with *S. schenckii* yeast cells [49]. However, it is not yet clear from these studies whether the elicited inflammatory response contributes to enhanced immunopathology or host protection.

Dectin-1 and dectin-2 are important PRRs that trigger Th-17 responses but currently there are only data for participation of dectin-1 in triggering Th-17 responses in an intraperitoneal mouse model of sporotrichosis [50]. Conversely, Zhang et al. showed that both dectin-1 and IL-17 production were dispensable for clearance of *S. schenckii* infection in a rat model [51]. There is also evidence from a mouse model of systemic infection that activation of the inflammasome exerts a transitory protective role, especially due to IL-1, IL-18 and caspase-1 [42,52,53], whose impairment reduced Th-17 and Th-1 mediated inflammatory responses leading to higher susceptibility to *S. schenckii* infection [52]. *S. schenckii* yeast cells can also activate the alternative (antibody independent) complement pathway in vitro but its relevance to in vivo host defenses was not defined [54].

Overall, these studies suggest that *Sporothrix* spp. can be recognized by different innate immunity receptors. Which particular set of these (and their signaling pathways) is involved in human infection, which could also be affected by the different infection routes (percutaneous or inhalatory), remains to be determined. Furthermore, with the identification of new *Sporothrix* spp., the involvement of immune receptors could be species-specific.

In fact, Arrillaga-Moncrieff et al. showed that pathogenicity differs among species: *S. brasiliensis* was the most pathogenic, followed by *S. schenckii*, when compared to *S. albicans*, *S. globosa* and *S. mexicana* [55]. Almeida-Paes et al. suggested those differences might even exist within a single species: *S. brasiliensis* isolates obtained from patients with more severe disease express more putative "virulence" factors, such as urease and melanin, and are able to cause a more disseminated disease [56,57]. In Venezuela, a retrospective study gathered isolates from patients and found that *S. globosa* is isolated mainly from patients with FC sporotrichosis while *S. schenckii* would be related to LC forms [58].

Fibronectin surface adhesins expressed by *S. schenckii* have also been described to increase pathogenicity in C57BL/6 mice. Although analyses of differences according to species were not available at that time, they did not find direct correlation between virulence and the clinical or environmental origin of the isolates: the lowest virulence was observed for an isolate recovered from a patient with meningeal sporotrichosis [59].

Recently, Martinez-Alvarez et al. showed that human peripheral blood mononuclear cells (PBMC) differentially recognize *S. brasiliensis* and *S. schenckii* [60]. The three *S. schenckii* morphologies stimulated higher levels of pro-inflammatory cytokines than *S. brasiliensis*, while the latter stimulated higher IL-10 levels. This finding could help to explain the apparent higher pathogenicity of *S. brasiliensis*. However, as we still do not know the first steps of the infection in humans, the contribution of each morphology to its successful occurrence remains to be established. The authors additionally showed that dectin-1 was a key receptor for cytokine production induced by *S. schenckii*, but was dispensable for *S. brasiliensis* germlings. TLR2 and TLR4 were also involved in sensing of *Sporothrix* cells, with a major role for the former during cytokine production. The mannose receptor had a minor contribution in *S. schenckii* yeast-like cells and germlings recognition, but *S. schenckii* conidia and *S. brasiliensis* yeast-like cells stimulated pro-inflammatory cytokines via this receptor.

Immunochemical studies in the 1970s already suggested that cell wall components elicited immediate and delayed immune responses [61]. Subsequent studies reinforced the important role played by cell-mediated mechanisms (i.e., TCD4+ lymphocytes and activated macrophages) in resistance to intravenous experimental sporotrichosis in athymic nude [62–65] and Swiss mice [66,67]. This research line was resumed more recently in the search for vaccine candidates. Live yeast cells and/or exoantigens were used, and Th-1 and Th-17 responses were, in general, generated [41,43,53], both of which appeared to be required for protection in these models. However, there was also evidence of an important participation of Th-2 responses at later stages and activation of macrophages with anti-inflammatory characteristics (defined as M2 macrophages) such as high levels of IL-10 secretion [47,48,60,68]. It has been suggested that isolates from cutaneous lesions were more potent to activate human monocyte-derived DCs to drive Th-1 responses than isolates from visceral lesions. However, this finding should be regarded with caution since the study was performed before reclassification of the *S. schenckii* complex into several species and thus the isolates' differences could rather reflect different species [69].

Sporothrix spp. components have also been studied with regard to human humoral responses. Sera from extracutaneous or more severe forms of sporotrichosis recognized a wider range of antigens and displayed higher antibody titers than sera from patients with cutaneous/less severe forms of sporotrichosis [56,70,71]. A protective role of antibodies, possibly through facilitation of phagocytosis, has been described in some experimental models [71–74].

Unfortunately, data stemming from patients are scarce and are represented mostly by histopathology studies of biopsies taken from patients, with a limited set of parameters analyzed due to limitations inherent to these methods. Moreover, some studies involved a rather small number of patients. Nonetheless, the cutaneous inflammatory process consists in most cases of a suppurative granulomatous response, with frequent presence of liquefaction and/or necrosis. Of note, the paucity of fungal elements (absent from 65% of the biopsies), associated with better granuloma formation (epithelioid granulomas, higher infiltration of lymphocytes, presence of fibrosis, absence of necrosis) suggests the ability of the human system in partially containing the fungal burden [75,76]. This may help to explain why (a) exposure does not necessarily result in development of illness and (b) some patients self-heal.

Immunohistochemistry studies showed the presence of CD4+ and CD8+ T-cells, CD83+ DC, macrophages and monocytes, and the expression of IFN-γ, but not of iNOS, within granulomas [38,77,78]. Compared to the FC form, LC patients had more intense signs of inflammation (higher infiltration of neutrophils and lymphocytes, and higher expression of nitric oxide synthase 2) and higher fungal burden. IFN-γ expression did not differ but IL-10 was more prominent in LC than FC lesions, consistent with the more intense inflammatory process in the former [77]. Interestingly, these authors also observed a higher ability of PBMC from patients than healthy individuals to release IFN-$\gamma$and IL-10 upon in vitro challenge with *S. schenckii* antigen [77]. An early report already noted a trend toward lower T-lymphocyte responsiveness in systemic disease as compared with the LC form [78]. Interestingly, of six systemic sporotrichosis patients, one had bone marrow aplasia and four reported daily consumption of variable amounts of alcohol, while none of the LC patients reported these conditions. Overall, these data reinforce the ability of the human immune response in limiting, at least partially, the disease caused by *S. schenckii*. However, this notion can be challenged by the report of severe extracutaneous sporotrichosis in apparently immunocompetent individuals [79–81]. A summarized, schematic view of the data obtained from experimental studies of the immune response in sporotrichosis is shown in Figure 1.

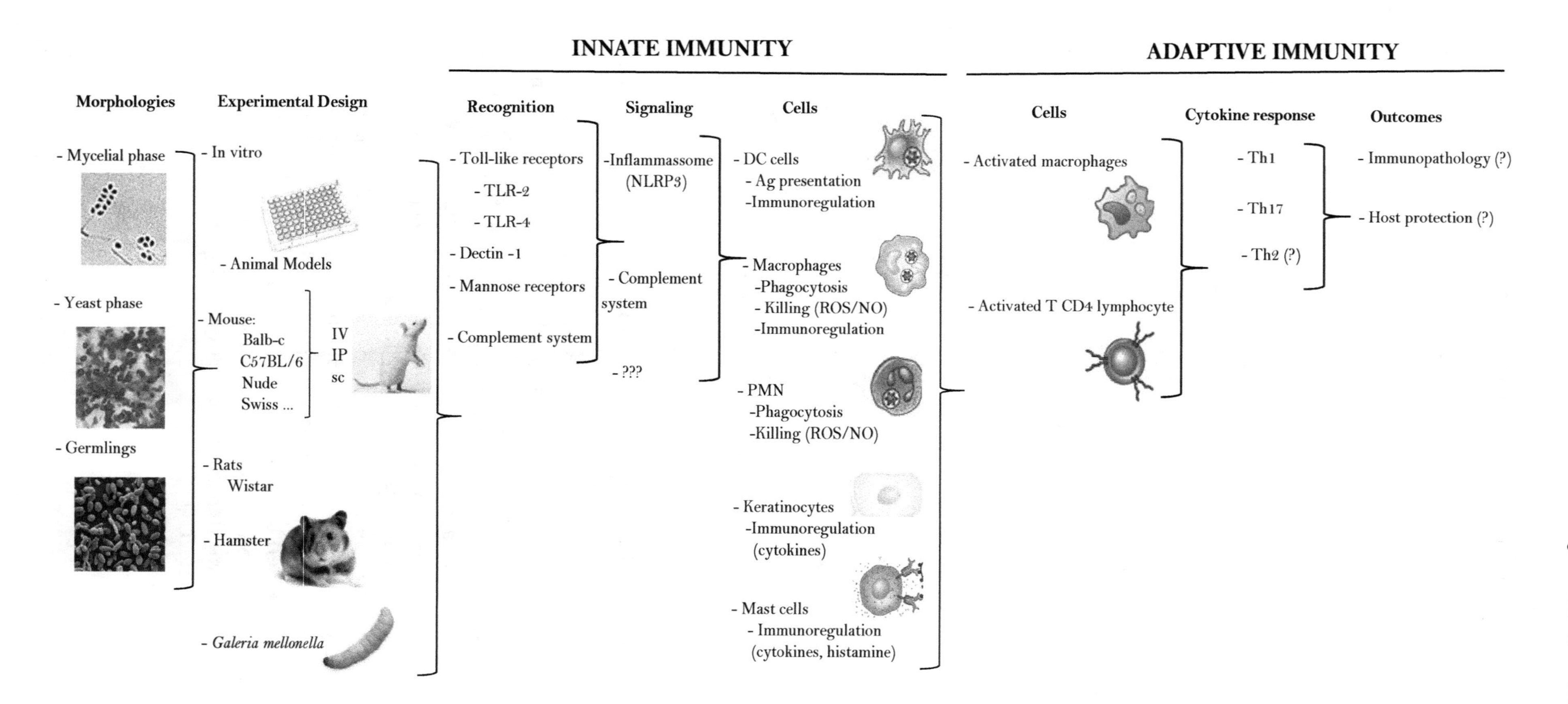

Figure 1. Schematic view of the data obtained from experimental studies of the immune response in sporotrichosis. IV: intravenous; IP: intraperitoneal; sc: subcutaneous; DC: dendritic cells; ROS: reactive oxygen species; NO: nitric oxide; PMN: polymorphonuclear cells; Th: T helper.

4. Sporotrichosis in AIDS Patients

The first case report of OS in HIV-infected individuals dates from 1985 [82]. In the last decade, the reemergence of the disease in Rio de Janeiro, Brazil, was followed by an increase in the number of cases in HIV co-infected patients. Nevertheless, to date, there are no more than 107 cases reported worldwide. Reflecting the rare occurrence of the disease in this group, our knowledge of the clinical features and management principles is based on expert opinions, case studies and retrospective cohort studies [12,83–85]. Still, it is worth noting that HIV/AIDS patients either had disseminated or cutaneous sporotrichosis, or did not become ill after exposure [12].

Data from the largest retrospective cohort study of 3618 cases of sporotrichosis revealed that 1.32% were HIV co-infected. Close to half (44%) were hospitalized over time, much more frequently than HIV negative patients (1%). Although the main cause for hospitalization in both groups was disseminated disease, this corresponded to 90.5% of the hospitalized HIV patients but only 43.2% of HIV negative subjects. In addition, hospitalized patients had a mean CD4 T lymphocyte count of 125 cells/μL and deaths attributed to sporotrichosis occurred 45 times more [83].

Clinical data from a systematic review showed the majority of HIV co-infected patients have cutaneous disease associated with involvement of other organs or systems. Their median CD4 T lymphocyte count was 97 cells/μL. In addition, unusual manifestations cannot be underestimated, with 17% of the cases presenting CNS involvement [84]. CNS disease has already been described in HIV-infected patients without clinical evidence of neurological symptoms [79,85,86]. Ten previous published cases with sufficient clinical details indicate poor prognosis for CNS involvement [85–93]. Thus, investigation of CNS disease in this specific population is strongly recommended [84] in order to provide early diagnosis and aggressive treatment. Most of these patients are males with a mean CD4 count of 101cells/μL. In all cases, skin lesions were present before or associated with the onset of meningeal symptoms. In addition, the concomitant involvement of lungs [85,89,92,93], mucosa [85,86,89], bone [86], kidney [93], testicles, epididymis, bone marrow, lymph nodes and pancreas [90] has also been described.

In the majority of cases, patients present positive cerebrospinal fluid (CSF) cultures on admission or during the follow-up. Only in one case, lumbar puncture was sterile and *Sporothrix* sp. was observed in tissue sections [90]. Most importantly, however, biopsies of skin lesions yielded growth of *Sporothrix* spp. in almost all the previous cases [85–88,91–93], allowing the diagnosis to be made before CSF cultures results were available or became positive. Treatment of sporotrichosis meningoencephalitis is by far the most challenging aspect in the disease management and the poor therapeutic response observed in these few cases is remarkable. Except in one patient [93], amphotericin B formulations were the initial therapeutic choice, although a significant rate of recurrence or relapse of neurological symptoms was observed [85,86,92]. Nine of the patients died; the single patient who survived resolved the infection without sequels [92].

Other unusual manifestations associated with cutaneous lesions were endocarditis, mucosal involvement (ocular and nasal), uveitis, endophthalmitis, and pulmonary and osteoarticular involvement [12,24]. Isolated involvement of the lungs and sinus has also already been described [94–97].

The prevalence of HIV co-infection in patients with less severe forms is not yet well established, because HIV routine testing is not recommended for every patient with sporotrichosis. Usually, only those with severe manifestations or a suspected HIV infection result in a laboratory investigation, which would overestimate the incidence of severe clinical presentations and poor prognosis in this population. Nevertheless, Freitas et al. evaluated the prevalence of HIV co-infection in the Rio de Janeiro epidemic by testing stored blood samples of 850 patients with benign forms of sporotrichosis from 2000 to 2008, and found only one positive result [83]. Moreover, only a few cases of LC and FC forms have been described so far in HIV-infected patients with a mean CD4 T lymphocyte count of 513 cells/μL [12,83,98]. Overall, these data support the notion that HIV co-infection modifies the clinical presentation, severity and outcome of the patients with sporotrichosis, in accordance to their immune status and degree of immunosuppression [83].

5. Sporotrichosis Associated with IRIS

Only six cases of sporotrichosis associated with immune reconstitution inflammatory syndrome (IRIS) have been published to date. There are two cases of paradoxical sporotrichosis meningitis IRIS in patients who exhibited cutaneous lesion and were under itraconazole treatment before the onset of neurological disease. Both patients had confirmation of virologic response to antiretroviral therapy (ART). Despite treatment with amphotericin B at 4–8 weeks, the patients presented recurrence of neurological symptoms during follow-up with itraconazole and/or amphotericin B 2–3 times a week. *Sporothrix* sp. was isolated from CSF at some point [92].

These cases are controversial due to the difficulties in making a definite diagnosis of sporotrichosis IRIS when cultures remain positive, properly excluding other causes of clinical deterioration as therapeutic antifungal failure. Difficulties in defining IRIS still apply to other endemic mycoses such as paracoccidioidomycosis and cryptococcosis [99,100]. Although some debate persists, an apparent consensus for diagnosis of paradoxical IRIS associated with opportunistic mycoses is the worsening or appearance of new clinical and/or radiological manifestations consistent with an inflammatory process occurring during appropriate antifungal therapy with sterile cultures for the initial fungal pathogen within 12 months of ART initiation [100,101].

Another questionable report from Brazil described one patient with disseminated sporotrichosis who was already under treatment for two months with itraconazole when ART started. After six weeks, the patient experienced reactivation of old lesions and development of new cutaneous and mucosal lesions. However, cultures of skin biopsy were positive for *Sporothrix* sp. and the patient recovered well with increased doses of itraconazole combined with amphotericin B [92].

Lyra et al. described two cases of disseminated cutaneous sporotrichosis whose clinical presentations were more consistent with IRIS than progressive fungal infection or failure of treatment. Both patients started antifungal therapy shortly followed by ART. However, after four and five weeks, the patients exhibited paradoxical clinical worsening with recurrence of the lesion as well as development of new lesions along with systemic inflammatory symptoms such as fever and arthralgia. Subsequent mycological examination did not reveal fungal growth. The patients were treated with prednisone, resulting in rapid improvement of arthralgia and fever, followed by resolution of skin lesions [102].

Finally, one case described a patient who had no cutaneous findings before ART, but experienced unmasking of disseminated cutaneous sporotrichosis after five weeks. Cultures of lesion exudates were positive for *Sporothrix* sp. The patient's cat had died of sporotrichosis one month before the patient started ART. He presented complete regression of the lesions after antifungal therapy [92].

Case Presentation

A 59-year-old Brazilian man presented cachexia and disseminated and ulcerated skin lesions with one-year evolution (Figure 2A). Before his illness, he worked as an agriculturist, truck driver and a sewerage system cleaner in his town. During his last professional activity, he was continuously exposed to polluted water. Eight months earlier, the diagnosis of leprosy was made without any microbiological evidence and he was unsuccessfully treated with rifampin, dapsone and clofazimine. Six months ago, HIV infection was detected and lamivudine, tenofovir and efavirenz were added. At admission, he was depressed, febrile and complaining of pain. His body weight was 40 kg, and, besides the cutaneous clinical manifestations, there were no signs of internal organ involvement. The main laboratory findings included anemia with hemoglobin of 9.1 g/dL, leukocytosis (12,100 cell/μL) and protein chain reaction (PCR) of 11mg/L. HIV test was positive with CD4 cell count of 584 cells/$mm^3$and viral load of 1558 copies/mL (log 3.1). A skin biopsy depicted a mixed exudative and granulomatous cellular infiltrate with a few round to elongated yeast cells (Figure 2B). The cultures of biopsy fragments yielded a dimorphic fungus phenotypically identified as *Sporothrix* sp., later identified by DNA sequence as *S. schenckii*. The anti-lepromatous therapy was stopped and the patient was treated with itraconazole, 400 mg per day, and cotrimoxazole 360mg/800mg per day for secondary bacterial infection. Because IRIS was suspected, prednisone at the daily dose of 20 mg per day was added and ART was changed to atazanavir/ritonavir due to probable drug-to-drug interactions between itraconazole and the previous antiretrovirals. He improved gradually and corticosteroid and cotrimoxazole were discontinued. After three months of therapy, itraconazole was reduced to 200 mg per day and discontinued after six months. The patient presented complete clinical and mycological responses. (Figure 2C).

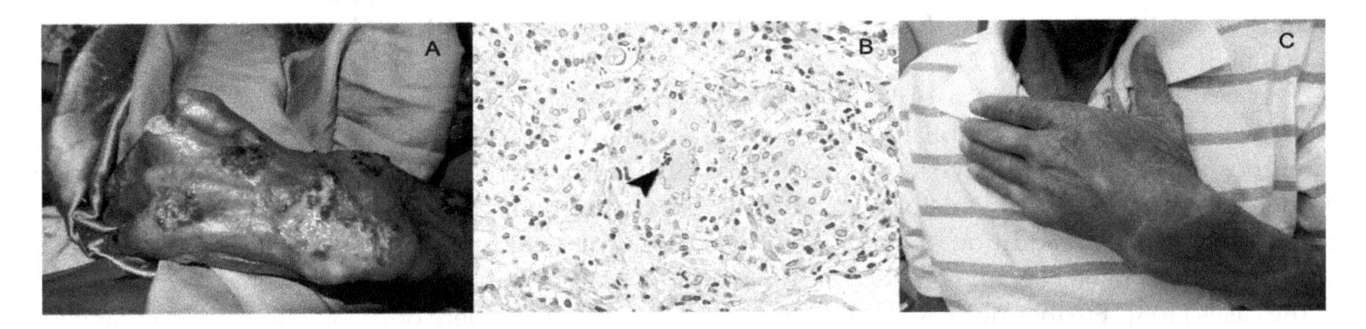

Figure 2. Ulcerated lesions in the hand and fist of a patient with human immunodeficiency virus (HIV) infection ad cutaneous disseminated sporotrichosis and immune reconstitution syndrome (**A**). A skin biopsy (**B**) depicted an exudative and granulomatous infiltrate with cigar shape and round yeast cells (arrow head), imbibed in multinucleate giant cells of the Langerhans type. Periodic Acid-Schiff stain $\times$ 400. The patient responded well to long course of continuous itraconazole intercalated with short courses of cotrimoxazole for secondary bacterial infection and prednisone for immune reconstitution inflammatory syndrome (IRIS) control (**C**).

6. Comorbidities as Risk Factors for Sporotrichosis

Diabetes Mellitus and Alcoholism

The main underlying disease reported in sporotrichosis outbreaks is diabetes mellitus, reaching up to 23% of the cases, followed by alcohol consumption, reaching from 5% to 8% [20,22,103]. Despite this observation, in hyperendemic areas of sporotrichosis, little is known about the contribution of these conditions to development of severe forms and there is limited understanding about the immunosuppressive mechanisms involved. To date, no large series or cohort studies described particularities of the clinical presentation in this population and most published reports of sporotrichosis in diabetes and alcoholism have focused on unusual manifestations, which are not necessarily the predominant forms observed in these populations.

Some of the first cases in alcoholic and diabetic patients date back to 1961 and 1970, respectively, and both patients presented with primary pulmonary sporotrichosis [104,105]. A recent systematic analysis of the literature addressing 86 cases of pulmonary sporotrichosis showed that diabetes mellitus was present in six patients and alcohol consumption in 34 cases. Most of these cases (75%) presented with primary pulmonary sporotrichosis. Of note, the cavitary pattern on radiology was the most common finding and 45% of the patients presented extrapulmonary involvement, the skin being the most affected site followed by joint involvement [4].

Putative differences in clinical presentation in diabetic and alcoholic patients occur. One presented disseminated cutaneous lesions and the other particularly severe/destructive localized lesions with a granulomatous aspect, denoting an enhanced pathogenicity in these localized cases. Nevertheless, all patients had a marked improvement with conventional antifungal treatment [106–109]. The most emblematic case of disseminated disease occurred in a diabetic and alcoholic patient who developed a fatal fungemia after 17 days of hospitalization [110]. The other cases corresponded to isolated monoarthritis, endophthalmitis and cutaneous disseminated sporotrichosis [111–113].

A retrospective study of 238 cases in the Peruvian highlands, where *S. schenckii* is hyperendemic, pointed to alcoholism and diabetes mellitus as significant underlying factors. The majority of patients had cutaneous or lymphocutaneous disease; only nine patients presented disseminated cutaneous disease and no cases of extracutaneous involvement were found [13]. In Brazil, clinical data from 178 patients with culture-positive sporotrichosis treated during the period of 1998–2001 showed that 80.9% of the cases presented LC or localized cutaneous forms. Systemic sporotrichosis was not diagnosed even in cases involving alcohol or diabetes as comorbidities, and in 29 patients (16.3%) with skin lesions at multiple locations, this was more likely due to repeated inoculations during persistent contact with sick animals [22].

In another Brazilian case of 24 patients with widespread cutaneous lesions, only two were associated with alcoholism and diabetes, and these conditions may have acted as an immunosuppressive factor favoring either the establishment of the infection and/or its dissemination. None of these presented a history of multiple exposures that could account for the widespread cutaneous lesions. In addition, none of the other patients showed any immunosuppressive condition and were found to be in good general condition, although fever and/or arthralgia were reported in 50% of the cases [103]. Furthermore, Rosa et al. described 304 patients where only four cases with cutaneous disseminated and extracutaneous forms were recognized, but data regarding their comorbidities were not provided [20].

7. Other Immunosuppressive Conditions

The literature reports several cases of cutaneous-disseminated and extra-cutaneous sporotrichosis in patients under immunosuppressive treatments for rheumatologic, autoimmune conditions, solid organ transplantation (SOT), hematologic cancers and primary immunodeficiencies.

Osteo-articular or disseminated sporotrichosis misdiagnosed as rheumatoid arthritis, presumed inflammatory arthritis or sarcoidosis illustrates the role of iatrogenic immunosuppressive regimen in severity and complicated outcome [114–116]. Immunosuppressive therapy included steroids, azathioprine, tocilizumab, tacrolimus and cyclophosphamide—in one case, after almost one year of inappropriate therapy with several immunosuppressives (including prednisolone, tocilizumab, tacrolimus and cyclophosphamide), the patient experienced fungemia and died of respiratory insufficiency due to pulmonary sporotrichosis [114]. Two patients also had delayed diagnosis and progressed to disseminate disease albeit clinical improvement was achieved after antifungal therapy (amphotericin B or itraconazole) was started in parallel to lessening the iatrogenic immunosuppression [115,116].

A retrospective review of 19 cases of sporotrichosis diagnosed at a single service in the United States showed that seven patients were misdiagnosed initially and four received immunosuppressive agents for other diagnoses, such as polyarteritis nodosa, sarcoidosis, pyoderma gangrenosum and vasculitis [117]. One additional patient had received immunosuppressive therapy for a pre-existing polyarthropathy before the development of his cutaneous lesion. In contrast, none of the patients presented extracutaneous disease. The index case was diagnosed as pyoderma gangrenosum for disseminated leg ulcerated lesions and received immunosuppressive treatment with aziatropine, prednisone and cyclosporine with further worsening of the lesions. Treatment required debridement of necrotic tissue, plastic surgery and subsequent staged skin grafting, together with an 18-month course of 600mg/day itraconazole and cessation of immunosuppression. However, no data regarding the response to treatment of these 19 patients were provided.

Similarly, cases of cutaneous disseminated and LC forms in immunosuppressive therapy with tacrolimus, anti-TNF-alpha and prednisone due to lupus nephritis, ankylosing spondylitis and sciatic pain, respectively, have been reported. All these patients had a good clinical response to antifungal therapy (potassium iodide or itraconazole) and discontinuation of immunosuppressive therapy [118–120].

The possible role of immunosuppressive drugs in atypical clinical presentation is reinforced by a systematic analysis of pulmonary sporotrichosis. In this study, of the 86 cases of pulmonary sporotrichosis included, 64 had primary pulmonary disease and 22 also had extra-pulmonary involvement. The only significant difference between the groups that could represent a risk factor for multifocal disease was the increased use of immunosuppressant drugs by the extra-pulmonary group [4].

Sporotrichosis in SOT recipients manifests as more severe disseminated forms than in immunocompetent hosts. Few exceptions respond well to antifungal treatment, being considered uncommon according to a prospective surveillance study of invasive fungal infections conducted in 15 SOT centers in United Sates, which did not identify any case of sporotrichosis [121]. However, in the Rio de Janeiro epidemic, sporotrichosis was retrospectively recognized in one subject, among 42 kidney transplant patients, with extracutaneous disease (LC and bone involvement) [122]. In addition, Caroti et al. followed 774 Italian kidney transplant patients for 18 years and subcutaneous nodules or cutaneous lesions were identified in seven [123]. One patient presented an erythematous papulonodular lesion with positive culture for *S. schenckii*. Despite treatment with fluconazole, seven years after renal transplantation, the patient developed acute osteomyelitis and gangrene in the left foot with ulcers. The patient was treated again with fluconazole together with interruption of the immunosuppressive agent mycophenolate mofetil, presenting gradual regression of the lesions. In India, during a period of two years, 40 renal transplants were performed and pulmonary sporotrichosis was diagnosed in one patient on triple drug immunosuppression [124]. Finally, there are three additional cases in kidney transplant recipients reported in the literature, one case of cutaneous disseminated and two of disseminated disease [23,125]. All patients were taking immunosuppressive agents and were successfully treated with antifungal therapy including amphotericin B deoxycholate, lipid amphotericin B formulations, fluconazole and itraconazole. One unusual case of urinary sporotrichosis after renal transplantation has also been described [25].

Sporotrichosis in SOT other than kidney transplantation is even more rare. Disseminated sporotrichosis with LC, articular and pulmonary involvement was described in a patient 10 years after liver transplantation still on immunosuppressive regimen (tacrolimus and prednisone) [23]. Despite antifungal treatment with itraconazole and reduction of immunosuppressant drugs, after 300 days of follow-up, the patient showed only partial improvement. Pulmonary sporotrichosis in a lung transplant recipient was also reported. On the second day of transplantation, while on induction of immunosuppression with high dose methylprednisolone, tacrolimus, and mycophenolate mofetil, he presented pulmonary diffuse patchy bilateral infiltrates and *S. schenckii* was isolated from bronchoalveolar lavage. Treatment with amphotericin B lipid formulation followed by itraconazole maintenance therapy was successful [126]. Rare cases of sporotrichosis in patients with hematologic cancer have also been described. One patient with multiple myeloma presented disseminated disease, successfully treated with amphotericin B [80]. Two Hodgkin's disease patients, one with fatal meningeal sporotrichosis [127] and the other with a disseminated form refractory to potassium iodide, also responded to a long course of amphotericin B [80]. Two other cases reported patients with hairy cell leukemia, both with disseminated disease, one whose difficult and life-threatening course required liposomal amphotericin B followed by posaconazole (taken indefinitely), and the other exhibited a good responded to itraconazole [128,129].

Finally, primary immunodeficiency has been recognized as a risk factor for severe forms of the disease. A fatal case of disseminated form of sporotrichosis was described in one patient with primary idiopathic CD4 lymphocytopenia [130]. Another unusual case of *S. schenckii* cervical lymphadenitis was identified in a 33-month-old male with X-linked CGD that was successfully treated with surgical excision and voriconazole [131].

Overall, based mostly on published case reports, we suggest that patients on immunosuppressive regimen due to SOT, rheumatologic disease or other comorbidities are at higher risk of more severe clinical presentations of sporotrichosis. However, most cases presented a good outcome when provided with more prolonged and higher doses of antifungal treatment than used in immunocompetent hosts. In this scenario, the drug of choice should be guided by the severity of the disease, with initial therapy with amphotericin B being frequently required [114,116,117,125–128,130,132].

8. Laboratory Diagnosis

The most relevant diagnostic tool for patients with suspicion of sporotrichosis is isolation of etiologic agent from clinical specimens such as secretions, abscess aspirates and biopsied tissue fragments. In extracutaneous clinical forms, synovial fluid, blood, CSF and sputum should be cultivated. Fungal cultures may be obtained in standard media such as Sabouraud dextrose agar with antibiotics, Mycosel, blood agar and brain heart infusion media [133]. After 5 to 10 days, the yeast-like colonies may be observed at 37 °C incubation, although this time may be extended up to 30 days. For phenotypic identification, the micromorphology of mycelial forms must be seen after incubation at room temperature, although it cannot distinguish individual species. Species determination requires molecular methods for definitive identification [10,133].

In contrast to immunocompetent patients, where direct mycologic or histopathologic exams show poor sensitivity, in immunocompromised hosts these methods may depict higher sensitivity, especially in AIDS patients with very low CD4 cell counts [83,84]. In severely immunocompromised AIDS patients, cutaneous and lymphatic lesions may depict a big fungal burden, similar to that observed in cats with *S. brasiliensis* infections [7,134] (Figure 3). Immunocompromised patients with cutaneous disseminated and extracutaneous clinical forms may depict *Sporothrix* spp. yeasts under immunofluorescence, Giemsa, Gram, Grocott-Gomori and periodic acid Schiff (PAS) stains. When observed, yeasts present round to oval and elongated "cigar shape" forms [2,10,135,136]. Non-microbiologic diagnostic tests such as immunoelectrophoresis, immunodiffusion, ELISA and DNA sequencing PCR methods are very important for typical and immune reactive forms but the only commercially available test for immunodiagnostic of sporotrichosis is the latex agglutination technique [10,137,138].

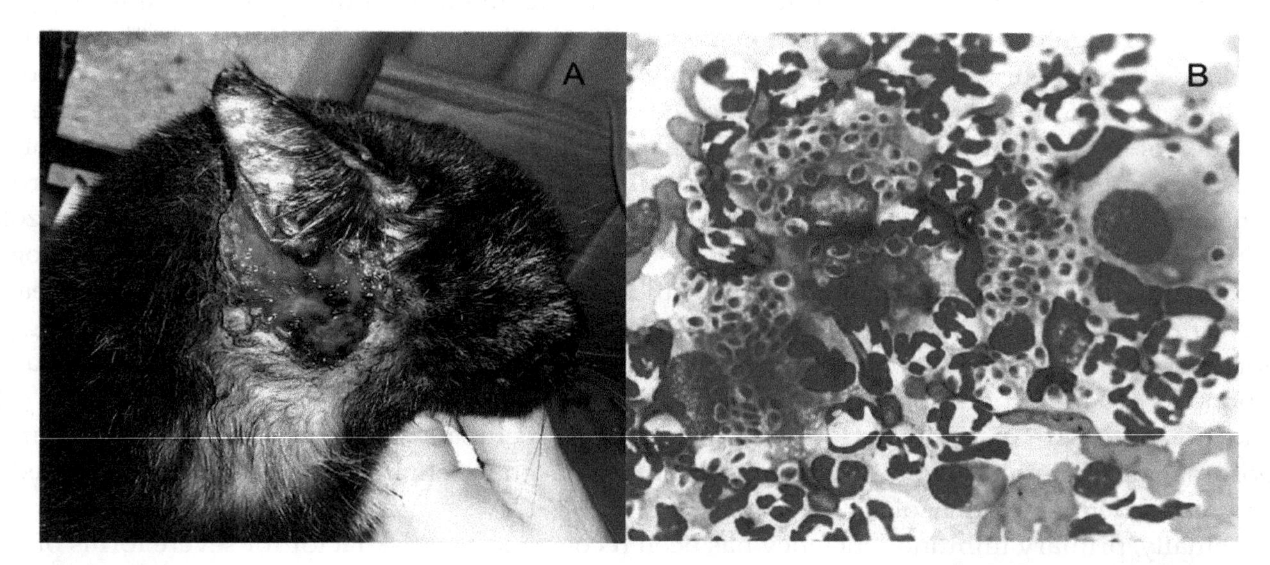

Figure 3. Ulcerated and papular vesicular lesions in the head and ear of a cat with proved *Sporothrix brasiliensis* infection (**A**). Cutaneous feline lesions are highly infective and harbor a great number of yeast cells of the fungus. Feline sporotrichosis can be easily diagnosed by secretion direct exam stained Giemsa × 1000 (**B**).

9. Treatment of Opportunistic Sporotrichosis

Therapy of patients with sporotrichosis associated with impaired host defenses does not differ from treatment modalities applied for immunocompetent individuals, except for inclusion of specific interventions for the underlying conditions leading to opportunistic disease. Although some experimental studies demonstrated different species might show variable in vitro sensitivity to systemic antifungal drugs, no clinical correlation in therapy of human sporotrichosis has been confirmed to date [133,138–141]. For patients with cutaneous or LC forms, itraconazole at the daily dose of 200–400 mg for 3–6 months is the therapy of choice. Exceptionally, if immunosuppression is maintained, a longer period of itraconazole therapy may be required [83,84]. Special attention is recommended for drug interaction between itraconazole and ART drugs such as efavirenz, ritonavir and darunavir [142]. If itraconazole is contraindicated due to intolerance, refractoriness or drug-to-drug interaction, 500 mg of terbinafine twice a day is the second option. Finally, for non-severe forms, super saturated potassium iodine solution, 40–50 drops three times per day, can be tried. Second generation triazoles as posaconazole and isavuconazole have not been evaluated yet.

Although infection with *Sporothrix* spp. is rarely life threatening, all forms of sporotrichosis require some kind of treatment. Unlike HIV/AIDS patients, diabetic and alcoholic subjects do not seem to have a worse prognosis and, in general, all patients show a satisfactory response even if in some cases it is necessary to increase the drug dose or the hospitalization stay due to the comorbidity itself or the severity of the lesions [22].

Regardless of the presence of any comorbidity, in pulmonary, severe or life-threatening disease, amphotericin B should be the initial therapy until the patient shows a favorable response, moving on to itraconazole [140]. In parallel, the management of those risk factors, such as control of chronic alcohol intake and steroid or anti TNF discontinuation should be carried out.

An issue in HIV-patient management concerns the best time to introduce ART. While some authors recommend its initiation should be delayed in patients with CNS disease in order to avoid IRIS [84], currently there is no sufficient evidence to support this recommendation for secondary prophylaxis. Data from AIDS patients and tuberculosis or *Cryptococcus* meningitis suggest that patients should start antifungals before ART [143,144]. Therefore, similar to indications already described in the literature for other opportunistic mycoses, long-term suppressive therapy should be considered in patients with severe forms or CNS infection after at least 1 year of successful treatment and then discontinued in patients with CD4 cells counts $\geq$200 cells/μL and who have undetectable viral loads on ART for >6 months [140,145]. Because the risk of relapse of meningeal sporotrichosis is high, lifelong suppressive therapy seems prudent and recommended in these cases [140].

10. What Can Immunocompromised Patients with Sporotrichosis Teach Us?

Many issues in our understanding of sporotrichosis remain unresolved. First, even though it is expected, a report of specific exposure (i.e., sick cats, gardening, lumbering, farming, hunting, etc.) for contracting this implantation mycosis, such as an epidemiological link, was not evident in most cases compiled in this review, except for CTS. Thus, we hypothesize that the inhalatory route would be the most likely mode of infection. Moreover, in many cases retrieved in this review, sporotrichosis was not suspected initially, delaying diagnosis and appropriate treatment initiation, which might have contributed to a higher severity of the disease.

Second, a yet unknown number of the exposed individuals do not develop disease, suggesting the full list of predisposing factors is completely unknown. Although deficiencies of the host immune response are indisputably a critical factor, reports of systemic or severe extracutaneous disease in individuals without clinical or laboratorial evidence of immunodeficiency [79–81] indicates many factors should be considered in infection pathogenesis.

Not surprisingly, a frequent association between atypical forms of sporotrichosis and HIV/AIDS, transplantation, hematological malignancies and iatrogenic immunosuppression for rheumatologic conditions was detected. As seen for other endemic mycoses such as paracoccidioidomycosis,

coccidioidomycosis and histoplasmosis, all these comorbidities, particularly HIV/AIDS, are high-risk factors for disseminated or atypical forms and can even change their natural history [146–148], mostly by promoting a defective T-cell immunity. This was suggested by the experimental works on nude mice reviewed here.

An interesting observation was identification of chronic alcohol abuse as a single predisposing factor to infection. The impact of chronic heavy alcohol consumption on the immune system is complex and time and dose dependent, typically resulting in a subclinical immunosuppression that becomes clinically significant only in the case of a secondary insult [149]. Innate immunity is affected, particularly by inhibiting cellular chemotaxis, phagocytosis (especially for alveolar macrophages) and production of growth factors [150]. Adaptive responses show severe compromise of T-cell function such as lymphopenia, increased cellular differentiation and activation, and reduced migration. The chronic activation of T-cell pool in alcoholic patients would alter its ability to expand and respond to pathogenic challenges or lead to their elimination through increased sensitivity to activation-induced cell death [149,151]. Although the higher risk of infections in alcoholic patients has been related mostly to bacterial (e.g., tuberculosis) and viral infections, our review points to chronic alcohol abuse as a risk factor of atypical, more severe, sporotrichosis. In addition, the higher susceptibility to infections by alcohol abuse may be in part related to behavioral changes that lead to enhanced exposure to pathogens.

Curiously, we only found two reports of primary immunodeficiencies patients with severe sporotrichosis (T CD4 lymphopenia and X-linked CGD). The latter corroborates the finding of increased susceptibility to sporotrichosis in the CGD mice model discussed earlier. Although studies of experimental sporotrichosis suggested that signaling via TLRs and other PRR would be crucial to recognition of the fungus and the mounting of effective immune responses, we did not find cases of OS associated with putative constitutive defects of these pathways. This may be due to the low frequency of these deficiencies in the general population compared with the immunosuppressed conditions aforementioned. We also did not find association between humoral immunodeficiencies and sporotrichosis, despite this subset of immunodeficiency being relatively more common. Thus, although antibodies are protective in some experimental models, they may not play a major role in human sporotrichosis.

Diabetes mellitus is an established risk factor for certain endemic mycoses (e.g., coccidioidomycosis, histoplasmosis, blastomycosis), but surprisingly not in others (e.g., paracoccidioidomycosis), and manifestations were more severe than in non-diabetic patients [152–154]. This was specially related to uncontrolled diabetes (chronic hyperglycemia). Decreased chemotaxis, phagocytic and killing activities of macrophages and neutrophils were described in uncontrolled diabetic patients [155]. However, as for chronic alcohol abuse, the (uncontrolled) diabetes mellitus induced alterations of both innate (neutrophils and macrophages) and adaptive immunity cells (mainly T cells) were related predominantly to enhanced susceptibility to tuberculosis [156–158]. Current evidence suggests underperforming innate immunity followed by a hyper-reactive T-cell-mediated immune response to *Mycobacterium tuberculosis* in patients with tuberculosis disease, but how these altered responses contribute to enhanced susceptibility or more adverse outcomes remains unclear. However, studies of latent tuberculosis individuals showed that diabetes leads to suboptimal induction of protective T-cell responses, thereby providing a possible mechanism for the increased susceptibility to active disease [158]. Conceivably, these alterations would apply to chronic granulomatous infections other than tuberculosis, such as those caused by fungal organisms. Thus, the observation of diabetes as the single underlying condition in an appreciable number of sporotrichosis patients should not be surprising. Further investigation is required to determine which of the immune dysfunctions presented by diabetic patients play a relevant role in the enhanced susceptibility to sporotrichosis.

11. Concluding Remarks and Future Perspectives

The data we gathered neither allow drawing definitive conclusions on several aspects of the opportunistic nature of sporotrichosis nor making a consensus on the management of these cases.

However, they suggest that OS, when not life-threatening, frequently progresses to larger and/or deeper lesions that usually require higher doses of the antifungals and more prolonged courses of therapy. Therapy was generally started with amphotericin B formulations, which were moved to itraconazole after the initial improvement, as judged by the assisting clinician. In general, the patients responded well to treatment, even if slowly. The few fatalities were mainly accounted for by delayed onset of antifungal therapy or by use of immunosuppressors due to misdiagnosis (rheumatologic disease, sarcoidosis, etc.). Potassium iodide was seldom used (mostly in the early case reports), with poorer responses, being then replaced by amphotericin B. Notably, in addition to the skin, the most affected sites were bones, joints, lungs and CNS, with diagnosis based on histopathology/mycological examination plus cultures of specimens such as biopsies, synovial liquid, cerebrospinal fluid, and blood. Unfortunately, non-microbiologic tests, such as antibody/antigen detection, PCR and other molecular methods are not routinely applied for diagnosis. We thus urgently need standardized and commercially available diagnostic tools to discover the deep part of the sporotrichosis iceberg.

Acknowledgments: We are grateful to Fabio Seiti for critical reading of the manuscript and English editing and Juliana Ruiz Fernandes for assistance with Figure 1. Figure 3B is a courtesy of Professor Marconi Rodrigues de Faria, Veterinary Catholic School of Curitiba, Paraná, Brazil.

References

1. Chakrabarti, A.; Bonifaz, A.; Gutierrez-Galhardo, M.C.; Mochizuki, T.; Li, S. Global epidemiology of sporotrichosis. *Med. Mycol.* **2015**, *53*, 3–14. [CrossRef] [PubMed]
2. Barros, M.B.; de Almeida Paes, R.; Schubach, A.O. Sporothrix schenckii and Sporotrichosis. *Clin. Microbiol. Rev.* **2011**, *24*, 633–654. [CrossRef] [PubMed]
3. Queiroz-Telles, F.; Nucci, M.; Colombo, A.L.; Tobón, A.; Restrepo, A. Mycoses of implantation in Latin America: An overview of epidemiology, clinical manifestations, diagnosis and treatment. *Med. Mycol.* **2011**, *49*, 225–236. [CrossRef] [PubMed]
4. Aung, A.K.; Teh, B.M.; McGrath, C.; Thompson, P.J. Pulmonary sporotrichosis: Case series and systematic analysis of literature on clinico-radiological patterns and management outcomes. *Med. Mycol.* **2013**, *51*, 534–544. [CrossRef]
5. Rodrigues, A.M.; de Hoog, G.S.; de Camargo, Z.P. Sporothrix Species Causing Outbreaks in Animals and Humans Driven by Animal-Animal Transmission. *PLoS Pathog.* **2016**, *12*, e1005638. [CrossRef] [PubMed]
6. Schubach, A.; Barros, M.B.; Wanke, B. Epidemic sporotrichosis. *Curr. Opin. Infect. Dis.* **2008**, *21*, 129–133. [CrossRef] [PubMed]
7. Gremião, I.D.; Menezes, R.C.; Schubach, T.M.; Figueiredo, A.B.; Cavalcanti, M.C.; Pereira, S.A. Feline sporotrichosis: Epidemiological and clinical aspects. *Med. Mycol.* **2015**, *53*, 15–21. [CrossRef]
8. Queiroz-Telles, F.; Fahal, A.H.; Falci, D.R.; Caceres, D.H.; Chiller, T.; Pasqualotto, A.C. Neglected endemic mycoses. *Lancet Infect. Dis.* **2017**, *17*, e367–e377. [CrossRef]
9. Fernández, N.; Iachini, R.; Farias, L.; Pozzi, N.; Tiraboschi, I. Esporotrichosis: Uma zoonosis em alerta. In Proceedings of the 13th Latin American Forum for Fungal Infections, Cordoba, Argentina, 5–7 November 2015; pp. 10–11.
10. Lopes-Bezerra, L.M.; Mora-Montes, H.M.; Zhang, Y.; Nino-Vega, G.; Rodrigues, A.M.; de Camargo, Z.P.; de Hoog, S. Sporotrichosis between 1898 and 2017: The evolution of knowledge on a changeable disease and on emerging etiological agents. *Med. Mycol.* **2018**, *56*, 126–143. [CrossRef]
11. Rios, M.E.; Suarez, J.; Moreno, J.; Vallee, J.; Moreno, J.P. Zoonotic Sporotrichosis Related to Cat Contact: First Case Report from Panama in Central America. *Cureus* **2018**, *10*, e2906. [CrossRef]
12. Freitas, D.F.; de Siqueira Hoagland, B.; do Valle, A.C.; Fraga, B.B.; de Barros, M.B.; de Oliveira Schubach, A.; de Almeida-Paes, R.; Cuzzi, T.; Rosalino, C.M.; Zancope-Oliveira, R.M.; et al. Sporotrichosis in HIV-infected patients: Report of 21 cases of endemic sporotrichosis in Rio de Janeiro, Brazil. *Med. Mycol.* **2012**, *50*, 170–178. [CrossRef] [PubMed]

13. Pappas, P.G.; Tellez, I.; Deep, A.E.; Nolasco, D.; Holgado, W.; Bustamante, B. Sporotrichosis in Peru: description of an area of hyperendemicity. *Clin. Infect. Dis.* **2000**, *30*, 65–70. [CrossRef] [PubMed]
14. Song, Y.; Li, S.S.; Zhong, S.X.; Liu, Y.Y.; Yao, L.; Huo, S.S. Report of 457 sporotrichosis cases from Jilin province, northeast China, a serious endemic region. *J. Eur. Acad. Dermatol. Venereol.* **2013**, *27*, 313–318. [CrossRef] [PubMed]
15. Yap, F.B. Disseminated cutaneous sporotrichosis in an immunocompetent individual. *Int. J. Infect. Dis.* **2011**, *15*, e727–e729. [CrossRef] [PubMed]
16. Romero-Cabello, R.; Bonifaz, A.; Romero-Feregrino, R.; Sánchez, C.J.; Linares, Y.; Zavala, J.T.; Romero, L.C.; Vega, J.T. Disseminated sporotrichosis. *BMJ Case Rep.* **2011**, *2011*. [CrossRef] [PubMed]
17. Fernandes, B.; Caligiorne, R.B.; Coutinho, D.M.; Gomes, R.R.; Rocha-Silva, F.; Machado, A.S.; Santrer, E.F.R.; Assuncao, C.B.; Guimaraes, C.F.; Laborne, M.S.; et al. A case of disseminated sporotrichosis caused by Sporothrix brasiliensis. *Med. Mycol. Case Rep.* **2018**, *21*, 34–36. [CrossRef] [PubMed]
18. Gandhi, N.; Chander, R.; Jain, A.; Sanke, S.; Garg, T. Atypical Cutaneous Sporotrichosis in an Immunocompetent Adult: Response to Potassium Iodide. *Indian J. Dermatol.* **2016**, *61*, 236. [CrossRef]
19. Hessler, C.; Kauffman, C.A.; Chow, F.C. The Upside of Bias: A Case of Chronic Meningitis Due to Sporothrix Schenckii in an Immunocompetent Host. *Neurohospitalist* **2017**, *7*, 30–34. [CrossRef]
20. da Rosa, A.C.; Scroferneker, M.L.; Vettorato, R.; Gervini, R.L.; Vettorato, G.; Weber, A. Epidemiology of sporotrichosis: A study of 304 cases in Brazil. *J. Am. Acad. Dermatol.* **2005**, *52*, 451–459. [CrossRef]
21. Itoh, M.; Okamoto, S.; Kariya, H. Survey of 200 cases of sporotrichosis. *Dermatologica* **1986**, *172*, 209–213. [CrossRef]
22. Barros, M.B.; Schubach, A.E.O.; do Valle, A.C.; Gutierrez Galhardo, M.C.; Conceição-Silva, F.; Schubach, T.M.; Reis, R.S.; Wanke, B.; Marzochi, K.B.; Conceição, M.J. Cat-transmitted sporotrichosis epidemic in Rio de Janeiro, Brazil: Description of a series of cases. *Clin. Infect. Dis.* **2004**, *38*, 529–535. [CrossRef] [PubMed]
23. da Silva, R.F.; Bonfitto, M.; da Silva Junior, F.I.M.; de Ameida, M.T.G.; da Silva, R.C. Sporotrichosis in a liver transplant patient: A case report and literature review. *Med. Mycol. Case Rep.* **2017**, *17*, 25–27. [CrossRef] [PubMed]
24. Silva-Vergara, M.L.; de Camargo, Z.P.; Silva, P.F.; Abdalla, M.R.; Sgarbieri, R.N.; Rodrigues, A.M.; dos Santos, K.C.; Barata, C.H.; Ferreira-Paim, K. Disseminated Sporothrix brasiliensis infection with endocardial and ocular involvement in an HIV-infected patient. *Am. J. Trop. Med. Hyg.* **2012**, *86*, 477–480. [CrossRef] [PubMed]
25. Agarwal, S.K.; Tiwari, S.C.; Dash, S.C.; Mehta, S.N.; Saxena, S.; Banerjee, U.; Kumar, R.; Bhunyan, U.N. Urinary sporotrichosis in a renal allograft recipient. *Nephron* **1994**, *66*, 485. [CrossRef] [PubMed]
26. Kauffman, C.A. Sporotrichosis. *Clin. Infect. Dis.* **1999**, *29*, 231–236, quiz 237. [CrossRef] [PubMed]
27. Alba-Fierro, C.A.; Pérez-Torres, A.; Toriello, C.; Pulido-Camarillo, E.; López-Romero, E.; Romo-Lozano, Y.; Gutiérrez-Sánchez, G.; Ruiz-Baca, E. Immune Response Induced by an Immunodominant 60 kDa Glycoprotein of the Cell Wall of Sporothrix schenckii in Two Mice Strains with Experimental Sporotrichosis. *J. Immunol. Res.* **2016**, *2016*, 6525831. [CrossRef] [PubMed]
28. Conceição-Silva, F.; Morgado, F.N. Immunopathogenesis of Human Sporotrichosis: What We Already Know. *J. Fungi* **2018**, *4*, 89. [CrossRef]
29. Castro, R.A.; Kubitschek-Barreira, P.H.; Teixeira, P.A.; Sanches, G.F.; Teixeira, M.M.; Quintella, L.P.; Almeida, S.R.; Costa, R.O.; Camargo, Z.P.; Felipe, M.S.; et al. Differences in cell morphometry, cell wall topography and gp70 expression correlate with the virulence of Sporothrix brasiliensis clinical isolates. *PLoS ONE* **2013**, *8*, e75656. [CrossRef]
30. de Almeida, J.R.F.; Jannuzzi, G.P.; Kaihami, G.H.; Breda, L.C.D.; Ferreira, K.S.; de Almeida, S.R. An immunoproteomic approach revealing peptides from Sporothrix brasiliensis that induce a cellular immune response in subcutaneous sporotrichosis. *Sci. Rep.* **2018**, *8*, 4192. [CrossRef]
31. Manente, F.A.; Quinello, C.; Ferreira, L.S.; de Andrade, C.R.; Jellmayer, J.A.; Portuondo, D.L.; Batista-Duharte, A.; Carlos, I.Z. Experimental sporotrichosis in a cyclophosphamide-induced immunosuppressed mice model. *Med. Mycol.* **2018**, *56*, 711–722. [CrossRef]
32. Kajiwara, H.; Saito, M.; Ohga, S.; Uenotsuchi, T.; Yoshida, S. Impaired host defense against Sporothrix schenckii in mice with chronic granulomatous disease. *Infect. Immun.* **2004**, *72*, 5073–5079. [CrossRef] [PubMed]

33. Cunningham, K.M.; Bulmer, G.S.; Rhoades, E.R. Phagocytosis and intracellular fate of Sporothrix schenckii. *J. Infect. Dis.* **1979**, *140*, 815–817. [CrossRef]
34. Schaffner, A.; Davis, C.E.; Schaffner, T.; Markert, M.; Douglas, H.; Braude, A.I. In vitro susceptibility of fungi to killing by neutrophil granulocytes discriminates between primary pathogenicity and opportunism. *J. Clin. Investig.* **1986**, *78*, 511–524. [CrossRef] [PubMed]
35. Fernandes, K.S.; Coelho, A.L.; Lopes Bezerra, L.M.; Barja-Fidalgo, C. Virulence of Sporothrix schenckii conidia and yeast cells, and their susceptibility to nitric oxide. *Immunology* **2000**, *101*, 563–569. [CrossRef] [PubMed]
36. Fernandes, K.S.; Neto, E.H.; Brito, M.M.; Silva, J.S.; Cunha, F.Q.; Barja-Fidalgo, C. Detrimental role of endogenous nitric oxide in host defence against Sporothrix schenckii. *Immunology* **2008**, *123*, 469–479. [CrossRef] [PubMed]
37. Morgado, F.N.; Schubach, A.O.; Barros, M.B.; Conceição-Silva, F. The in situ inflammatory profile of lymphocutaneous and fixed forms of human sporotrichosis. *Med. Mycol.* **2011**, *49*, 612–620. [CrossRef] [PubMed]
38. Morgado, F.N.; de Carvalho, L.M.V.; Leite-Silva, J.; Seba, A.J.; Pimentel, M.I.F.; Fagundes, A.; Madeira, M.F.; Lyra, M.R.; Oliveira, M.M.; Schubach, A.O.; et al. Unbalanced inflammatory reaction could increase tissue destruction and worsen skin infectious diseases—A comparative study of leishmaniasis and sporotrichosis. *Sci. Rep.* **2018**, *8*, 2898. [CrossRef] [PubMed]
39. Guzman-Beltran, S.; Perez-Torres, A.; Coronel-Cruz, C.; Torres-Guerrero, H. Phagocytic receptors on macrophages distinguish between different Sporothrix schenckii morphotypes. *Microbes Infect.* **2012**, *14*, 1093–1101. [CrossRef] [PubMed]
40. Madrid, I.M.; Xavier, M.O.; Mattei, A.S.; Fernandes, C.G.; Guim, T.N.; Santin, R.; Schuch, L.F.; Nobre, M.e.O.; Araújo Meireles, M.C. Role of melanin in the pathogenesis of cutaneous sporotrichosis. *Microbes Infect.* **2010**, *12*, 162–165. [CrossRef] [PubMed]
41. Verdan, F.F.; Faleiros, J.C.; Ferreira, L.S.; Monnazzi, L.G.; Maia, D.C.; Tansine, A.; Placeres, M.C.; Carlos, I.Z.; Santos-Junior, R.R. Dendritic cell are able to differentially recognize Sporothrix schenckii antigens and promote Th1/Th17 response in vitro. *Immunobiology* **2012**, *217*, 788–794. [CrossRef]
42. Goncalves, A.C.; Maia, D.C.; Ferreira, L.S.; Monnazzi, L.G.; Alegranci, P.; Placeres, M.C.; Batista-Duharte, A.; Carlos, I.Z. Involvement of major components from Sporothrix schenckii cell wall in the caspase-1 activation, nitric oxide and cytokines production during experimental sporotrichosis. *Mycopathologia* **2015**, *179*, 21–30. [CrossRef] [PubMed]
43. Ferreira, L.S.; Goncalves, A.C.; Portuondo, D.L.; Maia, D.C.; Placeres, M.C.; Batista-Duharte, A.; Carlos, I.Z. Optimal clearance of Sporothrix schenckii requires an intact Th17 response in a mouse model of systemic infection. *Immunobiology* **2015**, *220*, 985–992. [CrossRef] [PubMed]
44. Romo-Lozano, Y.; Hernandez-Hernandez, F.; Salinas, E. Sporothrix schenckii yeasts induce ERK pathway activation and secretion of IL-6 and TNF-alpha in rat mast cells, but no degranulation. *Med. Mycol.* **2014**, *52*, 862–868. [CrossRef] [PubMed]
45. Romo-Lozano, Y.; Hernandez-Hernandez, F.; Salinas, E. Mast cell activation by conidia of Sporothrix schenckii: Role in the severity of infection. *Scand. J. Immunol.* **2012**, *76*, 11–20. [CrossRef] [PubMed]
46. Sassa, M.F.; Saturi, A.E.; Souza, L.F.; Ribeiro, L.C.; Sgarbi, D.B.; Carlos, I.Z. Response of macrophage Toll-like receptor 4 to a Sporothrix schenckii lipid extract during experimental sporotrichosis. *Immunology* **2009**, *128*, 301–309. [CrossRef]
47. de, C.N.T.; Ferreira, L.S.; Arthur, R.A.; Alegranci, P.; Placeres, M.C.; Spolidorio, L.C.; Carlos, I.Z. Influence of TLR-2 in the immune response in the infection induced by fungus Sporothrix schenckii. *Immunol. Investig.* **2014**, *43*, 370–390. [CrossRef]
48. Negrini Tde, C.; Ferreira, L.S.; Alegranci, P.; Arthur, R.A.; Sundfeld, P.P.; Maia, D.C.; Spolidorio, L.C.; Carlos, I.Z. Role of TLR-2 and fungal surface antigens on innate immune response against Sporothrix schenckii. *Immunol. Investig.* **2013**, *42*, 36–48. [CrossRef] [PubMed]
49. Li, M.; Chen, Q.; Sun, J.; Shen, Y.; Liu, W. Inflammatory response of human keratinocytes triggered by Sporothrix schenckii via Toll-like receptor 2 and 4. *J. Dermatol. Sci.* **2012**, *66*, 80–82. [CrossRef]
50. Jellmayer, J.A.; Ferreira, L.S.; Manente, F.A.; Goncalves, A.C.; Polesi, M.C.; Batista-Duharte, A.; Carlos, I.Z. Dectin-1 expression by macrophages and related antifungal mechanisms in a murine model of Sporothrix schenckii sensu stricto systemic infection. *Microb. Pathog.* **2017**, *110*, 78–84. [CrossRef]

51. Zhang, Z.; Liu, X.; Lv, X.; Lin, J. Variation in genotype and higher virulence of a strain of Sporothrix schenckii causing disseminated cutaneous sporotrichosis. *Mycopathologia* **2011**, *172*, 439–446. [CrossRef]
52. Goncalves, A.C.; Ferreira, L.S.; Manente, F.A.; de Faria, C.; Polesi, M.C.; de Andrade, C.R.; Zamboni, D.S.; Carlos, I.Z. The NLRP3 inflammasome contributes to host protection during Sporothrix schenckii infection. *Immunology* **2017**, *151*, 154–166. [CrossRef] [PubMed]
53. Maia, D.C.; Gonçalves, A.C.; Ferreira, L.S.; Manente, F.A.; Portuondo, D.L.; Vellosa, J.C.; Polesi, M.C.; Batista-Duharte, A.; Carlos, I.Z. Response of Cytokines and Hydrogen Peroxide to Sporothrix schenckii Exoantigen in Systemic Experimental Infection. *Mycopathologia* **2016**, *181*, 207–215. [CrossRef] [PubMed]
54. Scott, E.N.; Muchmore, H.G.; Fine, D.P. Activation of the alternative complement pathway by Sporothrix schenckii. *Infect. Immun.* **1986**, *51*, 6–9. [PubMed]
55. Arrillaga-Moncrieff, I.; Capilla, J.; Mayayo, E.; Marimon, R.; Mariné, M.; Gené, J.; Cano, J.; Guarro, J. Different virulence levels of the species of Sporothrix in a murine model. *Clin. Microbiol. Infect.* **2009**, *15*, 651–655. [CrossRef] [PubMed]
56. Almeida-Paes, R.; Bailao, A.M.; Pizzini, C.V.; Reis, R.S.; Soares, C.M.; Peralta, J.M.; Gutierrez-Galhardo, M.C.; Zancope-Oliveira, R.M. Cell-free antigens of Sporothrix brasiliensis: Antigenic diversity and application in an immunoblot assay. *Mycoses* **2012**, *55*, 467–475. [CrossRef] [PubMed]
57. Almeida-Paes, R.; de Oliveira, L.C.; Oliveira, M.M.; Gutierrez-Galhardo, M.C.; Nosanchuk, J.D.; Zancope-Oliveira, R.M. Phenotypic characteristics associated with virulence of clinical isolates from the Sporothrix complex. *Biomed. Res. Int.* **2015**, *2015*, 212308. [CrossRef] [PubMed]
58. Camacho, E.; León-Navarro, I.; Rodríguez-Brito, S.; Mendoza, M.; Niño-Vega, G.A. Molecular epidemiology of human sporotrichosis in Venezuela reveals high frequency of Sporothrix globosa. *BMC Infect. Dis.* **2015**, *15*, 94. [CrossRef]
59. Teixeira, P.A.; de Castro, R.A.; Nascimento, R.C.; Tronchin, G.; Torres, A.P.; Lazéra, M.; de Almeida, S.R.; Bouchara, J.P.; Loureiro y Penha, C.V.; Lopes-Bezerra, L.M. Cell surface expression of adhesins for fibronectin correlates with virulence in Sporothrix schenckii. *Microbiology* **2009**, *155*, 3730–3738. [CrossRef]
60. Martínez-Álvarez, J.A.; Pérez-García, L.A.; Mellado-Mojica, E.; López, M.G.; Martínez-Duncker, I.; Lópes-Bezerra, L.M.; Mora-Montes, H.M. Sporothrix schenckii sensu stricto and Sporothrix brasiliensis Are Differentially Recognized by Human Peripheral Blood Mononuclear Cells. *Front. Microbiol.* **2017**, *8*, 843. [CrossRef]
61. Shimonaka, H.; Noguchi, T.; Kawai, K.; Hasegawa, I.; Nozawa, Y.; Ito, Y. Immunochemical studies on the human pathogen Sporothrix schenckii: Effects of chemical and enzymatic modification of the antigenic compounds upon immediate and delayed reactions. *Infect. Immun.* **1975**, *11*, 1187–1194.
62. Shiraishi, A.; Nakagaki, K.; Arai, T. Experimental sporotrichosis in congenitally athymic (nude) mice. *J. Reticuloendothel. Soc.* **1979**, *26*, 333–336.
63. Shiraishi, A.; Nakagaki, K.; Arai, T. Role of cell-mediated immunity in the resistance to experimental sporotrichosis in mice. *Mycopathologia* **1992**, *120*, 15–21. [CrossRef] [PubMed]
64. Tachibana, T.; Matsuyama, T.; Mitsuyama, M. Involvement of CD4+ T cells and macrophages in acquired protection against infection with Sporothrix schenckii in mice. *Med. Mycol.* **1999**, *37*, 397–404. [CrossRef] [PubMed]
65. Dickerson, C.L.; Taylor, R.L.; Drutz, D.J. Susceptibility of congenitally athymic (nude) mice to sporotrichosis. *Infect. Immun.* **1983**, *40*, 417–420. [PubMed]
66. Carlos, I.Z.; Sgarbi, D.B.; Angluster, J.; Alviano, C.S.; Silva, C.L. Detection of cellular immunity with the soluble antigen of the fungus Sporothrix schenckii in the systemic form of the disease. *Mycopathologia* **1992**, *117*, 139–144. [CrossRef] [PubMed]
67. Carlos, I.Z.; Sgarbi, D.B.; Placeres, M.C. Host organism defense by a peptide-polysaccharide extracted from the fungus Sporothrix schenckii. *Mycopathologia* **1998**, *144*, 9–14. [CrossRef] [PubMed]
68. Alegranci, P.; de Abreu Ribeiro, L.C.; Ferreira, L.S.; Negrini Tde, C.; Maia, D.C.; Tansini, A.; Goncalves, A.C.; Placeres, M.C.; Carlos, I.Z. The predominance of alternatively activated macrophages following challenge with cell wall peptide-polysaccharide after prior infection with Sporothrix schenckii. *Mycopathologia* **2013**, *176*, 57–65. [CrossRef] [PubMed]
69. Uenotsuchi, T.; Takeuchi, S.; Matsuda, T.; Urabe, K.; Koga, T.; Uchi, H.; Nakahara, T.; Fukagawa, S.; Kawasaki, M.; Kajiwara, H.; et al. Differential induction of Th1-prone immunity by human dendritic cells

activated with Sporothrix schenckii of cutaneous and visceral origins to determine their different virulence. *Int. Immunol.* **2006**, *18*, 1637–1646. [CrossRef]
70. Almeida-Paes, R.; Pimenta, M.A.; Pizzini, C.V.; Monteiro, P.C.; Peralta, J.M.; Nosanchuk, J.D.; Zancopé-Oliveira, R.M. Use of mycelial-phase Sporothrix schenckii exoantigens in an enzyme-linked immunosorbent assay for diagnosis of sporotrichosis by antibody detection. *Clin. Vaccine Immunol.* **2007**, *14*, 244–249. [CrossRef]
71. Portuondo, D.L.; Batista-Duharte, A.; Ferreira, L.S.; Martínez, D.T.; Polesi, M.C.; Duarte, R.A.; de Paula E Silva, A.C.; Marcos, C.M.; Almeida, A.M.; Carlos, I.Z. A cell wall protein-based vaccine candidate induce protective immune response against Sporothrix schenckii infection. *Immunobiology* **2016**, *221*, 300–309. [CrossRef]
72. Nascimento, R.C.; Espíndola, N.M.; Castro, R.A.; Teixeira, P.A.; Loureiro y Penha, C.V.; Lopes-Bezerra, L.M.; Almeida, S.R. Passive immunization with monoclonal antibody against a 70-kDa putative adhesin of Sporothrix schenckii induces protection in murine sporotrichosis. *Eur. J. Immunol.* **2008**, *38*, 3080–3089. [CrossRef] [PubMed]
73. Franco Dde, L.; Nascimento, R.C.; Ferreira, K.S.; Almeida, S.R. Antibodies Against Sporothrix schenckii Enhance TNF-alpha Production and Killing by Macrophages. *Scand. J. Immunol.* **2012**, *75*, 142–146. [CrossRef] [PubMed]
74. de Almeida, J.R.; Kaihami, G.H.; Jannuzzi, G.P.; de Almeida, S.R. Therapeutic vaccine using a monoclonal antibody against a 70-kDa glycoprotein in mice infected with highly virulent Sporothrix schenckii and Sporothrix brasiliensis. *Med. Mycol.* **2015**, *53*, 42–50. [CrossRef] [PubMed]
75. Zhang, Y.Q.; Xu, X.G.; Zhang, M.; Jiang, P.; Zhou, X.Y.; Li, Z.Z.; Zhang, M.F. Sporotrichosis: Clinical and histopathological manifestations. *Am. J. Dermatopathol.* **2011**, *33*, 296–302. [CrossRef] [PubMed]
76. Quintella, L.P.; Passos, S.R.; do Vale, A.C.; Galhardo, M.C.; Barros, M.B.; Cuzzi, T.; Reis, R.O.S.; de Carvalho, M.H.; Zappa, M.B.; Schubach, A.E.O. Histopathology of cutaneous sporotrichosis in Rio de Janeiro: A series of 119 consecutive cases. *J. Cutan. Pathol.* **2011**, *38*, 25–32. [CrossRef]
77. Morgado, F.N.; Schubach, A.O.; Pimentel, M.I.; Lyra, M.R.; Vasconcellos, É.; Valete-Rosalino, C.M.; Conceição-Silva, F. Is There Any Difference between the In Situ and Systemic IL-10 and IFN-γ Production when Clinical Forms of Cutaneous Sporotrichosis Are Compared? *PLoS ONE* **2016**, *11*, e0162764. [CrossRef]
78. Plouffe, J.F.; Silva, J.; Fekety, R.; Reinhalter, E.; Browne, R. Cell-mediated immune responses in sporotrichosis. *J. Infect. Dis.* **1979**, *139*, 152–157. [CrossRef]
79. Mialski, R.; de Oliveira, J.N., Jr.; da Silva, L.H.; Kono, A.; Pinheiro, R.L.; Teixeira, M.J.; Gomes, R.R.; de Queiroz-Telles, F.; Pinto, F.G.; Benard, G. Chronic Meningitis and Hydrocephalus due to Sporothrix brasiliensis in Immunocompetent Adults: A Challenging Entity. *Open Forum Infect. Dis.* **2018**, *5*, ofy081. [CrossRef]
80. Lynch, P.J.; Voorhees, J.J.; Harrell, E.R. Systemic sporotrichosis. *Ann. Intern. Med.* **1970**, *73*, 23–30. [CrossRef]
81. Almeida-Paes, R.; de Oliveira, M.M.; Freitas, D.F.; do Valle, A.C.; Zancopé-Oliveira, R.M.; Gutierrez-Galhardo, M.C. Sporotrichosis in Rio de Janeiro, Brazil: Sporothrix brasiliensis is associated with atypical clinical presentations. *PLoS Negl. Trop. Dis.* **2014**, *8*, e3094. [CrossRef]
82. Lipstein-Kresch, E.; Isenberg, H.D.; Singer, C.; Cooke, O.; Greenwald, R.A. Disseminated Sporothrix schenckii infection with arthritis in a patient with acquired immunodeficiency syndrome. *J. Rheumatol.* **1985**, *12*, 805–808. [PubMed]
83. Freitas, D.F.; Valle, A.C.; da Silva, M.B.; Campos, D.P.; Lyra, M.R.; de Souza, R.V.; Veloso, V.G.; Zancopé-Oliveira, R.M.; Bastos, F.I.; Galhardo, M.C. Sporotrichosis: An emerging neglected opportunistic infection in HIV-infected patients in Rio de Janeiro, Brazil. *PLoS Negl. Trop. Dis.* **2014**, *8*, e3110. [CrossRef]
84. Moreira, J.A.; Freitas, D.F.; Lamas, C.C. The impact of sporotrichosis in HIV-infected patients: A systematic review. *Infection* **2015**, *43*, 267–276. [CrossRef] [PubMed]
85. Donabedian, H.; O'Donnell, E.; Olszewski, C.; MacArthur, R.D.; Budd, N. Disseminated cutaneous and meningeal sporotrichosis in an AIDS patient. *Diagn. Microbiol. Infect. Dis.* **1994**, *18*, 111–115. [CrossRef]
86. Paixão, A.G.; Galhardo, M.C.G.; Almeida-Paes, R.; Nunes, E.P.; Gonçalves, M.L.C.; Chequer, G.L.; Lamas, C.D.C. The difficult management of disseminated Sporothrix brasiliensis in a patient with advanced AIDS. *AIDS Res. Ther.* **2015**, *12*, 16. [CrossRef] [PubMed]
87. Penn, C.C.; Goldstein, E.; Bartholomew, W.R. Sporothrix schenckii meningitis in a patient with AIDS. *Clin. Infect. Dis.* **1992**, *15*, 741–743. [CrossRef] [PubMed]

88. Hardman, S.; Stephenson, I.; Jenkins, D.R.; Wiselka, M.J.; Johnson, E.M. Disseminated Sporothix schenckii in a patient with AIDS. *J. Infect.* **2005**, *51*, e73–e77. [CrossRef]
89. Dong, J.A.; Chren, M.M.; Elewski, B.E. Bonsai tree: Risk factor for disseminated sporotrichosis. *J. Am. Acad. Dermatol.* **1995**, *33*, 839–840. [CrossRef]
90. Silva-Vergara, M.L.; Maneira, F.R.; De Oliveira, R.M.; Santos, C.T.; Etchebehere, R.M.; Adad, S.J. Multifocal sporotrichosis with meningeal involvement in a patient with AIDS. *Med. Mycol.* **2005**, *43*, 187–190. [CrossRef] [PubMed]
91. Vilela, R.; Souza, G.F.; Fernandes Cota, G.; Mendoza, L. Cutaneous and meningeal sporotrichosis in a HIV patient. *Rev. Iberoam. Micol.* **2007**, *24*, 161–163. [CrossRef]
92. Galhardo, M.C.; Silva, M.T.; Lima, M.A.; Nunes, E.P.; Schettini, L.E.; de Freitas, R.F.; Paes Rde, A.; Neves Ede, S.; do Valle, A.C. Sporothrix schenckii meningitis in AIDS during immune reconstitution syndrome. *J. Neurol. Neurosurg. Psychiatry* **2010**, *81*, 696–699. [CrossRef] [PubMed]
93. Rotz, L.D.; Slater, L.N.; Wack, M.F.; Boyd, A.L.; Nan, S.E.; Greenfield, R.A. Disseminated Sporotrichosis with Meningitis in a Patient with AIDS. *Infect. Dis. Clin. Pract.* **1996**, *5*, 566–568. [CrossRef]
94. Callens, S.F.; Kitetele, F.; Lukun, P.; Lelo, P.; Van Rie, A.; Behets, F.; Colebunders, R. Pulmonary Sporothrix schenckii infection in a HIV positive child. *J. Trop. Pediatr.* **2006**, *52*, 144–146. [CrossRef] [PubMed]
95. Losman, J.A.; Cavanaugh, K. Cases from the Osler Medical Service at Johns Hopkins University. Diagnosis: *P. carinii* pneumonia and primary pulmonary sporotrichosis. *Am. J. Med.* **2004**, *117*, 353–356. [CrossRef] [PubMed]
96. Gori, S.; Lupetti, A.; Moscato, G.; Parenti, M.; Lofaro, A. Pulmonary sporotrichosis with hyphae in a human immunodeficiency virus-infected patient. A case report. *Acta Cytol.* **1997**, *41*, 519–521. [CrossRef] [PubMed]
97. Morgan, M.; Reves, R. Invasive sinusitis due to Sporothrix schenckii in a patient with AIDS. *Clin. Infect. Dis.* **1996**, *23*, 1319–1320. [CrossRef] [PubMed]
98. Bustamante, B.; Lama, J.R.; Mosquera, C.; Soto, L. Sporotrichosis in Human Immunodeficiency Virus Infected Peruvian Patients: Two Case Reports and Literature Review. *Infect. Dis. Clin. Pract.* **2009**, *17*, 78–83. [CrossRef]
99. Buccheri, R.; Benard, G. Opinion: Paracoccidioidomycosis and HIV Immune Recovery Inflammatory Syndrome. *Mycopathologia* **2018**, *183*, 495–498. [CrossRef]
100. Haddow, L.J.; Colebunders, R.; Meintjes, G.; Lawn, S.D.; Elliott, J.H.; Manabe, Y.C.; Bohjanen, P.R.; Sungkanuparph, S.; Easterbrook, P.J.; French, M.A.; et al. Cryptococcal immune reconstitution inflammatory syndrome in HIV-1-infected individuals: Proposed clinical case definitions. *Lancet Infect. Dis.* **2010**, *10*, 791–802. [CrossRef]
101. Singh, N.; Perfect, J.R. Immune reconstitution syndrome associated with opportunistic mycoses. *Lancet Infect. Dis.* **2007**, *7*, 395–401. [CrossRef]
102. Lyra, M.R.; Nascimento, M.L.; Varon, A.G.; Pimentel, M.I.; Antonio, L.E.F.; Saheki, M.N.; Bedoya-Pacheco, S.J.; Valle, A.C. Immune reconstitution inflammatory syndrome in HIV and sporotrichosis coinfection: report of two cases and review of the literature. *Rev. Soc. Bras. Med. Trop.* **2014**, *47*, 806–809. [CrossRef] [PubMed]
103. de Lima Barros, M.B.; de Oliveira Schubach, A.; Galhardo, M.C.; Schubach, T.M.; dos Reis, R.S.; Conceição, M.J.; do Valle, A.C. Sporotrichosis with widespread cutaneous lesions: Report of 24 cases related to transmission by domestic cats in Rio de Janeiro, Brazil. *Int. J. Dermatol.* **2003**, *42*, 677–681. [CrossRef] [PubMed]
104. Scott, S.M.; Peasley, E.D.; Crymes, T.P. Pulmonary sporotrichosis. Report of two cases with cavitation. *N. Engl. J. Med.* **1961**, *265*, 453–457. [CrossRef] [PubMed]
105. Mohr, J.A.; Patterson, C.D.; Eaton, B.G.; Rhoades, E.R.; Nichols, N.B. Primary pulmonary sporotrichosis. *Am. Rev. Respir. Dis.* **1972**, *106*, 260–264. [CrossRef] [PubMed]
106. Zhang, Y.; Hagen, F.; Wan, Z.; Liu, Y.; Wang, Q.; de Hoog, G.S.; Li, R.; Zhang, J. Two cases of sporotrichosis of the right upper extremity in right-handed patients with diabetes mellitus. *Rev. Iberoam. Micol.* **2016**, *33*, 38–42. [CrossRef] [PubMed]
107. Mohamad, N.; Badrin, S.; Wan Abdullah, W.N.H. A Diabetic Elderly Man with Finger Ulcer. *Korean J. Fam. Med.* **2018**, *39*, 126–129. [CrossRef] [PubMed]
108. Ramirez-Soto, M.; Lizarraga-Trujillo, J. Granulomatous sporotrichosis: Report of two unusual cases. *Rev. Chil. Infectol.* **2013**, *30*, 548–553. [CrossRef]

109. Nassif, P.W.; Granado, I.R.; Ferraz, J.S.; Souza, R.; Nassif, A.E. Atypical presentation of cutaneous sporotrichosis in an alcoholic patient. *Dermatol. Online J.* **2012**, *18*, 12.
110. Castrejon, O.V.; Robles, M.; Zubieta Arroyo, O.E. Fatal fungaemia due to Sporothrix schenckii. *Mycoses* **1995**, *38*, 373–376. [CrossRef]
111. Solorzano, S.; Ramirez, R.; Cabada, M.M.; Montoya, M.; Cazorla, E. Disseminated cutaneous sporotrichosis with joint involvement in a woman with type 2 diabetes. *Rev. Peru. Med. Exp. Salud Publica* **2015**, *32*, 187–190. [CrossRef]
112. Agger, W.A.; Caplan, R.H.; Maki, D.G. Ocular sporotrichosis mimicking mucormycosis in a diabetic. *Ann. Ophthalmol.* **1978**, *10*, 767–771. [PubMed]
113. Benvegnu, A.M.; Stramari, J.; Dallazem, L.N.D.; Chemello, R.M.L.; Beber, A.A.C. Disseminated cutaneous sporotrichosis in patient with alcoholism. *Rev. Soc. Bras. Med. Trop.* **2017**, *50*, 871–873. [CrossRef] [PubMed]
114. Yamaguchi, T.; Ito, S.; Takano, Y.; Umeda, N.; Goto, M.; Horikoshi, M.; Hayashi, T.; Goto, D.; Matsumoto, I.; Sumida, T. A case of disseminated sporotrichosis treated with prednisolone, immunosuppressants, and tocilizumab under the diagnosis of rheumatoid arthritis. *Intern. Med.* **2012**, *51*, 2035–2039. [CrossRef]
115. Yang, D.J.; Krishnan, R.S.; Guillen, D.R.; Schmiege, L.M.; Leis, P.F.; Hsu, S. Disseminated sporotrichosis mimicking sarcoidosis. *Int. J. Dermatol.* **2006**, *45*, 450–453. [CrossRef] [PubMed]
116. Mauermann, M.L.; Klein, C.J.; Orenstein, R.; Dyck, P.J. Disseminated sporotrichosis presenting with granulomatous inflammatory multiple mononeuropathies. *Muscle Nerve* **2007**, *36*, 866–872. [CrossRef] [PubMed]
117. Byrd, D.R.; El-Azhary, R.A.; Gibson, L.E.; Roberts, G.D. Sporotrichosis masquerading as pyoderma gangrenosum: Case report and review of 19 cases of sporotrichosis. *J. Eur. Acad. Dermatol. Venereol.* **2001**, *15*, 581–584. [CrossRef] [PubMed]
118. Ursini, F.; Russo, E.; Leporini, C.; Calabria, M.; Bruno, C.; Tripolino, C.; Naty, S.; Grembiale, R.D. Lymphocutaneous Sporotrichosis during Treatment with Anti-TNF-Alpha Monotherapy. *Case Rep. Rheumatol.* **2015**, *2015*, 614504. [CrossRef]
119. Tochigi, M.; Ochiai, T.; Mekata, C.; Nishiyama, H.; Anzawa, K.; Kawasaki, M. Sporotrichosis of the face by autoinoculation in a patient undergoing tacrolimus treatment. *J. Dermatol.* **2012**, *39*, 796–798. [CrossRef]
120. Severo, L.C.; Festugato, M.; Bernardi, C.; Londero, A.T. Widespread cutaneous lesions due to Sporothrix schenckii in a patient under a long-term steroids therapy. *Rev. Inst. Med. Trop. Sao Paulo* **1999**, *41*, 59–62. [CrossRef]
121. Pappas, P.G.; Alexander, B.D.; Andes, D.R.; Hadley, S.; Kauffman, C.A.; Freifeld, A.; Anaissie, E.J.; Brumble, L.M.; Herwaldt, L.; Ito, J.; et al. Invasive fungal infections among organ transplant recipients: Results of the Transplant-Associated Infection Surveillance Network (TRANSNET). *Clin. Infect. Dis.* **2010**, *50*, 1101–1111. [CrossRef]
122. Guimaraes, L.F.; Halpern, M.; de Lemos, A.S.; de Gouvea, E.F.; Goncalves, R.T.; da Rosa Santos, M.A.; Nucci, M.; Santoro-Lopes, G. Invasive Fungal Disease in Renal Transplant Recipients at a Brazilian Center: Local Epidemiology Matters. *Transplant. Proc.* **2016**, *48*, 2306–2309. [CrossRef] [PubMed]
123. Caroti, L.; Zanazzi, M.; Rogasi, P.; Fantoni, E.; Farsetti, S.; Rosso, G.; Bertoni, E.; Salvadori, M. Subcutaneous nodules and infectious complications in renal allograft recipients. *Transplant. Proc.* **2010**, *42*, 1146–1147. [CrossRef] [PubMed]
124. Rao, K.H.; Jha, R.; Narayan, G.; Sinha, S. Opportunistic infections following renal transplantation. *Indian J. Med. Microbiol.* **2002**, *20*, 47–49.
125. Gewehr, P.; Jung, B.; Aquino, V.; Manfro, R.C.; Spuldaro, F.; Rosa, R.G.; Goldani, L.Z. Sporotrichosis in renal transplant patients. *Can. J. Infect. Dis. Med. Microbiol.* **2013**, *24*, e47–e49. [CrossRef] [PubMed]
126. Bahr, N.C.; Janssen, K.; Billings, J.; Loor, G.; Green, J.S. Respiratory Failure due to Possible Donor-Derived Sporothrix schenckii Infection in a Lung Transplant Recipient. *Case Rep. Infect. Dis.* **2015**, *2015*, 925718. [CrossRef] [PubMed]
127. Ewing, G.E.; Bosl, G.J.; Peterson, P.K. Sporothrix schenckii meningitis in a farmer with Hodgkin's disease. *Am. J. Med.* **1980**, *68*, 455–457. [CrossRef]
128. Bunce, P.E.; Yang, L.; Chun, S.; Zhang, S.X.; Trinkaus, M.A.; Matukas, L.M. Disseminated sporotrichosis in a patient with hairy cell leukemia treated with amphotericin B and posaconazole. *Med. Mycol.* **2012**, *50*, 197–201. [CrossRef]

129. Kumar, S.; Kumar, D.; Gourley, W.K.; Alperin, J.B. Sporotrichosis as a presenting manifestation of hairy cell leukemia. *Am. J. Hematol.* **1994**, *46*, 134–137. [CrossRef] [PubMed]
130. Yagnik, K.J.; Skelton, W.P.T.; Olson, A.; Trillo, C.A.; Lascano, J. A rare case of disseminated Sporothrix schenckii with bone marrow involvement in a patient with idiopathic CD4 lymphocytopenia. *IDCases* **2017**, *9*, 70–72. [CrossRef]
131. Trotter, J.R.; Sriaroon, P.; Berman, D.; Petrovic, A.; Leiding, J.W. Sporothrix schenckii lymphadentitis in a male with X-linked chronic granulomatous disease. *J. Clin. Immunol.* **2014**, *34*, 49–52. [CrossRef] [PubMed]
132. Gullberg, R.M.; Quintanilla, A.; Levin, M.L.; Williams, J.; Phair, J.P. Sporotrichosis: Recurrent cutaneous, articular, and central nervous system infection in a renal transplant recipient. *Rev. Infect. Dis.* **1987**, *9*, 369–375. [CrossRef] [PubMed]
133. Orofino-Costa, R.; Macedo, P.M.; Rodrigues, A.M.; Bernardes-Engemann, A.R. Sporotrichosis: An update on epidemiology, etiopathogenesis, laboratory and clinical therapeutics. *An. Bras. Dermatol.* **2017**, *92*, 606–620. [CrossRef] [PubMed]
134. Sanchotene, K.O.; Madrid, I.M.; Klafke, G.B.; Bergamashi, M.; Della Terra, P.P.; Rodrigues, A.M.; de Camargo, Z.P.; Xavier, M.O. Sporothrix brasiliensis outbreaks and the rapid emergence of feline sporotrichosis. *Mycoses* **2015**, *58*, 652–658. [CrossRef] [PubMed]
135. Bonifaz, A.; Tirado-Sánchez, A. Cutaneous Disseminated and Extracutaneous Sporotrichosis: Current Status of a Complex Disease. *J. Fungi* **2017**, *3*, 6. [CrossRef] [PubMed]
136. Tirado-Sanchez, A.; Bonifaz, A. Nodular Lymphangitis (Sporotrichoid Lymphocutaneous Infections). Clues to Differential Diagnosis. *J. Fungi* **2018**, *4*, 56. [CrossRef] [PubMed]
137. Bernardes-Engemann, A.R.; de Lima Barros, M.; Zeitune, T.; Russi, D.C.; Orofino-Costa, R.; Lopes-Bezerra, L.M. Validation of a serodiagnostic test for sporotrichosis: a follow-up study of patients related to the Rio de Janeiro zoonotic outbreak. *Med. Mycol.* **2015**, *53*, 28–33. [CrossRef] [PubMed]
138. Ottonelli Stopiglia, C.D.; Magagnin, C.M.; Castrillon, M.R.; Mendes, S.D.; Heidrich, D.; Valente, P.; Scroferneker, M.L. Antifungal susceptibilities and identification of species of the Sporothrix schenckii complex isolated in Brazil. *Med. Mycol.* **2014**, *52*, 56–64. [CrossRef]
139. Brilhante, R.S.; Rodrigues, A.M.; Sidrim, J.J.; Rocha, M.F.; Pereira, S.A.; Gremiao, I.D.; Schubach, T.M.; de Camargo, Z.P. In vitro susceptibility of antifungal drugs against Sporothrix brasiliensis recovered from cats with sporotrichosis in Brazil. *Med. Mycol.* **2016**, *54*, 275–279. [CrossRef]
140. Kauffman, C.A.; Bustamante, B.; Chapman, S.W.; Pappas, P.G. Infectious Diseases Society of America. Clinical practice guidelines for the management of sporotrichosis: 2007 update by the Infectious Diseases Society of America. *Clin. Infect. Dis.* **2007**, *45*, 1255–1265. [CrossRef]
141. de Lima Barros, M.B.; Schubach, A.O.; de Vasconcellos Carvalhaes de Oliveira, R.; Martins, E.B.; Teixeira, J.L.; Wanke, B. Treatment of cutaneous sporotrichosis with itraconazole—Study of 645 patients. *Clin. Infect. Dis.* **2011**, *52*, e200–e206. [CrossRef]
142. Shikanai-Yasuda, M.A.; Mendes, R.P.; Colombo, A.L.; Queiroz-Telles, F.; Kono, A.S.G.; Paniago, A.M.M.; Nathan, A.; Valle, A.; Bagagli, E.; Benard, G.; et al. Brazilian guidelines for the clinical management of paracoccidioidomycosis. *Rev. Soc. Bras. Med. Trop.* **2017**, *50*, 715–740. [CrossRef] [PubMed]
143. Torok, M.E.; Yen, N.T.; Chau, T.T.; Mai, N.T.; Phu, N.H.; Mai, P.P.; Dung, N.T.; Chau, N.V.; Bang, N.D.; Tien, N.A.; et al. Timing of initiation of antiretroviral therapy in human immunodeficiency virus (HIV)—Associated tuberculous meningitis. *Clin. Infect. Dis.* **2011**, *52*, 1374–1383. [CrossRef] [PubMed]
144. Boulware, D.R.; Meya, D.B.; Muzoora, C.; Rolfes, M.A.; Huppler Hullsiek, K.; Musubire, A.; Taseera, K.; Nabeta, H.W.; Schutz, C.; Williams, D.A.; et al. Timing of antiretroviral therapy after diagnosis of cryptococcal meningitis. *N. Engl. J. Med.* **2014**, *370*, 2487–2498. [CrossRef] [PubMed]
145. Kaplan, J.E.; Benson, C.; Holmes, K.K.; Brooks, J.T.; Pau, A.; Masur, H.; CDC; National Institutes of Health; HIV Medicine Association of the Infectious Diseases Society of America. Guidelines for prevention and treatment of opportunistic infections in HIV-infected adults and adolescents: Recommendations from CDC, the National Institutes of Health, and the HIV Medicine Association of the Infectious Diseases Society of America. *MMWR Recomm. Rep.* **2009**, *58*, 1–207, quiz CE201-204. [PubMed]
146. Benard, G.; Duarte, A.J. Paracoccidioidomycosis: A model for evaluation of the effects of human immunodeficiency virus infection on the natural history of endemic tropical diseases. *Clin. Infect. Dis.* **2000**, *31*, 1032–1039. [CrossRef] [PubMed]

147. Hajjeh, R.A. Disseminated histoplasmosis in persons infected with human immunodeficiency virus. *Clin. Infect. Dis.* **1995**, *21* (Suppl. 1), S108–S110. [CrossRef] [PubMed]
148. Galgiani, J.N. Coccidioidomycosis: Changes in clinical expression, serological diagnosis, and therapeutic options. *Clin. Infect. Dis.* **1992**, *14* (Suppl. 1), S100–S105. [CrossRef]
149. Szabo, G.; Saha, B. Alcohol's Effect on Host Defense. *Alcohol Res.* **2015**, *37*, 159–170.
150. Barr, T.; Helms, C.; Grant, K.; Messaoudi, I. Opposing effects of alcohol on the immune system. *Prog. Neuropsychopharmacol. Biol. Psychiatry* **2016**, *65*, 242–251. [CrossRef]
151. Pasala, S.; Barr, T.; Messaoudi, I. Impact of Alcohol Abuse on the Adaptive Immune System. *Alcohol Res.* **2015**, *37*, 185–197.
152. Pan, B.; Chen, M.; Pan, W.; Liao, W. Histoplasmosis: A new endemic fungal infection in China? Review and analysis of cases. *Mycoses* **2013**, *56*, 212–221. [CrossRef]
153. Santelli, A.C.; Blair, J.E.; Roust, L.R. Coccidioidomycosis in patients with diabetes mellitus. *Am. J. Med.* **2006**, *119*, 964–969. [CrossRef] [PubMed]
154. Lemos, L.B.; Baliga, M.; Guo, M. Blastomycosis: The great pretender can also be an opportunist. Initial clinical diagnosis and underlying diseases in 123 patients. *Ann. Diagn. Pathol.* **2002**, *6*, 194–203. [CrossRef] [PubMed]
155. Geerlings, S.E.; Hoepelman, A.I. Immune dysfunction in patients with diabetes mellitus (DM). *FEMS Immunol. Med. Microbiol.* **1999**, *26*, 259–265. [CrossRef] [PubMed]
156. Restrepo, B.I.; Schlesinger, L.S. Host-pathogen interactions in tuberculosis patients with type 2 diabetes mellitus. *Tuberculosis* **2013**, *93*, S10–S14. [CrossRef]
157. Martinez, N.; Kornfeld, H. Diabetes and immunity to tuberculosis. *Eur. J. Immunol.* **2014**, *44*, 617–626. [CrossRef]
158. Kumar Nathella, P.; Babu, S. Influence of diabetes mellitus on immunity to human tuberculosis. *Immunology* **2017**, *152*, 13–24. [CrossRef]

5

Generation of a *Mucor circinelloides* Reporter Strain—A Promising New Tool to Study Antifungal Drug Efficacy and Mucormycosis

Ulrike Binder [1,*], Maria Isabel Navarro-Mendoza [2], Verena Naschberger [1], Ingo Bauer [3], Francisco E. Nicolas [2], Johannes D. Pallua [4], Cornelia Lass-Flörl [1] and Victoriano Garre [2,*]

1 Division of Hygiene and Medical Microbiology, Medical University Innsbruck, Schöpfstrasse 41, 6020 Innsbruck, Austria; verena.naschberger@i-med.ac.at (V.N.); cornelia.lass-floerl@i-med.ac.at (C.L.-F.)
2 Departamento de Genética y Microbiología, Facultad de Biología, Universidad de Murcia, 30100 Murcia, Spain; mariaisabel.navarro3@um.es (M.I.N.-M.); fnicolas@um.es (F.E.N.)
3 Division of Molecular Biology, Biocenter, Medical University of Innsbruck, Innrain 80-82, 6020 Innsbruck, Austria; ingo.bauer@i-med.ac.at
4 Institute of Pathology, Neuropathology and Molecular Pathology, Medical University of Innsbruck, Müllerstraße 44, 6020 Innsbruck, Austria; johannes.pallua@i-med.ac.at
* Correspondence: ulrike.binder@i-med.ac.at (U.B.); vgarre@um.es (V.G.)

Abstract: Invasive fungal infections caused by Mucorales (mucormycosis) have increased worldwide. These life-threatening infections affect mainly, but not exclusively, immunocompromised patients, and are characterized by rapid progression, severe tissue damage and an unacceptably high rate of mortality. Still, little is known about this disease and its successful therapy. New tools to understand mucormycosis and a screening method for novel antimycotics are required. Bioluminescent imaging is a powerful tool for in vitro and in vivo approaches. Hence, the objective of this work was to generate and functionally analyze bioluminescent reporter strains of *Mucor circinelloides*, one mucormycosis-causing pathogen. Reporter strains were constructed by targeted integration of the firefly luciferase gene under control of the *M. circinelloides* promoter P*zrt1*. The luciferase gene was sufficiently expressed, and light emission was detected under several conditions. Phenotypic characteristics, virulence potential and antifungal susceptibility were indifferent to the wild-type strains. Light intensity was dependent on growth conditions and biomass, being suitable to determine antifungal efficacy in vitro. This work describes for the first time the generation of reporter strains in a basal fungus that will allow real-time, non-invasive infection monitoring in insect and murine models, and the testing of antifungal efficacy by means other than survival.

Keywords: *Mucor circinelloides*; mucormycosis; firefly luciferase; reporter strain; bioluminescence

---

1. Introduction

Mucor circinelloides, a member of the Mucoromycota, is ubiquitously found in the environment. It is thermotolerant, able to grow on a wide range of organic substrates and sporulates fast and abundantly [1,2]. It can cause mucormycosis—a severe animal and human disease. In recent decades, the incidence of mucormycosis has increased all over the world, becoming the second most common fungal disease in patients with haematological malignancies and transplant recipients [3–5]. Infections with mucormycetes are highly aggressive and destructive, resulting in tissue necrosis, invasion of blood vessels and subsequent thrombosis. The rapid progression, linked with shortcomings in diagnosis and therapy, results in high mortality rates which are estimated to range between 40–>90%, depending on the site of infection, the condition of the host and the therapeutic interventions [3,5–7]. The different types of mucormycosis are classified according to the anatomic site of infection, such as rhino-orbital-cerebral, pulmonary, cutaneous, gastrointestinal and disseminated infections [8]. Antifungal therapy is complicated by the limited treatment options that comprise lipid amphotericin B (AMB) as first-line therapeutic and posaconazole (POS) or isavuconazole (ISA) as salvage treatment [9,10].

The most common genera associated with human disease next to *Mucor* are *Rhizopus*, and *Lichtheimia* (formerly *Absidia*) and infections are associated with severe graft-versus-host disease, treatment with steroids, neutropenia, iron overload, diabetes and malnutrition [10]. *M. circinelloides* isolates have been associated with outbreaks of mucormycosis in the US, the UK and Europe and it poses a threat to public health by contaminating food and producing 3-nitropropionic acid [11–15].

Despite the growing relevance of mucormycetes in public health, little is known about the physiology and virulence factors associated with this group of fungi. The heterogeneity of this group and the difficulties in genetic manipulation are reasons thereof. However, *M. circinelloides* stands out among the rest of basal fungi offering the opportunity to carry out genetic manipulation by the development of an increasing number of molecular tools [16,17]. The intrinsic resistance of mucormycetes to drugs used as resistant markers in other fungi, leaves only the use of auxotrophic markers [18,19].

Bioluminescence imaging is a very useful technique to track microorganisms in living animals and has provided novel insights into the onset and progression of disease. The great advantage is the real-time monitoring of infection in one individual organism over time. Different enzymes exist in living organisms, which use different substrates and different cofactors to emit light. The most prominently used are the firefly (*Photinus pyralis*) and the copepod (*Gaussia princeps*) luciferase [20]. Both have already been successfully transformed into opportunistic fungal pathogens e.g., *Candida albicans* [21,22], *Aspergillus fumigatus* [23–25] and *A. terreus* [26]. Bioluminescence imaging with these strains has significantly enhanced our understanding of fungal infection. It revealed unexpected host sites in disseminated candidiasis, showing persistence of *Candida* cells in the gallbladder, even after antifungal treatment. Studies comparing *A. fumigatus* and *A. terreus* revealed delayed onset of disease in *A. terreus* infected mice and survival in 50% of *A. terreus* infected mice, although progression of disease was similar to those that died.

In this study, we generated bioluminescent strains in the opportunistic human pathogen *M. circinelloides* based on the expression of firefly luciferase and controlled under a highly expressed *M. circinelloides* promoter, for the first time. Light emission was correlated to fungal growth and concentration of the substrate luciferin. Phenotypic analysis and virulence potential in the alternative host *Galleria mellonella* revealed no differences to parental strains. Antifungal efficacy was determined successfully by the use of the obtained reporter strains. The strains generated in this study will be a useful tool to test novel antifungal agents both in vitro and in murine and insect models, in addition to shed light on the onset and progression of mucormycosis in animal models.

2. Material and Methods

2.1. Fungal Strains, Plasmids, Media and Growth Conditions

The strains and plasmids used in this study are listed in Table 1. All fungal strains used were *M. circinelloides f. lusitanicus*, referred to in this work as *M. circinelloides* for simplicity. To obtain spores, strains were grown on YPG (yeast peptone glucose agar; 3 g/L yeast extract, 10 g/L peptone, 20 g/L glucose, pH 4.5) medium at 26 °C in the light for 4–5 days. Spores were collected by scraping the plates with sterile spore solution buffer (0.9% NaCl, 0.01% Tween 80) and spore concentration was determined by hemocytometer. Media and growth conditions for the individual assays are given below, for most assays YNB (yeast nitrogen base; 1.5 g/L ammonium sulfate, 1.5 g/L glutamic acid, 0.5 g/L yeast nitrogen base (*w/o* ammonium sulfate and amino acids, Sigma-Aldrich, Steinheim, Germany, cat. no. Y1251), 10 g/L glucose, thiamine 1 µg/mL and niacin 1 µg/mL) was used [16]. All chemicals used were purchased from Sigma-Aldrich, Germany, unless otherwise stated.

Table 1. List of strains and plasmids used in this study.

	Name	Genotypes/Characteristics	Reference
Mucor circinelloides	R7B	*leuA1* (mutant alelle of *leuA* gene)	Roncero et al., 1984 [27]
	R7B_luc	*carRP::leu*	Obtained in this study
	R7B_luc1	*carRP::leu*	Obtained in this study
	MU402	*leuA1, pyrG*$^-$	Nicolas et al., 2007 [18]
	MU402_luc	*pyrG*$^-$, *carRP::leuA*	Obtained in this study
	MU402_luc1	*pyrG*$^-$, *carRP::leuA*	Obtained in this study
	DH5α	ampicillin resistance	Thermo Fisher (Germering, Germany)
Escherichia coli	pMAT1477	*Pzrt1, leuA*	Rodriguez-Frometa et al., 2013 [19]
	pGL3 basic vector	firefly luciferase reporter vector	Commercially available, Promega (Fitchburg, WI, USA)
Plasmids	pMAT1903	pMAT1477 + luciferase	Obtained in this study

Escherichia coli DH5α was used as a host for plasmid propagation and grown in lysogeny broth (LB) medium, supplemented with 0.1 mg/mL ampicillin if needed. Bacterial cultures were incubated at 37 °C overnight.

2.2. Cloning Procedures

2.2.1. Amplification of the Firefly Luciferase Gene

The plasmid pGL3 basic vector (Promega, Fitchburg, WI, USA) served as a template to amplify firefly luciferase by PCR using primers luc-FOWXhoI (AAACTCGAGATGGAAGACGCCAAAAAC ATAAAGAAAGG), luc-REVSacII (CGCCCCGCGGCTAGAATTACACGGCGATCTTTCC) and Herculase II fusion DNA polymerase (Agilent, Santa Clara, CA, UAS). This luciferase is an optimized version for the use in mammalian cells and does not contain the peroxisomal target sequence of the native firefly luciferase.

2.2.2. Plasmid Construction

The pMAT1477 plasmid carries the strong promoter of the *M. circinelloides zrt1* gene (P*zrt1*) and a functional *leuA* gene as a selective marker [19]. Amplified luciferase gene was digested by *Xho*I and *Sac*II and ligated into pMAT1477 to obtain pMAT1903 (Figure S1). By the targeted integration of the whole construct in the *M. circinelloides carRP* gene, which is involved in carotenoid biosynthesis, identification of clones with integrated gene was facilitated, as they formed albino colonies, while those without integration had a yellow phenotype [28].

2.3. *Transformation of* Mucor circinelloides *and Initial Screening*

Strain R7B (*leuA*$^-$) and MU402 (*leuA*$^-$, *pyrG*$^-$) were chosen as recipient strains. MU402 is derived from R7B. Transformation of *M. circinelloides* was performed by electroporation of protoplasts as described previously [16]. In brief, freshly harvested spores were incubated in YPG media (pH 4.5) for 2–4 h until spores were germinated and then transferred to a fresh tube for digestion of the cell wall by lysing enzymes (L-1412, Sigma-Aldrich, St. Louis, MO, USA) and chitosanase (C-0794, Sigma-Aldrich). Linearized DNA (5 μg) was used in each transformation reaction. Protoplasts were incubated on YNBS agar (pH 3.2; containing 0.2 g/L uridine for MU402 strains) and checked daily for colonies [16]. Colonies formed by protoplasts with correct gene targeting appeared white, because of disruption of the *carRP* gene by targeted integration. Albino colonies were repeatedly transferred to fresh selective agar plates (3–4 cycles) to obtain homokaryons. Several clones in each background were then chosen and tested for light emission. Therefore, spores were incubated in YPG in 6-well plates (Nunc GmbH, Langenselbold, Germany) overnight, then D-luciferin (10 mM, Synchem, Felsberg, Germany) was added to the cultures and light emission detected by a monochrome scientific grade CCD camera (BIO-Vision 3000 imaging system, Golden, CA, USA (Figure S3)). Clones that showed highest light emission were chosen for further experiments.

2.4. Genomic DNA Extraction and Southern Analysis

For the preparation of genomic DNA, lyophilized mycelia were ground to powder using a tungsten carbide ball in a Retsch MixerMill 400 and resuspended in 1 mL of DNA isolation buffer (50 mM Tris-HCl, 250 mM NaCl, 100 mM EDTA (Ethylenediaminetetraacetic acid), 1% (*w*/*v*) SDS (Sodium dodecyl sulfate), pH 8.0 at 25 °C) and 300 µL of PCI (Phenyl-Chloroform-Isoamylalcohol; Carl Roth, Karlsruhe, Germany). After incubation at room temperature for 5 min, the mixture was centrifuged for 10 min at 20,000× *g* and 4 °C. RNase A (10 µL; 10 mg/mL) were added to the supernatant and incubated for 10 min at 65 °C and further 30 min at 37 °C. After RNase digestion, DNA was extracted by addition of 1/3 volume of PCI, centrifuged and the resulting supernatant was precipitated by addition of 1 volume of isopropanol. The DNA pellet was washed with 180 µL of 70% ethanol, briefly air-dried and solubilized in 50 µL of a.d. Concentration and quality were determined by agarose gel electrophoresis. 2 µg of genomic DNA were digested overnight with 10–20 units of either *Bgl*II or *Pst*I and separated on 0.8% agarose gels. DIG-labeled marker VII (Roche, Basel, Switzerland) served as a marker for fragment size estimation. Capillary transfer of DNA onto nylon membrane was performed overnight. Membranes were hybridized with a probe for the luciferase coding sequence. Probe labeling with DIG-dUTP was performed by PCR amplification using primers luc1f (5′-TCGCATGCCAGAGATCCT) and luc1r (5′-CGCCCGGTTTATCATCCC).

2.5. Luciferase Activity

Luciferase activity was tested in vitro by measuring light emission of bioluminescent *M. circinelloides* strains in dependency of inoculum density and growth conditions by luminometer Tecan infinite 200 PRO plate reader (Tecan Group AG, Männedorf, Switzerland). First, two transformants of each background were chosen, grown in YNB (2×10^5 spores/mL) in 24-well microtiter plates (Nunc) for 24 h, 100 µL of luciferin (Roche, Luciferase Reporter Gene Assay) were added to the cultures and light emission was detected immediately (2 min after addition of luciferase), or 30 min after addition of substrate to check for stability of light emission.

To determine correlation to inoculum size and subsequent fungal growth/biomass, different spore concentrations were incubated (2×10^2–2×10^6/mL) and light emission was detected as previously described.

A dilution series of substrate luciferin (Roche)—undiluted, 1:2, 1:5 and 1:10—was tested to check for the optimal concentration needed for further experiments.

To determine if light emission is dependent upon growth conditions, strains (2×10^5 spores/mL) were pre-grown in YNB to the same amount of biomass overnight, before the medium was replaced by fresh YNB, YPG or $RPMI_{1640}$ (Sigma-Aldrich, Spittal/Drau, Austria), respectively. Substrate addition and measurement of emitted light were carried out as described above.

2.6. Phenotypic Analysis in Different Growth Conditions

Growth on different media of the recipient strains and the resulting luciferase expressing clone was compared on different media (YNB, YPG, $RPMI_{1640}$ and supplemented minimal agar, SUP [29]; with supplements added when necessary). 10^4 spores of the individual strains were dotted onto the respective agar plates and incubated at 30 °C and 37 °C, respectively. After 24 h colony diameters were measured and growth documented visually. Experiments were carried out with 3 parallels and repeated twice. For growth assays in hypoxic conditions cultures were grown at 1% O_2 (Biospherix, C-Chamber & Pro-Ox controller, Parish, NY, USA).

2.7. Antifungal Susceptibility Testing

Minimum inhibitory concentration (MIC) of AMB, POS, ISA, and itraconazole (ITRA) were determined for all strains according to the European Committee on Antimicrobial Susceptibility Testing (EUCAST) guidelines 9.2 [30]. MIC was defined as the lowest concentration that completely inhibited growth. Additionally, MICs were also determined in YNB medium, as this was used for luciferin activity assays.

To determine antifungal drug efficacy by correlation of fungal growth with light emission, R7B_luc was grown in YNB overnight in the presence of AMB (0.25 μg/mL and 1 μg/mL) or POS (2, 16 and 32 μg/mL), respectively, in 96 well plates. To test the growth inhibiting activity of AMB and POS on *M. circinelloides* hyphae, cultures were pre-grown in YNB overnight before antifungal drugs were added. After 4 h of drug exposure, luciferin was added and light emission detected as described above. All experiments were carried out in parallels and repeated twice.

2.8. Virulence Assay in Galleria mellonella

Sixth instar larvae of *G. mellonella* (SAGIP, Bagnacavallo, Italy), weighing 0.3–0.4 g, were selected for experimental use. Larvae, in groups of twenty, were injected through the last pro-leg into the hemocoel with 1×10^6 spores in a volume of 20 μL as described previously and incubated at 30 °C in the dark [29,31,32]. Untouched larvae and larvae injected with sterile insect physiological saline (IPS) served as controls. Survival was determined every 24 h over a period of 144 h. Experiments were repeated three times and the average survival rate was calculated. Significance was determined with log-rank (Mantel-Cox) test, utilizing GraphPad Prism 7 software (GraphPad Software, San Diego, CA, USA). Differences were considered significant at p-values < 0.05.

2.9. Histology of Larvae

Specimen were fixed in formalin for at least 15 days before being embedded in paraffin. Longitudinal tissue sections were carried out with a microtome at 3.0 μm thickness and stained with Grocott for histological validation. Slides were digitalized using a Pannoramic SCAN digital slide scanner (3DHISTECH, Budapest, Hungary) with plan-apochromat objective (magnification: 20×, Numerical aperture: 0.8). The histological evaluation and the scoring of the fungal infection were done by using the Pannoramic Viewer software (3DHISTECH).

3. Results and Discussion

3.1. Generation of Firefly Luciferase-Producing Mucor circinelloides Strains Resulted in Detectable Light Emission

For the generation of luciferase-producing *M. circinelloides* strains, we cloned the firefly (*P. pyralis*) luciferase gene, optimized for use in mammalian cells, under the control of a strong *M. circinelloides zrt1* promoter into plasmid pMAT1477 that contained the *leuA* gene, which was used as a selective marker in transformations. In the resulting plasmid, the luciferase gene and *leuA* are flanked by sequences of the *carRP* locus to favor targeted integration of the whole construct. The plasmid was linearized and then used to transform the leucine auxotroph strain R7B and the leucine and uridine auxotroph strain MU402. The double auxotroph was chosen to facilitate subsequent disruption or introduction of other genes in the bioluminescent strain. In both backgrounds, more than 50 colonies were obtained on selective transformation plates. Integration in the *carRP* locus renders albino colonies, hence fifteen independent transformants of each background with white appearance were selected. Due to the multinucleated nature of the protoplasts, they were repeatedly inoculated on selective agar until the transformants produced only white colonies, an indication that they were homokaryons. Five of these transformants per background were randomly selected and checked for luciferase production by observing light emission with the naked eye in the dark and visualization of light production (Figure S3). Two strains in each background, showing high light emission, were selected for Southern blot analysis using digoxigenin-labeled probes directed against the luciferase coding region (Figure S2). Restriction with *Bgl*II or *Pst*I confirmed correct insertion and single integration of the luciferase gene. The same four strains were chosen for further luminescence detection by microplate reader and CCD camera (Figure S3). As shown in Table 2, all transformants emitted more light than the parental strains, indicating that our approach of expressing firefly luciferase for the first time in *M. circinelloides* was successful. Measurements taken 30 min after substrate addition indicated that light emission was moderately stable and still significantly detectable after this time, which is essential for further use of the reporter strains. Furthermore, luciferase-harboring strains were stable over several generations because of the site- directed insertion and the selection for homokaryons. Light signals were lower for transformants in the MU402 background, which correlates to slower growth and less biomass of these strains compared to R7B. Therefore, mainly R7B_luc was used for further experiments. Regarding the difficulties with genetic engineering of basal filamentous fungi and in particular the opposition of mucormycetes to express foreign genes, this is an achievement that will be advantageous for further experimental work (e.g., in optimizing luciferase expression in *M. circinelloides* and other mucormycetes).

Table 2. Detection of luminescence signal by microplate reader. 10^5 spores/mL of the respective strains were grown in YNB medium (containing supplements where needed) for 24 h. Light emission was induced by the addition of D-luciferin (10 mM) and detected with a microplate reader (Tecan Group AG, Männedorf, Switzerland). Ten seconds were set as integration time. Measurements were carried out 2 min and 30 min after substrate addition. RLUs (relative light units) present the average of three experiments; SD represents standard deviation.

Strains	RLUs (2 min)	SD	RLUs (30 min)	SD
R7B	14	4	15	1
R7B_luc	4167	112	1444	129
R7B_luc1	3117	73	1194	95
MU402	10	1	7	3
MU402_luc	174	50	185	0
MU402_luc1	20	2	15	1

3.2. *In Vitro Characterization of Bioluminescent Mucor circinelloides Reporter Strains*

3.2.1. Radial Growth Is Not Altered by Insertion of the Luciferase Gene

To check growth ability of luciferase containing strains compared to their recipient strains, radial growth was determined on different media at 30 °C and 37 °C. Based on results shown in Table 2, R7B_luc and MU402_luc were chosen for the radial growth assays, because they showed higher light emission. R7B_luc is prototrophic, while MU402_luc is still auxotrophic for uridine. The aim to generate a reporter strain in the MU402 background was to have a tool in hand that can be used for further genetic manipulation, such as deletion of genes essential for virulence. Here, to rule out phenotypes resulting from luciferase integration, also the MU402_luc strains were used for growth characterization and light emission assays. However, these strains will not be used per se in future animal models. None of the strains displayed an obvious abnormal growth phenotype. At both temperatures, R7B_luc exhibits same growth and average colony diameter on each of the media tested compared to the parental strain R7B (Figure 1, Table S1). MU402_luc exhibited significantly smaller colony diameters on YNB (containing uridine) at both temperatures and on $RPMI_{1640}$ at 37 °C, but still, no differences were detected compared to the parental strain. As expected, colonies were smaller at 37 °C at this early time point, indicating difficulties of *M. circinelloides* with adaptation to high temperatures. Because oxygen levels are expected to be very low on site of infection in the human and animal body [33], and we aim to use the luciferase containing strains in animal models, growth was further evaluated in hypoxic conditions (1% oxygen). Even at this low oxygen concentration, all strains were able to grow and form hyphae, a pre-requisite of tissue invasion. Growth was reduced in all samples compared to normoxic conditions, especially on minimal media (YNB and RPMI). Surprisingly, MU402 and MU402_luc seemed to adapt better to the combination of low oxygen and elevated temperature than R7B_luc and its parental strain on SUP and YPG. At 48 h growth in hypoxia was restored at 30 °C and partially at 37 °C (Figure S4), indicating that *M. circinelloides* spores showed delayed germination in hypoxic conditions, but are able to adapt to low oxygen. This ensures that the luciferase strains will be suitable for use in infection models at a later time point.

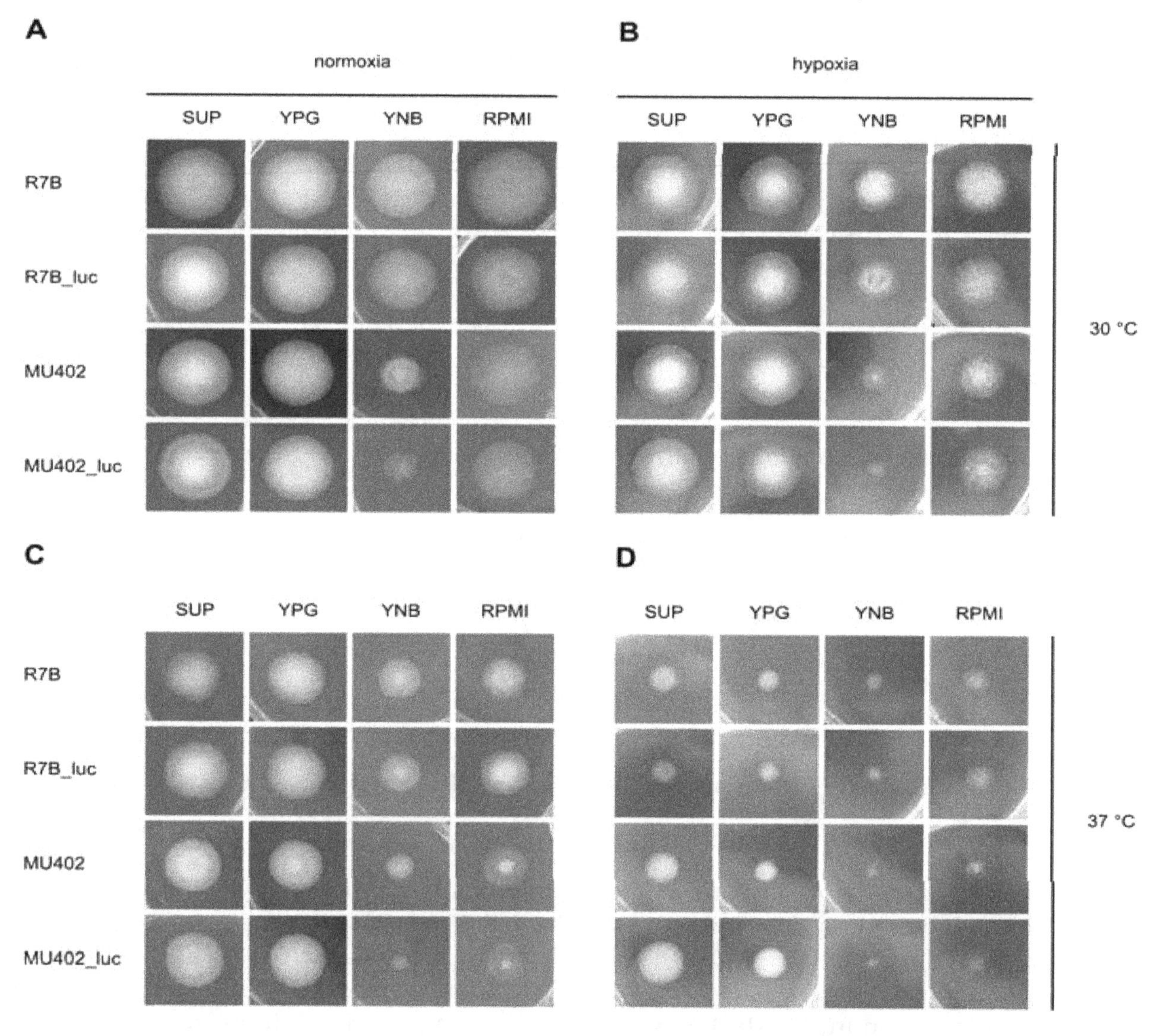

Figure 1. Growth phenotypes of recipient and luciferase-expressing strains grown for 24 h on different media at 30 °C (panels (**A**,**B**)), 37 °C (panels (**C**,**D**)) under normoxic (panels (**A**,**C**)) and hypoxic conditions (panels (**B**,**D**)). Hypoxia was induced by reducing the oxygen concentration in the incubator to 1%. SUP: supplemented minimal agar; YPG: yeast peptone glucose; YNB: yeast nitrogen base; RPMI: $RPMI_{1640}$.

3.2.2. Light Emission Correlates with Fungal Biomass and Amount of Available Luciferase Substrate

An important parameter for the use of bioluminescent reporter strains in animal infection is the detection limit of emitted light, which was determined by cultivation of R7B_luc spores at different inoculum densities and assessment of light emission after 24 h of growth. Light was detected in cultures inoculated with as low as 2×10^3 spores/mL compared to the controls without luciferin and increased with the number of spores used. Highest RLUs were observed at a spore concentration of 2×10^5/mL (Figure 2A, upper panel), the spore concentration that also led to the highest density of mycelia (Figure 2A, lower panel). All other spore concentrations led to significantly lower RLU measurement ($p < 0.05$). The highest inoculum concentration used (2×10^6/mL) did not result in highest light emission, which can be explained by lower growth rate and non-homogeneity of in the culture. Spores probably germinate but face a lack of nutrients at this high density. For all further in vitro experiments, 2×10^5 spores were used.

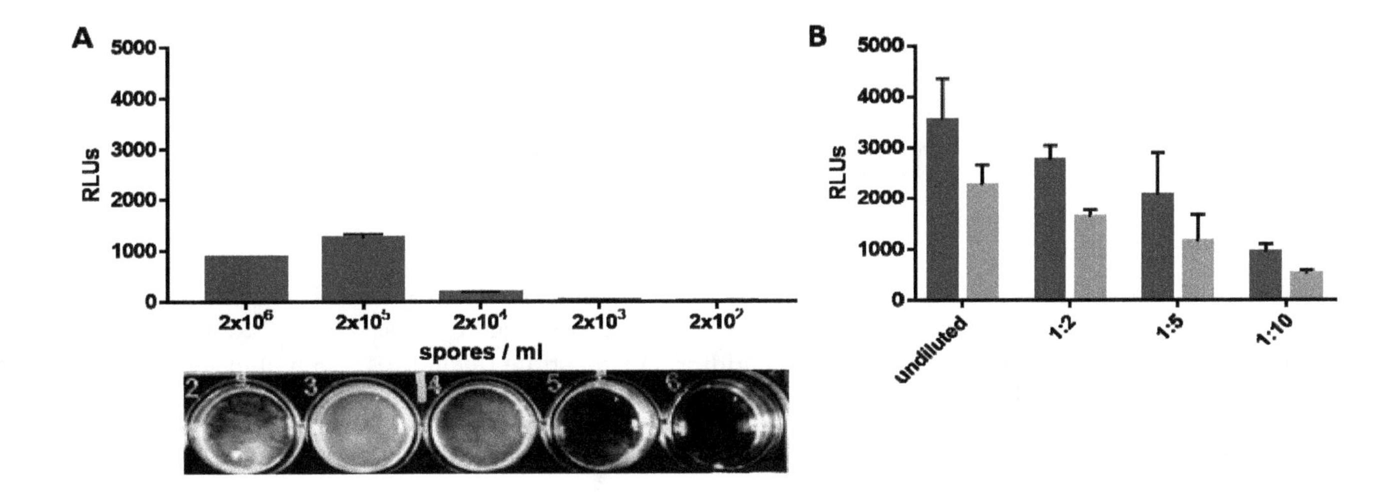

Figure 2. (**A**) Light emission in dependency of inoculum density. YNB medium was inoculated with different concentrations of R7B_luc spores, light emission was induced by addition of luciferin after 24 h and detected by plate reader (upper panel). Relative light units (RLUs) represent the average of three independent measurements. Error bars indicate standard deviation. The lower panel shows fungal growth at the various spore concentrations after 24 h of incubation. (**B**) Light emission in dependency of substrate concentration. 2×10^5 spores/mL were inoculated in YNB and light emission was induced by addition of different concentrations (undiluted, 1:2, 1:5; 1:10) of luciferin dissolved according to the manufacture's protocol (Roche, Basel, Switzerland). Light was detected immediately (dark grey bars) and 10 min after substrate addition (light grey bars). Error bars indicate standard deviation.

Different concentrations of the substrate luciferin were tested to evaluate the minimum amount necessary for R7B_luc to emit detectable light. As expected, light emission clearly correlated with amount of substrate added to the cultures (Figure 2B). As shown before, light emission decreased with time, with significantly reduced light emission at the later time point (t-test, $p < 0.05$), but was still detectable 10 min after substrate addition.

To evaluate the effect of growth media on luciferase expression, cultures were pre-grown in YNB overnight and the medium was replaced by fresh YNB, YPG or $RPMI_{1640}$ 3 h before addition of substrate and light detection. Measurement of RLUs revealed highest levels of emitted light in YNB medium and lowest in YPG (Figure 3). One possible explanation is the nature of the P*zrt1* promoter used for our construct. The gene *zrt1* codes for a zinc transporter whose expression is induced by reduced availability of zinc as is the case in minimal media such as YNB. In mammalian tissues such as human lung or blood, the concentration of zinc is very low; therefore, expression of luciferase driven by P*zrt1* should be high in vivo. Rich media, such as YPG contain higher concentrations of zinc, hypothetically resulting in downregulation of luciferase expression.

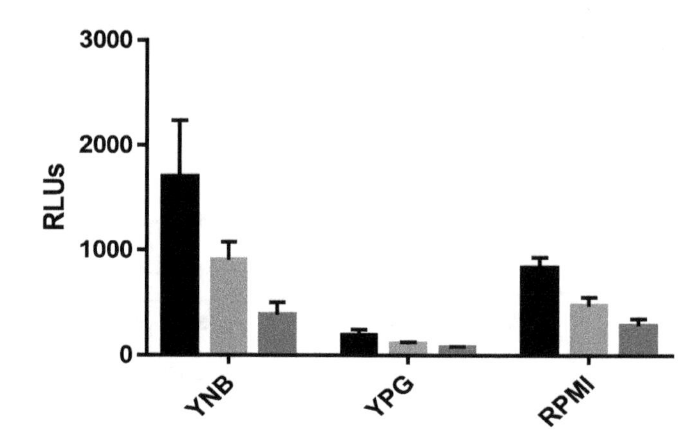

Figure 3. Light emission of R7B_luc in different growth media. YNB medium was inoculated with 2×10^5 spores/mL of R7B_luc, incubated for 16 h, and then replaced by fresh YNB, YPG, or RPMI, respectively. Light emission was induced by addition of substrate (luciferin 1:5) 3 h after medium exchange and detected by using a plate reader. RLUs were determined immediately after addition of substrate (black bars), 10 min (light grey bars) and 30 min (dark grey bars) after the addition of substrate. Error bars indicate standard deviation. RLUs emitted in each media were significantly different from the other media tested (Two-way analysis of variance (ANOVA), $p < 0.05$).

3.3. *Antifungal Susceptibility Testing*

3.3.1. Genetic Manipulation Does Not Influence Antifungal Susceptibility Patterns of *Mucor circinelloides* Strains

The expression of the luciferase gene does not affect the susceptibility patterns of *M. circinelloides* to commonly used antifungal agents, since strains expressing the luciferase gene showed the same susceptibility pattern as the recipient strains (Table 3). To better compare MIC results with results from light emission studies, MICs were also determined in YNB in addition to $RPMI_{1640}$ medium. All strains showed moderate susceptibility to AMB, resulting in complete growth inhibition at concentrations between 0.5 and 2 μg/mL in both media tested. This correlates well to other studies, and thus all 4 strains tested could be classified as susceptible to AMB according to the epidemiological cut-off value determined for *M. circinelloides* [34]. Despite of reducing growth, the azole concentrations applied were below the MIC in $RPMI_{1640}$, nevertheless, in YNB a posaconazole MIC could be determined for R7B and R7B_luc. Although POS is regarded as second-line treatment, it has been shown before that *M. circinelloides* isolates very often also exhibited resistance to this azole [35,36]. The fact that susceptibility patterns of luciferase-harboring and recipient strains were very similar, assures that the luciferase-harboring strains are suitable for the assessment of antifungal drug efficacy in vitro and in vivo.

Table 3. Minimal inhibitory concentrations (MICs; μg/mL) determined for amphotericin B (AMB) and azoles (posaconozale: POS; itraconazole: ITRA; isavuconazole: ISA) according to European Committee on Antimicrobial Susceptibility Testing (EUCAST) guidelines. MICs were determined after 24 h of incubation at 37 °C, except for MU402 and MU402_luc, where MICs were read after 48 h of growth in yeast nitrogen base (YNB).

	MIC (μg/mL)							
	$RPMI_{1640}$				YNB			
Strains	**AMB**	**POS**	**ITRA**	**ISA**	**AMB**	**POS**	**ITRA**	**ISA**
R7B	2	>16	>8	>4	1	2	>8	>4
R7B_luc	1	>16	>8	>4	0.5	4	>8	>4
MU402	2	>16	>8	>4	1	>16	>8	>4
MU402_luc	2	>16	>8	>4	1	>16	>8	>4

3.3.2. Bioluminescent Strains Can Be Used to Evaluate Efficacy of Antifungal Drugs

To test if the luciferase producing strain R7B_luc is suitable for monitoring the efficacy of antifungal substances, we used AMB and POS in two different experimental setups. First, the respective antifungal agent was added directly to spores of R7B_luc, mimicking the EUCAST protocol and light emission was determined after 24h of incubation. No light was detected in wells containing either AMB or POS; respectively (Figure 4A,B). Results obtained with 1 µg/mL AMB correlate well with the MIC determined (Table 3). At this concentration no growth was evident in the presence of AMB prior to germination and consequently no light is emitted. Although some spores could grow in the presence of 0.25 µg/mL AMB, also at this concentration biomass was too little to produce sufficient luciferase. Similarly, of all three POS concentrations tested, none resulted in the emission of detectable light units. Regarding a MIC of 4 mg/mL determined in YNB medium—the same medium that was used for the light emission assays—evidence of a correlation to standard MIC testing is observed. Even at a concentration of 2 µg/mL, growth (or at least fungal metabolism) was inhibited sufficiently to prevent production of luciferase and consequently, light.

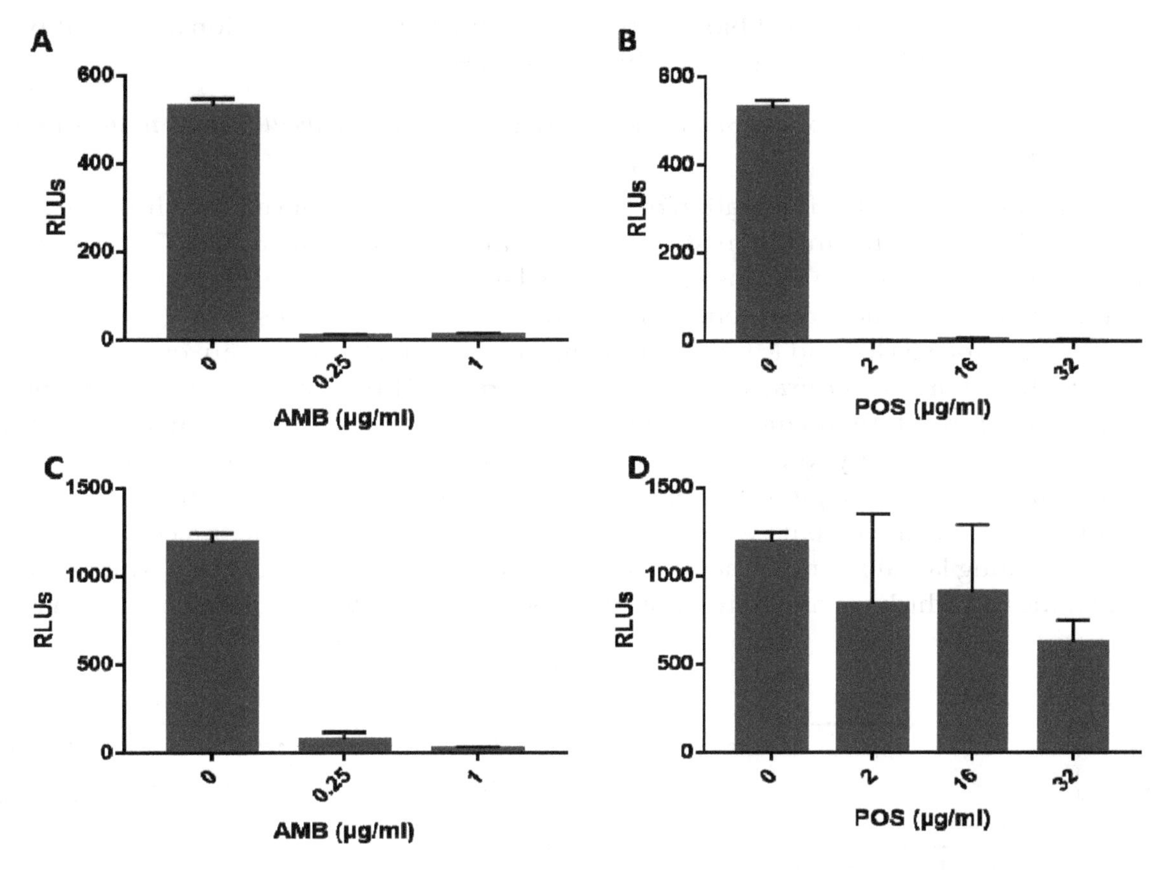

Figure 4. Graphical analysis of drug efficacy by detection of light emission. Luminescence was measured by a plate reader in wells containing 2 × 10^5 spores/mL of R7B_luc grown in YNB in the presence of amphotericin B (AMB) (**A**) or posaconazole (POS) (**B**), and with AMB (**C**) or POS (**D**) added to hyphae, respectively. Light emission was detected after 24 h of incubation (**A**,**B**) and subsequent luciferin (1:5, Roche) addition. Pre-grown cultures (16 h) were further incubated for 4 h once antifungals were added (**C**,**D**). Average values from three independent wells are given, error bars represent standard deviation.

The second approach was to test to what extent the production of luciferase is affected by the addition of antifungal agent, AMB and POS, to R7B_luc hyphae. AMB showed a tremendous effect on luciferase activity, resulting in no light emission when hyphae were incubated in 1 µg/mL AMB for 4 h, and only marginal light emission at 0.25 µg/mL (Figure 4C). This suggests a strong effect of AMB on hyphal metabolism, presumably resulting in decreased ATP levels within the hyphae, which consequently leads to reduced activity of the ATP-dependent luciferase. Another possible effect could be inhibition of substrate uptake due to AMB-induced metabolic changes or membrane dysfunction. Hyphae confronted with POS for 4 h exhibited reduced light emission, but with only a significant difference at 32 μg/mL compared to the untreated controls (Figure 4D). This correlated to previous data that showed reduced effect of POS on hyphae compared to AMB [37]. Furthermore, poor in vitro and in vivo POS efficacy against several *M. circinelloides* strains, due to a rather fungistatic than a fungicidal activity of POS, was shown by Salas et al. [38]. The results obtained with the luciferase-producing strains are in agreement with the MIC data shown before and studies undertaken by others [34,35]. Therefore, we can conclude that screening for antifungal drug efficacy is possible by using luciferase-expressing *M. circinelloides* strains and presents a valuable tool for testing novel antifungal drugs. For *M. circinelloides*, or Mucorales in general, this is of great importance, because many laboratories face difficulties applying and interpreting standard susceptibility test procedures, such as microbroth dilutions methods (EUCAST or clinical and laboratory standards institute (CLSI)) and especially Etest®, with this group of fungi. Often, results obtained with different methods do not correlate [39–41]; therefore, the use of bioluminescent strains will be an additional possibility to test the efficacy of (novel) antifungal drugs or combinations thereof.

3.4. Luciferase-Harboring Strains Exhibit Similar Virulence Potential as Recipient Strains in the Alternative Host Galleria mellonella

In order to test whether the integration of the luciferase gene influenced the virulence potential of the *M. circinelloides* strains, infection studies in the invertebrate host model *G. mellonella* were carried out. All strains were able to cause death to the larvae and no significant difference ($p > 0.05$) was detected between the luciferase-containing strains and the recipient strains (Figure 5). Lower mortality rates seen for MU402 and MU402_luc are most likely due to uridine auxotrophy, suggesting limited availability of uridine or uracil in *Galleria* hemolymph. This correlated to data obtained with *A. fumigatus*, that showed attenuated virulence potential of uridine or uracil auxotrophic strains in murine models [42]. Ability to cause disease in *Galleria* larvae and similarities in survival rates of luciferase-harboring and recipient strains confirms suitability of generated strains for future use in in vivo models. Fungal elements were found in tissue sections of larvae infected with R7B_luc (Figure S5), indicating larval killing by active fungal growth within the larval body. This is important for further studies, in which we aim to use this model system for in vivo bioluminescent imaging.

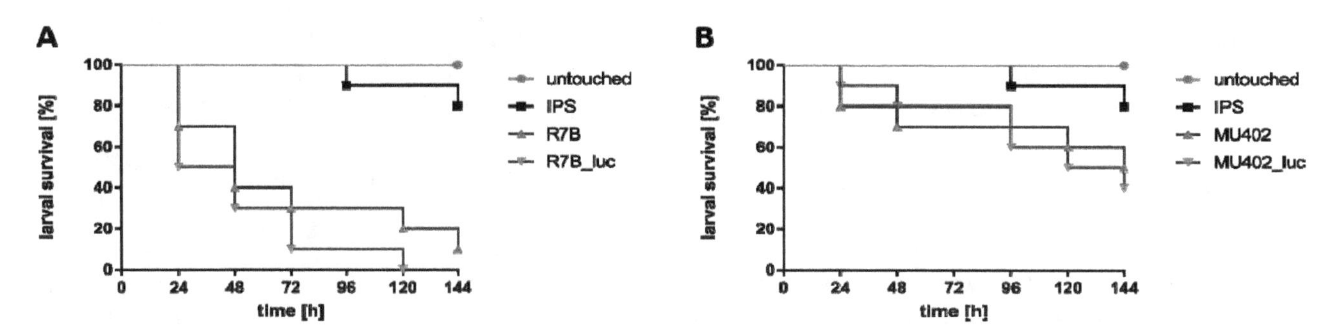

Figure 5. Survival of larvae infected with *M. circinelloides strains.* Larvae were infected with 10^6 spores of the respective strains and incubated at 30 °C. (**A**) represents Kaplan-Meier curves of larvae infected with R7B or R7B_luc strain. (**B**) represents Kaplan-Meier curves of larvae infected with MU402 or MU402_luc strain. Survival was monitored every 24 h up to 144 h. Untouched larvae and larvae injected with IPS buffer served as controls. Results are expressed as the mean of three independent experiments (60 larvae in total).

For other fungi, such as *A. fumigatus* and *C. albicans*, codon-optimized luciferase gene sequences were used and resulted in increased light emission, successfully detected in murine models [22,23,43].

Codon optimization of luciferase in *M. circinelloides* would probably increase luciferase expression and result in higher levels of detectable light, which would specifically be important in murine models, as light detection is quenched by tissue. Further, correlation between the copy number of luciferase and light emission was shown in *A. terreus* and *A. fumigatus* strains [23,26]. Although *M. circinelloides* is one of the few genetically tractable species among the basal fungi, it is not yet a robust genetic system, such as *A. nidulans* or *A. fumigatus*, and knowledge of other native, strong promoters besides *zrt1* is scarce. Nevertheless, using alternative promoters and/or integration of additional luciferase gene copies could further improve our model system.

4. Conclusions

The construction of bioluminescent reporter strains in the basal fungus *M. circinelloides* was successful and resulted in the first *M. circinelloides* strains expressing firefly luciferase, evident by detectable light emission. When comparing our newly generated reporter strains with their respective recipients, we obtained similar results regarding growth, antifungal susceptibility patterns, and virulence potential in the insect model *G. mellonella*. Further, luciferase-containing strains proved to be suitable for the evaluation of antifungal agents. The reporter strains obtained in this study represent a valuable tool for studies investigating the efficacy of novel antifungal agents and monitoring disease in a spatial and temporal manner in animal models in the future.

Supplementary Materials:
Figure S1: Plasmid pMAT1903 containing luciferase gene without peroxisomal target sequence under the strong zrt1 promoter and functional *leuA*, Figure S2: Gene targeting strategy and confirmation of homologous recombination. (A) Schematic presentation of gene targeting and Southern analysis strategy. For generation of the targeting cassette, a leucine auxotrophic marker followed by the gene coding for firefly luciferase were cloned in between 1 kb each of the 5′ UTR and the 3′ UTR of carRP, which were used as flanks for homologous recombination after linearization of the plasmid by *Cfr*9I (*Xma*I) digestion. Restriction sites (*Bgl*II, *Pst*I) and the site for probe hybridization are shown. Expected fragment length to be identified upon proper integration at the carRP locus are 4.6 kb and 5.3 kb. (B) Southern analysis. Genomic DNA of individual transformants was digested with *Bgl*II and *Pst*I, respectively, separated on a 0.8% agarose gel (lower panels) and blotted onto nylon membranes. To determine fragment size, DIG-labelled marker VII (Roche; M) was used; the length of selected marker bands is indicated, Figure S3: Visualisation of bioluminescence from *M. circinelloides* cultures. The *M. circinelloides* transformants (2 per strain) and the respective parental strains are shown. 10^5 spores/ml were inoculated in YNB medium and grown for 24 h. Light emission was induced by the addition of D-luciferin (10 mM) to the medium, and bioluminescence images of the cultures were acquired by a monochrome scientific grade CCD camera (BIO-VISION 3000 imaging system, right panel), Figure S4: Growth phenotypes of recipient and luciferase expressing strains grown for 48 h on different media at 30 °C (panels A and B) and 37 °C (panels C and D) under normoxic (panels A and C) and hypoxic conditions (panels B and D) are shown. Hypoxia was induced by reducing the oxygen concentration in the incubator to 1 %, Figure S5: Histological examination of *Mucor circinelloides* infected *Galleria mellonella* larvae. Specimen were fixed in formalin 72 h after infection with 10^6 spores of R7B_luc and embedded in paraffin. Tissue sections were prepared at a thickness of 3.0 µm and stained with Grocott silver stain to optimize visualisation of fungal elements, Table S1: Colony diameter of *M. circinelloides* strains on various growth media. Colony diameter was determined in triplicates after 24 h of incubation at 30 °C and 37 °C. Numbers given represent the average of two independent experiments. Significance was determined by calculating standard deviation (SD).

Author Contributions: Conceptualization, U.B., V.G.; Methodology, U.B., M.I.N.-M., I.B., V.N. and F.E.N.; Formal Analysis, U.B., M.I.N.-M., I.B., V.N.; Investigation, U.B., M.I.N.-M., I.B., V.N., J.D.P.; Resources, U.B., V.G. and C.L.-F.; Data Curation, U.B., M.I.N.-M., I.B., V.N.; Writing—Original Draft Preparation, U.B., V.G., I.B.; Writing—Review & Editing, U.B., M.I.N.-M., I.B., V.G., F.E.N., J.D.P., C.L.-F.; Supervision, U.B., C.L.-F.; V.G. and F.E.N.; Project Administration, U.B., C.L.-F.; Funding Acquisition, U.B., C.L.-F., V.G.

Acknowledgments: This work was financially supported by the EMBO short-term fellowship 6856 to U.B., the "Christian Doppler Forschungsgesellschaft" (CD-Labor Invasive Pilzinfektionen) to C.L.-F., The "Ministerio de Educación, Cultura y Deporte, Spain" (FPU-14/01832) to M.I.N.-M., "Ministerio de Economía y Competitividad, Spain" (RYC-2014-15844) to F.E.N., and "Ministerio de Economía y Competitividad, Spain" (BFU2015-65501-P co-financed by FEDER) and "Fundación Séneca-Agencia de Ciencia y Tecnología de la Región de Murcia, Spain" (19339/PI/14) to M.I.N.-M., F.E.N. and V.G. The authors are grateful to Fabio Gsaller for useful discussion regarding light imaging and Carmen Kandelbauer, Ines Brosch, Inge Jehart and Sabine Jöbstl for technical assistance.

References

1. Richardson, M. The ecology of the Zygomycetes and its impact on environmental exposure. *Clin. Microbiol. Infect.* **2009**, *15*, 2–9. [CrossRef] [PubMed]
2. Ingold, C.T. *The Biology of Mucor and Its Allies*; Studies in biology no. 88; E. Arnold: London, UK, 1978.
3. Kontoyiannis, D.P.; Lionakis, M.S.; Lewis, R.E.; Chamilos, G.; Healy, M.; Perego, C.; Safdar, A.; Kantarjian, H.; Champlin, R.; Walsh, T.J.; et al. Zygomycosis in a tertiary-care cancer center in the era of *Aspergillus*-active antifungal therapy: A case-control observational study of 27 recent cases. *J. Infect. Dis.* **2005**, *191*, 1350–1360. [CrossRef] [PubMed]
4. Lanternier, F.; Sun, H.Y.; Ribaud, P.; Singh, N.; Kontoyiannis, D.P.; Lortholary, O. Mucormycosis in organ and stem cell transplant recipients. *Clin. Infect. Dis.* **2012**, *54*, 1629–1636. [CrossRef] [PubMed]
5. Lewis, R.E.; Kontoyiannis, D.P. Epidemiology and treatment of mucormycosis. *Future Microbiol.* **2013**, *8*, 1163–1175. [CrossRef] [PubMed]
6. Chamilos, G.; Marom, E.M.; Lewis, R.E.; Lionakis, M.S.; Kontoyiannis, D.P. Predictors of pulmonary zygomycosis versus invasive pulmonary aspergillosis in patients with cancer. *Clin. Infect. Dis.* **2005**, *41*, 60–66. [CrossRef]
7. Hammond, S.P.; Baden, L.R.; Marty, F.M. Mortality in hematologic malignancy and hematopoietic stem cell transplant patients with mucormycosis, 2001 to 2009. *Antimicrob. Agents Chemother.* **2011**, *55*, 5018–5021. [CrossRef]
8. Spellberg, B.; Edwards, J., Jr.; Ibrahim, A. Novel perspectives on mucormycosis: Pathophysiology, presentation, and management. *Clin. Microbiol. Rev.* **2005**, *18*, 556–569. [CrossRef]
9. Caramalho, R.; Tyndal, J.D.A.; Monk, B.C.; Larentis, T.; Lass-Flörl, C.; Lackner, M. Intrinsic short-tailed azole resistance in mucormycetes is due to an evolutionary conserved aminoacid substitution of the lanosterol 14α-demethylase. *Sci. Rep.* **2017**, *7*, 15898. [CrossRef]
10. Skiada, A.; Pagano, L.; Groll, A.; Zimmerli, S.; Dupont, B.; Lagrou, K.; Lass-Florl, C.; Bouza, E.; Klimko, N.; Gaustad, P.; et al. Zygomycosis in Europe: Analysis of 230 cases accrued by the registry of the European Confederation of Medical Mycology (ECMM) working group on Zygomycosis between 2005 and 2007. *Clin. Microbiol. Infect.* **2011**, *17*, 1859–1867. [CrossRef]
11. Antoniadou, A. Outbreaks of zygomycosis in hospitals. *Clin. Microbiol. Infect.* **2009**, *15*, 55–59. [CrossRef]
12. Duffy, J.; Harris, J.; Gade, L.; Sehulster, L.; Newhouse, E.; O'Connell, H.; Noble-Wang, J.; Rao, C.; Balajee, S.A.; Chiller, T. Mucormycosis outbreak associated with hospital linens. *Pediatr. Infect. Dis. J.* **2014**, *33*, 472–476. [CrossRef] [PubMed]
13. Garcia-Hermoso, D.; Criscuolo, A.; Lee, S.C.; Legrand, M.; Chaouat, M.; Denis, B.; Lafaurie, M.; Rouveau, M.; Soler, C.; Schaal, J.-V.; et al. Outbreak of invasive wound mucormycosis in a burn unit due to multiple strains of *Mucor circinelloides* f. *circinelloides* resolved by whole-genome sequencing. *MBio* **2018**, *9*, e00573-18. [CrossRef] [PubMed]
14. Lee, S.C.; Billmyre, R.B.; Li, A.; Carson, S.; Sykes, S.M.; Huh, E.Y.; Mieczkowski, P.; Ko, D.C.; Cuomo, C.A.; Heitman, J. Analysis of a food-borne fungal pathogen outbreak: Virulence and genome of a *Mucor circinelloides* isolate from yogurt. *MBio* **2014**, *5*, e01390-14. [CrossRef] [PubMed]
15. Magan, N.; Olsen, M. *Mycotoxins in Food: Detection and Control*; Woodhead Publishing Ltd.: Cambridge, UK, 2004.
16. Torres-Martinez, S.; Ruiz-Vázquez, R.M.; Garre, V.; López-García, S.; Navarro, E.; Vila, A. Molecular tools for carotenogenesis analysis in the zygomycete *Mucor circinelloides*. *Methods Mol. Biol.* **2012**, *898*, 85–107. [PubMed]
17. Vellanki, S.; Navarro-Mendoza, M.I.; Garcia, A.; Murcia, L.; Perez-Arques, C.; Garre, V.; Nicolas, F.E.; Lee, S.C. *Mucor circinelloides*: Growth, maintenance, and genetic manipulation. *Curr. Protoc. Microbiol.* **2018**, *49*, e53. [CrossRef] [PubMed]
18. Nicolas, F.E.; de Haro, J.P.; Torres-Martínez, S.; Ruiz-Vázquez, R.M. Mutants defective in a *Mucor circinelloides dicer*-like gene are not compromised in siRNA silencing but display developmental defects. *Fungal Genet. Biol.* **2007**, *44*, 504–516. [CrossRef] [PubMed]
19. Rodriguez-Frometa, R.A.; Gutiérrez, A.; Torres-Martínez, S.; Garre, V. Malic enzyme activity is not the only bottleneck for lipid accumulation in the oleaginous fungus *Mucor circinelloides*. *Appl. Microbiol. Biotechnol.* **2013**, *97*, 3063–3072. [CrossRef] [PubMed]

20. Brock, M. Application of bioluminescence imaging for in vivo monitoring of fungal infections. *Int. J. Microbiol.* **2012**, *2012*, 956794. [CrossRef] [PubMed]
21. Delarze, E.; Ischer, F.; Sanglard, D.; Coste, A.T. Adaptation of a *Gaussia princeps* Luciferase reporter system in *Candida albicans* for in vivo detection in the *Galleria mellonella* infection model. *Virulence* **2015**, *6*, 684–693. [CrossRef]
22. Jacobsen, I.D.; Lüttich, A.; Kurzai, O.; Hube, B.; Brock, M. In vivo imaging of disseminated murine *Candida albicans* infection reveals unexpected host sites of fungal persistence during antifungal therapy. *J. Antimicrob. Chemother.* **2014**, *69*, 2785–2796. [CrossRef]
23. Brock, M.; Jouvion, G.; Droin-Bergère, S.; Dussurget, O.; Nicola, M.-A.; Ibrahim-Granet, O. Bioluminescent *Aspergillus fumigatus*, a new tool for drug efficiency testing and in vivo monitoring of invasive aspergillosis. *Appl. Environ. Microbiol.* **2008**, *74*, 7023–7035. [CrossRef] [PubMed]
24. Donat, S.; Hasenberg, M.; Schäfer, T.; Ohlsen, K.; Gunzer, M.; Einsele, H.; Löffler, J.; Beilhack, A.; Krappmann, S. Surface display of *Gaussia princeps* luciferase allows sensitive fungal pathogen detection during cutaneous aspergillosis. *Virulence* **2012**, *3*, 51–61. [CrossRef] [PubMed]
25. Ibrahim-Granet, O.; Jouvion, G.; Hohl, T.M.; Droin-Bergère, S.; Philippart, F.; Kim, O.Y.; Adib-Conquy, M.; Schwendener, R.; Cavaillon, J.M.; Brock, M.; et al. In vivo bioluminescence imaging and histopathopathologic analysis reveal distinct roles for resident and recruited immune effector cells in defense against invasive aspergillosis. *BMC Microbiol.* **2010**, *10*, 105. [CrossRef]
26. Slesiona, S.; Ibrahim-Granet, O.; Olias, P.; Brock, M.; Jacobsen, I.D. Murine infection models for *Aspergillus terreus* pulmonary aspergillosis reveal long-term persistence of conidia and liver degeneration. *J. Infect. Dis.* **2012**, *205*, 1268–1277. [CrossRef] [PubMed]
27. Roncero, M.I.G. Enrichment Method for the isolation of auxotrophic mutants of mucor using the polyene antibiotic *N*-glycosyl-polifungin. *Carlsberg Res. Commun.* **1984**, *49*, 685–690. [CrossRef]
28. Nicolas, F.E.; Navarro-Mendoza, M.I.; Pérez-Arques, C.; López-García, S.; Navarro, E.; Torres-Martínez, S.; Garre, V. Molecular tools for carotenogenesis analysis in the mucoral *Mucor circinelloides*. *Methods Mol. Biol.* **2018**, *1852*, 221–237.
29. Maurer, E.; Hörtnagl, C.; Lackner, M.; Grässle, D.; Naschberger, V.; Moser, P.; Segal, E.; Semis, M.; Lass-Flörl, C.; Binder, U. *Galleria mellonella* as a model system to study virulence potential of mucormycetes and evaluation of antifungal treatment. *Med. Mycol.* **2018**. [CrossRef]
30. Arendrup, M.C.; Hope, W.W.; Lass-Flörl, C.; Cuenca-Estrella, M.; Arikan, S.; Barchiesi, F.; Bille, J.; Chryssanthou, E.; Groll, A.; et al. EUCAST technical note on the EUCAST definitive document EDef 7.2: Method for the determination of broth dilution minimum inhibitory concentrations of antifungal agents for yeasts EDef 7.2 (EUCAST-AFST). *Clin. Microbiol. Infect.* **2012**, *18*, E246–E247. [CrossRef]
31. Fallon, J.; Kelly, J.; Kavanagh, K. *Galleria mellonella* as a model for fungal pathogenicity testing. *Methods Mol. Biol.* **2012**, *845*, 469–485.
32. Maurer, E.; Browne, N.; Surlis, C.; Jukic, E.; Moser, P.; Kavanagh, K.; Lass-Flörl, C.; Binder, U. *Galleria mellonella* as a host model to study *Aspergillus terreus* virulence and amphotericin B resistance. *Virulence* **2015**, *6*, 591–598. [CrossRef]
33. Grahl, N.; Shepardson, K.M.; Chung, D.; Cramer, R.A. Hypoxia and fungal pathogenesis: To air or not to air? *Eukaryot. Cell* **2012**, *11*, 560–570. [CrossRef] [PubMed]
34. Dannaoui, E. Antifungal resistance in mucorales. *Int. J. Antimicrob. Agents* **2017**, *50*, 617–621. [CrossRef] [PubMed]
35. Espinel-Ingroff, A.; Chakrabarti, A.; Chowdhary, A.; Cordoba, S.; Dannaoui, E.; Dufresne, P.; Fothergill, A.; Ghannoum, M.; Gonzalez, G.M.; Guarro, J. Multicenter evaluation of MIC distributions for epidemiologic cutoff value definition to detect amphotericin B, posaconazole, and itraconazole resistance among the most clinically relevant species of Mucorales. *Antimicrob. Agents Chemother.* **2015**, *59*, 1745–1750. [CrossRef] [PubMed]
36. Khan, Z.U.; Ahmad, S.; Brazda, A.; Chandy, R. *Mucor circinelloides* as a cause of invasive maxillofacial zygomycosis: an emerging dimorphic pathogen with reduced susceptibility to posaconazole. *J. Clin. Microbiol.* **2009**, *47*, 1244–1248. [CrossRef] [PubMed]
37. Perkhofer, S.; Locher, M.; Cuenca-Estrella, M.; Rüchel, R.; Würzner, R.; Dierich, M.P.; Lass-Flörl, C. Posaconazole enhances the activity of amphotericin B against hyphae of zygomycetes in vitro. *Antimicrob. Agents Chemother.* **2008**, *52*, 2636–2638. [CrossRef]

38. Salas, V.; Pastor, F.J.; Calvo, E.; Alvarez, E.; Sutton, D.A.; Mayayo, E.; Fothergill, A.W.; Rinaldi, M.G.; Guarro, J. In vitro and in vivo activities of posaconazole and amphotericin B in a murine invasive infection by *Mucor circinelloides*: poor efficacy of posaconazole. *Antimicrob. Agents Chemother.* **2012**, *56*, 2246–2250. [CrossRef]
39. Caramalho, R.; Maurer, E.; Binder, U.; Araújo, R.; Dolatabadi, S.; Lass-Flörl, C.; Lackner, M. Etest cannot be recommended for in vitro susceptibility testing of mucorales. *Antimicrob. Agents Chemother.* **2015**, *59*, 3663–3665. [CrossRef] [PubMed]
40. Vitale, R.G.; de Hoog, G.S.; Schwarz, P.; Dannaoui, E.; Deng, S.; Machouart, M.; Voigt, K.; van de Sande, W.W.; Dolatabadi, S.; Meis, J.F.; et al. Antifungal susceptibility and phylogeny of opportunistic members of the order mucorales. *J. Clin. Microbiol.* **2012**, *50*, 66–75. [CrossRef] [PubMed]
41. Rodriguez, M.M.; Pastor, F.J.; Sutton, D.A.; Calvo, E.; Fothergill, A.W.; Salas, V.; Rinaldi, M.G.; Guarro, J. Correlation between in vitro activity of posaconazole and in vivo efficacy against *Rhizopus oryzae* infection in mice. *Antimicrob. Agents Chemother.* **2010**, *54*, 1665–1669. [CrossRef]
42. D'Enfert, C.; Diaquin, M.; Delit, A.; Wuscher, N.; Debeaupuis, J.P.; Huerre, M.; Latge, J.P. Attenuated virulence of uridine-uracil auxotrophs of *Aspergillus fumigatus*. *Infect. Immun.* **1996**, *64*, 4401–4405.
43. Galiger, C.; Brock, M.; Jouvion, G.; Savers, A.; Parlato, M.; Ibrahim-Granet, O. Assessment of efficacy of antifungals against *Aspergillus fumigatus*: Value of real-time bioluminescence imaging. *Antimicrob. Agents Chemother.* **2013**, *57*, 3046–3059. [CrossRef] [PubMed]

6

Role of Homologous Recombination Genes in Repair of Alkylation Base Damage by *Candida albicans*

Toni Ciudad [†], Alberto Bellido [†], Encarnación Andaluz, Belén Hermosa and Germán Larriba [*]

Departamento de Microbiología, Facultad de Ciencias, Universidad de Extremadura, 06071 Badajoz, Spain; aciudad@unex.es (T.C.); abdiaz@unex.es (A.B.); eandaluz@unex.es (E.A.); belenh@unex.es (B.H.)

* Correspondence: glarriba@unex.es

† These authors contributed equally to this work.

Abstract: *Candida albicans* mutants deficient in homologous recombination (HR) are extremely sensitive to the alkylating agent methyl-methane-sulfonate (MMS). Here, we have investigated the role of HR genes in the protection and repair of *C. albicans* chromosomes by taking advantage of the heat-labile property (55 °C) of MMS-induced base damage. Acute MMS treatments of cycling cells caused chromosome fragmentation in vitro (55 °C) due to the generation of heat-dependent breaks (HDBs), but not in vivo (30 °C). Following removal of MMS wild type, cells regained the chromosome ladder regardless of whether they were transferred to yeast extract/peptone/dextrose (YPD) or to phosphate buffer saline (PBS); however, repair of HDB/chromosome restitution was faster in YPD, suggesting that it was accelerated by metabolic energy and further fueled by the subsequent overgrowth of survivors. Compared to wild type CAI4, chromosome restitution in YPD was not altered in a Ca*rad59* isogenic derivative, whereas it was significantly delayed in Ca*rad51* and Ca*rad52* counterparts. However, when post-MMS incubation took place in PBS, chromosome restitution in wild type and HR mutants occurred with similar kinetics, suggesting that the exquisite sensitivity of Ca*rad51* and Ca*rad52* mutants to MMS is due to defective fork restart. Overall, our results demonstrate that repair of HDBs by resting cells of *C. albicans* is rather independent of CaRad51, CaRad52, and CaRad59, suggesting that it occurs mainly by base excision repair (BER).

Keywords: homologous recombination; *Candida albicans*; alkylation damage; repair

---

1. Introduction

Methyl-methane-sulfonate (MMS) is used for the analysis of pathways involved in repair/tolerance to methylation [1]. Methyl-methane-sulfonate generates methylated bases on dsDNA whose repair can cause nicks, gaps, and, indirectly, double-strand breaks (DSBs) that can engage in homologous recombination (HR) directly or cause fork stalling during the next replication round [2–4]. Although methylated bases can be directly removed by DNA-methyl-transferases, the major repair pathway consists of a step-wise process known as base excision repair (BER) [1,5]. In *Saccharomyces cerevisiae,* BER is initiated by specific DNA-*N*-glycosylases that remove the damaged bases. The apurinic/apyrimidic (AP) sites generated are removed by redundant AP endonucleases Apn1 and Apn2, which cleave 5′ the AP-site to form nicks with a 5′ desoxyribose phosphate (5′dRP). Removal of 5′dRP is carried out by the coordinated action of DNA polymerase (δ or ε) and the flap endonuclease Rad27/Fen1, followed by ligation. Alternatively, AP-sites can be processed by unspecific Ntg1, Ntg2, or Ogg1 lyases to generate 3′AP sites (3′-dRP), which are then removed by the 3′-diesterase activity of Apn1/Apn2 or as part of an oligonucleotide generated by the endonuclease Rad1–Rad10. Finally, the gap is filled by DNA Pol and the backbone sealed by DNA ligase [1,4,6]. It is likely that similar enzymes and reactions account for BER in other fungi. However, only one APN endonuclease

(*APN1*) and one NTG lyase (*NTG1*), in addition to an *OGG1* homolog, have been so far identified and partially characterized in *Candida albicans* [7]. Importantly, null mutants in each of these *C. albicans* genes as well as the triple mutant exhibited wild type sensitivities to MMS suggesting the presence of redundant enzymes involved in repair of the methylated bases [7].

Importantly, in methylated DNA, AP-sites can arise from elimination of methylated bases by the action of glycosylase Mag1 [6] or, non-enzymatically, through the spontaneous depurination of methylated N3 and N7 purines, a process that is accelerated by heat [8]. It is well known that AP-sites are heat labile [9,10] and can be converted into single-strand breaks (SSBs) at 55 °C (referred to as HDB, for heat dependent breaks), a temperature generally used to prepare plugs for pulse-filed gel electrophoresis (PFGE) [11]. In fact, in *S. cerevisiae* stationary, or G1-arrested cells, DSBs and chromosome degradation was observed when methylated DNA or DNA carrying AP-sites was incubated at 55 °C [4,6,11]. Methyl-methane-sulfonate treatment can also cause opposed closely-spaced nicks in vivo (referred to as heat independent breaks (HIB) because they are detected when the same DNA samples are incubated at 30 °C), which can also result in secondary DSBs, and therefore in chromosome fragmentation [4,6,11]. For *S. cerevisiae* wild type, typical MMS treatments generate a large amount of HDBs and a low amount of HIB [12]. However, the latter are substantially increased in some BER-deficient cells as in *apn1/apn2* double mutants [4,6,12].

Rad51 and Rad52 are evolutionary conserved proteins that play crucial roles in HR. In yeast, recombinase Rad51 is required for recombination pathways involving strand invasion. These pathways also require Rad52, which is thought to mediate the Rad51-ssDNA nucleofilament assembly. In addition, Rad52 participates in all recombination processes that require single strand annealing. For this reason, the *rad52* mutation is epistatic to deletion of any other gene of the *RAD52* epistasis group. Rad59 is a yeast paralog of Rad52 that exhibits strand annealing activity but lacks the ability of loading Rad51 onto ssDNA (for reviews see References [3,13]). Several studies have indicated that, although *rad51* and *rad52* mutants are extremely sensitive to MMS, HR does not play any role in the repair of MMS-born HIB of haploid *S. cerevisiae* G1 stationary cells which do not have a partner for engaging in HR, and the same was true for diploid G2-arrested cells which do [4,6,11,14]. However, Rad52 was crucial for that repair of HIB in G2-arrested cells in the absence of Anp1 and Anp2 endonucleases, suggesting that HR acts as a backup to repair lesions produced by AP-lyases Ntg1/2 and Ogg1 in the absence of Anp1,2 [6]. Besides, Rad51 and Rad52 are required for replication of methylated DNA [15–17] as well as for repair of the gaps generated in the process when cells reach the G2 phase [17]. Methyl-methane-sulfonate lesions and BER intermediates that are not repaired before they encounter the replication fork may cause replication fork stalling and collapse, unless stalled forks are bypassed by translation synthesis (stalled forks) causing increased mutagenesis or faithfully repaired by HR (both stalled and collapsed forks) [6,16,17]. The importance of HR in repair of methylated DNA extends to *C. albicans* where mutants affected in either HR (Ca*rad52*) or resection of DSBs (Ca*rad50*, Ca*mre11*) were significantly more sensitive to MMS than single mutants in BER genes [7,18].

In addition to MMS, a number of both endogenous and environmental agents including anticancer drugs (i.e., 4-methyl-5-oxo-2,3,4,6,8-pentazabicyclo[4.3.0]nona-2,7,9-triene-9-carboxamide temozolomide) can cause methylation damage [8,19,20], and therefore may affect viability and virulence of commensal opportunistic pathogens such as *C. albicans*. We previously reported that Ca*rad52*-ΔΔ, and to a lesser extent Ca*rad51*-ΔΔ cells from *C. albicans*, exhibit increased sensitivity to MMS [21,22]. In the current study, we have determined methylation base damage and recovery in *C. albicans*, taking advantage of the secondary DSBs and subsequent chromosome fragmentation generated during preparation of samples for pulse-field gel electrophoresis (PFGE) at 55 °C. We also show that resting cells of *C. albicans* can repair HDBs in the absence of HR, whereas repair of HDBs (and other BER intermediates) by cycling cell was mostly dependent on efficient HR.

2. Materials and Methods

2.1. Strains

The *C. albicans* and *S. cerevisiae* strains used in this study are shown in Table 1. Strain CAF2-1 derives from reference strain SC5314 by deletion of one copy of *URA3* whereas CAI4 is the Uri$^-$ auxotrophic derivative of CAF2-1. All three strains are wild type for DNA recombination and repair. They were routinely grown in either yeast extract/peptone/dextrose (YPD) medium or synthetic complete (SC) medium supplemented with 33 mM uridine when necessary (i.e., Uri$^-$ strains) [23]. The haploid *S. cerevisiae* strain LSY0695-7D (Table 1) is a W303 derivative kindly provided by Lorraine Symington, from Columbia University [24]. Diploid W303 was obtained by standard genetic crosses [25].

Table 1. Strains used in this study.

Strains (Old Name)	Genotype	Parental	Reference
Candida albicans			
SC5314	Wild type		Gillum et al., 1984 [26]
CAF2-1	*Δura3::imm434/URA3*	SC5314	Fonzi and Irwin, 1993 [27]
CAI4	*Δura3::imm434/Δura3::imm434*	CAF2-1	Fonzi and Irwin, 1993 [27]
CAGL4 (TCR2.1)	*Δura3::imm434/Δura3::imm434 rad52::hisG/Δrad52::hisG-URA3-hisG*	CAI4	Ciudad et al., 2004 [28]
CAGL4.1 (TCR2.1.1)	*Δura3::imm434/Δura3::imm434 Δrad52::hisG/Δrad52::hisG*	CAGL4	Ciudad et al., 2004 [28]
CAGL17 (BNC1.1)	*Δura3::imm434/Δura3::imm434 Δrad59::hisG/Δrad59::hisG-URA3-hisG*	CAI4	García-Prieto et al., 2010 [21]
CAGL17.1 (BNC23.1)	*Δura3::imm434/Δura3::imm434 Δrad59::hisG/Δrad59::hisG*	CAGL17	Bellido et al., 2015 [22]
CAGL19 (JGR5)	*Δura3::imm434/Δura3::imm434 Δrad51::hisG/Δrad51::hisG-URA3-hisG*	CAI4	García-Prieto et al., 2010 [21]
CAGL19.1 (JGR5A)	*Δura3::imm434/Δura3::imm434 Δrad51::hisG/Δrad51::hisG*	CAGL19	García-Prieto et al., 2010 [21]
S. cerevisiae			
LSY0695-7D W303 haploid	*MAT*a; *ADE2 RAD5 met17-s ade2-1 trp1-1 his3-11,15 can1-100 ura3-1 leu2-3,112*		Bärtsch et al., 2000 [24]
W303 diploid	*MAT*a/α; *ADE2 RAD5 met17-s/met17-s ade2-1/ade2-1 trp1-1/trp1-1 his3-11,15/his3-11,15 can1-100/can1-100 ura3-1/ura3-1 leu2-3,112/leu2-3,112*		Bellido et al., 2015 [22]

2.2. DNA Extraction and Analysis and Cell Transformation

The DNA preparation for PFGE, as well as resolution of the samples, was carried out as reported before [29,30] using a Bio-Rad Chef Dry III. Gels were stained with ethidium bromide (0.5 μg/mL) for 2–4 h and imaged using a Molecular Imager (Bio-Rad Laboratories, Madrid, Spain). *C. albicans* cells were transformed using the lithium acetate method [31]. *C. albicans* chromosome fragments generated during incubation of methylated DNA with proteinase K at 55 °C were sized by using *S. cerevisiae* chromosomes as molecular weight (MW) markers (316–1091 kb) (see below).

A functional copy of the *C. albicans URA3* marker was obtained by digestion of pLUBP plasmid with Pst1-BglII [32]. The resulting 4.9 kb fragment containing the *URA3* gene was used to transform the Uri$^-$ strains CAI4 and its derivatives *rad59*-ΔΔ and *rad51*-ΔΔ. Uri$^+$ transformants were selected on minimal SC medium minus uridine and correct integration was verified by PCR using oligonucleotide URA3 left flank and URA3 right flank [32].

2.3. Sensitivity to DNA-Damaging Agents

For determination of survival following an acute short-term MMS treatment, about 5×10^5 exponentially growing cells suspended in 1 mL YPD were incubated with MMS (0.05%, final concentration) for 30 min. Incubation mixtures were diluted 10^3-fold with PBS, and 50 μL (containing 300–400 colony forming units (CFU) before treatment) were plated on YPD plates for 48 h to determine the number of colonies. All the assays were done in duplicate and repeated four or more times.

2.4. Generation of MMS-Induced Heat Labile Breaks

Alkylation was induced as described by Lundin et al. [11]. Briefly, the indicated yeast strain was grown overnight in YPD until $OD_{600} \approx 5–7$. MMS was added to a final concentration of 0.05% and cell suspensions were shaken at 30 °C for 15–30 min. The MMS was neutralized with 5% sodium thiosulfate by mixing 1:1 (*v*/*v*) ratio with 10% $Na_2S_2O_3$ and washed twice with phosphate buffer saline (PBS). An aliquot was processed for PFGE ($t = 0$) and the rest was incubated in MMS-free YPD medium at 30 °C. At regular intervals (1, 2, 4, 8 and 24 h) samples were taken for determination of OD_{600}, CFU, morphology, and repair of DNA damage (PFGE). Plugs for PFGE were prepared as described above and subsequently treated for 24 h with proteinase K (1 mg/mL) at 55 °C for analysis of HDBs or at 30 °C for analysis of HIBs as reported [4,6]. When indicated, thiosulfate neutralized cells were allowed to stand in PBS (resting cells) and samples were taken at the indicated times [4,6,11].

2.5. Calculations of Chromosome Fragments Sizes from Closely Spaced Single-Strand Breaks

Our calculations are based on the assumptions that SSBs (or heat-labile sites, in our assay) are distributed evenly between two DNA strands of a chromosome and that for a given SSB to form a DSB, a second SSB must appear on the opposite strand within an interval ≤S [6]. For further calculations of the number of DSBs, Ma et al. [6] used a circularized ChrIII, which only enters the gel following induction of a single DSB, to quantitatively determine the ratio ChrII to ChrIII using Southern blotting with a probe that hybridizes to both chromosomes. Since *C. albicans* does not maintain circular plasmids [28], we determined the range of sizes of the smear produced by the MMS treatment (0.05% MMS, 30 min) using the *S. cerevisiae* chromosomes as size markers. For *S. cerevisiae* diploid W303 strain (24 Mb), chromosome fragment sizes generated by MMS treatment ranged between 250 and 666 Kb, whose mean is 455 kb. Generation of uniform fragments of this size would require an average of 52 DSB per diploid genome or 26 DSB per haploid genome. For *C. albicans* (32 Mb), chromosome fragments sizes ranged between 316 and 1091 kb. By analogy, for a mean of 700kb we calculated 46 DSB per diploid genome or 23 DSB per haploid genome. Importantly, these values are not far from the range of 30 and 40 DSB per *S. cerevisae* haploid genome previously reported [6] for acute (30 min) 0.1% MMS treatments taking into account that we have used half MMS concentration (0.05% MMS) (Figure S1).

To quantify the extent of chromosome restitution during the post-MMS incubation, we estimated the intensities associated to Chr2, Chr5 bands (the same area for each chromosome) and smear from pulsed-field gels stained with ethidium bromide using the ImageJ software. Then, the ratio of the intensities Chr/smear at each time point was graphed versus the post-incubation time using Microsoft Excel. Data obtained with this approach agreed well with visual inspections of the gels. Chr2 and Chr5 were chosen as indicators of chromosome restitution because both are well resolved in PFGE and differ widely in size (2.23 Mb for Chr2 and 1.19 Mb for Chr5).

3. Results

3.1. The Role of HR Genes in Growth Polarization in Response to MMS

Genotoxic stress, including hydroxyurea and MMS treatments, triggers growth polarization of wild type *C. albicans* SC5314 and derivatives CAF2-1 (*URA3*/*ura3*) and CAI4 (*ura3*/*ura3*) (Table 1) generating elongated cells [33,34]. We have recently shown that the Uri^- strain CAI4 and its derivatives carrying additional auxotrophies (RM10, RM1000, BWP37 and SN strains) exhibit an enhanced growth polarization and susceptibility to 0.02% MMS compared to parental CAF2-1 [35]. This differential behavior is due to a spontaneous loss of heterozygosity (LOH) event during the generation of CAI4 on the right arm of chromosome 3 (Chr3R) that homozygosed *MBP1a*, which regulates expression of DNA repair genes at G1/S phase of the cell cycle [35]. To investigate if MMS-induced filamentation was further affected by HR mutations, we subjected wild type and mutant strains to 0.02% MMS for 16 h at 30 °C in liquid YPD (Figure 1). As expected, CAF2-1 displayed chains of elongated cells but no filaments, whereas its Uri^- derivative CAI4 showed long filaments. Reintegration of one copy of *URA3* in its own locus improved growth rate but did not affect filamentation of the CAI4 strain [35]. All three Ca*rad59*-ΔΔ strains filamented as CAI4, regardless of whether they were Uri^+ or Uri^-, or if one copy of *URA3* had been reintegrated into its own locus in the Uri^- version of the mutant (Figure 1). The same was true for Ca*rad51*-ΔΔ strains in its Uri^-, Uri^+, or *URA3*-reintegrated versions (Figure 1). For Ca*rad52*-ΔΔ, Uri^+ and Uri^- versions behaved also similarly (Figure 1); however, as described [28], it was not possible to reintegrate *URA3* into its own locus in *rad52*-ΔΔ strains. Importantly, in addition to the typical filamentous cells, Ca*rad51*-ΔΔ and Ca*rad52*-ΔΔ cultures also contain yeast cells [23]. In response to MMS, yeast cells also formed "germinative tubes" whose form and length were similar to their counterparts from Ca*rad59*-ΔΔ or CAI4 strains. We conclude that recombination mutants retain the ability to filament in response to MMS. It should be, finally, noted that cell elongation was a specific trait of *C. albicans* since it was not observed when diploid *S. cerevisiae* W303 or its Sc*rad52* derivative were incubated in MMS or during their post-MMS incubation in either PBS or YPD (not shown).

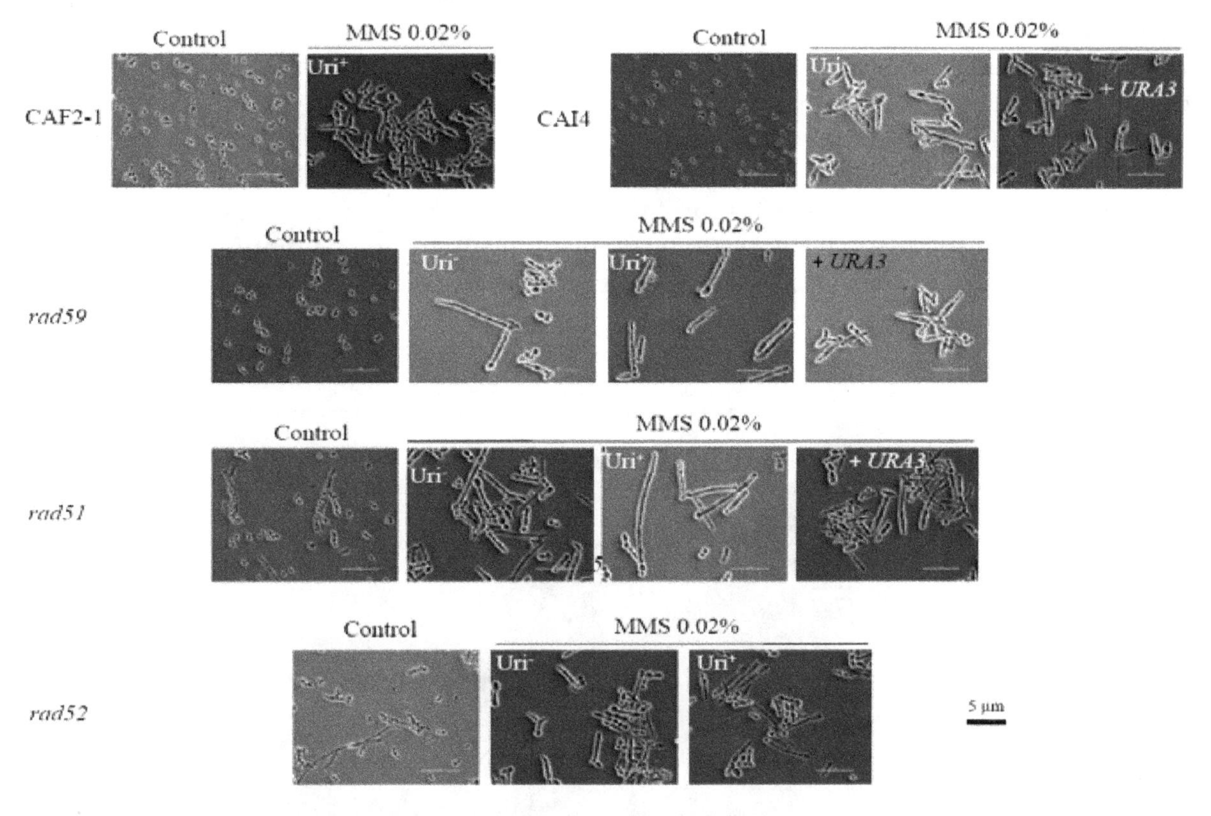

Figure 1. Methyl-methane-sulfonate (MMS) induces constitutive filamentous growth in wild type and HR mutants of *Candida albicans*. A yeast extract/peptone/dextrose (YPD) overnight culture of exponentially growing cells from the indicated strains was refreshed and adjusted to $OD_{600} = 1$. Following a further incubation for 2 h at 30 °C with shaking, one half was suspended in YPD supplemented or not (control) with 0.02% MMS. After 12 h at 30 °C, with gentle shaking, samples were photographed using a Nikon Eclipse 600 microscope with a 60× DIC objective. A CC-12 digital camera interfaced with Soft Imaging System software was used for imaging (Izasa Scientific, Alcobendas, Madrid, Spain). Each bar corresponds to 5 µm.

3.2. Generation and Repair of MMS-Induced Heat Labile Lesions on Growing C. albicans Wild Type Cells

Although MMS treatment of *S. cerevisiae* and human cells in vivo does not cause DSBs, methylated bases and AP-sites have been shown to be heat labile and converted into single strand breaks (SSBs) during the 55 °C proteinase K treatment used for preparation of PFGE plugs [6,9,11]. Furthermore, closely spaced SSBs located in opposite DNA strands may result in DSBs, and therefore chromosome fragmentation [6,9,11]. With these premises in mind, we investigated the extent to which MMS causes HDBs and HIBs in DNA from *C. albicans*.

Parental CAF2-1 strain exhibits the standard *C. albicans* karyotype. However, when medium-to-late exponentially growing cells (OD_{600} = 9; of note, when grown in YPD, *C. albicans* reaches OD_{600} up to 18) were subjected to an acute MMS treatment (0.05%, 15–30 min in YPD) and chromosome preparation for PFGE (which includes incubation with protease K) was conducted at 55 °C there was complete loss of chromosomal bands, which migrated now below Chr7 as a smear, indicating the presence of DSBs (Figure 2). Furthermore, the average size of the pool of degraded chromosomes obtained at 55 °C decreased with the incubation time in MMS, suggesting progressive chromosome fragmentation (Figure S2, lanes 1–5). By contrast, when following the acute MMS treatment (0.05%, 15–30 min in YPD) proteinase K incubation was conducted at 30 °C, strain CAF2-1 displayed the standard PFGE karyotype, and chromosome degradation was negligible (Figure S3, CAF2-1, lanes 1 and 2). Under these conditions, extension of the acute MMS treatment up to 120 min did not lead to a significant increase in chromosomal degradation (Figure S3, lanes 3–5). Some smear was likely due to the generation of a few closely opposed SSBs during manipulation of the samples, since it also was occasionally shown by preparations of untreated wild type cells incubated at 55 °C; however, the induction of small amounts of HIBs by MMS cannot be ruled out. As expected from previous reports [11,12], MMS-treated late-exponential phase cells of *S. cerevisiae* W303 subjected to the same process (i.e., preparation of samples for PFGE at 55 °C) displayed also chromosome fragmentation (Figure S1). These results indicate that almost all DSBs were generated during the incubation of *C. albicans* plugs with proteinase K at 55 °C and not in vivo, and accordingly provide an assessment of the overall number of HDBs [6]. Consistent with this interpretation, incubation of plugs at a lower temperature (50 °C) resulted in a reduced migration of the PFGE smear accompanied by the presence of vestiges of intact chromosomal bands (Figure S4, compare panels A and B, CAF2-1 lanes).

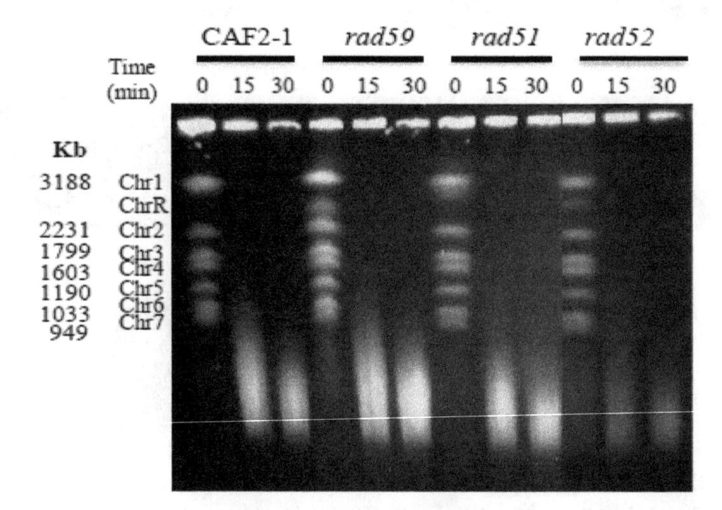

Figure 2. Determination of chromosome breaks (HDBs) following acute MMS treatments. Cells from wild type CAF2-1 and the several HR mutant derivatives were treated with 0.05% MMS for 15 and 30 min, followed by immediate DNA purification and pulse-field gel electrophoresis (PFGE). Proteinase K digestion was carried out at 55 °C to detect heat dependent breaks (HDBs). Chromosomes were visualized with ethidium bromide staining (see also Figure S2).

In *S. cerevisiae*, in vivo repair of the MMS-induced HDB prevented chromosome degradation in the subsequent 55 °C incubation with proteinase K [6]. In order to determine the time window required for repair of HDB in *C. albicans*, MMS-treated wild type CAF2-1 cells were post-incubated in either YPD or PBS lacking MMS and analyzed for chromosome restoration in time course experiments. Following a standard MMS treatment (0.05% MMS for 30 min in YPD and 55 °C incubations), no vestige of the chromosome ladder was apparent (Figure 3A, lanes 1 and 2). Traces of chromosomal bands were first seen after 2 h of recovery in YPD (Figure 3A, lane 4) and full restoration was accomplished by 4 h (Figure 3A, lane 5), with no significant changes being detected at later times (8–24 h) (Figure 3A, lanes 6 and 7). A similar conclusion was reached by quantification of Chr2 and Chr5 repair kinetics (Figure 3B). As expected from a true repair process, restoration of the chromosome ladder was paralleled by a significant reduction in the intensity of the smear (Figure 3A, compare lanes 2 to 7). It is worthy to mention that when PFGE samples were incubated at 50 °C "restitution" of the chromosomal ladder took only one hour (Figure S4A,B, lanes 1 to 3), as one could expect from the lower amount of SSBs generated at that temperature. Importantly, restitution of chromosomes was accelerated by the supply of metabolic energy (YPD) since it was significantly delayed when post-MMS incubation of CAF2-1 cells was carried out in PBS instead YPD (Figure 3A, lanes 8–12 and Figure 3B,C).

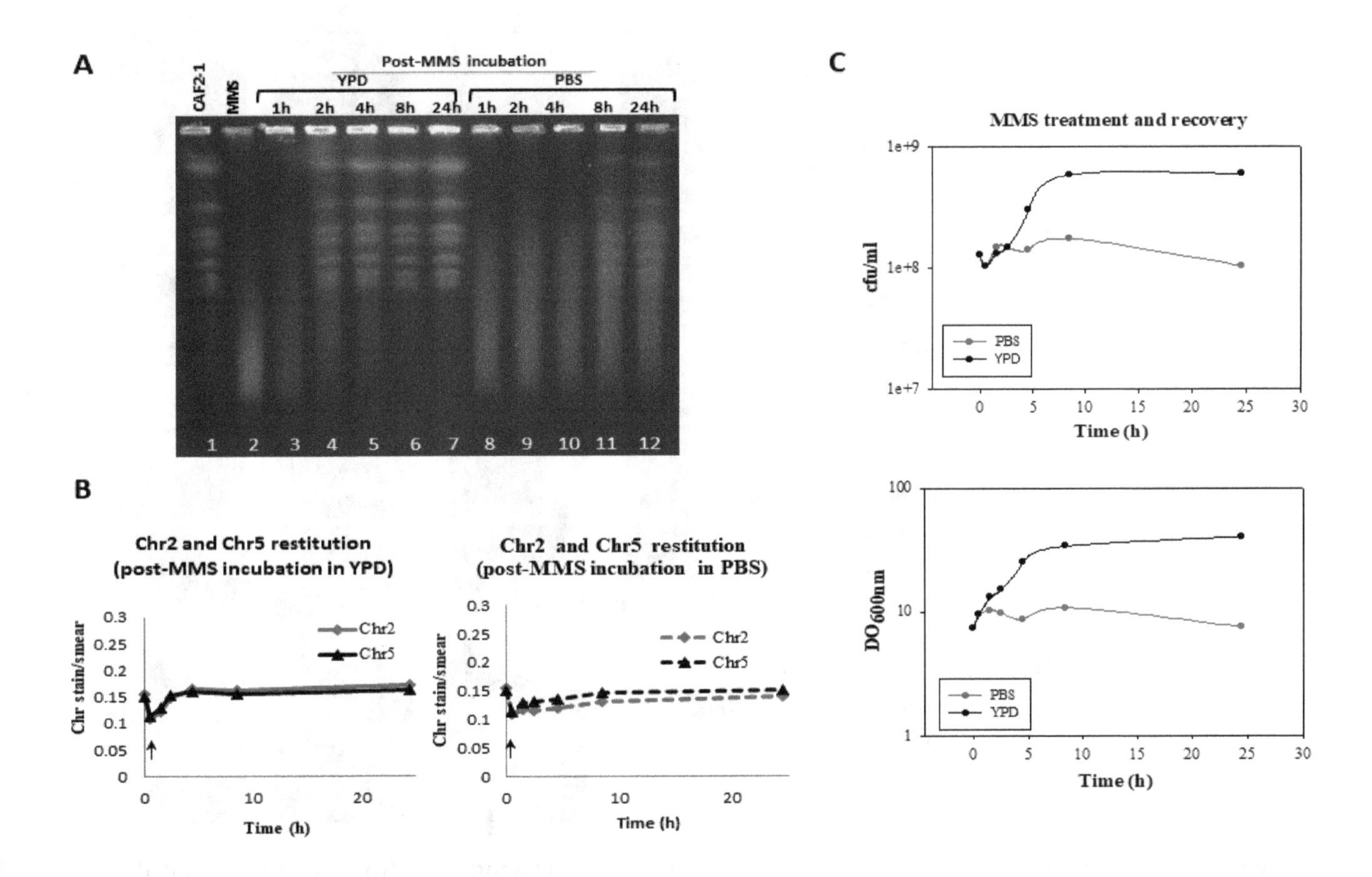

Figure 3. Induction and repair of HDBs by *C. albicans* cycling and resting wild type cells. *C. albicans* wild type (CAF2-1) was subjected to an acute MMS treatment (0.05% MMS for 30 min). Then cells were treated with 5% sodium thiosulphate, by mixing 1:1 (*v*/*v*) ratio with 10% $Na_2S_2O_3$, washed with PBS, and transferred to MMS-free YPD (cycling cells) or PBS (resting cells) at 30 °C. Samples were taken at the indicated times to determine PFGE profiles (**A**), to quantify Chr2 and Chr5 (**B**) and to calculate CFU and OD_{600} (**C**). For PFGE profiles, DNA plugs were incubated with proteinase K at 55 °C. Of note, in panels **B** and **C**, initial time (0) corresponds to samples before MMS treatment. Arrows in panel B shows the MMS-treated sample.

The CFU number of strain CAF2-1, that had been slightly reduced (4%) by the MMS-treatment, increased significantly throughout the post-MMS incubation in YPD (2–8 h) (Figure 3C). Importantly, by 4 h elongated cells carried apical and lateral buds further suggesting that they had undergone mitosis; later on, elongated branched cells steadily returned to the typical yeast form which became predominant by 24 h (Figure 4). In agreement with CFUs, the OD_{600} value steadily increased throughout the post-MMS incubation, including the first hour (Figure 3C), when few 55 °C -HDB had been repaired as indicated by the absence of chromosome "restitution" (see Figure 3A and Figure S4). It is likely that the initial increase in OD_{600} was mostly due to the formation of elongated cells (a retarded effect of the acute MMS treatment (Figure 1)), which reached maximal length by 2–4 h (Figure 4) whereas the late increase was due to cell division. Importantly, in contrast to the significant increase in cell number in YPD, neither cell division (Figure 3C) nor cell elongation (Figure 4) were detected when MMS-treated populations were transferred to PBS. Considering that under these conditions repair of HDB was significantly delayed, we conclude that overgrowth of survivors contributed significantly to the fast restitution of chromosomes during the post-MMS incubation in YPD (see Section 4, Discussion).

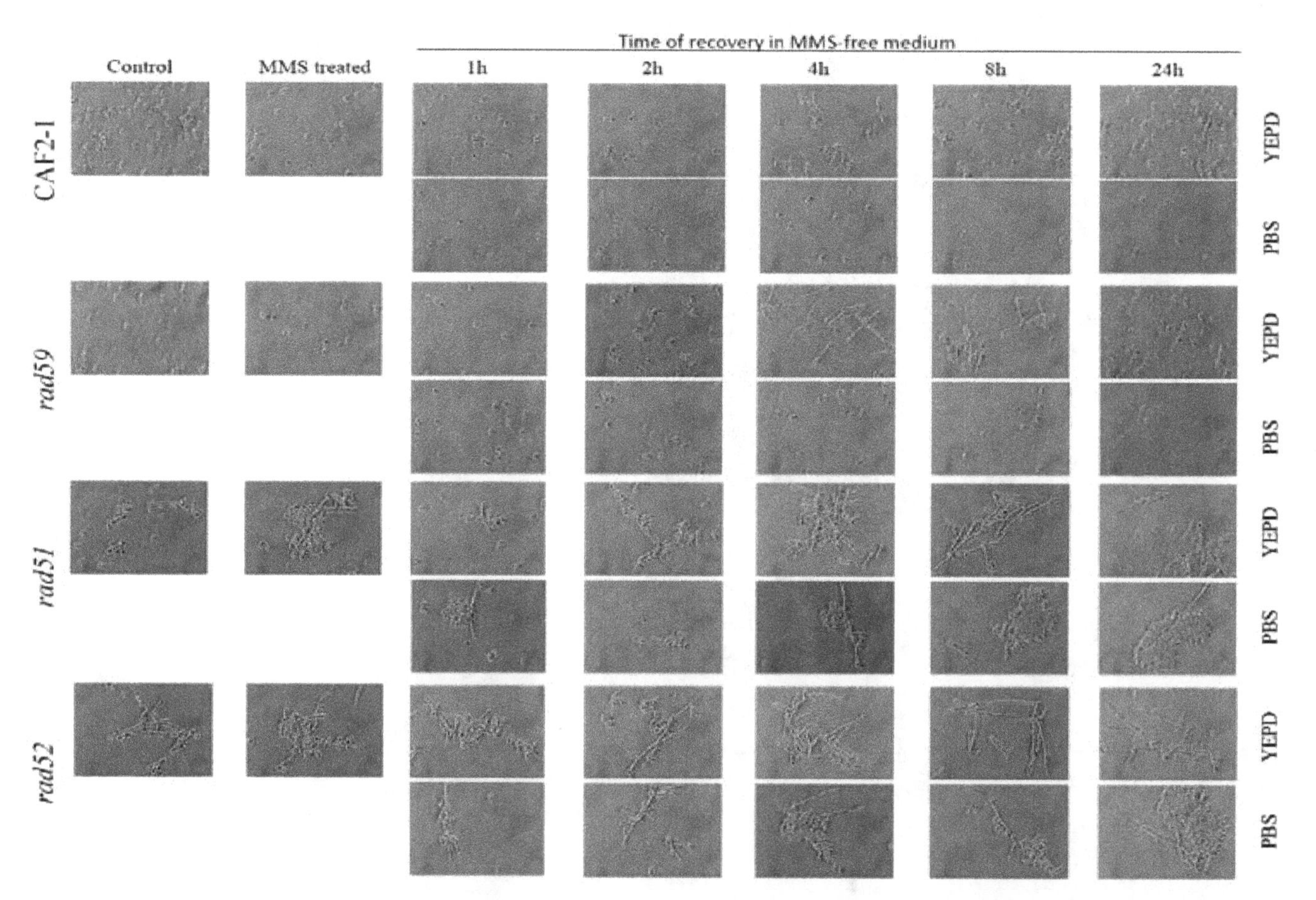

Figure 4. Cell morphology of MMS-treated exponentially growing cells from wild type and the indicated HR mutants during recovery in MMS-free YPD medium or PBS. For experimental conditions, see legend of Figure 3.

Because generation of CAI-4 from CAF2-1 resulted in increased sensitivity to low doses of MMS [35] (see above) and HR mutants (see below) were derived from CAI4, we compared chromosome recovery in MMS-treated CAF2-1 and CAI4 strains. As expected, CAI4 exhibited a small delay compared to CAF2-1 since no traces of chromosome bands could be detected in the former after 2 h of incubation in YPD. However, after 4 h both strains showed a full chromosome ladder (Figure S5).

3.3. Role of HR on the Repair of HDB

We have previously reported that *C. albicans* single mutants Ca*rad51*-ΔΔ, Ca*rad52*-ΔΔ, and Ca*rad59*-ΔΔ retained wild type PFGE karyotype profiles [21,22] (see also Figure 2 and Figure S3). When subjected to a standard acute MMS treatment and PFGE samples were incubated at 55 °C, all the mutants generated a smear similar to that shown by CAF2-1 cells (Figure 2). The average size of the pool of degraded chromosomes also decreased with the incubation time in MMS, suggesting a progressive accumulation of HDBs (Figure 2 and Figure S2). Besides, for each time-point (0–120 min in 0.05% MMS), the extent of chromosomal fragmentation was slightly higher for Ca*rad51*-ΔΔ and, to a larger extent, Ca*rad52*-ΔΔ strains compared to wild type suggesting that during the acute MMS treatment some repair in wild type is slowed down or blocked in Ca*rad51*-ΔΔ and Ca*rad52*-ΔΔ mutants (Figure S2). Finally, similarly to wild type, chromosomal fragmentation required in vitro incubations at high temperature since MMS-treated HR mutants also displayed standard chromosomal ladder when PFGE samples were incubated at 30 °C with proteinase K. Besides, prolongation of the MMS treatment up to 120 min did not result a significant increase in chromosomal degradation (Figure S3). We conclude that, as shown for wild type strain CAF2-1, treatment of HR mutants with MMS caused little or no induction of direct DSBs (HIBs).

Next, we examined the role of HR proteins in repair of HDBs during the post-MMS incubation at 30 °C in either YPD or PBS. In YPD, the Ca*rad59*-ΔΔ mutant showed a delay (4 h) in the repair of HDB/"restitution" of chromosomal bands (incubations of plugs conducted at 55 °C), compared to the wild type CAF2-1 control (2 h) (Figure S4). This delay was also detected in a parallel experiment in which samples were incubated at 50 °C but, under these conditions, repair of HDBs by Ca*rad59*-ΔΔ and wild type took only two and one hours respectively (Figure S4). However, no noticeable differences between CAI4-URA3 and Ca*rad59*-ΔΔ strains were detected regardless the post-MMS incubation took place in YPD or PBS (Figure 5A,B), suggesting that *MBP1a* homozygosis in Ca*rad59*-ΔΔ and not the absence of Rad59 itself could be responsible for the delay in chromosome restitution in YPD when compared to CAF2-1 (*MBP1a*/*MBP1b*). Importantly, as shown above for CAF2-1 (Figure 3), chromosome restitution in CAI4-URA3 and Ca*rad59*-ΔΔ was accelerated in YPD (Figures 5 and 6). In order to circumvent unwanted effects derived from the zygotic status of *MBP1*, CAI4-URA3 instead CAF2-1 was also used as a wild type control to investigate chromosome restitution in Ca*rad51*-ΔΔ and Ca*rad52*-ΔΔ mutants.

When post-incubated in YPD, MMS-treated Ca*rad51*-ΔΔ cultures exhibited a significant delay in chromosome restitution compared to CAI4-URA3 counterparts. As shown in Figures 5C and 6C, Ca*rad51*-ΔΔ exhibited a progressive increase in the average size of the smear followed by the appearance of some faint bands by 8 h and distinct clear chromosomal bands by 24 h (Figure 5C, lanes 3–7 and Figure 6C). A similar restitution pattern was shown by the Ca*rad52*-ΔΔ mutant (Figure 5D, lanes 3–7 and Figure 6D). However, when the post-MMS incubation took place in PBS chromosome restitution in both mutants occurred with kinetics similar to those of wild type and Ca*rad59*-ΔΔ (Figures 5 and 6). It is likely that replication of the few Ca*rad51*-ΔΔ and Ca*rad52*-ΔΔ survivors left by the MMS-treatment (12% and 7%, respectively) when transferred to YPD accounts by these differences.

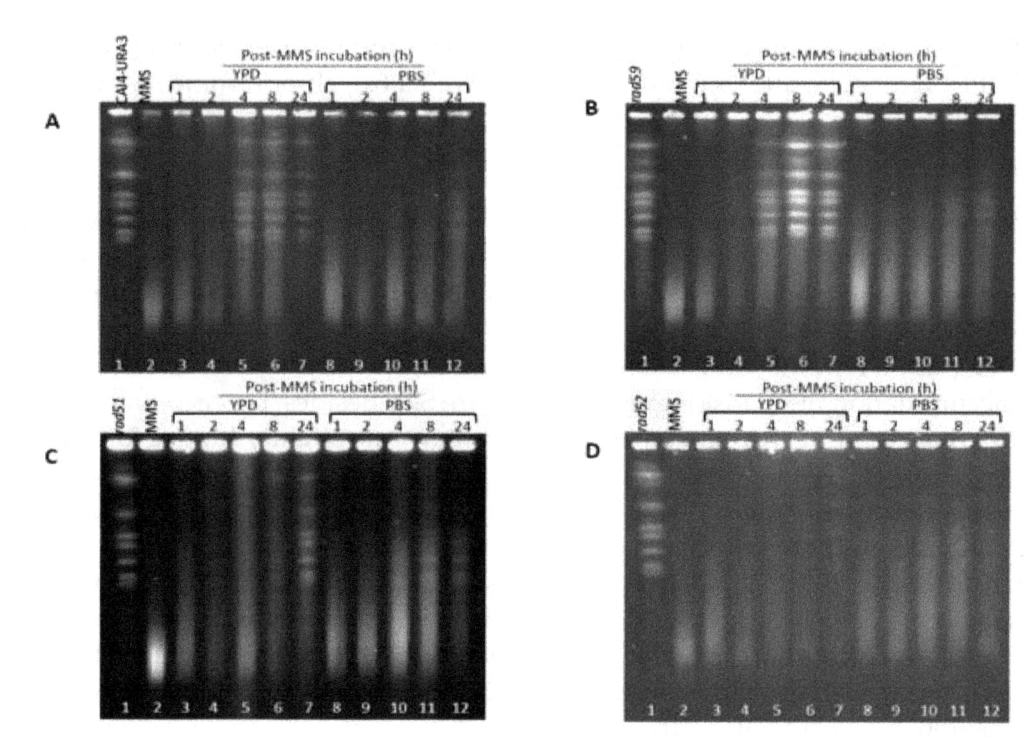

Figure 5. Induction and repair of HDBs by cycling and resting wild type and HR mutants cells of *C. albicans*. *C. albicans* wild type (CAI4-URA3) (**A**) and the indicated HR mutants (*rad59* –**B**-, *rad51* –**C**- and *rad52* –**D**-) were subjected to an acute MMS treatment (0.05% MMS for 30 min). Then cells were treated with 5% sodium thiosulphate, by mixing 1:1 (v/v) ratio with 10% $Na_2S_2O_3$, washed with PBS, and resuspended in MMS-free YPD (cycling cells) or PBS (resting cells) at 30 °C. Samples were taken at the indicated times to calculate OD_{600}, CFU (see Figure 7), and to determine PFGE profiles. For PFGE profiles, DNA plugs were incubated with proteinase K at 55 °C.

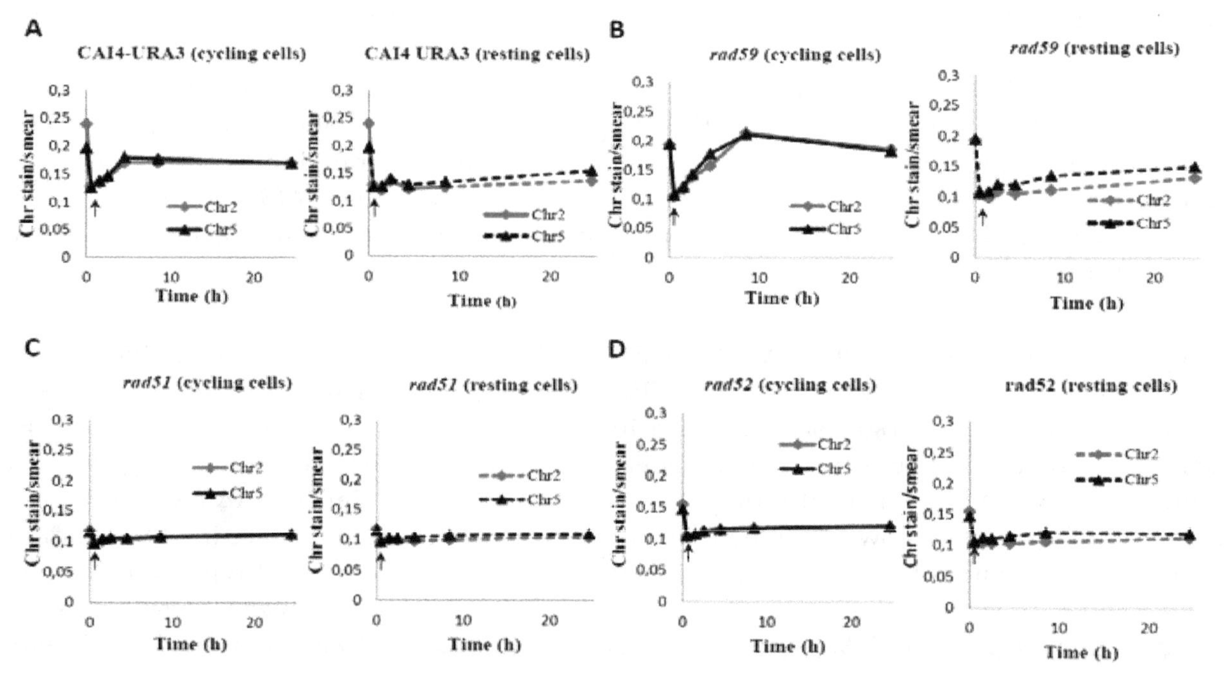

Figure 6. Quantification of Chr2 and Chr5 intensities during the MMS-treatment and the post-MMS incubation of wild type CAI4-URA3 (**A**) and HR Uri+ mutants Ca*rad59* (**B**), Ca*rad51* (**C**) and Ca*rad52* (**D**). Post-MMS incubation was carried out in YPD (cycling cells) or PBS (resting cells). PFGE gels shown in Figure 5 were quantified as described in Materials and Methods. Initial time (0) corresponds to samples before MMS treatment. Arrows within each panel shows the MMS-treated sample.

3.4. Effect of the MMS Treatment on Cell Number and Morphology during the Post-Incubation Recovery of Mutants Populations

Null Ca*rad59*-ΔΔ showed CFU and OD_{600} values similar to those of wild type CAF2-1 throughout the MMS-treatment and post-MMS incubation (Figure 7), but, as shown for CAI4 (Figure 1), mutant cells displayed a more elongated morphology due to the *MBP1a* homozygosis (Figure 4). As expected, the MMS treatment strongly reduced Ca*rad51*-ΔΔ and Ca*rad52*-ΔΔ survival to 12% and 7%, respectively, in terms of CFUs (Figure 7A). For both mutants, CFUs and OD_{600} continuously increased during the post-MMS recovery in YPD indicating replication of survivors (Figure 7B,C). It is worthy to notice the exacerbated elongation of Ca*rad51*-ΔΔ and Ca*rad52*-ΔΔ cells throughout the first 8 h of post-recovery as well as their stickiness and subsequent tendency to form aggregates (Figure 4). Importantly, as shown for wild type strains, no changes in morphology and a gentle decrease in OD_{600} were observed during the post-MMS incubation of CAF2-1, Ca*rad59*-ΔΔ, Ca*rad51*-ΔΔ and Ca*rad52*-ΔΔ cells in PBS (Figures 4 and 7E). However, in contrast to wild type and Ca*rad59*-ΔΔ, Ca*rad51*-ΔΔ and Ca*rad52*-ΔΔ CFUs increased significantly throughout post-MMS incubation in PBS despite the absence of cell replication (Figure 7D). It is likely that in the absence of nutrients damaged cells do not advance in the cell cycle and may repair HDBs by BER before being plated in YPD. Therefore, an increasing fraction of Ca*rad51*-ΔΔ and Ca*rad52*-ΔΔ cells can enter the S-phase with a repaired genome and give rise to healthy colonies on solid YPD.

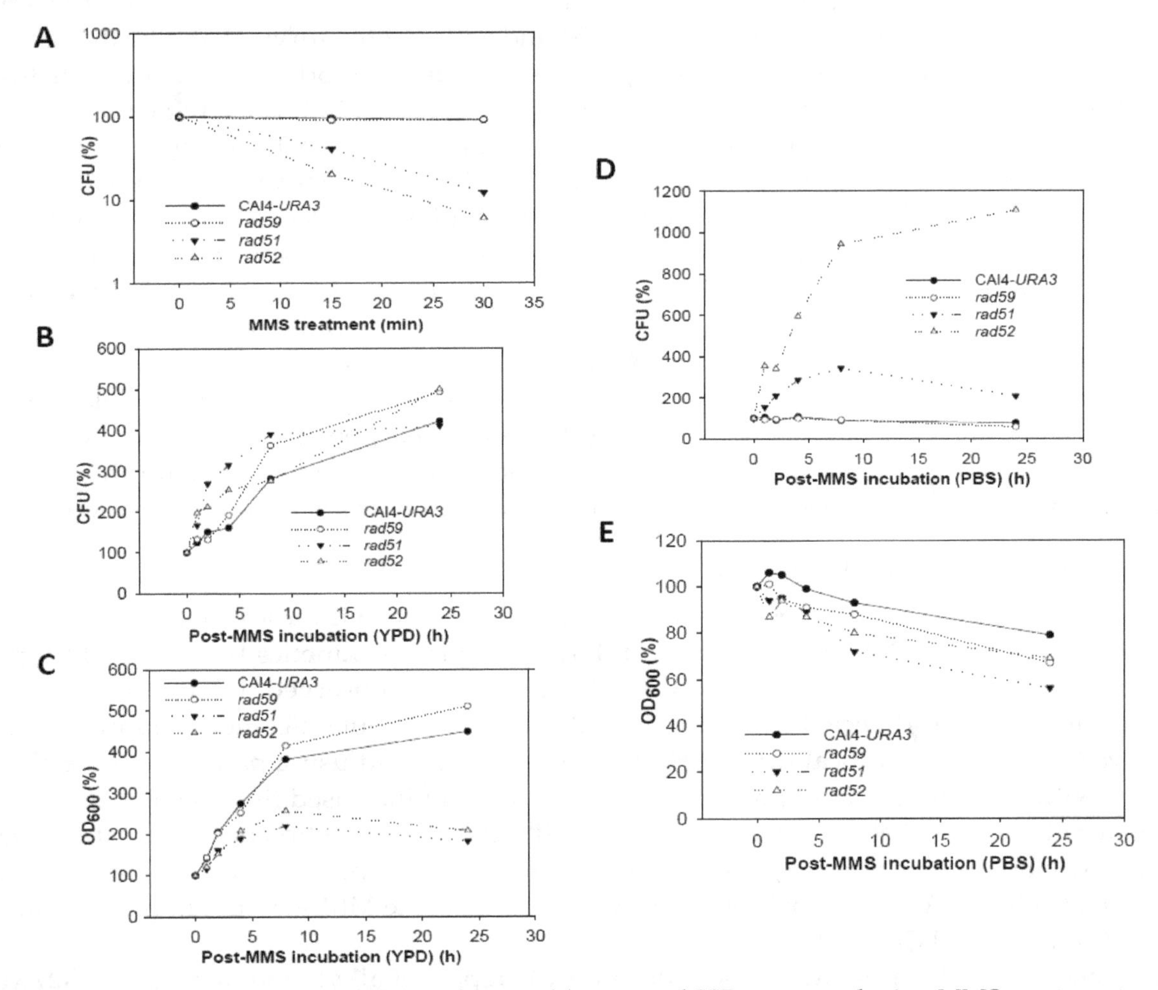

Figure 7. Changes in CFU and OD_{600} from wild type and HR mutants during MMS treatment and recovery. (**A**) Cell survival after the acute MMS-treatment. Survivability in terms of CFU was 95% for wild type (CAI4-URA3) and Ca*rad59*-ΔΔ, 12% for Ca*rad51*-ΔΔ, and 5% for Ca*rad52*-ΔΔ. (**B**,**C**) Variation of CFUs and OD_{600}, respectively, during recovery of MMS-treated cells in YPD. (**D**,**E**) Variation in CFU and OD_{600}, respectively during recovery in PBS.

4. Discussion

4.1. Role of HR in Repair of HDB by Proliferating C. Albicans Cells

In this study, we have analyzed the generation of HDBs by MMS in the genome of *C. albicans* as well as the role of HR genes Ca*RAD51*, Ca*RAD52*, and Ca*RAD59* in their repair. Importantly, we have used MMS concentrations reported not to cause direct DSBs in *S. cerevisiae* [9,12,36,37]. This was further confirmed by the absence of significant chromosome fragmentation in PFGE gels of MMS-treated *C. albicans* cells when incubation of plugs was conducted at 30 °C to detect exclusively HIB (Figure S3). As previously described [6], chromosome fragmentation was due to transformation of HDBs into SSBs at high temperature (55 °C). When close enough in opposite strands, SSBs are converted into secondary DSBs, which are manifested as genome shattering in PFGE gels. Our results are consistent with the observation that depurination of methylated DNA is a function of temperature [8]. When temperature dropped from 55 °C to 50 °C less HDBs/DSB were generated in vitro; this was reflected by the lower degree of chromosome fragmentation.

In *S. cerevisiae* repair of methylated bases by BER is constant and requires the action of the Mag1 glycosylase, which removes methylated bases leaving apurinic heat-sensitive sites throughout all the stages of the cell cycle [4,12,15]. However, Rad51 and Rad52 were also required for DNA replication in the presence of MMS [11,15,16] (see below). We found that for replicating cells of *C. albicans*, survival of an acute MMS-treatment was strongly dependent on HR proteins CaRad51 and, to a larger extent, CaRad52. In *S. cerevisiae*, the requirement of ScRad51 and ScRad52 for survival and further growth of survivors could be attributed to the role of both proteins in facilitating fork bypass through methylated DNA and subsequent repair of the resulting ssDNA gaps in G2/M [17]. It is likely that both functions are conserved in *C. albicans* where they are responsible for the rescue of >90% of damaged cells. In this scenario, the severe drop in survivability observed for cycling Ca*rad51*-ΔΔ and Ca*rad52*-ΔΔ cells is likely due to defective fork restart of damaged DNA. This S-phase block can potentially be bypassed by translesion synthesis which causes error-prone repair [38] and generates highly mutagenized survivors. In contrast to the high vulnerability of cells that transit S-phase, G1- or G2-cells present in our asynchronic cultures can potentially repair HDBs faithfully using BER and NER enzymes or other pathways before entering the S phase, thus providing additional non-mutagenized survivors [14,39]. In fact, the increase in CFU shown by Ca*rad51*-ΔΔ and Ca*rad52*-ΔΔ MMS-treated populations when incubated in YPD suggests that survivors have undergone at least one replication round (Figure 7), contributing in this way to restitution of the chromosome ladder. Consistent with this possibility, the relative amount of smear detected in PFGE gels by 8 and 24 h was significantly reduced.

4.2. Role of HR in Repair of HDB by Resting C. albicans Cells

Importantly, not only chromosomes were restituted, albeit with a lower kinetic, during the post-MMS incubation in PBS, but noticeable differences in repair kinetics between wild type and HR mutants Ca*rad51*-ΔΔ, Ca*rad52*-ΔΔ and Ca*rad59*-ΔΔ were not observed. Under these conditions, no cell proliferation was detected in wild type and Ca*rad59*-ΔΔ, but CFU increased in Ca*rad51*-ΔΔ and Ca*rad52*-ΔΔ mutants, indicating that HDBs are being repaired using pathways other than HR. Consistent with the absence of cell proliferation, OD_{600} did not increased (in fact some decrease was observed for all strains, likely as a consequence of cell lysis) and reduction of the smear throughout the post-MMS incubations was negligible compared to that observed in YPD. We conclude that in the absence of cell proliferation repair of HDBs caused by an acute MMS-treatment is independent of CaRad51, CaRad52, and CaRad59.

In *S. cerevisiae*, HR was shown to be crucial for the repair of alkylation damage by G2/M cells in the absence of Anp1 and Anp2 endonucleases, but not in its presence, suggesting that HR acts as a backup to repair lesions caused by unspecific Ntg1, Ntg2, and Ogg1 lyases [4]. Nothing is known on BER regulation in *C. albicans*. In this organism, only one *ANP* endonuclease (*APN1*) and one *NTG* lyase (*NTG1*), in addition to one *OGG1* lyase, have been reported. Furthermore, single (*anp1*, *ntg1*, *ogg1*) and double (*anp1 ntg1* and *ogg1 ntg1*) BER deletion mutants exhibited wild type sensitivity to MMS, while HR mutants were exquisitely sensitive, suggesting the existence of additional BER activities or that BER may be less important for repair of MMS damage in *C. albicans* compared to *S. cerevisiae* [7,18]. According to the present results, the single ANP endonuclease present in *C. albicans* seems to be able to repair methylation damage by resting cells whereas HR is mainly, if not exclusively, needed to allow replication of methylated DNA and further repair of the subsequent lesions when cells reach the G2/M phase. Work is in progress to determine if HR is needed for repair of methylation damage by G2/M-arrested *C. albicans* cells.

Figure S1. Determination of chromosome breaks (HDBs) following a MMS treatment. Figure S2. Kinetics of chromosome breaks (HDBs) throughout a prolonged MMS treatment. Figure S3. Determination of heat independent breaks (HIBs) following a prolonged MMS treatment. Figure S4. Effect of temperature on the induction and repair of HDBs by *C. albicans* cycling cells. *C. albicans* wild type (CAF2-1) and Carad59-mutant were subjected to an acute MMS treatment (0.05% MMS for 30 min). Figure S5. Influence of the MBP1 homozygosis on the repair of HDBs by *C. albicans* cycling cells.

Author Contributions: Conceived and designed the experiments: G.L. Performed the experiments: T.C., A.B., E.A. and B.H. Edited the manuscript: G.L. and T.C.

Acknowledgments: We thank to Andrés Aguilera (CABIMER and Universidad de Sevilla) for allowing us to use his laboratory for several assays and Wenjian Ma for critical reading of an early version of this manuscript.

References

1. Boiteux, S.; Jinks-Robertson, S. DNA repair mechanisms and the bypass of DNA damage in *Saccharomyces cerevisiae*. *Genetics* **2013**, *193*, 1025–1064. [CrossRef] [PubMed]
2. Budzowska, M.; Kanaar, R. Mechanisms of dealing with DNA damage-induced replication problems. *Cell Biochem. Biophys.* **2009**, *53*, 17–31. [CrossRef] [PubMed]
3. Heyer, W.D.; Ehmsen, K.T.; Liu, J. Regulation of homologous recombination in eukaryotes. *Annu. Rev. Genet.* **2010**, *44*, 113–139. [CrossRef] [PubMed]
4. Ma, W.; Westmoreland, J.W.; Gordenin, D.A.; Resnick, M.A. Alkylation base damage is converted into repairable double-strand breaks and complex intermediates in G2 cells lacking AP endonuclease. *PLoS Genet.* **2011**, *7*, e1002059. [CrossRef] [PubMed]
5. Sedgwick, B.; Bates, P.A.; Paik, J.; Jacobs, S.C.; Lindahl, T. Repair of alkylated DNA: Recent advances. *DNA Repair* **2007**, *6*, 429–442. [CrossRef] [PubMed]
6. Ma, W.; Resnick, M.A.; Gordenin, D.A. Apn1 and Apn2 endonucleases prevent accumulation of repair-associated DNA breaks in budding yeast as revealed by direct chromosomal analysis. *Nucleic Acids Res.* **2008**, *36*, 1836–1846. [CrossRef] [PubMed]
7. Legrand, M.; Chan, C.L.; Jauert, P.A.; Kirkpatrick, D.T. Analysis of base excision and nucleotide excision repair in *Candida albicans*. *Microbiology* **2008**, *154*, 2446–2456. [CrossRef] [PubMed]
8. Wyatt, M.D.; Pittman, D.L. Methylating agents and DNA repair responses: Methylated bases and sources of strand breaks. *Chem. Res. Toxicol.* **2006**, *19*, 1580–1594. [CrossRef] [PubMed]
9. Valenti, A.; Napoli, A.; Ferrara, M.C.; Nadal, M.; Rossi, M.; Ciaramella, M. Selective degradation of reverse gyrase and DNA fragmentation induced by alkylating agent in the archaeon *Sulfolobus solfataricus*. *Nucleic Acids Res.* **2006**, *34*, 98–108. [CrossRef] [PubMed]
10. Lindahl, T.; Andersson, A. Rate of chain breakage at apurinic sites in double-stranded deoxyribonucleic acid. *Biochemistry* **1972**, *11*, 3618–3623. [CrossRef] [PubMed]
11. Lundin, E.; North, M.; Erixon, K.; Walters, K.; Jenssen, D.; Goldman, A.S.; Helleday, T. Methyl methane sulfonate (MMS) produces heat-labile DNA damage but no detectable in vivo DNA double-strand breaks. *Nucleic Acids Res.* **2005**, *33*, 3799–3811. [CrossRef] [PubMed]
12. Ma, W.; Panduri, V.; Sterling, J.F.; Van Houten, B.; Gordenin, D.A.; Resnick, M.A. The transition of closely opposed lesions to double-strand breaks during long-patch base excision repair is prevented by the

coordinated action of DNA polymerase delta and Rad27/Fen1. *Mol. Cell. Biol.* **2009**, *29*, 1212–1221. [CrossRef] [PubMed]
13. Symington, L.S.; Rothstein, R.; Lisby, M. Mechanisms and regulation of mitotic recombination in *Saccharomyces cerevisiae*. *Genetics* **2014**, *198*, 795–835. [CrossRef] [PubMed]
14. Begley, T.J.; Rosenbach, A.S.; Ideker, T.; Samson, L.D. Damage recovery pathways in *Saccharomyces cerevisiae* revealed by genomic phenotyping and interactome mapping. *Mol. Cancer Res.* **2002**, *1*, 103–112. [PubMed]
15. Vázquez, M.V.; Rojas, V.; Tercero, J.A. sMultiple pathways cooperate to facilitate DNA replication fork progression through alkylated DNA. *DNA Repair* **2008**, *7*, 1693–1704. [CrossRef] [PubMed]
16. Tercero, J.A.; Diffley, J.F. Regulation of DNA replication fork progression through damaged DNA by the Mec1/Rad53 checkpoint. *Nature* **2001**, *412*, 553–557. [CrossRef] [PubMed]
17. González-Prieto, R.; Muñoz-Cabello, A.M.; Cabello-Lobato, M.J.; Prado, F. Rad51 replication fork recruitment is required for DNA damage tolerance. *EMBO J.* **2013**, *32*, 1307–1321. [CrossRef] [PubMed]
18. Legrand, M.; Chan, C.L.; Jauert, P.A.; Kirkpatrick, D.T. Role of DNA mismatch repair and double-strand break repair in genome stability and antifungal drug resistance in *Candida albicans*. *Eukaryot. Cell* **2007**, *6*, 2194–2205. [CrossRef] [PubMed]
19. Drablos, F.; Feyzi, E.; Aas, P.A.; Vaagbø, C.B.; Kavli, B.; Bratlie, M.S.; Peña-Diaz, J.; Otterlei, M.; Slupphaug, G.; Krokan, H.E. Alkylation damage in DNA and RNA-repair mechanisms and medical significance. *DNA Repair* **2004**, *3*, 1389–1407. [CrossRef] [PubMed]
20. Larson, K.; Sahm, J.; Shenkar, R.; Strauss, B. Methylation-induced blocks to in vitro DNA replication. *Mutat. Res.* **1985**, *150*, 77–84. [CrossRef]
21. García-Prieto, F.; Gómez-Raja, J.; Andaluz, E.; Calderone, R.; Larriba, G. Role of the homologous recombination genes *RAD51* and *RAD59* in the resistance of *Candida albicans* to UV light, radiomimetic and anti-tumor compounds and oxidizing agents. *Fungal Genet. Biol.* **2010**, *47*, 433–445. [CrossRef] [PubMed]
22. Bellido, A.; Andaluz, E.; Gómez-Raja, J.; Álvarez-Barrientos, A.; Larriba, G. Genetic interactions among homologous recombination mutants in *Candida albicans*. *Fungal Genet. Biol.* **2015**, *74*, 10–20. [CrossRef] [PubMed]
23. Andaluz, E.; Ciudad, T.; Gómez-Raja, J.; Calderone, R.; Larriba, G. Rad52 depletion in *Candida albicans* triggers both the DNA-damage checkpoint and filamentation accompanied by but independent of expression of hypha-specific genes. *Mol. Microbiol.* **2006**, *59*, 1452–1472. [CrossRef] [PubMed]
24. Bärtsch, S.; Kang, L.E.; Symington, L.S. *RAD51* is required for the repair of plasmid double-stranded gaps from either plasmid or chromosomal templates. *Mol. Cell. Biol.* **2000**, *20*, 1194–1205. [CrossRef] [PubMed]
25. Guthrie, C.; Fink, G.R. *Guide to Yeast Genetics and Molecular Biology*; Academic Press: San Diego, CA, USA, 1991.
26. Gillum, A.M.; Tsay, E.Y.; Kirsch, D.R. Isolation of the *Candida albicans* gene for orotidine-5′-phosphate decarboxylase by complementation of S. cerevisiae ura3 and E. coli pyrF mutations. *Mol. Gen. Genet.* **1984**, *198*, 179–182. [CrossRef] [PubMed]
27. Fonzi, W.A.; Irwin, M.Y. Isogenic strain construction and gene mapping in *Candida albicans*. *Genetics* **1993**, *134*, 717–728.
28. Ciudad, T.; Andaluz, E.; Steinberg-Neifach, O.; Lue, N.F.; Gow, N.A.; Calderone, R.A.; Larriba, G. Homologous recombination in *Candida albicans*: Role of CaRad52p in DNA repair, integration of linear DNA fragments and telomere length. *Mol. Microbiol.* **2004**, *53*, 1177–1194. [CrossRef] [PubMed]
29. Andaluz, E.; Gómez-Raja, J.; Hermosa, B.; Ciudad, T.; Rustchenko, E.; Calderone, R.; Larriba, G. Loss and fragmentation of chromosome 5 are major events linked to the adaptation of rad52-DeltaDelta strains of *Candida albicans* to sorbose. *Fungal Genet. Biol.* **2007**, *44*, 789–798. [CrossRef] [PubMed]
30. Andaluz, E.; Bellido, A.; Gómez-Raja, J.; Selmecki, A.; Bouchonville, K.; Calderone, R.; Berman, J.; Larriba, G. Rad52 function prevents chromosome loss and truncation in *Candida albicans*. *Mol. Microbiol.* **2011**, *79*, 1462–1482. [CrossRef] [PubMed]
31. Walther, A.; Wendland, J. An improved transformation protocol for the human fungal pathogen *Candida albicans*. *Curr. Genet.* **2003**, *42*, 339–343. [CrossRef] [PubMed]
32. Noble, S.M.; Johnson, A.D. Strains and strategies for large-scale gene deletion studies of the diploid human fungal pathogen *Candida albicans*. *Eukaryot. Cell* **2005**, *4*, 298–309. [CrossRef] [PubMed]

33. Shi, Q.M.; Wang, Y.M.; Zheng, X.D.; Lee, R.T.; Wang, Y. Critical role of DNA checkpoints in mediating genotoxic-stress-induced filamentous growth in *Candida albicans*. *Mol. Biol. Cell* **2007**, *18*, 815–826. [CrossRef] [PubMed]
34. Legrand, M.; Chan, C.L.; Jauert, P.A.; Kirkpatrick, D.T. The contribution of the S-phase checkpoint genes MEC1 and SGS1 to genome stability maintenance in *Candida albicans*. *Fungal Genet. Biol.* **2011**, *48*, 823–830. [CrossRef] [PubMed]
35. Ciudad, T.; Hickman, M.; Bellido, A.; Berman, J.; Larriba, G. Phenotypic consequences of a spontaneous loss of heterozygosity in a common laboratory strain of *Candida albicans*. *Genetics* **2016**, *203*, 1161–1176. [CrossRef] [PubMed]
36. Covo, S.; Westmoreland, J.W.; Gordenin, D.A.; Resnick, M.A. Cohesin is limiting for the suppression of DNA damage-induced recombination between homologous chromosomes. *PLoS Genet.* **2010**, *6*, e1001006. [CrossRef] [PubMed]
37. Redon, C.; Pilch, D.R.; Rogakou, E.P.; Orr, A.H.; Lowndes, N.F.; Bonner, W.M. Yeast histone 2A serine 129 is essential for the efficient repair of checkpoint-blind DNA damage. *EMBO Rep.* **2003**, *4*, 678–684. [CrossRef] [PubMed]
38. Auerbach, P.A.; Demple, B. Roles of Rev1, Pol zeta, Pol32 and Pol eta in the bypass of chromosomal abasic sites in *Saccharomyces cerevisiae*. *Mutagenesis* **2010**, *25*, 63–69. [CrossRef] [PubMed]
39. Fu, D.; Calvo, J.A.; Samson, L.D. Balancing repair and tolerance of DNA damage caused by alkylating agents. *Nat. Rev. Cancer* **2013**, *12*, 104–120. [CrossRef] [PubMed]

7

A Moldy Application of MALDI: MALDI-ToF Mass Spectrometry for Fungal Identification

Robin Patel [1,2]

[1] Department of Laboratory Medicine and Pathology, Division of Clinical Microbiology, Mayo Clinic, Rochester, MN 55905, USA; patel.robin@mayo.edu

[2] Department of Medicine, Mayo Clinic, Division of Infectious Diseases, Rochester, MN 55905, USA

Abstract: As a result of its being inexpensive, easy to perform, fast and accurate, matrix-assisted laser desorption ionization time-of-flight mass spectrometry (MALDI-ToF MS) is quickly becoming the standard means of bacterial identification from cultures in clinical microbiology laboratories. Its adoption for routine identification of yeasts and even dimorphic and filamentous fungi in cultures, while slower, is now being realized, with many of the same benefits as have been recognized on the bacterial side. In this review, the use of MALDI-ToF MS for identification of yeasts, and dimorphic and filamentous fungi grown in culture will be reviewed, with strengths and limitations addressed.

Keywords: MALDI-ToF MS; yeast; fungus

---

1. Background and Introduction

The concept of using mass spectrometry for bacterial identification was suggested by Catherine Fenselau and John Anhalt in 1975 [1], but at the time, intact proteins were not analyzable due to fragmentation during the mass spectrometry (MS) process, with mass spectrometric analysis of intact proteins only becoming possible a decade later. In 1985, Koichi Tanaka described a "soft desorption ionization" technique allowing mass spectrometry of biological macromolecules achieved using ultrafine metal powder and glycerol; for his discovery, he was awarded the Nobel Prize in Chemistry [2]. About the same time, Franz Hillenkamp and Michael Karas reported a soft desorption ionization using an organic compound matrix [3]; it was their approach for which the designation "matrix-assisted laser desorption ionization" or MALDI, was coined and on which subsequent clinical microbiology applications were based. Tied with time-of-flight or ToF analysis, this advancement made it possible to perform mass spectrometry on intact bacterial and, ultimately, fungal cells. However, it took until advances in informatics allowed the connection of microbial MALDI-ToF MS databases to automated computer-based analytics for MALDI-ToF MS to ultimately become usable for routine identification of bacteria and, eventually, fungi in clinical laboratories. In due course, these developments led to the commercialization and, ultimately, regulatory approval of MALDI-ToF MS systems for clinical microbiology laboratories. Although initial applications of MALDI-ToF MS for rapid, inexpensive identification of microorganisms in culture focused on bacteria, it was quickly realized that this new technology could be equally applied to yeasts and, with some caveats, dimorphic and filamentous fungi.

MALDI designates matrix that assists in desorption and ionization of highly abundant bacterial or fungal proteins through laser energy [4]. Like bacteria, fungi may be tested either by "direct transfer" to a MALDI-ToF MS target plate, with or without the addition of an on-plate formic acid treatment (to lyse cells, also referred to as "on-plate extraction" and "extended direct transfer"), or following a more formal (and time-consuming) off-plate protein extraction step. The former is most commonly used for yeasts and the latter for filamentous fungi. For direct transfer testing, whole cells from colonies are simply moved to the target plate using a loop, plastic or wooden stick, or pipette tip, to a "spot" on a MALDI-ToF MS target plate (a reusable or disposable plate with multiple test spots) (Figure 1). For on-plate formic acid treatment, a formic acid solution is incorporated, either by adding the formic acid solution prior to colony transfer or by overlaying the transferred colony with formic acid solution, followed by drying. Then, the microbial mass, either alone or after formic acid treatment, is overlain with matrix; following drying of the matrix, the target plate is moved into a mass spectrometer (Figure 2). After this point, the rest is automated vis-à-vis the described clinical microbiology applications. The matrix (e.g., α-cyano-4-hydroxycinnamic acid dissolved in 50% acetonitrile and 2.5% trifluoroacetic acid), which is used for bacteria and fungi alike, isolates microbial molecules from one another, protecting them from breaking up and allowing their desorption by laser energy; a majority of the energy is absorbed by the matrix, changing it to an ionized state. As a result of random impacts occurring in the gas phase, charge is moved from matrix to microbial molecules; ionized microbial molecules are then accelerated through a positively charged electrostatic field into a ToF, tube, which is under vacuum. In the tube, ions travel to an ion detector, with smaller analytes migrating fastest, followed by increasingly larger analytes; a mass spectrum is thereby generated, signifying the quantity of ions of a specified mass hitting the detector over time. The resultant mass spectrum represents the most abundant proteins, mainly ribosomal proteins, though with this application the specific proteins generating the mass spectrum are not separately identified. The overall mass spectrum is used as a signature profile of individual fungi (or bacteria), with peaks specific to groups, complexes, genera and/or species, depending on relatedness of the test organisms to other closely related ones. The mass spectrum of an individual isolate is compared to a database or library of reference spectra, producing a list of the most closely interrelated fungi (or bacteria) with numeric rankings (assessed as percentages or scores, depending on the system). As with any identification system, it is critical to have a comprehensive and well-curated database; this has been a notable limitation of historical fungal databases, especially those for dimorphic and filamentous fungi. Depending on relatedness of the test organism to the top match (and allowing for the next best matches), the organism is then identified at the group, complex, genus, species or subspecies-level. Usually, organisms are either appropriately identified or yield a low match, indicating that identification has not been attained; the latter suggests that the species being tested is not in the database, or that there is heterogeneity in individual species or genera, but may occur due to an insufficient amount of biomass being tested or poor technical preparation (in which case repeat testing, or testing after incubation for further growth, may be helpful). A Clinical and Laboratory Standards Institute guideline on MALDI-ToF MS was published in 2017 [5].

In the past, fungal identification has been a perplexing, multi-step process, tailored by organism-type. Clinical microbiology students were pedantically educated to interpret colony and microscopic morphology of fungi on solid media as a preface to choosing appropriate further testing, such as biochemical tests or sequencing. With MALDI-ToF MS, cultured yeasts may be correctly identified in minutes without a priori knowledge of organism-type; since it doesn't matter whether a bacterium or yeast is being tested, the decision-making procedure characteristically surrounding differentiation of bacteria or yeasts growing on solid media prior to selecting further testing is obviated. Filamentous fungi can also be identified, though usually their processing prior to MALDI-ToF MS analysis typically takes longer than yeasts. MALDI-ToF MS is enabling implementation of total laboratory automation in clinical microbiology laboratories, allowing automated specimen processing, plating, incubation, plate reading using digital imaging, and spotting to MALDI-ToF MS plates. Early growth detection by digital imaging, paired with MALDI-ToF MS may result in earlier detection of fungi than conventional techniques [6]. MALDI-ToF MS is also changing the educational needs of clinical microbiology laboratory management staff, medical technologists, as well as medical students, fellows and residents. For the curriculum of those who won't practice laboratory medicine, conventional biochemical-based identification is being deemphasized.

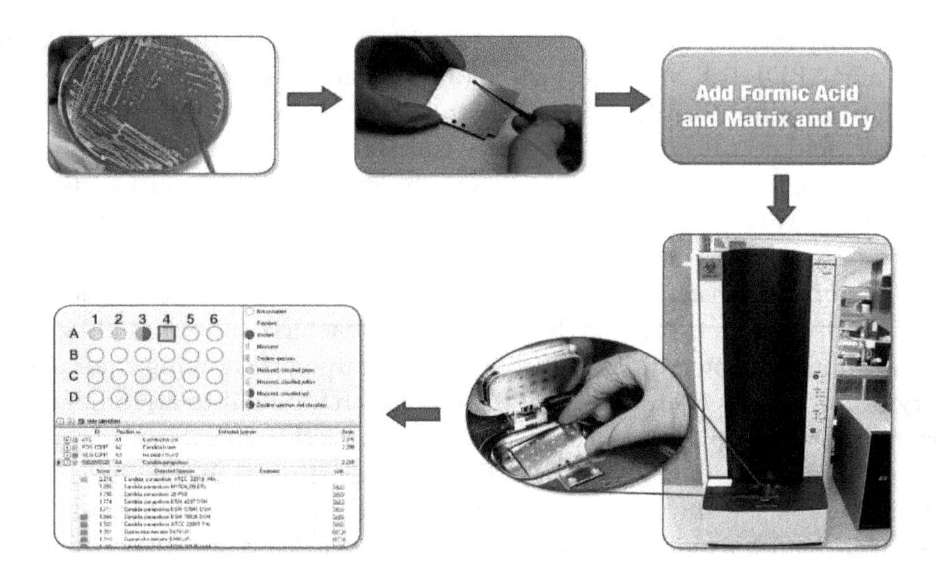

Figure 1. Process of matrix-assisted laser desorption ionization time-of-flight mass spectrometry (MALDI-ToF MS) for yeast identification [7,8]. A colony is picked from a culture plate to a spot on a MALDI-ToF MS target plate (a disposable or reusable plate with a number of spots, each of which may be used to test different colonies). For yeast applications, cells are typically treated with formic acid on the target plate, followed by drying. The spot is overlain with 1–2 µL of matrix and dried. The plate is placed in the ionization chamber of the mass spectrometer (Figure 2). A mass spectrum is produced and compared against a library of mass spectra by the software, resulting in identification of the yeast (*Candida parapsilosis* in position A4 in the example). Used with permission of the Mayo Foundation for Medical Education and Research. All rights reserved.

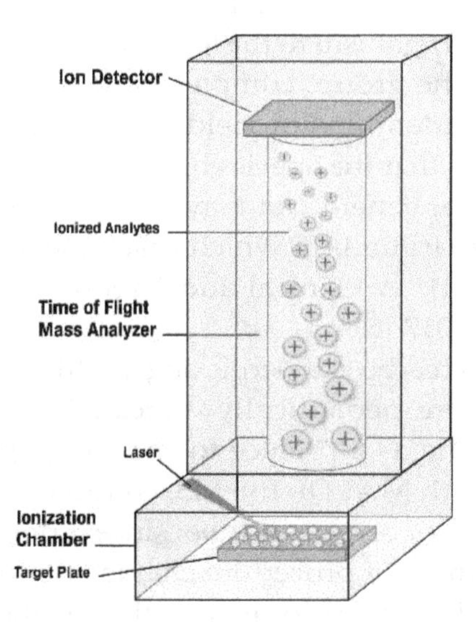

Figure 2. Mass spectrometer used for MALDI-ToF MS [7,8]. The MALDI-ToF MS plate is placed into the chamber of the instrument. Each spot to be analyzed is shot by a laser, resulting in desorption and ionization of bacterial or fungal and matrix molecules from the target plate. The cloud of ionized molecules is accelerated into the time-of-flight mass analyzer, toward a detector. Lighter molecules travel quicker, followed by progressively heavier ones. A mass spectrum is produced; it denotes the number of ions hitting the detector with time. Separation is by mass-to-charge ratio; because charge is typically single for this application, separation is by molecular weight. Used with permission of the Mayo Foundation for Medical Education and Research. All rights reserved.

MALDI-ToF MS instruments used in clinical microbiology laboratories are typically specific for clinical microbiology applications, though other testing may be performed on them and alternative instruments may be used for clinical microbiology purposes. For purposes of efficiency and biosafety, however, such instruments are typically located in (or near) clinical microbiology laboratories themselves, rather than in a centralized mass spectrometry core facility. Commercial MALDI-ToF MS systems for clinical microbiology laboratories are available from bioMérieux, Inc. (Durham, NC, USA) and Bruker Daltonics, Inc. (Billerica, MA, USA). (Other systems, such as Andromas (Paris, France), Clin-TOF (China) Quan TOF (China), Autof ms 1000 (China), and Microtyper MS (China) will not be discussed). In 2010, bioMérieux acquired a microbial database called Spectral Archiving and Microbial Identification System (SARAMIS) marketed by AnagnosTec (Zossen, Germany) and used with Shimadzu's AXIMA Assurance mass spectrometer (Shimadzu, Columbia, MD, USA), and transformed the label to VITEK MS research use only (RUO); bioMérieux then established a new database, software, and algorithms called VITEK MS IVD. bioMérieux's FDA-approved/cleared platform, available since 2013, is named Vitek MS. A RUO version, VITEK MS Plus is available, incorporating the VITEK MS and SARAMIS databases. Bruker began developing a system for identification of cultured microorganisms circa 2005, the so-called Biotyper system, obtaining FDA-approval/clearance shortly after bioMérieux in 2013, with a system referred to as the MALDI Biotyper CA System. Like bioMérieux, Bruker offers a more extensive RUO database. Bruker also has a specific RUO Filamentous Fungi Library. Bruker's mass spectrometer used for clinical microbiology testing is a desktop system, whereas bioMérieux's is a larger instrument that sits on the floor.

Yeasts and filamentous fungi claimed by the FDA cleared/approved versions of at least one commercial MALDI-ToF MS system are shown in Tables 1 and 2, respectively. Both companies' systems claim an extensive portfolio of yeasts commonly encountered in clinical practice, though there are some nomenclature differences and inclusion differences. In some cases, one system may use the teleomorph with the other using the anamorph name; for example, *Cyberlindnera jadinii* is officially claimed by the MALDI Biotyper CA system, whereas *Candida utilis* is officially claimed by the Vitek MS system. The MALDI Biotyper CA system claims *Trichosporon mucoides* group (which per the company's package insert includes *Trichosporon mucoides* and *Trichosporon dermatis*), whereas the Vitek MS system claims *T. mucoides*. Reporting of *Cryptococcus neoformans* and *Cryptococcus gattii* varies between the two systems (Table 1). Species uniquely claimed by the MALDI Biotyper CA system include *Candida boidinii*, *Candida duobushaemulonii*, *Candida metapsilosis*, *Candida orthopsilosis*, *Candida pararugosa*, *Candida valida*, and *Geotrichum candidum*, with *Candida rugosa* being uniquely claimed by the Vitek MS system. The Bruker and bioMérieux systems are different not just in databases, but also in database matching and relatedness reporting strategies. In most comparative studies, performance of the two has been similar, though not identical, assuming that the specific species being studied are represented in both databases [9–11]. Since there have been iterative and rapid growths and curations in both companies' databases over time, in reviewing the published literature, it is important to note not just the company whose system was studied, but also the specimen preparation method and the library version applied, alongside the cutoffs used for identification at the species-, genus-, group- or complex-level. The organism testing sets studied (i.e., supplemented with unusual organisms or not), and reference (i.e., comparator) identification procedures should also be considered. With both systems, users have the option to develop their own database(s), which can enhance performance; this, however, makes generalization to other users challenging. In addition, user-developed databases must be validated to meet regulatory requirements for clinical use. User-developed databases can be used in conjunction with commercial databases; alternatively or additionally, multiple databases from the same company can be used together. Success rates may be compromised if spectra in a particular library were not created from isolates prepared in the same way as they are being tested (e.g., on-plate formic acid preparation versus off-plate protein extraction) [12].

Table 1. Reportable yeasts for the FDA-approved/cleared Vitek MS and MALDI Biotyper CA systems as of October 2018 [13]. Those entries marked with "[V]" are FDA-approved/cleared for the Vitek MS system only and those marked with "[B]" are FDA-approved/cleared for the MALDI Biotyper CA system only. Those with marked with neither a "[V]" nor a "[B]" are FDA-approved/cleared on both systems.

Candida albicans	*Candida krusei*	*Candida tropicalis*	*Kodamaea/Pichia ohmeri* ***
Candida boidinii [B]	*Candida lambica*	*Candida utilis/Cyberlindnera jadinii* *	*Malassezia furfur*
Candida dubliniensis	*Candida lipolytica*	*Candida valida* [B]	*Malassezia pachydermatis*
Candida duobushaemulonii [B]	*Candida lusitaniae*	*Candida zeylanoides*	*Rhodotorula mucilaginosa*
Candida famata	*Candida metapsilosis* [B]	*Cryptococcus gattii* [B]	*Saccharomyces cerevisiae*
Candida glabrata	*Candida norvegensis*	*Cryptococcus neoformans* [V]	*Trichosporon asahii*
Candida guilliermondii	*Candida orthopsilosis* [B]	*Cryptococcus neoformans* var *grubii* [B]	*Trichosporon inkin*
Candida haemulonii	*Candida parapsilosis*	*Cryptococcus neoformans* var *neoformans* [B]	*Trichosporon mucoides* [V]
Candida inconspicua	*Candida pararugosa* [B]	*Geotrichum candidum* [B]	
Candida intermedia	*Candida pelliculosa*	*Geotrichum capitatum/Saprochaete capitate* **	*Trichosporon mucoides* group [B]
Candida kefyr	*Candida rugosa* [V]	*Kloeckera apiculata*	

* *Cyberlindnera jadinii* (teleomorph) is approved/cleared on the MALDI Biotyper CA system, whereas *Candida utilis* (anamorph) is approved/cleared on the Vitek MS system. ** *Geotrichum capitatum* is approved/cleared on the MALDI Biotyper CA system, whereas *Saprochaete capitate* is approved/cleared on the Vitek MS system **; *** *Kodamaea ohmeri* is approved/cleared on the Vitek MS system whereas *Pichia ohmeri* is approved/cleared on the MALDI Biotyper CA system.

Table 2. Reportable filamentous and dimorphic fungi for the FDA-approved/cleared Vitek MS system as of October 2018 [13].

Acremonium sclerotigenum	*Blastomyces dermatitidis*	*Histoplasma capsulatum*	*Rhizopus arrhizus* complex
Alternaria alternata	*Cladophialophora bantiana*	*Lecythophora hoffmannii*	*Rhizopus microsporus* complex
Aspergillus brasiliensis	*Coccidioides immitis/posadasii*	*Lichtheimia corymbifera*	*Sarocladium kiliense*
Aspergillus calidoustus	*Curvularia hawaiiensis*	*Microsporum audouinii*	*Scedosporium apiospermum*
Aspergillus flavus/oryzae	*Curvularia spicifera*	*Microsporum canis*	*Scedosporium prolificans*
Aspergillus fumigatus	*Epidermophyton floccosum*	*Microsporum gypseum*	*Sporothrix schenckii* complex
Aspergillus lentulus	*Exophiala dermatitidis*	*Mucor racemosus* complex	*Trichophyton interdigitale*
Aspergillus nidulans	*Exophiala xenobiotica*	*Paecilomyces variotii* complex	*Trichophyton rubrum*
Aspergillus niger complex	*Exserohilum rostratum*	*Penicillium chrysogenum*	*Trichophyton tonsurans*
Aspergillus sydowii	*Fusarium oxysporum* complex	*Pseudallescheria boydii*	*Trichophyton verrucosum*
Aspergillus terreus complex	*Fusarium proliferatum*	*Purpureocillium lilacinum*	*Trichophyton violaceum*
Aspergillus versicolor	*Fusarium solani* complex	*Rasamsonia argillacea* complex	

MALDI-ToF MS turnaround time is five or fewer minutes per isolate for direct target plate methods; the turnaround time is longer with off-plate protein extraction. Compared to standard methods, yeast and bacterial identification is achieved an average of 1.45 days faster [14], and since only a slight amount of organism is required, testing can be completed on small amounts of growth on primary culture plates without subculture. MALDI-ToF MS has a low reagent cost [14], being less expensive than biochemical- or sequencing-based identification. One study showed that a projected 87% of bacterial and yeast isolates may be identified on the first day using MALDI-ToF MS (versus 9% historically) [14]. Using MALDI-ToF MS, DNA sequencing expenses can be avoided, waste disposal reduced [15,16], and quality control and technologist labor/training for retired tests/replaced tests avoided.

2. Yeasts, with a Focus on *Candida* and *Cryptococcus* Species

MALDI-ToF MS has rapidly become a standard method for yeast identification, out-performing some historical phenotypic systems, and differentiating *Candida albicans* from *Candida dubliniensis*; *C. pararugosa* from *Candida rugosa*; *Candida krusei*, *Candida norvegensis*, and *Candida inconspicua* from one another; *C. orthopsilosis*, *C. metapsilosis* and *C. parapsilosis* from one another [17]; and *C. gattii* from *C. neoformans*, dependent on spectral database representation [18,19]. MALDI-ToF MS may outdo other identification systems for esoteric species, such as *C. famata* and *C. auris*, for example [12,17]. Although older studies used off-plate extraction for yeasts, on-plate extraction with formic acid is now favored for its simplicity; on-plate formic acid preparation yields higher identification rates than does direct transfer alone [20,21].

Dhiman et al. evaluated the Bruker system for identification of 138 common and 103 unusual yeast isolates, reporting 96% and 85% accurate species-level identification, respectively [22]. Westblade et al. assessed the Vitek MS v2.0 system for identification of 852 yeast isolates, including *Candida* species, *C. neoformans*, and other clinically relevant yeasts, using on-plate formic acid preparation, in a multicenter study, reporting 97% and 86% identification to the genus- and species-level, respectively [20]. Won et al. assessed the accuracy of yeast bloodstream isolate identification using the Vitek MS system; correct identification, misidentification and no identification were achieved in 96%, 1% and 3% of cases, respectively [23]. Mancini et al. compared the Bruker and Vitek MS systems for identification of yeasts; correct species-level identifications were comparable using the commercial databases (90% and 84%, respectively), with 100% identified using the Bruker system and a user-developed database [24]. More misidentifications were reported with the Vitek MS system compared to the Bruker system. Rosenvinge et al. studied the Bruker system with 200 yeast isolates, reporting 88% species-level identification (species cutoff of $\geq$1.700) using on-plate formic acid testing [25]. Lacroix et al. demonstrated that the Bruker system with protein extraction and using the manufacturer's species-level cutoff identified 97% of 1383 regularly isolated *Candida* isolates [26]. Pence et al. compared the VITEK MS (IVD Knowledgebase v.2.0) and Biotyper (software v3.1) for identification of 117 yeast isolates, showing correct identification of 95% and 83% of isolates, respectively, using on-plate formic acid testing [27]. Jamal et al. evaluated the Bruker and VITEK MS systems for identification of 188 clinically significant yeast isolates [28], reporting accurate identification of 93% of isolates with both. Three isolates were not identified by VITEK MS, while nine *C. orthopsilosis* were incorrectly identified as *C. parapsilosis*, which was not unexpected since *C. orthopsilosis* was not included in the database studied. Eleven isolates were not identified or misidentified by the Bruker system and although another 14 were identified correctly, their score was <1.700. Hamprecht et al. compared the VITEK MS (V2.0 knowledge base) and the Biotyper (v3.0 software, v3.0.10.0 database, using a species-level cutoff $\geq$2.000) systems for identification of 210 yeasts using on-plate formic acid testing, showing identification of 96% and 91%, respectively [29]. De Carolis et al. made an in-house database using spectra from 156 reference and clinical yeast isolates generated with a sample preparation procedure using suspension of a colony in 10% formic acid, and using 1 μL of the lysate for analysis [30]. Using their library and processing method, and the Bruker system (software v3.0) with a species-level cutoff of ≥2.000, they identified 96% of 4232 routinely isolated yeasts. Fatania et al. evaluated the Bruker system with 200 clinically significant yeasts, representing 19 species and five genera, showing agreement between MALDI-ToF MS and conventional methods for 91% [31]. Wang et al. evaluated 2683 yeast isolates comprising 41 species from the National China Hospital Invasive Fungal Surveillance Net program, reporting that the Bruker Biotyper MS system exhibited greater accuracy than the Vitek MS system for all isolates (99% and 95%, respectively) and for *Candida* and related species (99% and 96%, respectively) [32]. Fraser et al. evaluated MALDI-ToF MS using the Bruker system for identification of 6343 clinical isolates of yeasts representing 71 species using a user-developed simplified rapid extraction method, reporting correct identification of 94% of isolates, with a further 6% identified after full extraction [33]. Lee et al. compared the Bruker and VITEK MS systems for identification of 309 clinical isolates of four common *Candida* species, *C. neoformans*, as well as 37 uncommon yeast species, using on-plate formic acid preparation [34]. If "no identification" was obtained, isolates were retested using on-plate formic acid preparation and, for the Bruker system, tube-based extraction. Both systems accurately identified all 158 isolates of the common *Candida* species with initial analysis. The Bruker system correctly identified 9%, 30%, and 100% of 23 *C. neoformans* isolates after initial on-plate formic acid preparation, repeat on-plate formic acid preparation, and tube-based extraction, respectively; VITEK MS identified all *C. neoformans* isolates after initial on-plate formic acid preparation. Both systems had comparable identification rates for 37 uncommon yeast species following initial on-plate formic acid preparation (Bruker, 74%; VITEK MS, 73%) and repeat on-plate formic acid preparation (Bruker, 82%; VITEK MS, 73%). Marucco et al. compared identification of *Candida* species obtained by BD Phoenix (Becton Dickinson, Franklin Lakes, NJ, USA) and the Bruker system using 192 isolates from the strain collection of the Mycology Network of the Autonomous City of Buenos Aires, Argentina, reporting an observed concordance of 95%, with 5% of isolates not correctly identified by the BD Phoenix system [35]. Wilson et al. reported results of a multicenter assessment of the Bruker MALDI Biotyper CA system for identification of clinically significant bacteria and yeasts, including 815 yeast isolates evaluated using three processing methods [36]. The percentage identified and the percentage identified with a high level of confidence were 98% and 88%, respectively, with the extended direct transfer method being superior to the direct transfer method (74% and 49% success, respectively) [36]. Turhan et al. assessed the Bruker system with 117 yeasts, including 115 candidemia-associated *Candida* species, reporting 98% and 87% identification to the genus- and species-level, respectively [37]. Porte et al. compared the two commercial MALDI-ToF MS systems in a routine laboratory in Chile, in a study that included 47 yeasts; the bioMérieux system yielded higher rates of yeast identification to species-level than did the Bruker system (46 and 37 respectively) [38].

2.1. Malassezia *Species*

Malassezia furfur and *Malassezia pachydermatis* are included in both FDA-approved/cleared databases (Table 1). Denis et al. developed and evaluated a MALDI-ToF MS database for identifying *Malassezia* species using the Bruker system [39]. Forty-five isolates of *M. furfur*, *Malassezia slooffiae*, *Malassezia sympodialis*, *M. pachydermatis*, *Malassezia restricta* and *Malassezia globosa* were used to create a database, with 40 different isolates used to test the database; all isolates were identified with scores of >2.000.

2.2. Trichosporon *Species*

Trichosporon inkin and *Trichosporon asahii* are included in both FDA-approved/cleared databases, with *T. mucoides* additionally claimed by the Vitek MS database and *T. mucoides* group claimed by the MALDI Biotyper CA system (Table 1). de Almeida et al. subjected 16 *Trichosporon* species isolates to MALDI-ToF MS using the Bruker system, evaluating several extraction methods [40]. Overall, incubation for 30 min with 70% formic acid yielded spectra with the highest scores; among the six libraries studied, a library made of 18 strains plus seven clinical isolates yielded the best results, correctly identifying 99% of 68 clinical isolates.

3. Filamentous Fungi

Filamentous fungi demonstrate variable phenotypes as a result of which protein spectra may vary; heterogeneity can be affected by growth conditions and the zone of fungal mycelium examined. Nevertheless, filamentous fungi can be identified using MALDI-ToF MS [41]. The FDA-approved/cleared Vitek MS system claims 47 filamentous fungi, either species or complexes, including dimorphic pathogens, alongside dermatophytes. As mentioned above, Bruker has an RUO Filamentous Fungi Library. Sample preparation has varied from study-to-study, with sample preparation for molds recommended by companies having changed over time [41]; the FDA-approved/cleared Vitek MS system uses off-plate protein extraction.

McMullen et al. evaluated the Vitek MS using the Vitek MS Knowledge Base, v3.0 for identification of 319 mold isolates, representative of 43 genera, reporting 67% correct identification; when a modified SARAMIS database was used to supplement the v3.0 Knowledge Base, 77% were identified [42]. Rychert et al. reported correct species-level identification of 301/324 clinical isolates of various *Aspergillus* species tests as part of an FDA trial of the Vitek MS v3.0 system; species evaluated included *Aspergillus brasiliensis, Aspergillus calidoustus, Aspergillus flavus/oryzae, Aspergillus fumigatus, Aspergillus lentulus, Aspergillus nidulans, Aspergillus niger* complex, *Aspergillus sydowii, Aspergillus terreus* complex, and *Aspergillus versicolor* [43]. Rychert et al. also reported correct species identification of 205/325 clinical isolates of dematiaceous fungi in the same study, including *Alternaria alternata, Curvularia hawaiiensis, Curvularia spicifera, Exserohilum rostratum, Exophiala dermatitidis, Exophiala xenobiotica, Scedosporium boydii, Scedosporium apiospermum, Scedosporium prolificans* and *Cladophialophora bantiana* [43]. Finally, Rychert et al. reported correct species-level identification of 298/315 clinical isolates of "other potential pathogens", including *Fusarium oxysporum* complex, *Fusarium proliferatum, Fusarium solani* complex, *Paecilomyces variotii, Penicillium chrysogenum, Rasamsonia argillacea, Acremonium sclerotigenum, Lecythophora hoffmannii, Sarocladium kiliense* and *Purpureocillium lilacinum* [43].

De Carolis et al. established their own library of *Fusarium* species, *Aspergillus* species, and Mucorales using the Biotyper system and identified 97% of 94 isolates to the species-level [44]. Gautier et al. used an in-house database to assess the level to which MALDI-ToF MS performed using the Bruker platform enhanced identification; implementation of MALDI-ToF MS resulted in marked enhancement in mold identification at the species-level (to 98%) [45]. Lau et al. used a special extraction technique with a user-developed library representing 294 isolates of 76 genera and 152 species and the Bruker system, to test 421 mold isolates, achieving correct species- and genus-level identifications of 89% and 93% of isolates, respectively [46]. Zvezdanova et al. recently assessed the Bruker system with the Filamentous Fungi Library 1.0 for clinical mold identification using direct target plate testing and simplified processing consisting of mechanical lysis of molds preparatory to protein extraction [47]. They reported accurate species-level identification of 25/34 *Fusarium* species and all 10 *Mucor circinelloides* isolates tested. In addition, 1/21 *Pseudallescheria/Scedosporium and* 7/34 *Fusarium* species isolates were correctly identified to the genus level. The remaining 60 isolates were not identified using the commercial database. They then constructed an in-house database with 63 isolates, which allowed identification of 91% and 100% identification to the species- and genus-levels, respectively.

Normand et al. reported decision criteria for MALDI-ToF MS identification of molds and dermatophytes using the Bruker system [48]. They employed user-developed and Bruker databases as well as 422 isolates of 126 species to evaluate a number of thresholds and one to four spots. They found optimal results with a decision algorithm in which only the uppermost score of four spots was applied, with a 1.700 score threshold. Testing the complete panel enabled identification of 87% and 35% of isolates with the user-developed and Bruker databases, respectively. Applying the same rules to isolates with species represented by at least three strains in the database allowed identification of 92% and 47% of isolates with the user-developed and Bruker databases, respectively. Huang et al. described their findings using the Bruker system and 374 clinical filamentous fungal isolates with correct species and genus identification realized in 99% and 100% of isolates, respectively [49]. Riat et al. used the Bruker

Filamentous Fungi Library 1.0, reporting that an identification score of >1.700 was obtained for 92% of 48 mold isolates studied [50]. Using the Bruker system and a user-developed database, Masih et al. identified 95% of *Aspergillus* species [51]. Park et al. evaluated the Bruker's Filamentous Fungi Library 1.0 with 345 clinical *Aspergillus* isolates; compared with findings of internal transcribed spacer (ITS) sequencing, rates of accurate identification at the species-complex level were 95% and 99%, with cutoff values of 2.000 and 1.700, respectively [52]. Compared with β-tubulin gene sequencing, rates of accurate identification to the species-level were 96% (cutoff 2.000) and 100% (cutoff 1.700) for 303 *Aspergillus* isolates of five common species, but only 5% (cutoff 1.700) and 0% (cutoff 2.000) for 42 *Aspergillus* isolates of six rare species. Schulthess et al. evaluated Bruker's Filamentous Fungi Library 1.0, first studying 83 phenotypically- and molecularly-characterized, non-dermatophyte, non-dematiaceous molds from a clinical isolate collection [53]. Using manufacturer-recommended interpretative criteria, genus and species identification frequencies were 78% and 54%, respectively. Decreasing the species cutoff to 1.700 increased species identification to 71%, without impacting misidentification. In a follow-on prospective study, 200 successive clinical mold isolates were assessed; genus and species identification rates were 84% and 79%, respectively, with a species cutoff of 1.700. Sleiman et al. developed a database for identification of *Aspergillus*, *Fusarium* and *Scedosporium* species [54]. Using 117 isolates, species-level identification was enhanced when the user-developed database was used in conjunction with the Bruker Filamentous Fungi Library compared with the Bruker database alone (*Aspergillus* species, 93% versus 69%; *Fusarium* species, 84% versus 42%; and *Scedosporium* species, 94% versus 18%, respectively). Becker et al. employed a user-developed library and the Bruker system to evaluate 390 clinical isolates, reporting correct identification of 86% of isolates to the species-level using a cutoff of 1.700 [55]. Vidal-Acuña et al. created their own library using 42 clinical *Aspergillus* isolates and 11 strains, cultured in liquid medium, including 23 different species [56]. One hundred and ninety isolates cultured on solid media (179 clinical isolates identified by sequencing and the 11 strains) were studied, with species- and genus-level identifications of 87 and 100%, respectively. They then prospectively challenged their library with 200 *Aspergillus* clinical isolates grown on solid media; species identification was obtained in 96%. Stein et al. evaluated the Bruker system with clinical isolates and reference strains of molds using the Bruker mold, National Institutes of Health, and Mass Spectrometry Identification (MSI) online libraries, comparing results to morphological and molecular identification methods [57]. All libraries studied showed better accuracy in genus identification (≥95%) compared to conventional methods (86%), with 73% of isolates identified to the species-level. The MSI library showed the highest rate of species-level identification (72%) compared to National Institutes of Health (20%) and Bruker (14%) libraries. More than 20% of molds were unidentified to the species-level by all libraries studied, a finding attributed to library limitations and/or poor spectra. Triest et al. evaluated the Bruker system with a user-developed database for identification of 289 *Fusarium* isolates encompassing 40 species from the Belgian Coordinated Collections of Microorganisms/Institute of Hygiene and Epidemiology Mycology culture collection, observing no incorrect species complex identifications [58]. 83% of identifications were accurate to the species-level.

Rychert et al. reported correct species-level identification of clinical isolates of 24/30 *Mucor racemosus* complex, 22/28 *Rhizopus arrhizus* complex, 26/29 *Rhizopus microsporus* complex and 29/31 *Lichtheimia corymbifera*, as part of an FDA trial of the Vitek MS v3.0 system [43]. Dolatabadi et al. utilized the Bruker system with a user-developed database for identification of *R. arrhizus* and its varieties, *delemar* and *arrhizus*, as well as *R. microspores* [59]. Chen et al. assessed the Bruker system with 50 clinically encountered mold isolates, including *Talaromyces marneffei*, *Rhizopus* species, *Paecilomyces* species, *Fusarium solani*, and *Pseudallescheria boydii* [60]. The correct identification rate of *T. marneffei* (score ≥2.000) was 86% based on their user-developed library. Although all seven *P. variotii* isolates, two of the four *P. lilacinus*, four of the six *F. solani*, and two of the three isolates of *Rhizopus* species, and the *P. boydii* isolate had concordant identifications between MALDI-ToF MS and sequencing analysis, scores were all <1.700 [60]. Shao et al. studied 111 isolates of Mucorales belonging to six genera from the Research Center for Medical Mycology of Peking University, initially using the Bruker

Filamentous Fungi library (v1.0), showing 50% and 67% identification to species- and genus-levels, respectively [61]. They then created an in-house library, the Beijing Medical University database, using [11] strains of *Mucor hiemalis, Mucor racemosus, Mucor irregularis, Cunninghamella phaeospora, Cunninghamella bertholletiae,* and *Cunninghamella echinulate.* Using the Beijing Medical University and Bruker databases together, all 111 isolates were identified, 81% and 100% to the species- and genus-levels, respectively.

Singh et al. analyzed 72 melanized clinical fungal isolates from patients in 19 Indian medical centers using the Bruker system and a user-developed database, reporting 100% identification [62]. Paul et al. created an in-house database of 59 melanized fungi using a modified protein extraction protocol, and tested 117 clinical isolates using the database [63]. Whereas using the Bruker database only 29 (25%) molds were identified, all were accurately identified accurately by supplementing the Bruker database with the in-house library.

Dermatophytes

With appropriate databases, dermatophytes may be identified using MALDI-ToF MS [64–66]. *Microsporum audouinii, Microsporum canis, Microsporum gypseum, Epidermophyton floccosum, Trichophyton rubrum, Trichophyton interdigitale, Trichophyton tonsurans, Trichophyton verrucosum,* and *Trichophyton violaceum* are included in the FDA-approved/cleared Vitek MS system, with no dermatophytes in the FDA-approved/cleared Bruker system. Rychert et al. reported correct species-level identification of clinical isolates of 30/33 *M. audouinii*, 30/31 *M. canis*, 32/35 *M. gypseum*, 30/31 *E. floccosum*, 31/31 *T. rubrum*, 29/30 *T. interdigitale*, 30/33 *T. tonsurans*, 18/31 *T. verrucosum*, and 13/34 *T. violaceum*, as part of an FDA trial of the Vitek MS v3.0 system [43].

Packeu et al. evaluated the Bruker system with a user-developed library for the identification of 176 clinical dermatophyte isolates [67]. MALDI-ToF MS yielded accurate identifications of 97 and 90% of isolates with lowered scores and application of the user-supplemented database, respectively, versus 52% and 14% correct identifications with the unmodified library and recommended scores at the genus- and species-levels, respectively. Calderaro et al. determined the ability of a user-developed database with the Bruker system to identify 64 clinical isolates; all were correctly identified (score of >2.000 for 47 isolates, and 1.700 to 2.000 for the other 17 isolates) [68]. An on-plate procedure after 3 days of incubation produced 40% accurate identification; prolonging incubation time and using an extraction procedure both yielded 100% accurate identification. Karabicak et al. evaluated the Bruker system using a user-developed database with 126 dermatophytes, including 115 clinical isolates and [9] strains; using a combination of the user-developed database and lowered cutoff scores, genus and species identifications were achieved for 97% and 90% of the isolates [69]. L'Ollivier et al. appraised ten studies published between 2008 and 2015 showing accuracy of MALDI-ToF MS-based identification of dermatophytes to vary between 14 and 100% [70]; they ascribed inconsistencies, in part, to processing variability. Use of a tube-based extraction step and a manufacturer database augmented with user-developed spectra were helpful for accurate species identification. Da Cunha et al. assessed whether the direct transfer method can be used with dermatophytes [71]. They built their own library using the Bruker system and evaluated its performance with a panel of mass spectra produced with molecularly-identified isolates and, compared MALDI-ToF MS to morphology-based identification. Although dermatophyte identification using the Bruker library was poor, their database yielded 97% concordance between ITS sequencing and MALDI-ToF MS with 276 isolates. The direct transfer method using unpolished target plates permitted the correct identification of 85% of the clinical dermatophyte isolates.

4. Dimorphic Fungi

The Vitek MS database includes *Blastomyces dermatitidis, Coccidioides immitis/posadasii, Histoplasma capsulatum,* and *Sporothrix schenckii* complex (Table 2); Rychert et al. evaluated 40, 38, 32 and 31 of these, respectively, as part of an FDA trial of the Vitek MS v3.0 system, reporting 100% identification [43].

Lau et al. assessed the Bruker system for identification of 39 isolates of *T. marneffei* [72]. Using the Filamentous Fungi Library 1.0, MALDI-ToF MS did not identify the isolates; when the database was expanded by including spectra from 21 *T. marneffei* isolates, all isolates in the mold or yeast phase were identified to the species-level. De Almeida et al. showed that the Bruker system with a user-developed database, could identify *Paracoccidioides brasiliensis* and *Paracoccidioides lutzii* [73]. Valero et al. established their own *H. capsulatum* Bruker database using six strains [74]. Then, 30 *H. capsulatum* isolates from the Collection of the Spanish National Centre for Microbiology were studied and correctly identified, 87% with scores above 1.700. The created database was able to identify both growth phases of the fungus, with the most reliable results for the mycelial phase.

5. Limitations

MALDI-ToF MS has limitations. Unlike publicly available sequence databases, such as GenBank, commercial MALDI-ToF MS databases are typically exclusive to companies. Although low identification rates for some organisms may be enhanced by user addition of mass spectral entries of underrepresented species or strains (to cover intraspecies variability), or even re-addition of reference strain spectra to the library, especially those created using parallel growth conditions and preparation methods, doing so may be beyond the know-how of some laboratories. Because of low scores/percentages, repeat testing of isolates may be required [14]. Growth on some media may yield low scores/percentages [75], and small or mucoid colonies may fail. Using experimental capsule size manipulation, it was demonstrated that capsule size of *C. neoformans* and *C. gattii* can compromise identification by the Bruker system [76]. Refined interpretive criteria may be needed to discriminate closely related species and distinguish them from the next best taxon match. For some species of fungi, genus- or species-specific (including lowered) cutoffs may be needed. Mistakes that may occur include testing mixed colonies, spreading amongst spots, spotting into incorrect target plate positions, not properly cleaning re-usable target plates, and wrongly entering results into laboratory information systems. There is a learning curve to depositing ideal biomass onto target plates [77]. Although results are normally reproducible, sources of variability include the technologist, mass spectrometer and especially laser age, matrix and solvent make-up, biological variability, and culture conditions [48]. Instrument (e.g., laser) and software failure may happen. As a result of the simplicity of MALDI-ToF MS, technologists may lose or never develop fine-tuned abilities to visually identify fungi, macroscopically and microscopically.

6. Conclusions

In summary, MALDI-ToF MS has become a routine method for the identification of yeasts and is also being applied to filamentous and dimorphic fungi. Although databases are slowly becoming more complete with regards to clinically-relevant fungi, due to evolving nomenclature and constant description of new species/genera, systematic and continuing library updates will be needed to deliver quality fungal identification into the anticipatable future.

This manuscript has its limitation as the appreciation of MALDI-ToF MS to fungi is rapidly evolving such that some of the cited studies, even if recently published, may rapidly be antiquated. This paper is based in part on [4,7,8,13,78].

Conflicts of Interest: Patel reports grants from CD Diagnostics, BioFire, Curetis, Merck, Contrafect, Hutchison Bioflim Medical Solutions, Accelerate Diagnostics, Allergan, EnBiotix, Contrafect and The Medicines Company. Patel is or has been a consultant to Curetis, Specific Technologies, Selux Dx, Genmark Diagnotics, PathoQuest, Heraeus Medical, and Qvella; monies are paid to Mayo Clinic. In addition, Patel has a patent on *Bordetella pertussis/parapertussis* PCR issued, a patent on a device method for sonication with royalties paid by Samsung to Mayo Clinic, and a patent on an anti-biofilm substance issued. Patel received travel reimbursmnet from ASM and IDSA and an editor's stipend from ASM and IDSA, as well as honoraria from the NBME, Up-to-Date and the Infectious Diseases Board Review Course.

References

1. Anhalt, J.; Fenselau, C. Identification of bacteria using mass spectrometry. *Anal. Chem.* **1975**, *47*, 219–225. [CrossRef]
2. Tanaka, K. The origin of macromolecule ionization by laser irradiation (Nobel Lecture). *Angew. Chem.* **2003**, *42*, 3860–3870. [CrossRef] [PubMed]
3. Karas, M.; Hillenkamp, F. Laser desorption ionization of proteins with molecular masses exceeding 10,000 Daltons. *Anal. Chem.* **1988**, *60*, 2299–2301. [CrossRef] [PubMed]
4. Patel, R. MALDI-TOF mass spectrometry: Transformative proteomics for clinical microbiology. *Clin. Chem.* **2013**, *59*, 340–342. [CrossRef] [PubMed]
5. Clinical and Laboratory Standards Institute. *Methods for the Identification of Cultured Microorganisms Using Matrix-Assisted Laser Desorption/Ionization Time-of-Flight Mass Spectrometry*, 1st ed.; Clinical and Laboratory Standards Institute: Wayne, PA, USA, 2017; Volume M58.
6. Mutters, N.T.; Hodiamont, C.J.; de Jong, M.D.; Overmeijer, H.P.; van den Boogaard, M.; Visser, C.E. Performance of Kiestra total laboratory automation combined with MS in clinical microbiology practice. *Ann. Lab. Med.* **2014**, *34*, 111–117. [CrossRef] [PubMed]
7. Patel, R. Matrix-assisted laser desorption ionization-time of flight mass spectrometry in clinical microbiology. *Clin. Infect. Dis.* **2013**, *57*, 564–572. [CrossRef]
8. Patel, R. MALDI-TOF MS for the diagnosis of infectious diseases. *Clin. Chem.* **2015**, *61*, 100–111. [CrossRef]
9. McElvania TeKippe, E.; Burnham, C.A. Evaluation of the Bruker Biotyper and VITEK MS MALDI-TOF MS systems for the identification of unusual and/or difficult-to-identify microorganisms isolated from clinical specimens. *Eur. J. Clin. Microbiol. Infect. Dis.* **2014**, *33*, 2163–2171. [CrossRef]
10. Bilecen, K.; Yaman, G.; Ciftci, U.; Laleli, Y.R. Performances and reliability of Bruker Microflex LT and VITEK MS MALDI-TOF mass spectrometry systems for the identification of clinical microorganisms. *BioMed Res. Int.* **2015**, *2015*, 516410. [CrossRef]
11. Levesque, S.; Dufresne, P.J.; Soualhine, H.; Domingo, M.C.; Bekal, S.; Lefebvre, B.; Tremblay, C. A side by side comparison of Bruker Biotyper and VITEK MS: Utility of MALDI-TOF MS technology for microorganism identification in a public health reference laboratory. *PLoS ONE* **2015**, *10*, e0144878. [CrossRef]
12. Bao, J.R.; Master, R.N.; Azad, K.N.; Schwab, D.A.; Clark, R.B.; Jones, R.S.; Moore, E.C.; Shier, K.L. Rapid, accurate identification of *Candida auris* by using a novel matrix-assisted laser desorption ionization-time of flight mass spectrometry (MALDI-TOF MS) database (library). *J. Clin. Microbiol.* **2018**, *56*. [CrossRef] [PubMed]
13. Carroll, K.; Patel, R. Systems for identification of bacteria and fungi. In *Manual of Clinical Microbiology*, 12th ed.; ASM Press: Washington, DC, USA, 2018; Volume 1, in press.
14. Tan, K.E.; Ellis, B.C.; Lee, R.; Stamper, P.D.; Zhang, S.X.; Carroll, K.C. Prospective evaluation of a matrix-assisted laser desorption ionization-time of flight mass spectrometry system in a hospital clinical microbiology laboratory for identification of bacteria and yeasts: A bench-by-bench study for assessing the impact on time to identification and cost-effectiveness. *J. Clin. Microbiol.* **2012**, *50*, 3301–3308. [CrossRef] [PubMed]
15. Ge, M.C.; Kuo, A.J.; Liu, K.L.; Wen, Y.H.; Chia, J.H.; Chang, P.Y.; Lee, M.H.; Wu, T.L.; Chang, S.C.; Lu, J.J. Routine identification of microorganisms by matrix-assisted laser desorption ionization time-of-flight mass spectrometry: Success rate, economic analysis, and clinical outcome. *J. Microbiol. Immunol. Infect.* **2016**. [CrossRef] [PubMed]
16. Sparbier, K.; Weller, U.; Boogen, C.; Kostrzewa, M. Rapid detection of *Salmonella* sp. by means of a combination of selective enrichment broth and MALDI-TOF MS. *Eur. J. Clin. Microbiol. Infect. Dis.* **2012**, *31*, 767–773. [CrossRef] [PubMed]
17. Castanheira, M.; Woosley, L.N.; Diekema, D.J.; Jones, R.N.; Pfaller, M.A. *Candida guilliermondii* and other species of candida misidentified as *Candida famata*: Assessment by vitek 2, DNA sequencing analysis, and matrix-assisted laser desorption ionization-time of flight mass spectrometry in two global antifungal surveillance programs. *J. Clin. Microbiol.* **2013**, *51*, 117–124. [CrossRef] [PubMed]
18. Firacative, C.; Trilles, L.; Meyer, W. MALDI-TOF MS enables the rapid identification of the major molecular types within the *Cryptococcus neoformans/C. gattii* species complex. *PLoS ONE* **2012**, *7*, e37566. [CrossRef]

19. Sendid, B.; Ducoroy, P.; Francois, N.; Lucchi, G.; Spinali, S.; Vagner, O.; Damiens, S.; Bonnin, A.; Poulain, D.; Dalle, F. Evaluation of MALDI-TOF mass spectrometry for the identification of medically-important yeasts in the clinical laboratories of Dijon and Lille hospitals. *Med. Mycol.* **2013**, *51*, 25–32. [CrossRef]
20. Westblade, L.F.; Jennemann, R.; Branda, J.A.; Bythrow, M.; Ferraro, M.J.; Garner, O.B.; Ginocchio, C.C.; Lewinski, M.A.; Manji, R.; Mochon, A.B.; et al. Multicenter study evaluating the Vitek MS system for identification of medically important yeasts. *J. Clin. Microbiol.* **2013**, *51*, 2267–2272. [CrossRef]
21. Theel, E.S.; Schmitt, B.H.; Hall, L.; Cunningham, S.A.; Walchak, R.C.; Patel, R.; Wengenack, N.L. Formic acid-based direct, on-plate testing of yeast and *Corynebacterium* species by Bruker Biotyper matrix-assisted laser desorption ionization-time of flight mass spectrometry. *J. Clin. Microbiol.* **2012**, *50*, 3093–3095. [CrossRef]
22. Dhiman, N.; Hall, L.; Wohlfiel, S.L.; Buckwalter, S.P.; Wengenack, N.L. Performance and cost analysis of matrix-assisted laser desorption ionization-time of flight mass spectrometry for routine identification of yeast. *J. Clin. Microbiol.* **2011**, *49*, 1614–1616. [CrossRef]
23. Won, E.J.; Shin, J.H.; Lee, K.; Kim, M.N.; Lee, H.S.; Park, Y.J.; Joo, M.Y.; Kim, S.H.; Shin, M.G.; Suh, S.P.; et al. Accuracy of species-level identification of yeast isolates from blood cultures from 10 university hospitals in South Korea by use of the matrix-assisted laser desorption ionization-time of flight mass spectrometry-based Vitek MS system. *J. Clin. Microbiol.* **2013**, *51*, 3063–3065. [CrossRef]
24. Mancini, N.; De Carolis, E.; Infurnari, L.; Vella, A.; Clementi, N.; Vaccaro, L.; Ruggeri, A.; Posteraro, B.; Burioni, R.; Clementi, M.; et al. Comparative evaluation of the Bruker Biotyper and Vitek MS matrix-assisted laser desorption ionization-time of flight (MALDI-TOF) mass spectrometry systems for identification of yeasts of medical importance. *J. Clin. Microbiol.* **2013**, *51*, 2453–2457. [CrossRef] [PubMed]
25. Rosenvinge, F.S.; Dzajic, E.; Knudsen, E.; Malig, S.; Andersen, L.B.; Lovig, A.; Arendrup, M.C.; Jensen, T.G.; Gahrn-Hansen, B.; Kemp, M. Performance of matrix-assisted laser desorption-time of flight mass spectrometry for identification of clinical yeast isolates. *Mycoses* **2013**, *56*, 229–235. [CrossRef]
26. Lacroix, C.; Gicquel, A.; Sendid, B.; Meyer, J.; Accoceberry, I.; Francois, N.; Morio, F.; Desoubeaux, G.; Chandenier, J.; Kauffmann-Lacroix, C.; et al. Evaluation of two matrix-assisted laser desorption ionization-time of flight mass spectrometry (MALDI-TOF MS) systems for the identification of *Candida* species. *Clin. Microbiol. Infect.* **2014**, *20*, 153–158. [CrossRef]
27. Pence, M.A.; McElvania Tekippe, E.; Wallace, M.A.; Burnham, C.A. Comparison and optimization of two MALDI-TOF MS platforms for the identification of medically relevant yeast species. *Eur. J. Clin. Microbiol. Infect. Dis.* **2014**. [CrossRef] [PubMed]
28. Jamal, W.Y.; Ahmad, S.; Khan, Z.U.; Rotimi, V.O. Comparative evaluation of two matrix-assisted laser desorption/ionization time-of-flight mass spectrometry (MALDI-TOF MS) systems for the identification of clinically significant yeasts. *Int. J. Infect. Dis.* **2014**, *26*, 167–170. [CrossRef] [PubMed]
29. Hamprecht, A.; Christ, S.; Oestreicher, T.; Plum, G.; Kempf, V.A.; Gottig, S. Performance of two MALDI-TOF MS systems for the identification of yeasts isolated from bloodstream infections and cerebrospinal fluids using a time-saving direct transfer protocol. *Med. Microbiol. Immunol.* **2014**, *203*, 93–99. [CrossRef] [PubMed]
30. De Carolis, E.; Vella, A.; Vaccaro, L.; Torelli, R.; Posteraro, P.; Ricciardi, W.; Sanguinetti, M.; Posteraro, B. Development and validation of an in-house database for matrix-assisted laser desorption ionization-time of flight mass spectrometry-based yeast identification using a fast protein extraction procedure. *J. Clin. Microbiol.* **2014**, *52*, 1453–1458. [CrossRef] [PubMed]
31. Fatania, N.; Fraser, M.; Savage, M.; Hart, J.; Abdolrasouli, A. Comparative evaluation of matrix-assisted laser desorption ionisation-time of flight mass spectrometry and conventional phenotypic-based methods for identification of clinically important yeasts in a UK-based medical microbiology laboratory. *J. Clin. Pathol.* **2015**, *68*, 1040–1042. [CrossRef]
32. Wang, H.; Fan, Y.Y.; Kudinha, T.; Xu, Z.P.; Xiao, M.; Zhang, L.; Fan, X.; Kong, F.; Xu, Y.C. A comprehensive evaluation of the Bruker Biotyper MS and Vitek MS matrix-assisted laser desorption ionization-time of flight mass spectrometry systems for identification of yeasts, part of the national China hospital invasive fungal surveillance net (CHIF-NET) study, 2012 to 2013. *J. Clin. Microbiol.* **2016**, *54*, 1376–1380. [CrossRef]
33. Fraser, M.; Brown, Z.; Houldsworth, M.; Borman, A.M.; Johnson, E.M. Rapid identification of 6328 isolates of pathogenic yeasts using MALDI-ToF MS and a simplified, rapid extraction procedure that is compatible with the Bruker Biotyper platform and database. *Med. Mycol.* **2016**, *54*, 80–88. [CrossRef] [PubMed]

34. Lee, H.S.; Shin, J.H.; Choi, M.J.; Won, E.J.; Kee, S.J.; Kim, S.H.; Shin, M.G.; Suh, S.P. Comparison of the Bruker Biotyper and VITEK MS matrix-assisted laser desorption/ionization time-of-flight mass spectrometry systems using a formic acid extraction method to identify common and uncommon yeast isolates. *Ann. Lab. Med.* **2017**, *37*, 223–230. [CrossRef] [PubMed]
35. Marucco, A.P.; Minervini, P.; Snitman, G.V.; Sorge, A.; Guelfand, L.I.; Moral, L.L.; Integrantes de la Red de Micologia CABA. Comparison of the identification results of *Candida* species obtained by BD Phoenix and Maldi-TOF (Bruker Microflex LT Biotyper 3.1). *Rev. Argent. Microbiol.* **2018**. [CrossRef] [PubMed]
36. Wilson, D.A.; Young, S.; Timm, K.; Novak-Weekley, S.; Marlowe, E.M.; Madisen, N.; Lillie, J.L.; Ledeboer, N.A.; Smith, R.; Hyke, J.; et al. Multicenter evaluation of the Bruker MALDI Biotyper CA system for the identification of clinically important bacteria and yeasts. *Am. J. Clin. Pathol.* **2017**, *147*, 623–631. [CrossRef] [PubMed]
37. Turhan, O.; Ozhak-Baysan, B.; Zaragoza, O.; Er, H.; Saritas, Z.E.; Ongut, G.; Ogunc, D.; Colak, D.; Cuenca-Estrella, M. Evaluation of MALDI-TOF-MS for the identification of yeast isolates causing bloodstream infection. *Clin. Lab.* **2017**, *63*, 699–703. [CrossRef] [PubMed]
38. Porte, L.; Garcia, P.; Braun, S.; Ulloa, M.T.; Lafourcade, M.; Montana, A.; Miranda, C.; Acosta-Jamett, G.; Weitzel, T. Head-to-head comparison of Microflex LT and Vitek MS systems for routine identification of microorganisms by MALDI-TOF mass spectrometry in Chile. *PLoS ONE* **2017**, *12*, e0177929. [CrossRef]
39. Denis, J.; Machouart, M.; Morio, F.; Sabou, M.; Kauffmann-LaCroix, C.; Contet-Audonneau, N.; Candolfi, E.; Letscher-Bru, V. Performance of matrix-assisted laser desorption ionization-time of flight mass spectrometry for identifying clinical *Malassezia* isolates. *J. Clin. Microbiol.* **2017**, *55*, 90–96. [CrossRef]
40. de Almeida, J.N.; Figueiredo, D.S.; Toubas, D.; Del Negro, G.M.; Motta, A.L.; Rossi, F.; Guitard, J.; Morio, F.; Bailly, E.; Angoulvant, A.; et al. Usefulness of matrix-assisted laser desorption ionisation-time-of-flight mass spectrometry for identifying clinical *Trichosporon* isolates. *Clin. Microbiol. Infect.* **2014**, *20*, 784–790. [CrossRef]
41. Sanguinetti, M.; Posteraro, B. Identification of molds by matrix-assisted laser desorption ionization-time of flight mass spectrometry. *J. Clin. Microbiol.* **2017**, *55*, 369–379. [CrossRef]
42. McMullen, A.R.; Wallace, M.A.; Pincus, D.H.; Wilkey, K.; Burnham, C.A. Evaluation of the Vitek MS matrix-assisted laser desorption ionization-time of flight mass spectrometry system for identification of clinically relevant filamentous fungi. *J. Clin. Microbiol.* **2016**, *54*, 2068–2073. [CrossRef]
43. Rychert, J.; Slechta, E.S.; Barker, A.P.; Miranda, E.; Babady, N.E.; Tang, Y.W.; Gibas, C.; Wiederhold, N.; Sutton, D.; Hanson, K.E. Multicenter Evaluation of the Vitek MS v3.0 System for the Identification of Filamentous Fungi. *J. Clin. Microbiol.* **2018**, *56*, e01353-17. [CrossRef] [PubMed]
44. De Carolis, E.; Posteraro, B.; Lass-Florl, C.; Vella, A.; Florio, A.R.; Torelli, R.; Girmenia, C.; Colozza, C.; Tortorano, A.M.; Sanguinetti, M.; et al. Species identification of *Aspergillus*, *Fusarium* and Mucorales with direct surface analysis by matrix-assisted laser desorption ionization time-of-flight mass spectrometry. *Clin. Microbiol. Infect.* **2012**, *18*, 475–484. [CrossRef] [PubMed]
45. Gautier, M.; Ranque, S.; Normand, A.C.; Becker, P.; Packeu, A.; Cassagne, C.; L'Ollivier, C.; Hendrickx, M.; Piarroux, R. Matrix-assisted laser desorption ionization time-of-flight mass spectrometry: Revolutionizing clinical laboratory diagnosis of mould infections. *Clin. Microbiol. Infect.* **2014**, *20*, 1366–1371. [CrossRef] [PubMed]
46. Lau, A.F.; Drake, S.K.; Calhoun, L.B.; Henderson, C.M.; Zelazny, A.M. Development of a clinically comprehensive database and a simple procedure for identification of molds from solid media by matrix-assisted laser desorption ionization-time of flight mass spectrometry. *J. Clin. Microbiol.* **2013**, *51*, 828–834. [CrossRef]
47. Zvezdanova, M.E.; Escribano, P.; Ruiz, A.; Martinez-Jimenez, M.C.; Pelaez, T.; Collazos, A.; Guinea, J.; Bouza, E.; Rodriguez-Sanchez, B. Increased species-assignment of filamentous fungi using MALDI-TOF MS coupled with a simplified sample processing and an in-house library. *Med. Mycol.* **2018**. [CrossRef]
48. Normand, A.C.; Cassagne, C.; Gautier, M.; Becker, P.; Ranque, S.; Hendrickx, M.; Piarroux, R. Decision criteria for MALDI-TOF MS-based identification of filamentous fungi using commercial and in-house reference databases. *BMC Microbiol.* **2017**, *17*, 25. [CrossRef]
49. Huang, Y.; Zhang, M.; Zhu, M.; Wang, M.; Sun, Y.; Gu, H.; Cao, J.; Li, X.; Zhang, S.; Wang, J.; et al. Comparison of two matrix-assisted laser desorption ionization-time of flight mass spectrometry systems for the identification of clinical filamentous fungi. *World J. Microbiol. Biotechnol.* **2017**, *33*, 142. [CrossRef]

50. Riat, A.; Hinrikson, H.; Barras, V.; Fernandez, J.; Schrenzel, J. Confident identification of filamentous fungi by matrix-assisted laser desorption/ionization time-of-flight mass spectrometry without subculture-based sample preparation. *Int. J. Infect. Dis.* **2015**, *35*, 43–45. [CrossRef]
51. Masih, A.; Singh, P.K.; Kathuria, S.; Agarwal, K.; Meis, J.F.; Chowdhary, A. Identification by molecular methods and matrix-assisted laser desorption ionization-time of flight mass spectrometry and antifungal susceptibility profiles of clinically significant rare *Aspergillus* species in a referral chest hospital in Delhi, India. *J. Clin. Microbiol.* **2016**, *54*, 2354–2364. [CrossRef]
52. Park, J.H.; Shin, J.H.; Choi, M.J.; Choi, J.U.; Park, Y.J.; Jang, S.J.; Won, E.J.; Kim, S.H.; Kee, S.J.; Shin, M.G.; et al. Evaluation of matrix-assisted laser desorption/ionization time-of-fight mass spectrometry for identification of 345 clinical isolates of *Aspergillus* species from 11 Korean hospitals: Comparison with molecular identification. *Diagn. Microbiol. Infect. Dis.* **2017**, *87*, 28–31. [CrossRef]
53. Schulthess, B.; Ledermann, R.; Mouttet, F.; Zbinden, A.; Bloemberg, G.V.; Bottger, E.C.; Hombach, M. Use of the Bruker MALDI Biotyper for identification of molds in the clinical mycology laboratory. *J. Clin. Microbiol.* **2014**, *52*, 2797–2803. [CrossRef]
54. Sleiman, S.; Halliday, C.L.; Chapman, B.; Brown, M.; Nitschke, J.; Lau, A.F.; Chen, S.C. Performance of matrix-assisted laser desorption ionization-time of flight mass spectrometry for identification of *Aspergillus*, Scedosporium, and *Fusarium* spp. in the Australian clinical setting. *J. Clin. Microbiol.* **2016**, *54*, 2182–2186. [CrossRef]
55. Becker, P.T.; de Bel, A.; Martiny, D.; Ranque, S.; Piarroux, R.; Cassagne, C.; Detandt, M.; Hendrickx, M. Identification of filamentous fungi isolates by MALDI-TOF mass spectrometry: Clinical evaluation of an extended reference spectra library. *Med. Mycol.* **2014**, *52*, 826–834. [CrossRef]
56. Vidal-Acuna, M.R.; Ruiz-Perez de Pipaon, M.; Torres-Sanchez, M.J.; Aznar, J. Identification of clinical isolates of *Aspergillus*, including cryptic species, by matrix assisted laser desorption ionization time-of-flight mass spectrometry (MALDI-TOF MS). *Med. Mycol.* **2018**, *56*, 838–846. [CrossRef] [PubMed]
57. Stein, M.; Tran, V.; Nichol, K.A.; Lagace-Wiens, P.; Pieroni, P.; Adam, H.J.; Turenne, C.; Walkty, A.J.; Normand, A.C.; Hendrickx, M.; et al. Evaluation of three MALDI-TOF mass spectrometry libraries for the identification of filamentous fungi in three clinical microbiology laboratories in Manitoba, Canada. *Mycoses* **2018**, *61*, 743–753. [CrossRef]
58. Triest, D.; Stubbe, D.; De Cremer, K.; Pierard, D.; Normand, A.C.; Piarroux, R.; Detandt, M.; Hendrickx, M. Use of matrix-assisted laser desorption ionization-time of flight mass spectrometry for identification of molds of the *Fusarium* genus. *J. Clin. Microbiol.* **2015**, *53*, 465–476. [CrossRef] [PubMed]
59. Dolatabadi, S.; Kolecka, A.; Versteeg, M.; de Hoog, S.G.; Boekhout, T. Differentiation of clinically relevant Mucorales *Rhizopus microsporus* and *R. arrhizus* by matrix-assisted laser desorption ionization time-of-flight mass spectrometry (MALDI-TOF MS). *J. Med. Microbiol.* **2015**, *64*, 694–701. [CrossRef]
60. Chen, Y.S.; Liu, Y.H.; Teng, S.H.; Liao, C.H.; Hung, C.C.; Sheng, W.H.; Teng, L.J.; Hsueh, P.R. Evaluation of the matrix-assisted laser desorption/ionization time-of-flight mass spectrometry Bruker Biotyper for identification of *Penicillium marneffei*, *Paecilomyces* species, *Fusarium solani*, *Rhizopus* species, and *Pseudallescheria Boydii*. *Front. Microbiol.* **2015**, *6*, 679. [CrossRef] [PubMed]
61. Shao, J.; Wan, Z.; Li, R.; Yu, J. Species identification and delineation of pathogenic Mucorales by matrix-assisted laser desorption ionization-time of flight mass spectrometry. *J. Clin. Microbiol.* **2018**, *56*. [CrossRef]
62. Singh, A.; Singh, P.K.; Kumar, A.; Chander, J.; Khanna, G.; Roy, P.; Meis, J.F.; Chowdhary, A. Molecular and matrix-assisted laser desorption ionization-time of flight mass spectrometry-based characterization of clinically significant melanized fungi in India. *J. Clin. Microbiol.* **2017**, *55*, 1090–1103. [CrossRef]
63. Paul, S.; Singh, P.; Sharma, S.; Prasad, G.S.; Rudramurthy, S.M.; Chakrabarti, A.; Ghosh, A.K. MALDI-TOF MS-based identification of melanized fungi is faster and reliable after the expansion of in-house database. *Proteom. Clin. Appl.* **2018**. [CrossRef] [PubMed]
64. Theel, E.S.; Hall, L.; Mandrekar, J.; Wengenack, N.L. Dermatophyte identification using matrix-assisted laser desorption ionization-time of flight mass spectrometry. *J. Clin. Microbiol.* **2011**, *49*, 4067–4071. [CrossRef] [PubMed]
65. Nenoff, P.; Erhard, M.; Simon, J.C.; Muylowa, G.K.; Herrmann, J.; Rataj, W.; Graser, Y. MALDI-TOF mass spectrometry—A rapid method for the identification of dermatophyte species. *Med. Mycol.* **2013**, *51*, 17–24. [CrossRef] [PubMed]

66. de Respinis, S.; Tonolla, M.; Pranghofer, S.; Petrini, L.; Petrini, O.; Bosshard, P.P. Identification of dermatophytes by matrix-assisted laser desorption/ionization time-of-flight mass spectrometry. *Med. Mycol.* **2013**, *51*, 514–521. [CrossRef] [PubMed]
67. Packeu, A.; De Bel, A.; l'Ollivier, C.; Ranque, S.; Detandt, M.; Hendrickx, M. Fast and accurate identification of dermatophytes by matrix-assisted laser desorption ionization-time of flight mass spectrometry: Validation in the clinical laboratory. *J. Clin. Microbiol.* **2014**, *52*, 3440–3443. [CrossRef] [PubMed]
68. Calderaro, A.; Motta, F.; Montecchini, S.; Gorrini, C.; Piccolo, G.; Piergianni, M.; Buttrini, M.; Medici, M.C.; Arcangeletti, M.C.; Chezzi, C.; et al. Identification of dermatophyte species after implementation of the in-house MALDI-TOF MS database. *Int. J. Mol. Sci.* **2014**, *15*, 16012–16024. [CrossRef]
69. Karabicak, N.; Karatuna, O.; Ilkit, M.; Akyar, I. Evaluation of the Bruker matrix-assisted laser desorption-ionization time-of-flight mass spectrometry (MALDI-TOF MS) system for the identification of clinically important dermatophyte species. *Mycopathologia* **2015**, *180*, 165–171. [CrossRef] [PubMed]
70. L'Ollivier, C.; Ranque, S. MALDI-TOF-based dermatophyte identification. *Mycopathologia* **2017**, *182*, 183–192. [CrossRef] [PubMed]
71. da Cunha, K.C.; Riat, A.; Normand, A.C.; Bosshard, P.P.; de Almeida, M.T.G.; Piarroux, R.; Schrenzel, J.; Fontao, L. Fast identification of dermatophytes by MALDI-TOF/MS using direct transfer of fungal cells on ground steel target plates. *Mycoses* **2018**. [CrossRef] [PubMed]
72. Lau, S.K.; Lam, C.S.; Ngan, A.H.; Chow, W.N.; Wu, A.K.; Tsang, D.N.; Tse, C.W.; Que, T.L.; Tang, B.S.; Woo, P.C. Matrix-assisted laser desorption ionization time-of-flight mass spectrometry for rapid identification of mold and yeast cultures of *Penicillium Marneffei*. *BMC Microbiol.* **2016**, *16*, 36. [CrossRef] [PubMed]
73. de Almeida, J.N.; Del Negro, G.M.; Grenfell, R.C.; Vidal, M.S.; Thomaz, D.Y.; de Figueiredo, D.S.; Bagagli, E.; Juliano, L.; Benard, G. Matrix-assisted laser desorption ionization-time of flight mass spectrometry for differentiation of the dimorphic fungal species *Paracoccidioides brasiliensis* and *Paracoccidioides Lutzii*. *J. Clin. Microbiol.* **2015**, *53*, 1383–1386. [CrossRef] [PubMed]
74. Valero, C.; Buitrago, M.J.; Gago, S.; Quiles-Melero, I.; Garcia-Rodriguez, J. A matrix-assisted laser desorption/ionization time of flight mass spectrometry reference database for the identification of *Histoplasma Capsulatum*. *Med. Mycol.* **2018**, *56*, 307–314. [CrossRef] [PubMed]
75. Anderson, N.W.; Buchan, B.W.; Riebe, K.M.; Parsons, L.N.; Gnacinski, S.; Ledeboer, N.A. Effects of solid-medium type on routine identification of bacterial isolates by use of matrix-assisted laser desorption ionization-time of flight mass spectrometry. *J. Clin. Microbiol.* **2012**, *50*, 1008–1013. [CrossRef] [PubMed]
76. Thomaz, D.Y.; Grenfell, R.C.; Vidal, M.S.; Giudice, M.C.; Del Negro, G.M.; Juliano, L.; Benard, G.; de Almeida Junior, J.N. Does the capsule interfere with performance of matrix-assisted laser desorption ionization-time of flight mass spectrometry for identification of *Cryptococcus neoformans* and *Cryptococcus gattii*? *J. Clin. Microbiol.* **2016**, *54*, 474–477. [CrossRef] [PubMed]
77. Harris, P.; Winney, I.; Ashhurst-Smith, C.; O'Brien, M.; Graves, S. Comparison of Vitek MS (MALDI-TOF) to standard routine identification methods: An advance but no panacea. *Pathology* **2012**, *44*, 583–585. [CrossRef] [PubMed]
78. Heaton, P.; Patel, R. Mass spectrometry applications in infectious disease and pathogens identification. In *Principles and Applications of Clinical Mass Spectrometry, Small Molecules, Peptides, and Pathogens*; Elsevier: Amsterdam, The Netherlands, 2018; pp. 93–114.

8

Strengthening the One Health Agenda: The Role of Molecular Epidemiology in *Aspergillus* Threat Management

Eta E. Ashu [1] and **Jianping Xu** [1,2,*]

1 Department of Biology, McMaster University, 1280 Main St. W, Hamilton, Ontario, ON L8S 4K1, Canada; ashue@mcmaster.ca
2 Public Research Laboratory, Hainan Medical University, Haikou, Hainan 571199, China
* Correspondence: jpxu@mcmaster.ca

Abstract: The United Nations' One Health initiative advocates the collaboration of multiple sectors within the global and local health authorities toward the goal of better public health management outcomes. The emerging global health threat posed by *Aspergillus* species is an example of a management challenge that would benefit from the One Health approach. In this paper, we explore the potential role of molecular epidemiology in *Aspergillus* threat management and strengthening of the One Health initiative. Effective management of *Aspergillus* at a public health level requires the development of rapid and accurate diagnostic tools to not only identify the infecting pathogen to species level, but also to the level of individual genotype, including drug susceptibility patterns. While a variety of molecular methods have been developed for *Aspergillus* diagnosis, their use at below-species level in clinical settings has been very limited, especially in resource-poor countries and regions. Here we provide a framework for *Aspergillus* threat management and describe how molecular epidemiology and experimental evolution methods could be used for predicting resistance through drug exposure. Our analyses highlight the need for standardization of loci and methods used for molecular diagnostics, and surveillance across *Aspergillus* species and geographic regions. Such standardization will enable comparisons at national and global levels and through the One Health approach, strengthen *Aspergillus* threat management efforts.

Keywords: molecular epidemiology; One Health; *Aspergillus fumigatus*; invasive fungal diseases; threat management

---

1. The Genus *Aspergillus*

Fungal infections affect over a billion people and cause approximately 1.5 million deaths each year worldwide [1]. Regrettably, due to increases in the number of at-risk populations, fungal infections are projected to rise [1,2]. It is estimated that death can be averted in over 80% of fungal disease patients through improved diagnostics, treatment surveillance, and effective antifungal therapies [1]. However, to achieve such success, an inter-disciplinary approach is needed. An emerging example of the inter-disciplinary approach is the One Health initiative. The World Health Organization defines One Health as 'an approach to designing and implementing programs, polices, legislation, and research in which multiple sectors communicate and work together to achieve better public health outcomes'.

Species in the ascomycete genus *Aspergillus* have emerged as key agents of the fungal infections around the world [3–19]. For example, *Aspergilli* are the leading cause of chronic severe and allergic fungal infections, and the second leading cause of acute invasive fungal infections [1]. The genus *Aspergillus* was first described at the end of the 18th century by a Catholic priest and botanist named Pier Antonio Micheli. Viewing the microscopic spore-bearing structure of *Aspergillus*, Micheli was reminded of a holy water sprinkler—an aspergillum [20–22]. Since then, the number of species in genus *Aspergillus* has grown to encompass eight subgenera and over 250 species [23,24]. Of these species, approximately 15% are of known clinical importance [25,26]. DNA sequence-based methods are revealing an increasing number of cryptic species of *Aspergillus* associated with human diseases [27–30]. For example, surveys carried out in the United States, Brazil, and Spain revealed the percentage of phylogenetically divergent lineages representing cryptic *Aspergilli* species among clinical samples to be between 11–19%, a percentage which is notably higher than those seen in other clinically important filamentous fungi, including those belonging to the orders Mucorales, Microascales, and Hypocreales [31–33]. This is particularly important given that, in addition to being pathogenic, up to 40% of these cryptic *Aspergilli* can be resistant to antifungal drugs [30,32]. Of greater importance is the fact that some of these cryptic *Aspergilli* are resistant to multiple antifungals which can exacerbate infections caused by these species. Indeed, fungal infections, including those caused by *Aspergilli*, have become a menace to global public health.

Aspergilli cause a wide range of infections, commonly referred to as aspergillosis. Allergic bronchopulmonary, chronic pulmonary, and invasive aspergillosis (IA) are the three most common types of *Aspergillus* infections. Allergic bronchopulmonary and chronic pulmonary aspergillosis results from immune hypersensitivity and scarring due to an *Aspergillus* respiratory tract infection. On the other hand, IA can affect a wider range of body organs belonging to the urinary, digestive, and nervous systems. A significant proportion of *Aspergillus* infections are asymptomatic. However, in patients with symptomatic infections, most symptoms are non-specific, and include low-grade fevers, generalized malaise, wheezing, headaches, and haemoptysis [34,35]. Approximately eight million people world-wide are estimated to have aspergillosis [1]. Invasive aspergillosis is the most lethal type of aspergillosis and is estimated to affect >300,000 people globally every year, with a mortality rate as high as 90% in at-risk populations [1,36]. Allergic bronchopulmonary and chronic pulmonary aspergillosis affect approximately 4.8 and 3 million people annually, respectively [1].

Although aspergillosis cases are predominantly sporadic, outbreaks are not uncommon. Specifically, there have been at least 75 documented aspergillosis outbreaks between January 1966 and December 2015 [37–45]. Interestingly, a recent study showed a non-construction-related outbreak that was associated with high airborne spore concentrations in hospital areas with low efficiency air filters [44]. These results highlight the threat posed if environmental spore concentrations reach critical levels within community or home settings. Multiple *Aspergillus* species including *A. fumigatus*, *A. flavus*, *A. terreus*, *A. niger*, *A. glaucus*, *A. oryzae*, and *A. ustus* are known to have caused outbreaks. Among these, *A. fumigatus*, and *A. flavus* are the most frequently identified species [38], and are responsible for approximately 87% of all aspergillosis case reports [26].

Changes in the antifungal susceptibility patterns of these two species have further increased the threat posed by *Aspergilli.* Since its emergence in 1997, resistance to triazole in *A. fumigatus* has steadily increased and is a current global health menace [46,47]. Furthermore, in *A. fumigatus,* there are emerging reports of increased resistance to polyenes such as amphotericin B (AMB), an antifungal to which very little resistance has been reported thus far [48,49]. For example, a recent study carried out in Brazil showed that 27% of a clinical sample of *A. fumigatus* isolates was resistant to AMB (minimum inhibitory concentration (MIC) $\geq$ 2 mg/L) [48]. Similarly, our group also very recently found that 96% of a combined environmental and clinical sample from Hamilton, Canada was resistant to amphotericin. The recent emergence of voriconazole (VRC) resistance in *A. flavus* will likely cause significant problems in the management of aspergillosis caused by *A. flavus.* Triazoles, especially VRC, are first-line drugs used in the treatment of aspergillosis [50–53]. Although not yet reported, multi-drug resistant aspergillosis outbreaks similar to those caused by *Candida auris* and *Acinetobacter baumannii* will likely emerge in the near future [54–58].

With the increasing number of clinically important *Aspergillus* species and the changing antifungal susceptibility patterns of key *Aspergilli* such *A. fumigatus*, and *A. flavus*, there is a pressing need to develop novel and effective *Aspergillus* threat management strategies. Below, we propose a framework that can be used in *Aspergillus* threat management. When put in context, this framework can also be used in the management of *Candida* and other clinically relevant fungi. We encourage essential stakeholders to engage in discussions aimed at *Aspergillus* threat management.

2. Molecular Epidemiology in *Aspergillus* Threat Management

A variety of host, pathogen, host-pathogen interaction, and environmental factors have been identified as contributors to the increased threat caused by *Aspergilli*. Of interest among pathogen-related factors is the recent global rise in resistance to antifungal drugs. Triazole resistance in *A. fumigatus* has now been reported in every continent but Antarctica [47]. Generally speaking, pathogen threat management has three interdependent components: preparedness, response, and prevention. In Figure 1, we suggest a non-exclusive framework that could be used in the management of *Aspergillus* threats, including those caused by resistant strains. This review however only focuses on the molecular epidemiology components of preparedness and prevention; specifically, on molecular diagnostics and surveillance (Figure 1).

2.1. The Usefulness of Molecular Epidemiology in Aspergillus Surveillance

In epidemiology, surveillance is defined as the collection and analysis of data necessary to develop, implement, and evaluate preventative health measures. Over the last two decades, molecular epidemiology has emerged as a very important tool in the surveillance of diverse human pathogens [59,60]. This burgeoning branch of epidemiology merges traditional epidemiology and molecular biology in order to better characterize virulence, pathogen transmission patterns, and outbreak incidence. In molecular epidemiology, marker genes are used to elucidate the genotypes and the relationship between strains and populations. In addition, some of these marker genes are becoming indispensable for understanding virulence determinants and the distribution of aspergillosis.

Over the years, a wide range of molecular markers has been used to study the molecular epidemiology of *Aspergilli*. These markers include multilocus sequences, microsatellites, PCR-restriction fragment length polymorphisms, Southern hybridization of restriction enzyme-digested DNA, randomly amplified polymorphic DNA, and mating type genes [61,62]. For instance, using microsatellite markers, Guinea and colleagues investigated an aspergillosis outbreak in a major heart surgery unit of a hospital in Spain and showed that such markers were a valuable tool in IA outbreak source investigation [41]. It is however, important to note that aspergillosis outbreaks most often do not have a single source and can consist of a series of unrelated events. As such, pinpointing the source of aspergillosis outbreaks can be difficult [63]. In contrast, molecular markers have been used with more success in determining the sources of infections in non-outbreak cases. For example, a recent study highlighted that the home environment can be an important source of infection for isolated cases of triazole-resistant *A. fumigatus* [64].

In addition to its value in infection source investigation, molecular epidemiology can be used to track *Aspergilli* transmission patterns. For example, using microsatellite markers, a recent global study showed that resistant populations of *A. fumigatus* are significantly differentiated geographically [65]. This result suggests that it may be possible to track triazole resistant *A. fumigatus* strains across national and regional borders. However, significant caution should be applied here. Compared to the large global population of *Aspergilli*, relatively few isolates and genotypes have been analyzed to date, and our current understanding of the molecular variation between and within these populations may not be representative of the true global diversity. In regard to tracking *Aspergilli*, geographic sub-structuring can vary by country and region, hence tracking transmission patterns of clinically relevant *Aspergilli* within or between certain countries might prove to be easier than in others [39,66]. For example, little to no geographic population structuring has been reported in *A. fumigatus* samples from India and Netherlands, whereas Cameroonian *A. fumigatus* samples show significant evidence for geographic sub-structuring [66]. As a result, tracking clinically relevant *A. fumigatus* strains would be a more feasible task in Cameroon than it would be in India or Netherlands.

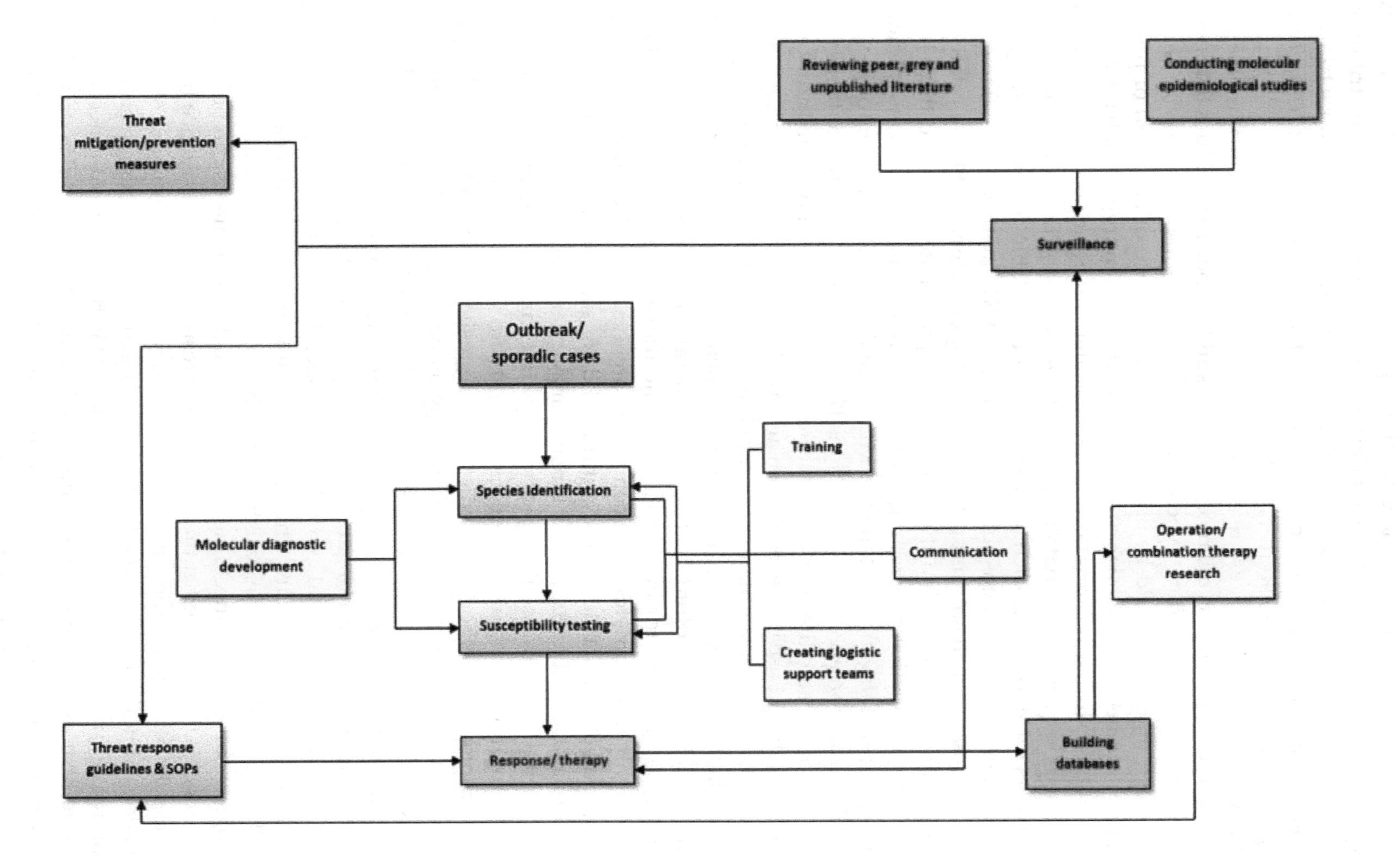

Figure 1. Framework for the management of *Aspergillus* outbreaks and critically important sporadic cases. Activities pertaining to threat preparedness are highlighted in yellow, while those pertaining to prevention are in green. Training within the context of this review refers to training laboratory technicians to perform rapid and accurate diagnosis of *Aspergilli*, including resistant strains and cryptic species. It also entails cross-training other laboratory technicians to carry out the designated emergency technician's routine duties. Communication here covers having a standardized plan to act as quickly as possible in relaying information on species and susceptibility diagnosis to the appropriate people, in order to ensure an adequate response and the safety of groups or persons at risk. Logistic support teams will be responsible for ensuring diagnostic supplies, admission documentation, travel arrangements, quality control, and all components essential for rapid and accurate diagnosis are in place in the event of an *Aspergillus* threat. SOP: standard operating procedure.

Aside from transmission pattern tracking and outbreak incidence investigation, molecular markers can also be used in virulence and antifungal drug-resistance characterization. Mating type loci genotyping has been successfully used to characterize the degree of disease severity caused by *A. fumigatus* strains [67–69], making it possible for these markers to be used in mapping the distribution of hyper-virulent strains. Similarly, certain mutations in the *cyp51A* gene are known to cause triazole drug resistance. Being able to map the distribution of hyper-virulent and drug resistant *Aspergilli* is essential for emergency preparedness and infection control efforts.

With reference to the wide range of markers used in genotyping *Aspergilli*, whilst molecular epidemiology holds promise for *Aspergilli* surveillance, there are still two pertinent issues that need to be addressed in order for the surveillance data to be fully implemented at the global scale. Firstly, we need a consensus on which markers should be used for the surveillance of *Aspergilli* of public health importance. Secondly, we need to establish and curate a reliable genotype database on the consensus gene markers to which similar information on new isolates can be added and compared. At present, there is a multilocus sequence type database for *A. fumigatus* based on seven loci (https://pubmlst.org/afumigatus/) [70]. However, such a database is not available for other *Aspergillus* species. In addition, while the current multilocus sequence typing (MLST) method for *A. fumigatus* is highly reproducible and can distinguish strains at the genus and species levels, it has a relatively low discriminatory power among strains within the same species [70]. Instead, due to their high polymorphism and greater discrimination power, a group of nine microsatellite markers is more commonly used for genotyping *A. fumigatus*, making the MLST database of *A. fumigatus* of limited use in epidemiological and population genetic investigations [71]. However, a shared database for strain genotypes based on the microsatellite markers is not available. In the short/medium term, we recommend that future work should use the seven consensus MLST loci for genotyping all *Aspergilli* strains, and for *A. fumigatus*, the additional nine microsatellite loci should be used for genotyping. Furthermore, a publicly available database should be set up to include both types of genetic data. Ultimately, with increasing availability and affordability of next-generation sequencing, whole-genome sequencing (WGS) should be considered for epidemiological monitoring for the long term. Data from WGS will not only be highly discriminatory and commonly archived, but also can help identify genetic polymorphisms associated with virulence and antifungal drug susceptibility. Such data will significantly strengthen national and global *Aspergillus* threat management efforts.

Despite the rather successful use of molecular epidemiology in characterizing hyper-virulence, pathogen transmission patterns, and outbreak incidences of *Aspergilli*, very little has been done in leveraging molecular epidemiological methods in predicting drug resistance in endemic wild-type genotypes. For example, wild-type genotypes that are strongly associated with the emergence of resistance have been shown in the human immunodeficiency virus type 1 (HIV-1) [72]. This is particularly important, as infections with wild-type genotypes with higher propensities to become drug resistant could result in inappropriate antifungal therapy, and ultimately lead to treatment failure. In Figure 2, we suggest an in-vitro experimental evolutionary model that can be used in predicting drug resistance likelihood through drug exposure. The above experiment is aimed at measuring two primary outcomes: the likelihood of becoming resistant and the time it takes to attain resistance breakpoint or epidemiological cutoff (resistance time). Data from such experiments can be used to develop statistical learning models that can predict the likelihood of resistance and resistance time in a matter of minutes on a laptop. For example, supposing that acquiring resistance is through a stochastic process related to spontaneous mutation(s), then the likelihood of a genotype becoming resistant can be obtained using the following function $L(G/O) = P(O/G)$, where O is the observed outcome, G is the parameter set (genotype) that define the stochastic process, L is likelihood, and P is probability. Information on a wild-type genotype's propensity to become resistant and resistance time can be very useful in determining a treatment course and formulating disease control strategies. Similarly, given the role of sexual reproduction in the emergence of genotypes of clinical importance [63], developing tools capable of predicting highly fit and super mating genotypes is critical, as such genotypes can easily spread resistance genes through sexual reproduction and rapidly expanding their distribution across geographic areas. Indeed, a highly fit multi-triazole resistant *A. fumigatus* genotype which very likely acquired resistance through sexual reproduction has expanded across thousands of kilometers in India [73,74].

In conclusion, from a practical perspective, researchers involved in *Aspergillus* surveillance should monitor the following: (i) genetic structure of the local, regional, national, and global populations of human pathogenic *Aspergillus*; (ii) the distribution and overall prevalence of hyper-virulent strains; (iii) the distribution and overall prevalence of antifungal drug resistant strains including those with a higher propensity to become resistant; and (iv) the transmission patterns of the latter two categories of genotypes. Furthermore, given current increases in non-*fumigatus* aspergillosis, significant attention should be paid to monitoring non-*fumigatus Aspergilli*, especially cryptic species.

2.2. The Usefulness of Molecular Epidemiology in Threat Preparedness

Recent increases in globalization have led to more frequent movement of goods and people than ever before in human history. The constant movement of people and goods across geographic scales has significant implications for the spread of clinically important pathogens like *Aspergilli*. Indeed, both anthropogenic and non-anthropogenic activities have impacted the spread of clinically important *Aspergilli* across geographic boarders [65,76]. Such impacts highlight the need for preparedness even in countries with low aspergillosis incidences. There is an age-old adage that says "Luck favors the prepared mind". Threat preparedness in the context of this review implies being able to effectively anticipate and take the right steps to manage threats caused by *Aspergilli*. Among other things, this entails being able to accurately and rapidly diagnose *Aspergilli*, especially drug-resistant *Aspergilli*. Thus far, multiple diagnostic methods—including electrospray ionization and matrix-assisted laser desorption ionization mass spectrometry, nucleic acid sequence-based amplification, polymerase chain reactions (real-time, multiplex and nested), microsphere-based Luminex, loop-mediated isothermal amplification, and enzyme linked immunosorbent assays—have been used to identify *Aspergilli* in a wide range of specimens including whole blood, serum, plasma, bronchoalveolar lavages, and exhale breath condensate [77–81]. These methods, including their respective advantages and disadvantages, have been extensively reviewed and are not discussed in detail here. Instead, our focus is on providing suggestions necessary for leveraging molecular diagnostics in preparedness against the threat caused by *Aspergilli*.

Specifically, we would like to highlight two critical issues that need to be addressed by many countries in their efforts to manage the threat caused by *Aspergilli*. Firstly, we note that more research needs to be done in developing rapid diagnostic molecular markers for clinically important non-*fumigatus Aspergilli*, including the divergent lineages/cryptic species. This is particularly important because although some of these species represent only a small proportion of clinically diagnosed *Aspergilli*, they form significant proportions of drug-resistant *Aspergilli*. For example, although *A. lentulus* accounted for only ~3% (3/86) of a set of 86 *Aspergillus* isolates obtained from Italy and Netherlands, they represented approximately 21% (3/14) of all ($\geq$ 2 mg/L) voriconazole-resistant strains [82]. Another important example to note is *A. calidoustus*, an emerging pathogen in lung transplants which can also colonize water distribution systems, including those in health care settings [83,84]. Thus, the need for accurate and rapid diagnosis of non-*fumigatus Aspergilli* and cryptic species is becoming increasingly important.

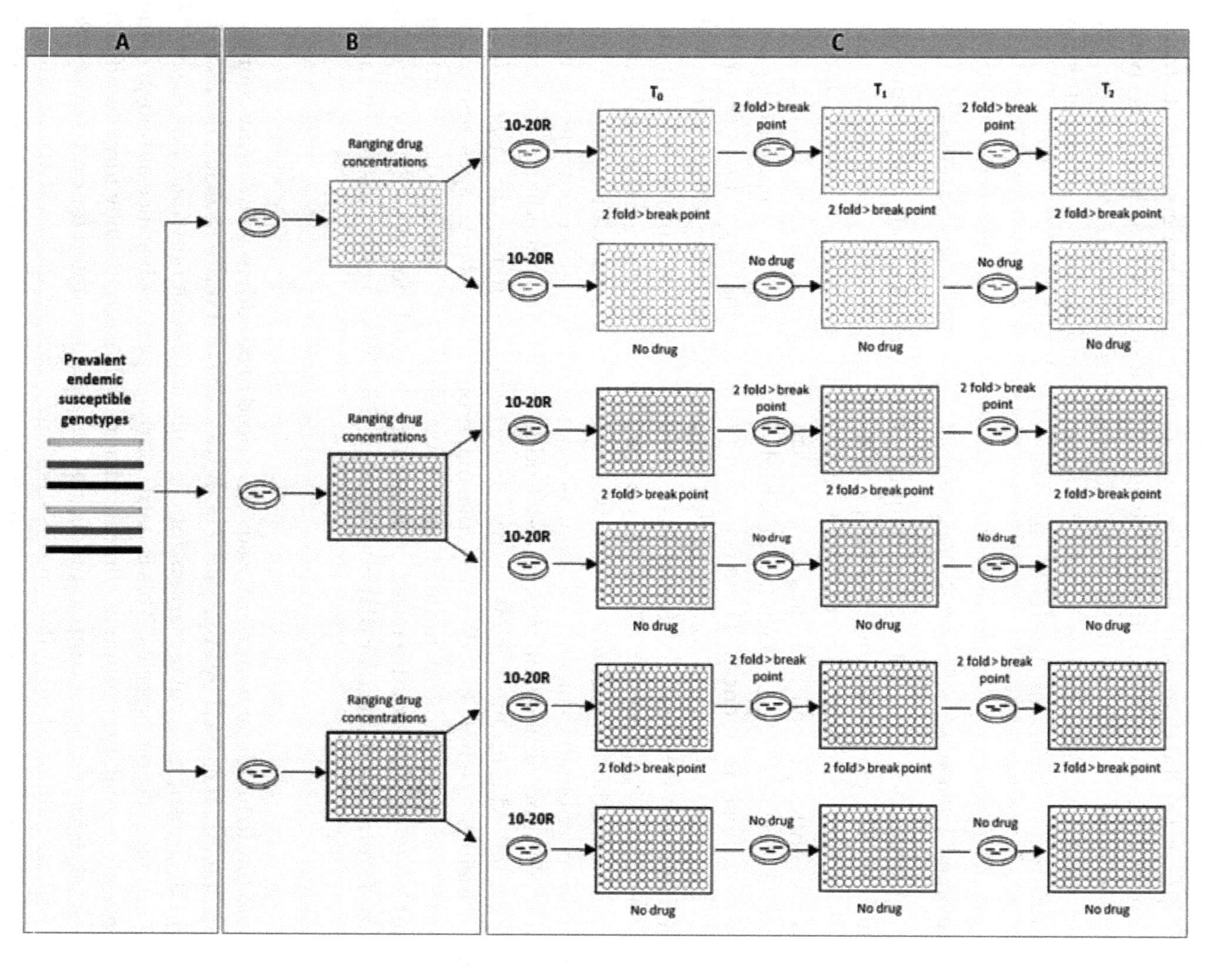

Figure 2. Experimental design to predict resistance through drug exposure in susceptible genotypes. The procedures depicted in panels A and B entail identifying prevalent endemic susceptible genotypes and determining their minimum inhibitory concentrations (MICs) to specific antifungal drugs. All genotypes used for the evolutionary experiment in panel C should have similar starting MICs, preferably very low. The evolutionary experiment consists of continuously exposing 10–20 replicates (R) of the selected genotypes to an antifungal drug as described in the Clinical & Laboratory Standards Institute (CLSI) protocol while including a no-drug control group [75]. At the end of 48 h, MICs are recorded, and new petri dishes are inoculated with microtiter plate content from the previous round of selection. The steps are repeated until genotypes reach resistance break point or epidemiological cutoff values.

Secondly, we highlight the need for standardization of molecular diagnostic procedures and markers at country and regional levels. Generally speaking, of all molecular diagnostic platforms, PCR platforms seem to show the best potential as they are able to identify *Aspergilli* to the species level while also being able to help identify antifungal susceptibility patterns. Thus far, plasma samples have resulted in the highest sensitivity (91%), while whole blood produced the highest specificity (96%) [85]. The highest diagnostic sensitivity and specificity in bronchoalveolar lavages samples thus far were 91% and 92%, respectively [86]. In cerebrospinal fluid, an *Aspergillus*-specific nested PCR assay showed sensitivity and specificity values of 100% and 93%, respectively [87]. Despite the great potential of PCR platforms, the specificity and sensitivity of PCR can be notably affected by two key factors: (i) method of DNA extraction; and (ii) specimen type. It was recently shown that factors including bead beating, white cell lysis, elution, and specimen volume can notably affect the quality and quantity of extracted DNA, and consequently PCR sensitivity [88]. In spite of the significant progress made by the European *Aspergillus* PCR initiative in quelling variation associated with DNA extraction from whole blood [89], the blood fraction best suited for PCR assays is still disputed. Furthermore, PCR performance is still to be evaluated in a wide range of patients, with other types of immunosuppressive conditions as most specimens used for PCR assays this far were performed on samples from hematology-oncology patients. Similarly, the differences in bronchoalveolar lavage procedures between patients and medical centers can affect PCR interpretation [90]. Indeed, there is room for improvements in all molecular diagnostic platforms.

Regardless of these shortfalls, molecular diagnostic platforms such as PCR can still be leveraged in the management of *Aspergilli* threats in two key ways. Firstly, molecular diagnostic platforms can be used in combinations to obtain optimal results. For example, a recent study showed that combining a lateral flow device with PCR yielded 100% diagnostic sensitivity and specificity [91]. Similarly, an earlier diagnosis and a lower incidence of IA were associated with a combination of galactomannan (GM) and PCR-based *Aspergillus* detection [92]. Secondly, PCR and similar molecular diagnostic platforms can be preemptively used in the surveillance of high risk patients in order to provide rapid usable results if needed. For instance, an improved 30-day survival rate was observed in a group that received PCR-guided prophylactic treatment compared to that for those treated on the basis of symptoms alone [93]. Similarly, combined surveillance of serum GM and PCR was shown to help decrease the incidence of IA in at-risk populations [92]. Surveillance of at-risk populations, health care facilities and adjacent environmental areas using molecular diagnostic platforms depicts how interdependent components of pathogen threat management can be. However, combining different platforms for diagnosis, prevention, and treatment requires standardization of procedures across different immunosuppressed populations and the mastery of diverse mycological procedures. Similarly, routine molecular diagnostic surveillance in high risk populations will require a significant amount of human resources. In the context of *Aspergillus* threat management, more specifically preparedness, standard operating procedures for the rapid and accurate diagnosis of both common and uncommon clinically important *Aspergilli* still need to be developed and adopted in many countries. Furthermore, the necessary human resources, cost effectiveness, and research capacity required to achieve the latter should be considered in most clinical microbiology laboratories.

Within individual *Aspergillus* species, having a good understanding of the molecular epidemiology is vital in formulating integrative molecular diagnostic plans. For instance, given that almost all multi-triazole resistant *A. fumigatus* strains in India belonged to a single microsatellite genotype (14/20/9/31/9/10/8/10/28) [74], screening for this genotype in addition to the common *A. fumigatus* species-level gene markers discussed here could diagnose an *A. fumigatus* isolate to a below-species level, including whether it is multi-triazole resistant. Such a plan could also be used in other geographic areas if one or a few genotypes dominate the drug-resistant population.

3. Conclusions

In this paper, we highlight the importance of surveillance and molecular diagnostics in *Aspergillus* threat preparedness and prevention. Given how interconnected that the world has become, standardization of loci and methods used for molecular diagnostics and surveillance is critically important for managing the current global *Aspergillus* threat. Although not discussed in detail here, combination therapy recommendations and vaccine research are two other key *Aspergillus* threat preparedness and prevention components worth mentioning.

Effectively managing the current threat caused by *Aspergilli* will require the significant incorporation of the molecular epidemiology component into *Aspergillus* threat preparedness and prevention. From a public health policy perspective, incorporation of genetic information into *Aspergillus* threat management is still a relatively new concept. For example, the European Organization for Research and Treatment of Cancer/Invasive Fungal Infections Cooperative Group and the National Institute of Allergy and Infectious Diseases Mycoses Study Group (EORTC/MSG) only recently suggested the inclusion of molecular diagnostics for the case definition of IA [94]. However, the recommended diagnosis can only identify the infecting pathogen to species level. A recent study identified that different genetic populations within *A. fumigatus* have different rates of triazole resistance [65]. Thus, identifying infecting pathogens to the genotype level will have significant treatment value. Furthermore, molecular markers targeting drug resistance mutations are being developed. The adoption by clinical microbiology labs around the globe of these fine-scale molecular methods will lead to better-targeted treatments and improved patient outcome at the population level.

Author Contributions: J.X. conceived the paper. E.E.A. produced the initial draft. Both E.E.A. and J.X. contributed to finalizing the text.

Acknowledgments: We thank Heather Yoell for proofreading the manuscript.

References

1. Bongomin, F.; Gago, S.; Oladele, R.O.; Denning, D.W. Global and multi-national prevalence of fungal diseases—Estimate precision. *J. Fungi* **2017**, *3*, 57. [CrossRef] [PubMed]
2. Vallabhaneni, S.; Mody, R.K.; Walker, T.; Chiller, T. The global burden of fungal diseases. *Infect. Dis. Clin.* **2016**, *30*, 1–11. [CrossRef] [PubMed]
3. Pegorie, M.; Denning, D.W.; Welfare, W. Estimating the burden of invasive and serious fungal disease in the United Kingdom. *J. Infect.* **2017**, *74*, 60–71. [CrossRef] [PubMed]
4. Gamaletsou, M.N.; Drogari-Apiranthitou, M.; Denning, D.W.; Sipsas, N.V. An estimate of the burden of serious fungal diseases in Greece. *Eur. J. Clin. Microbiol. Infect. Dis.* **2016**, *35*, 1115–1120. [CrossRef] [PubMed]
5. Gangneux, J.-P.; Bougnoux, M.-E.; Hennequin, C.; Godet, C.; Chandenier, J.; Denning, D.W.; Dupont, B. LIFE program, the Société française de mycologie médicale SFMM-study group. An estimation of burden of serious fungal infections in France. *J. Mycol. Med.* **2016**, *26*, 385–390. [CrossRef] [PubMed]
6. Oladele, R.O.; Denning, D.W. Burden of serious fungal infection in Nigeria. *West Afr. J. Med.* **2014**, *33*, 107–114. [PubMed]
7. Ben, R.; Denning, D.W. Estimating the burden of fungal diseases in Israel. *Isr. Med. Assoc. J. IMAJ* **2015**, *17*, 374–379. [PubMed]
8. Taj-Aldeen, S.J.; Chandra, P.; Denning, D.W. Burden of fungal infections in Qatar. *Mycoses* **2015**, *58* (Suppl. 5), 51–57. [CrossRef] [PubMed]
9. Lagrou, K.; Maertens, J.; Van Even, E.; Denning, D.W. Burden of serious fungal infections in Belgium. *Mycoses* **2015**, *58* (Suppl. 5), 1–5. [CrossRef] [PubMed]
10. Osmanov, A.; Denning, D.W. Burden of serious fungal infections in Ukraine. *Mycoses* **2015**, *58* (Suppl. 5), 94–100. [CrossRef] [PubMed]

11. Mortensen, K.L.; Denning, D.W.; Arendrup, M.C. The burden of fungal disease in Denmark. *Mycoses* **2015**, *58* (Suppl. 5), 15–21. [CrossRef] [PubMed]
12. Khwakhali, U.S.; Denning, D.W. Burden of serious fungal infections in Nepal. *Mycoses* **2015**, *58* (Suppl. 5), 45–50. [CrossRef] [PubMed]
13. Badiane, A.S.; Ndiaye, D.; Denning, D.W. Burden of fungal infections in Senegal. *Mycoses* **2015**, *58* (Suppl. 5), 63–69. [CrossRef] [PubMed]
14. Klimko, N.; Kozlova, Y.; Khostelidi, S.; Shadrivova, O.; Borzova, Y.; Burygina, E.; Vasilieva, N.; Denning, D.W. The burden of serious fungal diseases in Russia. *Mycoses* **2015**, *58* (Suppl. 5), 58–62. [CrossRef] [PubMed]
15. Faini, D.; Maokola, W.; Furrer, H.; Hatz, C.; Battegay, M.; Tanner, M.; Denning, D.W.; Letang, E. Burden of serious fungal infections in Tanzania. *Mycoses* **2015**, *58* (Suppl. 5), 70–79. [CrossRef] [PubMed]
16. Sinkó, J.; Sulyok, M.; Denning, D.W. Burden of serious fungal diseases in Hungary. *Mycoses* **2015**, *58* (Suppl. 5), 29–33. [CrossRef] [PubMed]
17. Mandengue, C.E.; Denning, D.W. The burden of serious fungal infections in Cameroon. *J. Fungi* **2018**, *4*, 44. [CrossRef] [PubMed]
18. Guto, J.A.; Bii, C.C.; Denning, D.W. Estimated burden of fungal infections in Kenya. *J. Infect. Dev. Ctries.* **2016**, *10*, 777–784. [CrossRef] [PubMed]
19. Corzo-León, D.E.; Armstrong-James, D.; Denning, D.W. Burden of serious fungal infections in Mexico. *Mycoses* **2015**, *58* (Suppl. 5), 34–44. [CrossRef] [PubMed]
20. George Agrios. *Plant Pathology*, 5th ed.; Academic Press: New York, NY, USA, 2005; ISBN 978-0-12-044565-3.
21. Bennett, J.W. An overview of the genus *Aspergillus*. In *Aspergillus: Molecular Biology and Genomics*; Katsuya, G., Masayuki, M., Eds.; Horizon Scientific Press: Norfolk, UK, 2010; pp. 1–17.
22. Ashu, E.; Forsythe, A.; Vogan, A.; Xu, J. Filamentous Fungi in Fermented Foods. In *Fermented Foods, Part I: Biochemistry and Biotechnology*; Didier, M., Ramesh, R., Eds.; CRC Press: Boca Raton, FL, USA, 2016; pp. 60–90, ISBN 978-1-4987-4081-4.
23. Dyer, P.S.; O'Gorman, C.M. A fungal sexual revolution: *Aspergillus* and *Penicillium* show the way. *Curr. Opin. Microbiol.* **2011**, *14*, 649–654. [CrossRef] [PubMed]
24. Geiser, D.M.; Klich, M.A.; Frisvad, J.C.; Peterson, S.W.; Varga, J.; Samson, R.A. The current status of species recognition and identification in *Aspergillus*. *Stud. Mycol.* **2007**, *59*, 1–10. [CrossRef] [PubMed]
25. Sugui, J.A.; Kwon-Chung, K.J.; Juvvadi, P.R.; Latgé, J.-P.; Steinbach, W.J. *Aspergillus fumigatus* and related species. *Cold Spring Harb. Perspect. Med.* **2015**, *5*, a019786. [CrossRef] [PubMed]
26. Guarro, J.; Xavier, M.O.; Severo, L.C. Differences and similarities amongst pathogenic *Aspergillus* species. In *Aspergillosis: From Diagnosis to Prevention*; Alessandro, C.P., Ed.; Springer: Dordrecht, The Netherlands, 2009; pp. 7–32. ISBN 978-90-481-2407-7.
27. Varga, J.; Houbraken, J.; Van Der Lee, H.A.L.; Verweij, P.E.; Samson, R.A. *Aspergillus calidoustus* sp. nov., causative agent of human infections previously assigned to *Aspergillus ustus*. *Eukaryot. Cell* **2008**, *7*, 630–638. [CrossRef] [PubMed]
28. Balajee, S.A.; Gribskov, J.L.; Hanley, E.; Nickle, D.; Marr, K.A. *Aspergillus lentulus* sp. nov., a new sibling species of *A. fumigatus*. *Eukaryot. Cell* **2005**, *4*, 625–632. [CrossRef] [PubMed]
29. Gautier, M.; Normand, A.-C.; Ranque, S. Previously unknown species of *Aspergillus*. *Clin. Microbiol. Infect.* **2016**, *22*, 662–669. [CrossRef] [PubMed]
30. Alastruey-Izquierdo, A.; Alcazar-Fuoli, L.; Cuenca-Estrella, M. Antifungal susceptibility profile of cryptic species of *Aspergillus*. *Mycopathologia* **2014**, *178*, 427–433. [CrossRef] [PubMed]
31. Balajee, S.A.; Kano, R.; Baddley, J.W.; Moser, S.A.; Marr, K.A.; Alexander, B.D.; Andes, D.; Kontoyiannis, D.P.; Perrone, G.; Peterson, S.; et al. Molecular identification of *Aspergillus* species collected for the transplant-associated infection surveillance network. *J. Clin. Microbiol.* **2009**, *47*, 3138–3141. [CrossRef] [PubMed]
32. Alastruey-Izquierdo, A.; Mellado, E.; Peláez, T.; Pemán, J.; Zapico, S.; Alvarez, M.; Rodríguez-Tudela, J.L.; Cuenca-Estrella, M.; Group, F.S. Population-based survey of filamentous fungi and antifungal resistance in Spain (FILPOP Study). *Antimicrob. Agents Chemother.* **2013**, *57*, 3380–3387. [CrossRef] [PubMed]
33. Negri, C.E.; Gonçalves, S.S.; Xafranski, H.; Bergamasco, M.D.; Aquino, V.R.; Castro, P.T.O.; Colombo, A.L. Cryptic and rare *Aspergillus* species in Brazil: Prevalence in clinical samples and in vitro susceptibility to triazoles. *J. Clin. Microbiol.* **2014**, *52*, 3633–3640. [CrossRef] [PubMed]
34. Maturu, V.N.; Agarwal, R. Itraconazole in chronic pulmonary aspergillosis: In whom, for how long, and at what dose? *Lung India Off. Organ Indian Chest Soc.* **2015**, *32*, 309–312. [CrossRef]

35. Agarwal, R. Allergic bronchopulmonary aspergillosis. *Chest* **2009**, *135*, 805–826. [CrossRef] [PubMed]
36. Dagenais, T.R.T.; Keller, N.P. Pathogenesis of *Aspergillus fumigatus* in invasive aspergillosis. *Clin. Microbiol. Rev.* **2009**, *22*, 447–465. [CrossRef] [PubMed]
37. Vonberg, R.-P.; Gastmeier, P. Nosocomial aspergillosis in outbreak settings. *J. Hosp. Infect.* **2006**, *63*, 246–254. [CrossRef] [PubMed]
38. Weber, D.J.; Peppercorn, A.; Miller, M.B.; Sickbert-Benett, E.; Rutala, W.A. Preventing healthcare-associated *Aspergillus* infections: Review of recent CDC/HICPAC recommendations. *Med. Mycol.* **2009**, *47*, S199–S209. [CrossRef] [PubMed]
39. Balajee, S.A.; Tay, S.T.; Lasker, B.A.; Hurst, S.F.; Rooney, A.P. Characterization of a novel gene for strain typing reveals substructuring of *Aspergillus fumigatus* across North America. *Eukaryot. Cell* **2007**, *6*, 1392–1399. [CrossRef] [PubMed]
40. Chang, C.C.; Cheng, A.C.; Devitt, B.; Hughes, A.J.; Campbell, P.; Styles, K.; Low, J.; Athan, E. Successful control of an outbreak of invasive aspergillosis in a regional haematology unit during hospital construction works. *J. Hosp. Infect.* **2008**, *69*, 33–38. [CrossRef] [PubMed]
41. Guinea, J.; García de Viedma, D.; Peláez, T.; Escribano, P.; Muñoz, P.; Meis, J.F.; Klaassen, C.H.W.; Bouza, E. Molecular epidemiology of *Aspergillus fumigatus*: An in-depth genotypic analysis of isolates involved in an outbreak of invasive aspergillosis. *J. Clin. Microbiol.* **2011**, *49*, 3498–3503. [CrossRef] [PubMed]
42. Peláez, T.; Muñoz, P.; Guinea, J.; Valerio, M.; Giannella, M.; Klaassen, C.H.W.; Bouza, E. Outbreak of invasive aspergillosis after major heart surgery caused by spores in the air of the intensive care unit. *Clin. Infect. Dis.* **2012**, *54*, e24–e31. [CrossRef] [PubMed]
43. Pettit, A.C.; Kropski, J.A.; Castilho, J.L.; Schmitz, J.E.; Rauch, C.A.; Mobley, B.C.; Wang, X.J.; Spires, S.S.; Pugh, M.E. The index case for the fungal meningitis outbreak in the United States. *N. Engl. J. Med.* **2012**, *367*, 2119–2125. [CrossRef] [PubMed]
44. Vena, A.; Muñoz, P.; Pelaez, T.; Guinea, J.; Valerio, M.; Bouza, E. Non-construction related *Aspergillus* outbreak in non-hematological patients related to high concentrations of airborne spores in non-HEPA filtered areas. *Open Forum Infect. Dis.* **2015**, *2*, 352. [CrossRef]
45. Kabbani, D.; Goldraich, L.; Ross, H.; Rotstein, C.; Husain, S. Outbreak of invasive aspergillosis in heart transplant recipients: The role of screening computed tomography scans in asymptomatic patients and universal antifungal prophylaxis. *Transpl. Infect. Dis.* **2017**, *20*, e12808. [CrossRef] [PubMed]
46. Rivero-Menendez, O.; Alastruey-Izquierdo, A.; Mellado, E.; Cuenca-Estrella, M. Triazole resistance in *Aspergillus* spp.: A worldwide problem? *J. Fungi* **2016**, *2*, 21. [CrossRef] [PubMed]
47. Garcia-Rubio, R.; Cuenca-Estrella, M.; Mellado, E. Triazole resistance in *Aspergillus* species: An emerging problem. *Drugs* **2017**, *77*, 599–613. [CrossRef] [PubMed]
48. Reichert-Lima, F.; Lyra, L.; Pontes, L.; Moretti, M.L.; Pham, C.D.; Lockhart, S.R.; Schreiber, A.Z. Surveillance for azoles resistance in *Aspergillus* spp. highlights a high number of amphotericin B-resistant isolates. *Mycoses* **2018**, *61*. [CrossRef] [PubMed]
49. Ashu, E.; Korfanty, G.; Samarasinghe, H.; Pum, N.; Man, Y.; Yamamura, D.; Xu, J. Widespread presence of amphotericin B resistant *Aspergillus fumigatus* in Hamilton, Canada. *Infect. Drug Resist.* **2018**, under review.
50. Paul, R.A.; Rudramurthy, S.M.; Meis, J.F.; Mouton, J.W.; Chakrabarti, A. A novel Y319H substitution in CYP51C associated with azole resistance in *Aspergillus flavus*. *Antimicrob. Agents Chemother.* **2015**, *59*, 6615–6619. [CrossRef] [PubMed]
51. Liu, W.; Sun, Y.; Chen, W.; Liu, W.; Wan, Z.; Bu, D.; Li, R. The T788G mutation in the *CYP51C* gene confers voriconazole resistance in *Aspergillus flavus* causing aspergillosis. *Antimicrob. Agents Chemother.* **2012**, *56*, 2598–2603. [CrossRef] [PubMed]
52. Sharma, C.; Kumar, R.; Kumar, N.; Masih, A.; Gupta, D.; Chowdhary, A. Investigation of multiple resistance mechanisms in voriconazole-resistant *Aspergillus flavus* clinical isolates from a chest hospital surveillance in Delhi, India. *Antimicrob. Agents Chemother.* **2018**, *62*. [CrossRef] [PubMed]
53. Nami, S.; Baradaran, B.; Mansoori, B.; Kordbacheh, P.; Rezaie, S.; Falahati, M.; Mohamed Khosroshahi, L.; Safara, M.; Zaini, F. The utilization of RNA silencing technology to mitigate the voriconazole resistance of *Aspergillus Flavus*; lipofectamine-based delivery. *Adv. Pharm. Bull.* **2017**, *7*, 53–59. [CrossRef] [PubMed]
54. Chowdhary, A.; Sharma, C.; Meis, J.F. *Candida auris*: A rapidly emerging cause of hospital-acquired multidrug-resistant fungal infections globally. *PLoS Pathog.* **2017**, *13*, e1006290. [CrossRef] [PubMed]

55. Sarma, S.; Upadhyay, S. Current perspective on emergence, diagnosis and drug resistance in *Candida auris*. *Infect. Drug Resist.* **2017**, *10*, 155–165. [CrossRef] [PubMed]
56. Dettori, M.; Piana, A.; Deriu, M.G.; Lo Curto, P.; Cossu, A.; Musumeci, R.; Cocuzza, C.; Astone, V.; Contu, M.A.; Sotgiu, G. Outbreak of multidrug-resistant *Acinetobacter baumannii* in an intensive care unit. *New Microbiol.* **2014**, *37*, 185–191. [PubMed]
57. Ghaith, D.M.; Zafer, M.M.; Al-Agamy, M.H.; Alyamani, E.J.; Booq, R.Y.; Almoazzamy, O. The emergence of a novel sequence type of MDR *Acinetobacter baumannii* from the intensive care unit of an Egyptian tertiary care hospital. *Ann. Clin. Microbiol. Antimicrob.* **2017**, *16*, 34. [CrossRef] [PubMed]
58. Zarrilli, R.; Casillo, R.; Di Popolo, A.; Tripodi, M.-F.; Bagattini, M.; Cuccurullo, S.; Crivaro, V.; Ragone, E.; Mattei, A.; Galdieri, N.; et al. Molecular epidemiology of a clonal outbreak of multidrug-resistant *Acinetobacter baumannii* in a university hospital in Italy. *Clin. Microbiol. Infect.* **2007**, *13*, 481–489. [CrossRef] [PubMed]
59. Eybpoosh, S.; Haghdoost, A.A.; Mostafavi, E.; Bahrampour, A.; Azadmanesh, K.; Zolala, F. Molecular epidemiology of infectious diseases. *Electron. Phys.* **2017**, *9*, 5149–5158. [CrossRef] [PubMed]
60. Villari, P.; Iacuzio, L.; Torre, I.; Scarcella, A. Molecular epidemiology as an effective tool in the surveillance of infections in the neonatal intensive care unit. *J. Infect.* **1998**, *37*, 274–281. [CrossRef]
61. De Valk, H.A.; Klaassen, C.H.W.; Meis, J.F.G.M. Molecular typing of *Aspergillus* species. *Mycoses* **2008**, *51*, 463–476. [CrossRef] [PubMed]
62. Varga, J. Molecular typing of *Aspergilli*: Recent developments and outcomes. *Med. Mycol.* **2006**, *44*, 149–161. [CrossRef]
63. Ashu, E.E.; Xu, J. The roles of sexual and asexual reproduction in the origin and dissemination of strains causing fungal infectious disease outbreaks. *Infect. Genet. Evol.* **2015**, *36*, 199–209. [CrossRef] [PubMed]
64. Lavergne, R.-A.; Chouaki, T.; Hagen, F.; Toublanc, B.; Dupont, H.; Jounieaux, V.; Meis, J.F.; Morio, F.; Le Pape, P. Home environment as a source of life-threatening azole-resistant *Aspergillus fumigatus* in immunocompromised patients. *Clin. Infect. Dis.* **2017**, *64*, 76–78. [CrossRef] [PubMed]
65. Ashu, E.E.; Hagen, F.; Chowdhary, A.; Meis, J.F.; Xu, J. Global population genetic analysis of *Aspergillus fumigatus*. *MSphere* **2017**, *2*, e00019-17. [CrossRef] [PubMed]
66. Ashu, E.E.; Korfanty, G.A.; Xu, J. Evidence of unique genetic diversity in *Aspergillus fumigatus* isolates from Cameroon. *Mycoses* **2017**, *60*, 739–748. [CrossRef] [PubMed]
67. Alvarez-Perez, S.; Blanco, J.L.; Alba, P.; Garcia, M.E. Mating type and invasiveness are significantly associated in *Aspergillus fumigatus*. *Med. Mycol.* **2010**, *48*, 273–277. [CrossRef] [PubMed]
68. Cheema, M.S.; Christians, J.K. Virulence in an insect model differs between mating types in *Aspergillus fumigatus*. *Med. Mycol.* **2011**, *49*, 202–207. [CrossRef] [PubMed]
69. Monteiro, M.C.; Garcia-Rubio, R.; Alcazar-Fuoli, L.; Peláez, T.; Mellado, E. Could the determination of *Aspergillus fumigatus* mating type have prognostic value in invasive aspergillosis? *Mycoses* **2018**, *61*, 172–178. [CrossRef] [PubMed]
70. Bain, J.M.; Tavanti, A.; Davidson, A.D.; Jacobsen, M.D.; Shaw, D.; Gow, N.A.; Odds, F.C. Multilocus sequence typing of the pathogenic fungus *Aspergillus fumigatus*. *J. Clin. Microbiol.* **2007**, *45*, 1469–1477. [CrossRef] [PubMed]
71. De Valk, H.A.; Meis, J.F.G.M.; Curfs, I.M.; Muehlethaler, K.; Mouton, J.W.; Klaassen, C.H.W. Use of a novel panel of nine short tandem repeats for exact and high-resolution fingerprinting of *Aspergillus fumigatus* isolates. *J. Clin. Microbiol.* **2005**, *43*, 4112–4120. [CrossRef] [PubMed]
72. García-Lerma, J.G.; Nidtha, S.; Blumoff, K.; Weinstock, H.; Heneine, W. Increased ability for selection of zidovudine resistance in a distinct class of wild-type HIV-1 from drug-naive persons. *Proc. Natl. Acad. Sci. USA* **2001**, *98*, 13907–13912. [CrossRef] [PubMed]
73. Abdolrasouli, A.; Rhodes, J.; Beale, M.A.; Hagen, F.; Rogers, T.R.; Chowdhary, A.; Meis, J.F.; Armstrong-James, D.; Fisher, M.C. Genomic context of azole resistance mutations in *Aspergillus fumigatus* determined using whole-genome sequencing. *MBio* **2015**, *6*, e00536-15. [CrossRef] [PubMed]
74. Chang, H.; Ashu, E.; Sharma, C.; Kathuria, S.; Chowdhary, A.; Xu, J. Diversity and origins of Indian multi-triazole resistant strains of *Aspergillus fumigatus*. *Mycoses* **2016**, *59*, 450–466. [CrossRef] [PubMed]
75. Clinical and Laboratory Standards Institute. *Reference Method for Broth Dilution Antifungal Susceptibility Testing of Filamentous Fungi*, 2nd ed.; Approved Standard, CLSI M38-A2; Clinical and Laboratory Standards Institute: Wayne, PA, USA, 2008.

76. Dunne, K.; Hagen, F.; Pomeroy, N.; Meis, J.F.; Rogers, T.R. Intercountry transfer of triazole-resistant *Aspergillus fumigatus* on Plant Bulbs. *Clin. Infect. Dis.* **2017**, *65*, 147–149. [CrossRef] [PubMed]
77. Lamoth, F. *Aspergillus fumigatus*-related species in clinical practice. *Front. Microbiol.* **2016**, *7*, 683. [CrossRef] [PubMed]
78. Tang, Q.; Tian, S.; Yu, N.; Zhang, X.; Jia, X.; Zhai, H.; Sun, Q.; Han, L. Development and evaluation of a loop-mediated isothermal amplification method for rapid detection of *Aspergillus fumigatus*. *J. Clin. Microbiol.* **2016**, *54*, 950–955. [CrossRef] [PubMed]
79. Powers-Fletcher, M.V.; Hanson, K.E. Molecular diagnostic testing for *Aspergillus*. *J. Clin. Microbiol.* **2016**, *54*, 2655–2660. [CrossRef] [PubMed]
80. Etienne, K.A.; Kano, R.; Balajee, S.A. Development and validation of a microsphere-based luminex assay for rapid identification of clinically relevant *Aspergilli*. *J. Clin. Microbiol.* **2009**, *47*, 1096–1100. [CrossRef] [PubMed]
81. Bhimji, A.; Bhaskaran, A.; Singer, L.G.; Kumar, D.; Humar, A.; Pavan, R.; Lipton, J.; Kuruvilla, J.; Schuh, A.; Yee, K.; et al. *Aspergillus* galactomannan detection in exhaled breath condensate compared to bronchoalveolar lavage fluid for the diagnosis of invasive aspergillosis in immunocompromised patients. *Clin. Microbiol. Infect.* **2018**, *24*, 640–645. [CrossRef] [PubMed]
82. Mello, E.; Posteraro, B.; Vella, A.; De Carolis, E.; Torelli, R.; D'Inzeo, T.; Verweij, P.E.; Sanguinetti, M. Susceptibility testing of common and uncommon *Aspergillus* species against posaconazole and other mold-active antifungal azoles using the Sensititre method. *Antimicrob. Agents Chemother.* **2017**, *61*, e00168-17. [CrossRef] [PubMed]
83. Hageskal, G.; Kristensen, R.; Fristad, R.F.; Skaar, I. Emerging pathogen *Aspergillus calidoustus* colonizes water distribution systems. *Med. Mycol.* **2011**, *49*, 588–593. [CrossRef] [PubMed]
84. Egli, A.; Fuller, J.; Humar, A.; Lien, D.; Weinkauf, J.; Nador, R.; Kapasi, A.; Kumar, D. Emergence of *Aspergillus calidoustus* infection in the era of post-transplantation azole prophylaxis. *Transplantation* **2012**, *94*, 403–410. [CrossRef] [PubMed]
85. Springer, J.; White, P.L.; Hamilton, S.; Michel, D.; Barnes, R.A.; Einsele, H.; Löffler, J. Comparison of performance characteristics of *Aspergillus* PCR in testing a range of blood-based samples in accordance with international methodological recommendations. *J. Clin. Microbiol.* **2016**, *54*, 705–711. [CrossRef] [PubMed]
86. Sun, W.; Wang, K.; Gao, W.; Su, X.; Qian, Q.; Lu, X.; Song, Y.; Guo, Y.; Shi, Y. Evaluation of PCR on bronchoalveolar lavage fluid for diagnosis of invasive aspergillosis: A bivariate meta-analysis and systematic review. *PLoS ONE* **2011**, *6*, e28467. [CrossRef] [PubMed]
87. Reinwald, M.; Buchheidt, D.; Hummel, M.; Duerken, M.; Bertz, H.; Schwerdtfeger, R.; Reuter, S.; Kiehl, M.G.; Barreto-Miranda, M.; Hofmann, W.-K.; et al. Diagnostic performance of an *Aspergillus*-specific nested PCR assay in cerebrospinal fluid samples of immunocompromised patients for detection of central nervous system aspergillosis. *PLoS ONE* **2013**, *8*, e56706. [CrossRef] [PubMed]
88. White, P.L.; Bretagne, S.; Klingspor, L.; Melchers, W.J.G.; McCulloch, E.; Schulz, B.; Finnstrom, N.; Mengoli, C.; Barnes, R.A.; Donnelly, J.P.; et al. European *Aspergillus* PCR initiative *Aspergillus* PCR: One step closer to standardization. *J. Clin. Microbiol.* **2010**, *48*, 1231–1240. [CrossRef] [PubMed]
89. White, P.L.; Perry, M.D.; Loeffler, J.; Melchers, W.; Klingspor, L.; Bretagne, S.; McCulloch, E.; Cuenca-Estrella, M.; Finnstrom, N.; Donnelly, J.P.; et al. European *Aspergillus* PCR initiative critical stages of extracting DNA from *Aspergillus fumigatus* in whole-blood specimens. *J. Clin. Microbiol.* **2010**, *48*, 3753–3755. [CrossRef] [PubMed]
90. Alanio, A.; Bretagne, S. Challenges in microbiological diagnosis of invasive *Aspergillus* infections. *F1000Research* **2017**, *6*. [CrossRef] [PubMed]
91. White, P.L.; Parr, C.; Thornton, C.; Barnes, R.A. Evaluation of real-time PCR, galactomannan enzyme-linked immunosorbent assay (ELISA), and a novel lateral-flow device for diagnosis of invasive aspergillosis. *J. Clin. Microbiol.* **2013**, *51*, 1510–1516. [CrossRef] [PubMed]
92. Aguado, J.M.; Vázquez, L.; Fernández-Ruiz, M.; Villaescusa, T.; Ruiz-Camps, I.; Barba, P.; Silva, J.T.; Batlle, M.; Solano, C.; Gallardo, D.; et al. PCRAGA Study Group; Spanish Stem Cell Transplantation Group; Study Group of Medical Mycology of the Spanish Society of Clinical Microbiology and Infectious Diseases; Spanish Network for Research in Infectious Diseases. Serum galactomannan versus a combination of galactomannan and polymerase chain reaction-based *Aspergillus* DNA detection for early therapy of invasive aspergillosis in high-risk hematological patients: A randomized controlled trial. *Clin. Infect. Dis.* **2015**, *60*, 405–414. [CrossRef] [PubMed]

93. Hebart, H.; Klingspor, L.; Klingebiel, T.; Loeffler, J.; Tollemar, J.; Ljungman, P.; Wandt, H.; Schaefer-Eckart, K.; Dornbusch, H.J.; Meisner, C.; et al. A prospective randomized controlled trial comparing PCR-based and empirical treatment with liposomal amphotericin B in patients after allo-SCT. *Bone Marrow Transplant.* **2009**, *43*, 553–561. [CrossRef] [PubMed]
94. De Pauw, B.; Walsh, T.J.; Donnelly, J.P.; Stevens, D.A.; Edwards, J.E.; Calandra, T.; Pappas, P.G.; Maertens, J.; Lortholary, O.; Kauffman, C.A.; et al. Revised definitions of invasive fungal disease from the European Organization for Research and Treatment of Cancer/Invasive Fungal Infections Cooperative Group and the National Institute of Allergy and Infectious Diseases Mycoses Study Group (EORTC/MSG) consensus group. *Clin. Infect. Dis.* **2008**, *46*, 1813–1821. [CrossRef] [PubMed]

9

From the Clinical Mycology Laboratory: New Species and Changes in Fungal Taxonomy and Nomenclature

Nathan P. Wiederhold * and Connie F. C. Gibas

Fungus Testing Laboratory, Department of Pathology and Laboratory Medicine, University of Texas Health Science Center at San Antonio, San Antonio, TX 78229, USA; gibas@uthscsa.edu

* Correspondence: wiederholdn@uthscsa.edu

Abstract: Fungal taxonomy is the branch of mycology by which we classify and group fungi based on similarities or differences. Historically, this was done by morphologic characteristics and other phenotypic traits. However, with the advent of the molecular age in mycology, phylogenetic analysis based on DNA sequences has replaced these classic means for grouping related species. This, along with the abandonment of the dual nomenclature system, has led to a marked increase in the number of new species and reclassification of known species. Although these evaluations and changes are necessary to move the field forward, there is concern among medical mycologists that the rapidity by which fungal nomenclature is changing could cause confusion in the clinical literature. Thus, there is a proposal to allow medical mycologists to adopt changes in taxonomy and nomenclature at a slower pace. In this review, changes in the taxonomy and nomenclature of medically relevant fungi will be discussed along with the impact this may have on clinicians and patient care. Specific examples of changes and current controversies will also be given.

Keywords: taxonomy; fungal nomenclature; phylogenetics; species complex

---

1. Introduction

Kingdom Fungi is a large and diverse group of organisms for which our knowledge is rapidly expanding. This kingdom includes numerous species that are capable of causing disease in humans, animals and plants. Infections caused by fungi are highly prevalent in humans, as it is estimated that greater than 1 billion people worldwide have infections caused by these organisms [1,2]. However, the full extent of fungi capable of causing infections in humans remains unknown. Although only several hundred species have been reported to cause disease in humans [3], it is estimated that there are between 1.5 million to 5 million fungal species and only approximately 100,000 species have been identified [4,5]. The potential clinical relevance of yet to be discovered species is highlighted by the nearly 10-fold increase in reports of newly described fungal pathogens in plants, animals and humans since 1995 [6], as well as by outbreaks of infections caused by fungi previously not associated with severe disease in humans [7–10]. Those that are capable of causing systemic infections in humans often have key attributes that make this possible (e.g., growth at 37 °C, penetrate or circumvent host barriers, digest and absorb components of human tissue, withstand immune responses of host) [11]. Many are also capable of persisting in the environment due to saprobic potential (i.e., the ability to grow on dead or decaying material) [12]. In addition, many species may be generalist pathogens with little host specificity and have dynamic genomes allowing for rapid adaption and evolution [11–13]. Thus, the number of fungal species that are etiologic agents of human infections will continue to grow. As the number of pathogenic species continues to grow, many of which are opportunists, new classifications and nomenclature will be introduced. In addition, revisions to current taxonomy will continue to be made based on our increased understanding of the diversity of this kingdom. In this review, changes in taxonomy and nomenclature of clinically relevant fungi will be discussed as will the challenges posed to clinicians and clinical microbiology laboratories by these changes.

2. Changes in Fungal Taxonomy and Nomenclature

Over the last several years significant changes have occurred in fungal taxonomy and nomenclature, as new fungi are discovered and the relationships of individual species to others and within larger taxonomic groups have been re-evaluated and redefined. Although the discovery of new fungal species and their classification has been a continuous process since the advent of the field of mycology, the pace of discovery and re-evaluation of taxonomic status has increased with the introduction of molecular and proteomic tools. Historically, morphologic characteristics and other phenotypic traits (e.g., growth on different media at different temperature, biochemical analysis) have been used for both taxonomic evaluation and species identification in clinical settings. However, the phenotypic traits that are observed may vary under different conditions and are thus subjective. Errors in species identification may occur because of this. DNA sequence analysis is now considered the gold standard for fungal species identification and has been a driving force for the increased pace of the discovery of new species and changes in fungal taxonomy and nomenclature [14–16]. Phylogenetic analysis based on the sequences of multiple loci within fungal DNA is often used for taxonomic designation of new species and in the re-evaluation of previous classifications that had been based solely on phenotypic characteristics. An advantage of phylogenetic analysis for taxonomic purposes is that close relatives become grouped together regardless of differences in morphology and these relationships may be useful for predicting pathogenicity and susceptibility to antifungal drugs [17]. These methods have led to the discovery of numerous cryptic species, which are indistinguishable from closely related species based on morphologic characteristics but can be identified by molecular means [18]. However, the use of phylogenetic analysis for taxonomic re-evaluation is not without its flaws, as the relationships created may be subject to change with increased understanding of fungal diversity since phylogenetic trees are highly subject to sampling effects [17]. In addition, no delimitation criteria exist above the species level [19]. Newer technologies, such as matrix assisted laser desorption ionization time-of-flight mass spectrometry (MALDI-TOF MS), are also being used with increased frequency for rapid species identification in clinical settings as well as for the taxonomic evaluation of fungi [20–23]. It should be noted that clinical laboratories may need to exercise caution in the adoption of these technologies for the identification of all fungal isolates until appropriately validated in the literature. Some examples of new and clinically relevant fungal species are listed in Table 1. Clearly, the description and recognition of new species helps to advance the field of medical mycology by increasing our understanding of the epidemiology of various fungal infections, the geographic distribution of species that cause these infections and how infections caused by different species may respond differently to treatment [24–28].

In addition to new tools for fungal identification and taxonomic re-evaluation, changes in fungal nomenclature have also been brought about by the elimination of the dual nomenclature system. When fungal taxonomy was based solely on morphologic characteristics, many fungi were forced to have multiple names describing either their sexual (teleomorph) or asexual (anamorph) life cycle stages under Article 59 of the International Code for Botanical Nomenclature. However, this dual nomenclature system became obsolete with the introduction of molecular tools since different morphologic stages are identical at the genetic level [17,19,29]. Thus, the system was abolished under the newly named International Code of Nomenclature of algae, fungi and plants in which fungi are now only to have one name [30]. However, decisions regarding which names to use have not always been straightforward. Some examples of clinically relevant changes in nomenclature for yeasts and molds are shown in Table 2.

Table 1. Examples of recently described and medically relevant fungi.

Species	Family	Order	Sites & Infections in Humans	Reference
Apophysomyces mexicanus	Saksenaeaceae	Mucorales	Necrotizing fasciitis	[31]
Aspergillus citrinoterreus	Aspergillaceae	Eurotiales	Pulmonary infection	[32]
Aspergillus suttoniae	Aspergillaceae	Eurotiales	Human sputum	[33]
Aspergillus tanneri	Aspergillaceae	Eurotiales	Lung, gastric abscess	[34]
Candida auris	Incertae sedis	Saccharomycetales	Various sites, candidemia	[35]
Curvularia americana	Pleosporaceae	Pleosporales	Nasal sinus, bone marrow	[36]
Curvularia chlamydospora	Pleosporaceae	Pleosporales	Nasal sinus, nail	[36]
Emergomyces canadensis	Ajellomycetaceae	Onygenales	Pneumonia, fungemia	[37,38]
Exophiala polymorpha	Herpotrichiellaceae	Chaetothyriales	Subcutaneous & cutaneous infections	[39]
Paracoccidioides lutzi	Ajellomycetaceae	Onygenales	Various	[40]
Rasamsonia aegroticola	Aspergillaceae	Eurotiales	Pulmonary infections	[41,42]
Spiromastigoides albida	Spiromastigaceae	Onygenales	Lung biopsy	[43]

Table 2. Examples of fungal nomenclature changes in medically relevant fungi.

New Name	Previous Name	Family	Order	Reference
Yeasts				
Apiotrichum mycotoxinivorans	*Trichosporon mycotoxinivorans*	Trichosporonaceae	Trichosporonales	[44]
Candida duobushaemulonii	*Candida haemulonii* group II	Incertae sedis	Saccharomycetales	[45]
Kluyveromyces marxianus	*Candida kefyr*	Saccharomycetaceae	Saccharomycetales	[46]
Magnusiomyces capitatus	*Blastoschizomyces capitatus/Geotrichum capitatum*	Dipodascaceae	Saccharomycetales	[47]
Meyerozyma guilliermondii	*Candida guilliermondii*	Debaryomycetaceae	Saccharomycetales	[48]
Moulds				
Blastomyces helicus	*Emmonsia helica*	Onygenaceae	Onygenalses	[33]
Blastomyces parvus	*Emmonsia parva*	Onygenaceae	Onygenales	[33]
Curvularia australiensis	*Bipolaris australiensis*	Pleosporaceae	Pleosporales	[49]
Cuvularia hawaiiensis	*Bipolaris hawaiiensis*	Pleosporaceae	Pleosporales	[49]
Curvularia spicifera	*Bipolaris spicifera*	Pleosporaceae	Pleosporales	[49]
Lichtheimia corymbifera	*Absidia corymbifera*	Lichtheimiaceae	Mucorales	[50]
Neocosmospora solani	*Fusarium solani*	Nectriaceae	Hypocreales	[51]
Purpureocillium lilacinum	*Paecilomyces lilacinus*	Ophiocrodycipitaceae	Hypocreales	[52]
Aspergillus thermomutatus	*Neosartorya pseudofischeri*	Aspergillaceae	Eurotiales	[53,54]
Aspergillus udagawae	*Neosartorya udagawae*	Aspergillaceae	Eurotiales	[54]
Rasamsonia argillacea	*Geosmithia argilacea*	Aspergillaceae	Eurotiales	[37]
Scedosporium boydii	*Pseudallescheria boydii*	Microascaceae	Microascales	[55]
Verruconis gallopava	*Ochroconis gallopava*	Sympoventuriaceae	Venturiales	[56]
Talaromyces marneffei	*Penicillium marneffei*	Aspergillaceae	Eurotiales	[57]

3. Implications of Changes in Nomenclature for Medical Mycology

The abolishment of the dual nomenclature system and the introduction of molecular tools for species identification have implications for medical mycology. There is concern that these changes may lead to confusion in the clinical literature regarding the names of the organisms or the diseases they cause among clinicians who do not closely follow taxonomic changes but are still responsible for navigating the medical publications to find clinically useful information regarding invasive mycoses and their etiologic agents in order to optimize patient care [17]. In addition, there is no single source that can be used to stay abreast of changes in fungal taxonomy and literature, as descriptions of new species or revised classifications are published in various scientific journals [58], many of which lack clinical scope. Websites that serve as useful online repositories include Mycobank (http://www.mycobank.org) and Index Fungorum (http://www.indexfungorum.org). Other useful resources include the Westerdijk Fungal Biodiversity Institute (http://www.westerdijkinstitute.nl/), the Atlas of Clinical Fungi, (http://www.clinicalfungi.org/), The Yeasts website (http://theyeasts.org/) and the International Commission of *Penicillium* and *Aspergillus* (https://www.aspergilluspenicillium.org/).

4. Recommendations for Nomenclature Changes in Medical Mycology

The relevance of nomenclature changes to medical mycology is often unknown at first and only later once cryptic or sibling species have been further evaluated in in vitro studies, animal models, or with the publications of case reports, does the clinical significance, or lack thereof, become better understood [17,19]. Because of this and the confusion that may be present in the literature due to differences in fungal nomenclature used between the clinical and purely mycologic literature, the International Society for Human and Animal Mycology Working Group on Nomenclature of Medical Fungi has made recommendations on the adoption of new fungal names. In general, this group has proposed that the clinical arena be allowed to follow and adopt changes in nomenclature at a slower pace [17,19]. At the genus level and higher, the taxa with similar medical attributes/characteristics would be maintained and changes should be made once validated and a consensus is reached regarding new classifications and nomenclature. Taxa should not be too large as this could possibly conceal phenotypic differences of clinical importance. Conversely, taxa should not be too small as this would reduce the distinction between genus and species; however, monotypic genera do exist (e.g., *Epidermophyton* and *Lophophyton*) [19,59].

At the species level, the term species complex should be used to cover the name used in medical practice for a group of similar organisms when there is a lack of evidence of the clinical relevance of cryptic species. Once the significance of the cryptic species becomes known to the medical community, the new name can be adopted and used by clinical laboratories and medical mycologists.

5. Species Complexes

In the mycology literature there has been an increased use of the term species complex. However, there is no clear taxonomic definition/statute for this term and various authors have used it in different contexts [60]. Some have used it to describe a selected group of organisms that are difficult to differentiate between based on standard diagnostic means, including classic morphologic and other phenotypic characteristics and in some cases DNA barcode analysis using single targets [60]. An example of this is the *Aspergillus viridinutans* species complex within *Aspergillus* section *Fumigati*, which includes 10 closely related species, including the human and animal pathogens *A. udagawae, A. felis, A. pseudofelis, A. parafelis, A. pseudoviridinutans and A. wyomingensis* [61].

In contrast, others have used the term species complex as a substitute for the subgenus term section. Examples of this use can be found within the *Fusarium, Aspergillus and Trichoderma* genera [62,63]. Still others have used species complexes to group together well-described species for which there are no known or insignificant differences in clinical parameters. An example of this is the *Aspergillus niger* species complex, in which there is a lack in differences in antifungal susceptibility profiles between the various species [64]. Other examples of the use of this term in this fashion may include the *Coccidioides immitis* species complex [65,66], which is now recognized to consist of the separate species *C. immitis* and *C. posadasii* [67], the *Candida albicans* species complex, which known to consist of *C. albicans*, *C. africana* and *C. stellatoidea* [66,68–71] and the *Candida glabrata* species complex, consisting of *C. glabrata*, *C. nivariensis* and *C. bracarensis* [66,71–74]. Although *C. immitis* and *C. posadasii* may differ in their geographic distributions [75,76], no clinically relevant differences appear to exist between these two species. Interestingly, the term species complex has been used in the literature for the examples listed above, even when it is known that these consist of distinct species [60,66,71,74].

Lastly, species complex has also been employed to group together species when the taxonomy is unsettled or under debate in the literature. This has been proposed for seven separate *Cryptococcus* species (*Cryptococcus neoformans* species complex) [23], although this is still under debate and different groups have different opinions as to the lumping or splitting of these species [66,77]. Although there may be important phenotypic differences among the species, the clinical relevance of these is not fully understood. Another example where key phenotypic differences may exist but the clinical relevance is not fully known is the *Candida parapsilosis* species complex (*C. parapsilosis*, *C. metapsilosis* and *C. orthopsilosis*). The original species, *C. parapsilosis*, is known to have reduced in vitro susceptibility to the echinocandins [78], although patients often respond well to therapy [79–81]. It is now known that *C. metapsilosis* and *C. orthopsilosis* are hybrids and this may be of clinical relevance [82,83].

One way that clinical laboratories can use species complexes is in the reporting of preliminary microbiologic test results. If a preliminary identification of an isolate can be reported to a clinician at the species complex level, this information may be useful in making treatment decisions while further studies are performed to identify the exact species. Once the species is known, the final results should then be provided. However, clinicians should also be made aware that all species within a particular complex may not have the same antifungal susceptibility profiles. Thus, a full species identification should be provided if available. This information will then be available for clinicians, epidemiologists and other mycologists for further study.

6. Clinically Relevant Changes in Fungal Nomenclature and Current Controversies

Acute invasive aspergillosis and chronic pulmonary aspergillosis are primarily caused by the species *A. fumigatus, A. flavus, A. nidulans, A. niger and A. terreus* [84–86]. However, surveillance studies that have used molecular means of species identification have reported higher rates of cryptic species than previously appreciated. In the TRANSNET study, which included solid organ and hematopoietic stem cell transplant recipients in U.S. centers, 11% of the 218 *Aspergillus* species isolated were found to be cryptic species, including *A. lentulus* (1.8%) and *A. udagawae* (1.4%) from section *Fumigati*, *A. tubingensis* (2.8%) from section *Nigri* and *A. calidoustus* (2.8%) from section *Usti* [87]. Similarly, in the FILPOP study, a population-based survey study conducted in Spain, 14.5% of the *Aspergillus* isolates were considered to be cryptic species [86]. This may be of clinical importance as several cryptic species have reduced susceptibility to the azoles or multiple classes of clinically available antifungals. For example, section *Fumigati*, there are currently at least 63 phylogenetically distinct species, of which at least 19 have been reported to cause disease in humans and animals [88–90]. This includes several that were previously known as *Neosartorya* species, including *A. fischeri* (formerly *N. fischeri*), *A. hiratsukae* (formerly *N. hiratsukae*), *A. thermomutatus* (formerly *N. pseudofischeri*) and *A. udagawae* (formerly *N. udagawae*) [91–95]. The previously discussed *A. viridinutans* species complex also falls into section *Fumigati*. Although the importance of distinguishing between members of this complex in the clinical setting is unknown, it is important to know that the species causing infection falls within this complex as these species are often associated with chronic infections as well as reduced antifungal susceptibility and thus may be refractory to therapy [61].

Another group of clinically important fungi that has undergone major taxonomic and nomenclature changes over the last decade is that of *Scedosporium*. Previously, *Pseudallescheria boydii* and *Scedosporium apiospermum* were considered to be the same species and were identified by morphology in clinical microbiology laboratories as *P. boydii* (teleomorph) or *S. apiospermum* (anamorph) based on their ability to develop sexual structures on routine culture media. This changed when it was determined that *P. boydii* (anamorph *Scedosporium boydii*) and *Pseudallescheria apiosperma* (anamorph *S. apiospermum)* were separate species based on phylogenetic analysis [96,97]. Subsequently, other species that are morphologically identical but genetically different have been discovered through the use of molecular phylogenetics [55,96,98]. The *Scedosporium apiospermum* species complex is composed of *S. apiospermum*, *S. boydii* and *Pseudallescheria angusta* [60]. However, other *Scedosporium* species, including *S. aurantiacum, S. dehoogii and S. minutisporum*, have not been placed within this species complex due to clear phylogenetic differences among the species and those that comprise this group, as well as differences in antifungal susceptibility patterns [60,99]. The morphologically distinct species previously known as *Scedosporium prolificans* has been renamed *Lomentospora prolificans* based on significant phylogenetic differences [55]. As *L. prolificans* is highly resistant to multiple antifungals [99–104] and infections caused by this organism are extremely difficult to treat [100], the distinction between this species and those in the genus *Scedosporium* species is clinically relevant.

Recently, a revision to the taxonomy of *Cryptococcus* species that frequently cause disease in humans was proposed. In a study that included 115 isolates, *Cryptococcus neoformans* var. *grubii* and *Cryptococcus neoformans* var. *neoformans* were split into the separate species *Cryptococcus neoformans* and *Cryptococcus deneoformans*, respectively [23], while *Cryptococcus gattii* was proposed to be split into 5 distinct species (Table 3). This was based on the results from multi-locus sequence typing (MLST) based phylogenetic analysis using 11 different loci, differences in phenotypic characteristics and other means. Phenotypic characteristics that were evaluated in this study and others have included temperature, melanin content, virulence in a *Drosophila melanogaster* model, sensitivity to mycophenolic acid and growth on L-canavanine glycine bromothymol blue (CGB) agar and creatinine dextrose bromothymol blue thymine (CDBT) agar [23,66]. The authors also evaluated MALDI-TOF MS and reported that this technology could also readily distinguish between the different *Cryptococcus* species.

Table 3. Proposed names for *Cryptococcus neoformans* and *C. gattii* species [23].

Current Name	Molecular Type	Proposed Name
Cryptococcus neoformans var. *grubii*	VNI, VNII, VNB	*Cryptococcus neoformans*
Cryptococcus neoformans var. *neoformans*	VNIV	*Cryptococcus deneoformans*
Cryptococcus gattii	VGI	*Cryptococcus gattii*
	VGIII	*Cryptococcus bacillisporus*
	VGII	*Cryptococcus deuterogattii*
	VGIV	*Cryptococcus tetragattii*
	VGIV/VGIIIc	*Cryptococcus decagattii*
Serotypes AD hybrid	VNIII	*Cryptococcus neoformans* x *Cryptococcus deneoformans* hybrid
Serotypes DB hybrid	AFLP8	*Cryptococcus deneoformans* x *Cryptococcus gattii* hybrid
Serotypes AB hybrid	AFLP9	*Cryptococcus neoformans* x *Cryptococcus gattii* hybrid
Serotypes AB hybrid	AFLP11	*Cryptococcus neoformans* x *Cryptococcus deuterogattii* hybrid

This proposal to divide the *Cryptococcus neoformans/gattii* species complex into different species has not been without criticism. In an editorial, Kwon-Chung et al. argued that the proposed division was premature as an insufficient number of isolates were used to make this taxonomic change [77]. A previous, larger analysis including over 2000 isolates, had showed greater genetic diversity and the possibility of even more species [77]. In addition, since loci from only 6 of the 14 chromosomes in *Cryptococcus* were used in the MLST-based phylogenetic analysis, the true extent of diversity and recombination events remains unknown. It was also argued that the proposed division is impractical for routine use in clinical microbiology laboratories. Eleven concatenated loci were used in the phylogenetic analysis that supported separating the species and even the most commonly used MLST scheme of seven concatenated loci recommended by the ISHAM Genotyping Working Group of *C. neoformans* and *C. gattii* is too complicated for clinical microbiology laboratories and even reference laboratories, especially since the loci commonly used for molecular identification of fungal species (i.e., ITS and D1/D2) are not included [77,105]. In addition, the MALDI-TOF MS score threshold used was somewhat different than the usual score cutoff value for species recognition [23,77,106] and the newly proposed species are not currently available in databases cleared by regulatory agencies for use in clinical microbiology laboratories. It should be noted that other studies have reported that lower score thresholds can be used to reliably identify fungal species [107–109]. However, many clinical microbiology laboratories may be reluctant to use lower score thresholds without internal validation studies. Concern was also raised regarding the possible creation of confusion between the taxonomic and clinical literature. Specifically, under the proposed nomenclature the Vancouver *C. gattii* epidemic reference strain R265 would no longer be *C. gattii* but instead would be reclassified as *C. deuterogattii*. Although it was recognized that the designation of seven separate species would be an important step for the formal recognition of the biodiversity of pathogenic *Cryptococcus* species, Kwon-Chung et al. instead proposed the use of *Cryptococcus neoformans* species complex and *Cryptococcus gattii* species complex based on these issues and our current insufficient understanding of the clinical differences among the various proposed *Cryptococcus* species. In a rebuttal, Hagen et al. defended the nomenclature changes and noted that the main advantage will be the advancement of the field through stimulation of further studies to assess for similarities and differences between the recognized species [66]. Additional work has subsequently reported that the newly proposed *Cryptococcus* species may indeed have clinically significant differences [23,66,110,111]. However, many clinical microbiology laboratories may not be able to adapt to the new nomenclature into their routine workflow in the near future, as the new species are not yet incorporated into commercially available assays and databases cleared for clinical use by regulatory agencies for diagnosis or species identification.

Fusarium species are significant causes of invasive infections in highly immunocompromised hosts [112,113]. In addition, infections including keratitis and onychomycosis, can also occur in immunocompetent patients [27,114]. Human infections can be caused by species grouped within 8 different species complexes, including: *Fusarium solani* species complex, *Fusarium oxysporum* species complex, *Fusarium fujikuroi* species complex, *Fusarium chlamydosporum* species complex, *Fusarium dimerum* species complex, *Fusarium incarnatum-equiseti* species complex, *Fusarium sambucinum* species complex and *Fusarium tricinctum* species complex [25], although most infections are caused by members of the *F. solani* and *F. oxysporum* species complexes [112]. The *F. solani* species complex encompass at least 60 phylogenetically distinct species and in addition to causing disease in humans and animals, also includes a number of important agricultural pathogens [115]. Traditionally, clinical microbiology laboratories have identified and reported these isolates as *F. solani* species complex and some reference laboratories also report the specific haplotype based on MLST of the translation elongation factor 1α and RNA polymerase II gene. Although some members of this complex have received formal species names (e.g., *F. petroliphilum* [halplotype 1], *F. keratoplasticum* [haplotype 2], *F. falciforme* [haplotype 3+4] and *F. solani* [haplotype 5]) many others have not [116,117].

Recently, it has been proposed that members of the *F. solani* species complex be moved to the genus *Neocosmospora* based on the results of phylogenetic analysis [53] and new species previously classified only as haplotypes have been described [118]. This includes the species *N. petroliphila* (*F. petroliphilum*), *N. keratoplastica* (*F. keratoplasticum*), *N. falciformis* (*F. falciforme*) and *N. solani* (*F. solani*), along with the new species *N. gamsii* (haplotype 7), *N. suttoniana* (haplotype 20) and *N. catenata* (haplotype 43). Others have argued against renaming members of the *F. solani* species complex based on the long-standing, historical concept of this genus [119]. In order to provide up-to-date information to both clinicians and clinical microbiologists, as well as facilitate their ability to find relevant information in the medical literature, the reports generated by our reference mycology laboratory provide both the name commonly used in the medical literature as well as the new nomenclature. For example, for a recent species identification of an isolate cultured from the cornea of a patient, the name frequently found in the clinical literature, *Fusarium falciforme*, was provided along with a statement that the species is now known as *Neocosmospora falciformis*.

7. Summary

The field of medical mycology is rapidly changing due to the introduction of new molecular and proteomic technologies. New species are rapidly being discovered in the environment, as are new etiologic agents of disease in humans and animals. The adoption of these technologies, along with the abandonment of the dual nomenclature system, has led to marked changes in fungal taxonomy and nomenclature, as organisms previously thought to be unrelated are now recognized as being genetically similar. Conversely, we are now learning that species that were previously considered to be related are in fact very different from each other. The rapidity of these changes has caused concern among some medical mycologists and clinicians that the nomenclature changes may lead to negative clinical consequences, as the ever-changing literature could cause confusion among those who are responsible direct patient care. To mitigate this possibility, it has been proposed that medical mycology, specifically clinicians and clinical microbiology laboratories, may need to adopt changes in fungal nomenclature and taxonomy at a more measured pace. In addition, clinical microbiology and reference laboratories should provide useful information that will aide clinicians in this endeavor. This includes keeping abreast of changes in taxonomy and nomenclature, serving as a resource for clinicians as to previous names that may be published in the literature, as well as the clinical significance of the new classifications. An example of how the clinical laboratories may provide up-to-date information as well as species names that are prevalent in the clinical literature is provided above (*Neocosmospora falciformis* and *Fusarium falciforme*). The discovery of new fungal species capable of causing disease in humans and animals and the reclassification of various groups will continue as our knowledge of fungal diversity increases. However, this should not impede clinicians in their treatment of patients with fungal infections.

Author Contributions: Writing—original draft preparation, N.P.W; review and editing, N.P.W and C.F.C.G.

References

1. Bongomin, F.; Gago, S.; Oladele, R.O.; Denning, D.W. Global and multi-national prevalence of fungal diseases-estimate precision. *J. Fungi* **2017**, *3*, 57. [CrossRef] [PubMed]
2. Havlickova, B.; Czaika, V.A.; Friedrich, M. Epidemiological trends in skin mycoses worldwide. *Mycoses* **2008**, *51* (Suppl. 4), 2–15. [CrossRef] [PubMed]
3. Taylor, L.H.; Latham, S.M.; Woolhouse, M.E. Risk factors for human disease emergence. *Philos. Trans. R. Soc. Lond. B Biol. Sci.* **2001**, *356*, 983–989. [CrossRef] [PubMed]
4. Jones, N. Planetary disasters: It could happen one night. *Nature* **2013**, *493*, 154–156. [CrossRef] [PubMed]
5. O'Brien, H.E.; Parrent, J.L.; Jackson, J.A.; Moncalvo, J.M.; Vilgalys, R. Fungal community analysis by large-scale sequencing of environmental samples. *Appl. Environ. Microbiol.* **2005**, *71*, 5544–5550. [CrossRef] [PubMed]
6. Fisher, M.C.; Henk, D.A.; Briggs, C.J.; Brownstein, J.S.; Madoff, L.C.; McCraw, S.L.; Gurr, S.J. Emerging fungal threats to animal, plant and ecosystem health. *Nature* **2012**, *484*, 186–194. [CrossRef]
7. Kauffman, C.A.; Pappas, P.G.; Patterson, T.F. Fungal infections associated with contaminated methylprednisolone injections. *N. Engl. J. Med.* **2013**, *368*, 2495–2500. [CrossRef]

8. Nucci, M.; Akiti, T.; Barreiros, G.; Silveira, F.; Revankar, S.G.; Sutton, D.A.; Patterson, T.F. Nosocomial fungemia due to *Exophiala jeanselmei* var. *jeanselmei and a Rhinocladiella species: Newly described causes of bloodstream infection. J. Clin. Microbiol.* **2001**, *39*, 514–518.
9. Nucci, M.; Akiti, T.; Barreiros, G.; Silveira, F.; Revankar, S.G.; Wickes, B.L.; Sutton, D.A.; Patterson, T.F. Nosocomial outbreak of Exophiala jeanselmei fungemia associated with contamination of hospital water. *Clin. Infect. Dis.* **2002**, *34*, 1475–1480. [CrossRef]
10. Saracli, M.A.; Mutlu, F.M.; Yildiran, S.T.; Kurekci, A.E.; Gonlum, A.; Uysal, Y.; Erdem, U.; Basustaoglu, A.C.; Sutton, D.A. Clustering of invasive *Aspergillus ustus* eye infections in a tertiary care hospital: A molecular epidemiologic study of an uncommon species. *Med. Mycol.* **2007**, *45*, 377–384. [CrossRef]
11. Kohler, J.R.; Casadevall, A.; Perfect, J. The spectrum of fungi that infects humans. *Cold Spring Harb. Perspect. Med.* **2014**, *5*, a019273. [CrossRef] [PubMed]
12. Casadevall, A. Determinants of virulence in the pathogenic fungi. *Fungal Biol. Rev.* **2007**, *21*, 130–132. [CrossRef] [PubMed]
13. Odds, F.C. Ecology and epidemiology of Candida species. *Zentralbl. Bakteriol. Mikrobiol. Hyg. A* **1984**, *257*, 207–212. [PubMed]
14. Petti, C.A. Detection and identification of microorganisms by gene amplification and sequencing. *Clin. Infect. Dis.* **2007**, *44*, 1108–1114. [PubMed]
15. Schoch, C.L.; Seifert, K.A.; Huhndorf, S.; Robert, V.; Spouge, J.L.; Levesque, C.A.; Chen, W.; Bolchacova, E.; Voigt, K.; Crous, P.W.; et al. Nuclear ribosomal internal transcribed spacer (ITS) region as a universal DNA barcode marker for Fungi. *Proc. Natl. Acad. Sci. USA* **2012**, *109*, 6241–6246. [CrossRef] [PubMed]
16. Seifert, K.A. Progress towards DNA barcoding of fungi. *Mol. Ecol. Resour.* **2009**, *9* (Suppl. 1), 83–89. [CrossRef] [PubMed]
17. De Hoog, G.S.; Chaturvedi, V.; Denning, D.W.; Dyer, P.S.; Frisvad, J.C.; Geiser, D.; Graser, Y.; Guarro, J.; Haase, G.; Kwon-Chung, K.J.; et al. Name changes in medically important fungi and their implications for clinical practice. *J. Clin. Microbiol.* **2015**, *53*, 1056–1062. [CrossRef]
18. Howard, S.J. Multi-resistant aspergillosis due to cryptic species. *Mycopathologia* **2014**, *178*, 435–439. [CrossRef]
19. De Hoog, G.S.; Haase, G.; Chaturvedi, V.; Walsh, T.J.; Meyer, W.; Lackner, M. Taxonomy of medically important fungi in the molecular era. *Lancet Infect. Dis.* **2013**, *13*, 385–386. [CrossRef]
20. Rychert, J.; Slechta, E.S.; Barker, A.P.; Miranda, E.; Babady, N.E.; Tang, Y.W.; Gibas, C.; Wiederhold, N.; Sutton, D.; Hanson, K.E. Multicenter Evaluation of the Vitek MS v3.0 System for the Identification of Filamentous Fungi. *J. Clin. Microbiol.* **2018**, *56*, e01353-17. [CrossRef]
21. Lau, A.F.; Drake, S.K.; Calhoun, L.B.; Henderson, C.M.; Zelazny, A.M. Development of a clinically comprehensive database and a simple procedure for identification of molds from solid media by matrix-assisted laser desorption ionization-time of flight mass spectrometry. *J. Clin. Microbiol.* **2013**, *51*, 828–834. [CrossRef] [PubMed]
22. Brun, S.; Madrid, H.; Gerrits Van Den Ende, B.; Andersen, B.; Marinach-Patrice, C.; Mazier, D.; De Hoog, G.S. Multilocus phylogeny and MALDI-TOF analysis of the plant pathogenic species *Alternaria dauci* and relatives. *Fungal Biol.* **2013**, *117*, 32–40. [CrossRef] [PubMed]
23. Hagen, F.; Khayhan, K.; Theelen, B.; Kolecka, A.; Polacheck, I.; Sionov, E.; Falk, R.; Parnmen, S.; Lumbsch, H.T.; Boekhout, T. Recognition of seven species in the *Cryptococcus gattii/Cryptococcus neoformans* species complex. *Fungal Genet. Biol.* **2015**, *78*, 16–48. [CrossRef] [PubMed]
24. Zhang, Y.; Hagen, F.; Stielow, B.; Rodrigues, A.M.; Samerpitak, K.; Zhou, X.; Feng, P.; Yang, L.; Chen, M.; Deng, S.; et al. Phylogeography and evolutionary patterns in *Sporothrix* spanning more than 14000 human and animal case reports. *Persoonia* **2015**, *35*, 1–20. [CrossRef] [PubMed]
25. O'Donnell, K.; Sarver, B.A.; Brandt, M.; Chang, D.C.; Noble-Wang, J.; Park, B.J.; Sutton, D.A.; Benjamin, L.; Lindsley, M.; Padhye, A.; et al. Phylogenetic diversity and microsphere array-based genotyping of human pathogenic Fusaria, including isolates from the multistate contact lens-associated U.S. keratitis outbreaks of 2005 and 2006. *J. Clin. Microbiol.* **2007**, *45*, 2235–2248. [CrossRef] [PubMed]
26. Short, D.P.; O'Donnell, K.; Thrane, U.; Nielsen, K.F.; Zhang, N.; Juba, J.H.; Geiser, D.M. Phylogenetic relationships among members of the *Fusarium solani* species complex in human infections and the descriptions of *F. keratoplasticum* sp. nov. and *F. petroliphilum* stat. nov. *Fungal Genet. Biol.* **2013**, *53*, 59–70. [CrossRef] [PubMed]

27. Chang, D.C.; Grant, G.B.; O'Donnell, K.; Wannemuehler, K.A.; Noble-Wang, J.; Rao, C.Y.; Jacobson, L.M.; Crowell, C.S.; Sneed, R.S.; Lewis, F.M.; et al. Multistate outbreak of Fusarium keratitis associated with use of a contact lens solution. *JAMA* **2006**, *296*, 953–963. [CrossRef]
28. Marimon, R.; Cano, J.; Gene, J.; Sutton, D.A.; Kawasaki, M.; Guarro, J. *Sporothrix brasiliensis*, *S. globosa*, and *S. mexicana*, three new *Sporothrix* species of clinical interest. *J. Clin. Microbiol.* **2007**, *45*, 3198–3206. [CrossRef]
29. Hawksworth, D.L. A new dawn for the naming of fungi: Impacts of decisions made in Melbourne in July 2011 on the future publication and regulation of fungal names. *IMA Fungus* **2011**, *2*, 155–162. [CrossRef]
30. Norvell, L.L. Melbourne approves a new CODE. *Mycotaxon* **2011**, *116*, 481–490. [CrossRef]
31. Bonifaz, A.; Stchigel, A.M.; Guarro, J.; Guevara, E.; Pintos, L.; Sanchis, M.; Cano-Lira, J.F. Primary cutaneous mucormycosis produced by the new species *Apophysomyces mexicanus*. *J. Clin. Microbiol.* **2014**, *52*, 4428–4431. [CrossRef] [PubMed]
32. Guinea, J.; Sandoval-Denis, M.; Escribano, P.; Pelaez, T.; Guarro, J.; Bouza, E. Aspergillus citrinoterreus, a new species of section Terrei isolated from samples of patients with nonhematological predisposing conditions. *J. Clin. Microbiol.* **2015**, *53*, 611–617. [CrossRef] [PubMed]
33. Siqueira, J.P.Z.; Wiederhold, N.; Gene, J.; Garcia, D.; Almeida, M.T.G.; Guarro, J. Cryptic *Aspergillus* from clinical samples in the USA and description of a new species in section *Flavipedes*. *Mycoses* **2018**, *61*, 814–825. [CrossRef] [PubMed]
34. Sugui, J.A.; Peterson, S.W.; Clark, L.P.; Nardone, G.; Folio, L.; Riedlinger, G.; Zerbe, C.S.; Shea, Y.; Henderson, C.M.; Zelazny, A.M.; et al. *Aspergillus tanneri* sp. nov., a new pathogen that causes invasive disease refractory to antifungal therapy. *J. Clin. Microbiol.* **2012**, *50*, 3309–3317. [CrossRef] [PubMed]
35. Satoh, K.; Makimura, K.; Hasumi, Y.; Nishiyama, Y.; Uchida, K.; Yamaguchi, H. *Candida auris* sp. nov., a novel ascomycetous yeast isolated from the external ear canal of an inpatient in a Japanese hospital. *Microbiol. Immunol.* **2009**, *53*, 41–44. [CrossRef]
36. Madrid, H.; da Cunha, K.C.; Gene, J.; Dijksterhuis, J.; Cano, J.; Sutton, D.A.; Guarro, J.; Crous, P.W. Novel Curvularia species from clinical specimens. *Persoonia* **2014**, *33*, 48–60. [CrossRef]
37. Jiang, Y.; Dukik, K.; Muñoz, J.F.; Sigler, L.; Schwartz, I.S.; Govender, N.P.; Kenyon, C.; Feng, P.; van den Ende, B.G.; Stielow, J.B.; et al. Phylogeny, ecology and taxonomy of systemic pathogens and their relatives in *Ajellomycetaceae* (Onygenales): *Blastomyces*, *Emergomyces*, *Emmonsia*, *Emmonsiellopsis*. *Fungal Divers.* **2018**, *90*, 245–291. [CrossRef]
38. Schwartz, I.S.; Sanche, S.; Wiederhold, N.P.; Patterson, T.F.; Sigler, L. *Emergomyces canadensis*, a Dimorphic Fungus Causing Fatal Systemic Human Disease in North America. *Emerg. Infect. Dis.* **2018**, *24*, 758–761. [CrossRef]
39. Yong, L.K.; Wiederhold, N.P.; Sutton, D.A.; Sandoval-Denis, M.; Lindner, J.R.; Fan, H.; Sanders, C.; Guarro, J. Morphological and Molecular Characterization of *Exophiala polymorpha* sp. nov. Isolated from Sporotrichoid Lymphocutaneous Lesions in a Patient with Myasthenia Gravis. *J. Clin. Microbiol.* **2015**, *53*, 2816–2822. [CrossRef]
40. Teixeira, M.M.; Theodoro, R.C.; Nino-Vega, G.; Bagagli, E.; Felipe, M.S. Paracoccidioides species complex: Ecology, phylogeny, sexual reproduction, and virulence. *PLoS Pathog.* **2014**, *10*, e1004397. [CrossRef]
41. Houbraken, J.; Giraud, S.; Meijer, M.; Bertout, S.; Frisvad, J.C.; Meis, J.F.; Bouchara, J.P.; Samson, R.A. Taxonomy and antifungal susceptibility of clinically important Rasamsonia species. *J. Clin. Microbiol.* **2013**, *51*, 22–30. [CrossRef] [PubMed]
42. Hong, G.; White, M.; Lechtzin, N.; West, N.E.; Avery, R.; Miller, H.; Lee, R.; Lovari, R.J.; Massire, C.; Blyn, L.B.; et al. Fatal disseminated Rasamsonia infection in cystic fibrosis post-lung transplantation. *J. Cyst. Fibros* **2017**, *16*, e3–e7. [CrossRef] [PubMed]
43. Stchigel, A.M.; Sutton, D.A.; Cano-Lira, J.F.; Wiederhold, N.; Guarro, J. New Species Spiromastigoides albida from a Lung Biopsy. *Mycopathologia* **2017**, *182*, 967–978. [CrossRef] [PubMed]
44. Liu, X.Z.; Wang, Q.M.; Goker, M.; Groenewald, M.; Kachalkin, A.V.; Lumbsch, H.T.; Millanes, A.M.; Wedin, M.; Yurkov, A.M.; Boekhout, T.; et al. Towards an integrated phylogenetic classification of the Tremellomycetes. *Stud. Mycol.* **2015**, *81*, 85–147. [CrossRef]

45. Cendejas-Bueno, E.; Kolecka, A.; Alastruey-Izquierdo, A.; Theelen, B.; Groenewald, M.; Kostrzewa, M.; Cuenca-Estrella, M.; Gomez-Lopez, A.; Boekhout, T. Reclassification of the *Candida haemulonii* complex as *Candida haemulonii* (*C. haemulonii* group I), *C. duobushaemulonii* sp. nov. (*C. haemulonii* group II), and *C. haemulonii* var. vulnera var. nov.: Three multiresistant human pathogenic yeasts. *J. Clin. Microbiol.* **2012**, *50*, 3641–3651. [CrossRef]
46. Van der Walt, J.P. The emendation of the genus *Kluyveromyces* v. d. Walt. *Antonie Van Leeuwenhoek* **1965**, *31*, 341–348. [CrossRef]
47. De Hoog, G.S.; Smith, M.T. Ribsomal gene phylogeny and species delimitation in Geotrichum and its teleomorphs. *Stud. Mycol.* **2004**, *50*, 489–515.
48. Kurtzman, C.P.; Suzuki, M. Phylogenetic analysis of ascomycete yeasts that form coenzyme Q-9 and the proposal of the new genera *Babjeviella, Meyerozyma, Millerozyma, Priceomyces,* and *Scheffersomyces. Mycoscience* **2010**, *51*, 2–14. [CrossRef]
49. Manamgoda, D.S.; Cai, L.; McKenzie, H.C.; Crous, P.W.; Madrid, H.; Chikeatirote, E.; Shivas, R.G.; Tan, Y.P.; Hyde, K.D. A phylogenetic and taxonomic re-evaluation of the *Bipolaris-Cochliobolus-Curvularia* complex. *Fungal Divers.* **2012**, *56*, 131–144. [CrossRef]
50. Alastruey-Izquierdo, A.; Hoffmann, K.; de Hoog, G.S.; Rodriguez-Tudela, J.L.; Voigt, K.; Bibashi, E.; Walther, G. Species recognition and clinical relevance of the zygomycetous genus *Lichtheimia* (syn. *Absidia pro parte, Mycocladus). J. Clin. Microbiol.* **2010**, *48*, 2154–2170. [CrossRef]
51. Lombard, L.; van der Merwe, N.A.; Groenewald, J.Z.; Crous, P.W. Generic concepts in Nectriaceae. *Stud. Mycol.* **2015**, *80*, 189–245. [CrossRef] [PubMed]
52. Luangsa-Ard, J.; Houbraken, J.; van Doorn, T.; Hong, S.B.; Borman, A.M.; Hywel-Jones, N.L.; Samson, R.A. Purpureocillium, a new genus for the medically important *Paecilomyces lilacinus. FEMS Microbiol. Lett.* **2011**, *321*, 141–149. [CrossRef] [PubMed]
53. Peterson, S.W. *Neosartorya pseudofischeri* sp. nov. and its relationship to other species in *Aspergillus* section *Fumigati. Mycol. Res.* **1992**, *96*, 547–554. [CrossRef]
54. Samson, R.A.; Visagie, C.M.; Houbraken, J.; Hong, S.B.; Hubka, V.; Klaassen, C.H.; Perrone, G.; Seifert, K.A.; Susca, A.; Tanney, J.B.; et al. Phylogeny, identification and nomenclature of the genus *Aspergillus. Stud. Mycol.* **2014**, *78*, 141–173. [CrossRef] [PubMed]
55. Lackner, M.; de Hoog, G.S.; Yang, L.; Moreno, L.F.; Ahmed, S.A.; Andreas, F.; Kaltseis, J.; Nagl, M.; Lass-Florl, C. Proposed nomenclature for *Pseudallescheria, Scedosporium* and related genera. *Fungal Divers.* **2014**, *67*, 1–10. [CrossRef]
56. Samerpitak, K.; Van der Linde, E.; Choi, H.J.; Van den Ende, B.G.; Machouart, M.; Gueidan, C.; De Hoog, G.S. Taxonomy of Ochroconis, genus including opportunistic pathogens on humans and animals. *Fungal Divers.* **2014**, *65*, 89–126. [CrossRef]
57. Samson, R.A.; Yilmaz, N.; Houbraken, J.; Spierenburg, H.; Seifert, K.A.; Peterson, S.W.; Varga, J.; Frisvad, J.C. Phylogeny and nomenclature of the genus *Talaromyces* and taxa accommodated in *Penicillium* subgenus *Biverticillium. Stud. Mycol.* **2011**, *70*, 159–183. [CrossRef]
58. Warnock, D.W. Name Changes for Fungi of Medical Importance, 2012 to 2015. *J. Clin. Microbiol.* **2017**, *55*, 53–59. [CrossRef]
59. De Hoog, G.S.; Dukik, K.; Monod, M.; Packeu, A.; Stubbe, D.; Hendrickx, M.; Kupsch, C.; Stielow, J.B.; Freeke, J.; Goker, M.; et al. Toward a Novel Multilocus Phylogenetic Taxonomy for the Dermatophytes. *Mycopathologia* **2017**, *182*, 5–31. [CrossRef]
60. Chen, M.; Zeng, J.; De Hoog, G.S.; Stielow, B.; Gerrits Van Den Ende, A.H.; Liao, W.; Lackner, M. The 'species complex' issue in clinically relevant fungi: A case study in *Scedosporium apiospermum. Fungal Biol.* **2016**, *120*, 137–146. [CrossRef]
61. Talbot, J.J.; Barrs, V.R. One-health pathogens in the *Aspergillus viridinutans* complex. *Med. Mycol.* **2018**, *56*, 1–12. [CrossRef]
62. Balajee, S.A.; Houbraken, J.; Verweij, P.E.; Hong, S.B.; Yaghuchi, T.; Varga, J.; Samson, R.A. *Aspergillus* species identification in the clinical setting. *Stud. Mycol.* **2007**, *59*, 39–46. [CrossRef]
63. Chaverri, P.; Branco-Rocha, F.; Jaklitsch, W.; Gazis, R.; Degenkolb, T.; Samuels, G.J. Systematics of the *Trichoderma harzianum* species complex and the re-identification of commercial biocontrol strains. *Mycologia* **2015**, *107*, 558–590. [CrossRef] [PubMed]

64. Howard, S.J.; Harrison, E.; Bowyer, P.; Varga, J.; Denning, D.W. Cryptic species and azole resistance in the *Aspergillus niger* complex. *Antimicrob. Agents Chemother.* **2011**, *55*, 4802–4809. [CrossRef]
65. Koufopanou, V.; Burt, A.; Szaro, T.; Taylor, J.W. Gene genealogies, cryptic species, and molecular evolution in the human pathogen *Coccidioides immitis* and relatives (Ascomycota, Onygenales). *Mol. Biol. Evol.* **2001**, *18*, 1246–1258. [CrossRef] [PubMed]
66. Hagen, F.; Lumbsch, H.T.; Arsic Arsenijevic, V.; Badali, H.; Bertout, S.; Billmyre, R.B.; Bragulat, M.R.; Cabanes, F.J.; Carbia, M.; Chakrabarti, A.; et al. Importance of Resolving Fungal Nomenclature: The Case of Multiple Pathogenic Species in the *Cryptococcus* Genus. *mSphere* **2017**, *2*, e00238-17. [CrossRef] [PubMed]
67. Fisher, M.C.; Koenig, G.L.; White, T.J.; Taylor, J.W. Molecular and phenotypic description of *Coccidioides posadasii* sp. nov., previously recognized as the non-California population of *Coccidioides immitis*. *Mycologia* **2002**, *94*, 73–84. [CrossRef] [PubMed]
68. Chowdhary, A.; Hagen, F.; Sharma, C.; Al-Hatmi, A.M.S.; Giuffre, L.; Giosa, D.; Fan, S.; Badali, H.; Felice, M.R.; de Hoog, S.; et al. Whole Genome-Based Amplified Fragment Length Polymorphism Analysis Reveals Genetic Diversity in *Candida africana*. *Front. Microbiol.* **2017**, *8*, 556. [CrossRef] [PubMed]
69. Tietz, H.J.; Hopp, M.; Schmalreck, A.; Sterry, W.; Czaika, V. *Candida africana* sp. nov., a new human pathogen or a variant of *Candida albicans*? *Mycoses* **2001**, *44*, 437–445. [CrossRef]
70. Romeo, O.; Criseo, G. First molecular method for discriminating between *Candida africana*, *Candida albicans*, and *Candida dubliniensis* by using *hwp1* gene. *Diagn. Microbiol. Infect. Dis.* **2008**, *62*, 230–233. [CrossRef]
71. Arastehfar, A.; Fang, W.; Pan, W.; Liao, W.; Yan, L.; Boekhout, T. Identification of nine cryptic species of *Candida albicans*, *C. glabrata*, and *C. parapsilosis* complexes using one-step multiplex PCR. *BMC Infect. Dis.* **2018**, *18*, 480. [CrossRef] [PubMed]
72. Alcoba-Florez, J.; Mendez-Alvarez, S.; Cano, J.; Guarro, J.; Perez-Roth, E.; del Pilar Arevalo, M. Phenotypic and molecular characterization of *Candida nivariensis* sp. nov., a possible new opportunistic fungus. *J. Clin. Microbiol.* **2005**, *43*, 4107–4111. [CrossRef] [PubMed]
73. Correia, A.; Sampaio, P.; James, S.; Pais, C. *Candida bracarensis* sp. nov., a novel anamorphic yeast species phenotypically similar to *Candida glabrata*. *Int. J. Syst. Evol. Microbiol.* **2006**, *56*, 313–317. [CrossRef] [PubMed]
74. Lockhart, S.R.; Messer, S.A.; Gherna, M.; Bishop, J.A.; Merz, W.G.; Pfaller, M.A.; Diekema, D.J. Identification of *Candida nivariensis* and *Candida bracarensis* in a large global collection of *Candida glabrata* isolates: Comparison to the literature. *J. Clin. Microbiol.* **2009**, *47*, 1216–1217. [CrossRef] [PubMed]
75. Hirschmann, J.V. The early history of coccidioidomycosis: 1892–1945. *Clin. Infect. Dis.* **2007**, *44*, 1202–1207. [CrossRef] [PubMed]
76. Brown, J.; Benedict, K.; Park, B.J.; Thompson, G.R., III. Coccidioidomycosis: Epidemiology. *Clin. Epidemiol.* **2013**, *5*, 185–197.
77. Kwon-Chung, K.J.; Bennett, J.E.; Wickes, B.L.; Meyer, W.; Cuomo, C.A.; Wollenburg, K.R.; Bicanic, T.A.; Castaneda, E.; Chang, Y.C.; Chen, J.; et al. The Case for Adopting the "Species Complex" Nomenclature for the Etiologic Agents of Cryptococcosis. *mSphere* **2017**, *2*, e00357-16. [CrossRef]
78. Trevino-Rangel Rde, J.; Garza-Gonzalez, E.; Gonzalez, J.G.; Bocanegra-Garcia, V.; Llaca, J.M.; Gonzalez, G.M. Molecular characterization and antifungal susceptibility of the *Candida parapsilosis* species complex of clinical isolates from Monterrey, Mexico. *Med. Mycol.* **2012**, *50*, 781–784. [CrossRef]
79. Mora-Duarte, J.; Betts, R.; Rotstein, C.; Colombo, A.L.; Thompson-Moya, L.; Smietana, J.; Lupinacci, R.; Sable, C.; Kartsonis, N.; Perfect, J.; et al. Comparison of caspofungin and amphotericin B for invasive candidiasis. *N. Engl. J. Med.* **2002**, *347*, 2020–2029. [CrossRef]
80. Kuse, E.R.; Chetchotisakd, P.; da Cunha, C.A.; Ruhnke, M.; Barrios, C.; Raghunadharao, D.; Sekhon, J.S.; Freire, A.; Ramasubramanian, V.; Demeyer, I.; et al. Micafungin versus liposomal amphotericin B for candidaemia and invasive candidosis: A phase III randomised double-blind trial. *Lancet* **2007**, *369*, 1519–1527. [CrossRef]
81. Reboli, A.C.; Rotstein, C.; Pappas, P.G.; Chapman, S.W.; Kett, D.H.; Kumar, D.; Betts, R.; Wible, M.; Goldstein, B.P.; Schranz, J.; et al. Anidulafungin versus fluconazole for invasive candidiasis. *N. Engl. J. Med.* **2007**, *356*, 2472–2482. [CrossRef]
82. Pryszcz, L.P.; Nemeth, T.; Gacser, A.; Gabaldon, T. Genome comparison of *Candida orthopsilosis* clinical strains reveals the existence of hybrids between two distinct subspecies. *Genome Biol. Evol.* **2014**, *6*, 1069–1078. [CrossRef] [PubMed]

83. Pryszcz, L.P.; Nemeth, T.; Saus, E.; Ksiezopolska, E.; Hegedusova, E.; Nosek, J.; Wolfe, K.H.; Gacser, A.; Gabaldon, T. The Genomic Aftermath of Hybridization in the Opportunistic Pathogen *Candida metapsilosis*. *PLoS Genet.* **2015**, *11*, e1005626. [CrossRef]
84. Pappas, P.G.; Alexander, B.D.; Andes, D.R.; Hadley, S.; Kauffman, C.A.; Freifeld, A.; Anaissie, E.J.; Brumble, L.M.; Herwaldt, L.; Ito, J.; et al. Invasive fungal infections among organ transplant recipients: Results of the Transplant-Associated Infection Surveillance Network (TRANSNET). *Clin. Infect. Dis.* **2010**, *50*, 1101–1111. [CrossRef]
85. Kontoyiannis, D.P.; Marr, K.A.; Park, B.J.; Alexander, B.D.; Anaissie, E.J.; Walsh, T.J.; Ito, J.; Andes, D.R.; Baddley, J.W.; Brown, J.M.; et al. Prospective surveillance for invasive fungal infections in hematopoietic stem cell transplant recipients, 2001–2006: Overview of the Transplant-Associated Infection Surveillance Network (TRANSNET) Database. *Clin. Infect. Dis.* **2010**, *50*, 1091–1100. [CrossRef] [PubMed]
86. Alastruey-Izquierdo, A.; Mellado, E.; Pelaez, T.; Peman, J.; Zapico, S.; Alvarez, M.; Rodriguez-Tudela, J.L.; Cuenca-Estrella, M.; Group, F.S. Population-based survey of filamentous fungi and antifungal resistance in Spain (FILPOP Study). *Antimicrob. Agents Chemother.* **2013**, *57*, 3380–3387. [CrossRef]
87. Balajee, S.A.; Kano, R.; Baddley, J.W.; Moser, S.A.; Marr, K.A.; Alexander, B.D.; Andes, D.; Kontoyiannis, D.P.; Perrone, G.; Peterson, S.; et al. Molecular identification of *Aspergillus* species collected for the Transplant-Associated Infection Surveillance Network. *J. Clin. Microbiol.* **2009**, *47*, 3138–3141. [CrossRef]
88. Frisvad, J.C.; Larsen, T.O. Extrolites of *Aspergillus fumigatus* and Other Pathogenic Species in *Aspergillus* Section *Fumigati*. *Front. Microbiol.* **2015**, *6*, 1485. [CrossRef] [PubMed]
89. Barrs, V.R.; Beatty, J.A.; Dhand, N.K.; Talbot, J.J.; Bell, E.; Abraham, L.A.; Chapman, P.; Bennett, S.; van Doorn, T.; Makara, M. Computed tomographic features of feline sino-nasal and sino-orbital aspergillosis. *Vet. J.* **2014**, *201*, 215–222. [CrossRef] [PubMed]
90. Sugui, J.A.; Peterson, S.W.; Figat, A.; Hansen, B.; Samson, R.A.; Mellado, E.; Cuenca-Estrella, M.; Kwon-Chung, K.J. Genetic relatedness versus biological compatibility between *Aspergillus fumigatus* and related species. *J. Clin. Microbiol.* **2014**, *52*, 3707–3721. [CrossRef] [PubMed]
91. Balajee, S.A.; Gribskov, J.; Brandt, M.; Ito, J.; Fothergill, A.; Marr, K.A. Mistaken identity: *Neosartorya pseudofischeri* and its anamorph masquerading as *Aspergillus fumigatus*. *J. Clin. Microbiol.* **2005**, *43*, 5996–5999. [CrossRef] [PubMed]
92. Matsumoto, N.; Shiraga, H.; Takahashi, K.; Kikuchi, K.; Ito, K. Successful treatment of *Aspergillus* peritonitis in a peritoneal dialysis patient. *Pediatr. Nephrol.* **2002**, *17*, 243–245. [CrossRef]
93. Jarv, H.; Lehtmaa, J.; Summerbell, R.C.; Hoekstra, E.S.; Samson, R.A.; Naaber, P. Isolation of *Neosartorya pseudofischeri* from blood: First hint of pulmonary Aspergillosis. *J. Clin. Microbiol.* **2004**, *42*, 925–928. [CrossRef] [PubMed]
94. Zbinden, A.; Imhof, A.; Wilhelm, M.J.; Ruschitzka, F.; Wild, P.; Bloemberg, G.V.; Mueller, N.J. Fatal outcome after heart transplantation caused by *Aspergillus lentulus*. *Transpl. Infect. Dis. Off. J. Transplant. Soc.* **2012**, *14*, E60–E63. [CrossRef] [PubMed]
95. Ghebremedhin, B.; Bluemel, A.; Neumann, K.H.; Koenig, B.; Koenig, W. Peritonitis due to *Neosartorya pseudofischeri* in an elderly patient undergoing peritoneal dialysis successfully treated with voriconazole. *J. Med. Microbiol.* **2009**, *58*, 678–682. [CrossRef] [PubMed]
96. Gilgado, F.; Cano, J.; Gene, J.; Sutton, D.A.; Guarro, J. Molecular and phenotypic data supporting distinct species statuses for *Scedosporium apiospermum* and *Pseudallescheria boydii* and the proposed new species *Scedosporium dehoogii*. *J. Clin. Microbiol.* **2008**, *46*, 766–771. [CrossRef]
97. Gilgado, F.; Gene, J.; Cano, J.; Guarro, J. Heterothallism in *Scedosporium apiospermum* and description of its teleomorph *Pseudallescheria apiosperma* sp. nov. *Med. Mycol.* **2010**, *48*, 122–128. [CrossRef] [PubMed]
98. Gilgado, F.; Cano, J.; Gene, J.; Guarro, J. Molecular phylogeny of the *Pseudallescheria boydii* species complex: Proposal of two new species. *J. Clin. Microbiol.* **2005**, *43*, 4930–4942. [CrossRef] [PubMed]
99. Lackner, M.; de Hoog, G.S.; Verweij, P.E.; Najafzadeh, M.J.; Curfs-Breuker, I.; Klaassen, C.H.; Meis, J.F. Species-specific antifungal susceptibility patterns of *Scedosporium* and *Pseudallescheria* species. *Antimicrob. Agents Chemother.* **2012**, *56*, 2635–2642. [CrossRef] [PubMed]
100. Cortez, K.J.; Roilides, E.; Quiroz-Telles, F.; Meletiadis, J.; Antachopoulos, C.; Knudsen, T.; Buchanan, W.; Milanovich, J.; Sutton, D.A.; Fothergill, A.; et al. Infections caused by *Scedosporium* spp. *Clin. Microbiol. Rev.* **2008**, *21*, 157–197. [CrossRef]

101. Walsh, T.J.; Groll, A.; Hiemenz, J.; Fleming, R.; Roilides, E.; Anaissie, E. Infections due to emerging and uncommon medically important fungal pathogens. *Clin. Microbiol. Infect. Off. Publ. Eur. Soc. Clin. Microbiol. Infect. Dis.* **2004**, *10* (Suppl. 1), 48–66. [CrossRef]
102. Lackner, M.; Hagen, F.; Meis, J.F.; Gerrits van den Ende, A.H.; Vu, D.; Robert, V.; Fritz, J.; Moussa, T.A.; de Hoog, G.S. Susceptibility and diversity in the therapy-refractory genus scedosporium. *Antimicrob. Agents Chemother.* **2014**, *58*, 5877–5885. [CrossRef] [PubMed]
103. Lewis, R.E.; Wiederhold, N.P.; Klepser, M.E. In vitro pharmacodynamics of amphotericin B, itraconazole, and voriconazole against *Aspergillus*, *Fusarium*, and *Scedosporium* spp. *Antimicrob. Agents Chemother.* **2005**, *49*, 945–951. [CrossRef] [PubMed]
104. Wiederhold, N.P.; Lewis, R.E. Antifungal activity against *Scedosporium* species and novel assays to assess antifungal pharmacodynamics against filamentous fungi. *Med. Mycol.* **2009**, *47*, 422–432. [CrossRef] [PubMed]
105. Meyer, W.; Aanensen, D.M.; Boekhout, T.; Cogliati, M.; Diaz, M.R.; Esposto, M.C.; Fisher, M.; Gilgado, F.; Hagen, F.; Kaocharoen, S.; et al. Consensus multi-locus sequence typing scheme for *Cryptococcus neoformans* and *Cryptococcus gattii*. *Med. Mycol.* **2009**, *47*, 561–570. [CrossRef]
106. Firacative, C.; Trilles, L.; Meyer, W. MALDI-TOF MS enables the rapid identification of the major molecular types within the *Cryptococcus neoformans/C. gattii* species complex. *PLoS ONE* **2012**, *7*, e37566. [CrossRef] [PubMed]
107. Normand, A.C.; Cassagne, C.; Gautier, M.; Becker, P.; Ranque, S.; Hendrickx, M.; Piarroux, R. Decision criteria for MALDI-TOF MS-based identification of filamentous fungi using commercial and in-house reference databases. *BMC Microbiol.* **2017**, *17*, 25. [CrossRef]
108. Van Herendael, B.H.; Bruynseels, P.; Bensaid, M.; Boekhout, T.; De Baere, T.; Surmont, I.; Mertens, A.H. Validation of a modified algorithm for the identification of yeast isolates using matrix-assisted laser desorption/ionisation time-of-flight mass spectrometry (MALDI-TOF MS). *Eur. J. Clin. Microbiol. Infect. Dis.* **2012**, *31*, 841–848. [CrossRef]
109. Vlek, A.; Kolecka, A.; Khayhan, K.; Theelen, B.; Groenewald, M.; Boel, E.; Multicenter Study, G.; Boekhout, T. Interlaboratory comparison of sample preparation methods, database expansions, and cutoff values for identification of yeasts by matrix-assisted laser desorption ionization-time of flight mass spectrometry using a yeast test panel. *J. Clin. Microbiol.* **2014**, *52*, 3023–3029. [CrossRef]
110. Herkert, P.F.; Dos Santos, J.C.; Hagen, F.; Ribeiro-Dias, F.; Queiroz-Telles, F.; Netea, M.G.; Meis, J.F.; Joosten, L.A.B. Differential In Vitro Cytokine Induction by the Species of *Cryptococcus gattii* Complex. *Infect. Immun.* **2018**, *86*. [CrossRef]
111. Nyazika, T.K.; Hagen, F.; Meis, J.F.; Robertson, V.J. *Cryptococcus tetragattii* as a major cause of cryptococcal meningitis among HIV-infected individuals in Harare, Zimbabwe. *J. Infect.* **2016**, *72*, 745–752. [CrossRef] [PubMed]
112. Nucci, M.; Marr, K.A.; Queiroz-Telles, F.; Martins, C.A.; Trabasso, P.; Costa, S.; Voltarelli, J.C.; Colombo, A.L.; Imhof, A.; Pasquini, R.; et al. *Fusarium* infection in hematopoietic stem cell transplant recipients. *Clin. Infect. Dis.* **2004**, *38*, 1237–1242. [CrossRef] [PubMed]
113. Nucci, M.; Garnica, M.; Gloria, A.B.; Lehugeur, D.S.; Dias, V.C.; Palma, L.C.; Cappellano, P.; Fertrin, K.Y.; Carlesse, F.; Simoes, B.; et al. Invasive fungal diseases in haematopoietic cell transplant recipients and in patients with acute myeloid leukaemia or myelodysplasia in Brazil. *Clin. Microbiol. Infect. Off. Publ. Eur. Soc. Clin. Microbiol. Infect. Dis.* **2013**, *19*, 745–751. [CrossRef] [PubMed]
114. Nucci, M.; Anaissie, E. Cutaneous infection by *Fusarium* species in healthy and immunocompromised hosts: Implications for diagnosis and management. *Clin. Infect. Dis.* **2002**, *35*, 909–920. [CrossRef] [PubMed]
115. Aoki, T.; O'Donnell, K.; Homma, Y.; Lattanzi, A.R. Sudden-death syndrome of soybean is caused by two morphologically and phylogenetically distinct species within the *Fusarium solani* species complex-*F. virguliforme* in North America and *F. tucumaniae* in South America. *Mycologia* **2003**, *95*, 660–684. [PubMed]
116. O'Donnell, K.; Rooney, A.P.; Proctor, R.H.; Brown, D.W.; McCormick, S.P.; Ward, T.J.; Frandsen, R.J.; Lysoe, E.; Rehner, S.A.; Aoki, T.; et al. Phylogenetic analyses of *RPB1* and *RPB2* support a middle Cretaceous origin for a clade comprising all agriculturally and medically important fusaria. *Fungal Genet. Biol.* **2013**, *52*, 20–31. [CrossRef] [PubMed]

117. O'Donnell, K.; Humber, R.A.; Geiser, D.M.; Kang, S.; Park, B.; Robert, V.A.; Crous, P.W.; Johnston, P.R.; Aoki, T.; Rooney, A.P.; et al. Phylogenetic diversity of insecticolous fusaria inferred from multilocus DNA sequence data and their molecular identification via FUSARIUM-ID and Fusarium MLST. *Mycologia* **2012**, *104*, 427–445. [CrossRef] [PubMed]
118. Sandoval-Denis, M.; Crous, P.W. Removing chaos from confusion: Assigning names to common human and animal pathogens in Neocosmospora. *Persoonia* **2018**, *41*, 109–129. [CrossRef]
119. Geiser, D.M.; Aoki, T.; Bacon, C.W.; Baker, S.E.; Bhattacharyya, M.K.; Brandt, M.E.; Brown, D.W.; Burgess, L.W.; Chulze, S.; Coleman, J.J.; et al. One fungus, one name: Defining the genus *Fusarium* in a scientifically robust way that preserves longstanding use. *Phytopathology* **2013**, *103*, 400–408. [CrossRef] [PubMed]

10

Portrait of Matrix Gene Expression in *Candida glabrata* Biofilms with Stress Induced by Different Drugs

Célia F. Rodrigues and Mariana Henriques *

Laboratório de Investigação em Biofilmes Rosário Oliveira (LIBRO), Centre of Biological Engineering, University of Minho, 4710-057 Braga, Portugal; rodriguescf@ceb.uminho.pt

* Correspondence: mcrh@deb.uminho.pt

Abstract: (1) Background: *Candida glabrata* is one of the most significant *Candida* species associated with severe cases of candidiasis. Biofilm formation is an important feature, closely associated with antifungal resistance, involving alterations of gene expression or mutations, which can result in the failure of antifungal treatments. Hence, the main goal of this work was to evaluate the role of a set of genes, associated with matrix production, in the resistance of *C. glabrata* biofilms to antifungal drugs. (2) Methods: the determination of the expression of *BGL2, XOG1, FKS1, FKS2, GAS2, KNH1, UGP1*, and *MNN2* genes in 48-h biofilm's cells of three *C. glabrata* strains was performed through quantitative real-time PCR (RT-qPCR), after contact with Fluconazole (Flu), Amphotericin B (AmB), Caspofungin (Csf), or Micafungin (Mcf). (3) Results: Mcf induced a general overexpression of the selected genes. It was verified that the genes related to the production of β-1,3-glucans (*BGL2, XOG1, GAS2*) had the highest expressions. (4) Conclusion: though β-1,6-glucans and mannans are an essential part of the cell and biofilm matrix, *C. glabrata* biofilm cells seem to contribute more to the replacement of β-1,3-glucans. Thus, these biopolymers seem to have a greater impact on the biofilm matrix composition and, consequently, a role in the biofilm resistance to antifungal drugs.

Keywords: *Candida*; biofilms; matrix; drug resistance; gene expression; *Candida glabrata*

1. Introduction

Fungal infections continue to increase worldwide, particularly among immunosuppressed patients, individuals under prolonged hospitalization, catheterization, or continued antimicrobial treatments [1–3]. *Candida* spp. are the commonest fungal species involved in these diseases. *Candida albicans* is the most isolated species, but *Candida glabrata* and *Candida parapsilosis* are the second most isolated species in the United States of America and Europe, respectively [1,4,5]. Though *C. glabrata* does not have the capacity to form hyphae and pseudohyphae or to secret proteases, this species has other virulence factors, such as the ability to secrete phospholipases, lipases, and haemolysins and, importantly, the capacity to form biofilms [6–8]. These factors highly contribute to a high aggressiveness, resulting in a low therapeutic response and severe cases of recurrent candidiasis [8,9]. Biofilms are communities of microorganisms that colonize tissues and indwelling medical devices, embedded in an extracellular matrix [10,11]. These heterogeneous structures provide high resistance to antifungal therapy and strong host immune responses [7,8,12]. *C. glabrata* has shown to form a compact biofilm structure in different multilayers [6,7], with proteins, carbohydrates, and ergosterol into their matrices [6,7,13].

Various reports have shown the presence of β-1,3 glucans in the biofilm matrices of *C. albicans* [14–17]. Interestingly, it has been demonstrated that an increase in cell wall glucan was associated with biofilm growth [14] and, more recently, β-1,3 glucans were shown to be also present in the matrices of *C. glabrata* biofilms [13,18,19]. This specific carbohydrate has been associated with a general increase of extracellular matrix delivery, which is critical for securing biofilm cells to a surface and crucial to develop an antifungal drug resistance phenotype [14,19–23]. Several genes are involved in the delivery and accumulation of extracellular matrix. It is recognized that, in *C. albicans*, the major β-1,3 glucan synthases are encoded mainly by *FKS1* but also by *FKS2* [24]. The *BGL2* and *XOG1* genes also have important roles in glucan matrix delivery by encoding glucanosyltransferases and β-1,3 exoglucanase, respectively [25,26]. These genes play an important part in cell wall remodeling, however, the influence of the corresponding enzymes in matrix glucan delivery does not appear to affect cell wall ultrastructure or β-1,3 glucan concentration, suggesting that these enzymes function specifically for matrix delivery [17,19,26–28]. Identical to *Saccharomyces cerevisiae*, in *C. glabrata*, the *GAS* gene family is a regulator of the production of β-1,3 glucan [29]. *Gas2*, a glycosylphosphatidylinositol (GPI)-anchored cell surface protein [30,31], is a putative carbohydrate-active enzyme that may change cell wall polysaccharides [29,32].

Another carbohydrate of *C. glabrata* cell wall is β-1,6-glucan, present as a polymer covalently attached to glycoproteins [33–36], β-1,3-glucan, and chitin [37]. Nagahashi et al. [36] reported the isolation of *KNH1* homologs (genes encoding cell surface *O*-glycoproteins), suggesting the evolutionary conservation of these molecules as essential components of β-1,6-glucan synthesis in *C. glabrata*, which was also discussed before [35,38]. Additionally, the *UGP1* gene is a putative uridine diphosphate (UDP)-glucose pyrophosphorylase related to the general β-1,6-D-glucan biosynthetic process [39,40]. During stress conditions, several *S. cerevisiae* orthologous genes are induced in *C. glabrata*. In glucose starvation stress, *UGP1* is induced [39].

The external layer cell wall of *Candida* spp. also consists of highly glycosylated mannoproteins [41–43], which play a major role in host recognition, adhesion, and cell wall integrity [44–56]. These proteins have both *N*- and *O*-linked sugars, predominantly mannans, which are also known to be present in the biofilm matrices of *C. albicans* [57–59]. The *MNN2* gene is one putative element of *N*-linked glycosylation, directly responsible for mannans production for both cell and biofilm matrices of *C. glabrata* [57–59].

The goal of this work was to determine the expression profile of selected genes (Tables 1 and 2) related to the production of biofilm matrix components in response to stress caused by drugs from the most important antifungal classes: azoles (Fluconazole, Flu), polyenes (Amphotericin B, AmB), and echinocandins (Caspofungin, Csf, and Micafungin, Mcf).

2. Materials and Methods

2.1. Organisms

Three strains of *C. glabrata* were used in the course of this study: One reference strain (*C. glabrata* ATCC 2001) from the American Type Culture Collection (Manassas, VA, USA), one strain recovered from the urinary tract (*C. glabrata* 562123) of a patient, and one strain recovered from the vaginal tract of a patient (*C. glabrata* 534784) in the Hospital Escala, Braga, Portugal. The identity of all isolates was confirmed using CHROMagar^{TM} *Candida* (CHROMagar^{TM}, Paris, France) and by PCR-based sequencing using specific primers (*ITS1* and *ITS4*) against the 5.8 s subunit gene reference [60]. The PCR products were sequenced using the ABI-PRISM Big Dye terminator cycle sequencing kit (Perkin Elmer, Applied Biosystems, Warrington, UK).

2.2. Growth Conditions

For each experiment, *C. glabrata* ATCC2001, *C. glabrata* 534784, and *C. glabrata* 562123 strains were subcultured on Sabouraud dextrose agar (SDA) (Merck, Darmstadt, Germany) for 24 h at 37 °C. The cells were then inoculated in Sabouraud dextrose broth (SDB) (Merck) and incubated for 18 h at 37 °C under agitation at 120 rpm. After incubation, the cells were harvested by centrifugation at 3000× *g* (Thermo Scientific, CL10, Hampton, NH, USA) for 10 min at 4 °C and washed twice with phosphate buffered saline (PBS, pH = 7.5). The cell pellets were then suspended in Roswell Park memorial institute (RPMI), and the cellular density was adjusted to 1×10^5 cells/mL, using a Neubauer counting chamber.

2.3. Antifungal Drugs

Flu, Csf, and Mcf were kindly provided by Pfizer® (New York, NY, USA), MSD® (Kenilworth, NJ, USA) and Astellas Pharma, Ltd., (Tokyo, Japan), respectively, in their pure form. AmB was purchased from Sigma® (Sigma-Aldrich, Buffalo, NY, USA). Aliquots of 5000 mg/L were prepared using dimethyl sulfoxide (DMSO). The final concentrations used were prepared in RPMI-1640 (Sigma-Aldrich).

2.4. Biofilm Formation

The minimum biofilm eradicatory concentration (MBEC) values were previously determined by the group, according to the European committee on antimicrobial susceptibility testing (EUCAST) guidelines [61,62]. For biofilm formation, standardized cell suspensions (1000 μL) were placed into selected wells of 24-wells polystyrene microtiter plates (Orange Scientific, Braine-l'Alleud, Belgium). At 24 h, 500 μL of RPMI-1640 was removed, and an equal volume of fresh RPMI-1640 plus the antifungal solution was added, on the basis of the MBEC values determined and indicated in bold in Table 1 (2× concentrated). The plates were incubated at 37 °C for additional 24 h at 120 rpm. RPMI-1640 containing only the antifungal agent was used as a negative control. As a positive control, cell suspensions were tested in the absence of the antifungal agent [18].

Table 1. Minimum biofilm eradicatory concentrations (MBEC) for the *Candida glabrata* strains of fluconazole (Flu), amphotericin B (AmB), caspofungin (Csf), and micafungin (Mcf) (mg/L).

Origin	Strain	Flu	AmB	Csf	Mcf
Reference (Wild Type)	ATCC2001	**>1250**	**4**	2.5–**3**	16–**17**
Urinary Tract	562123	**625**	**2**	0.5–**1**	16–**17**
Vaginal Tract	534784	**>1250**	**2**	2.5–**3**	5.5–**6**

Bold: concentrations applied to the pre-formed biofilms.

2.5. Gene Expression Analysis

2.5.1. Gene Selection and Primer Design for Quantitative Real-Time PCR

Genes related to the production of biofilm matrix components (β-1,3, β-1,6 glucans, and mannans)—*BGL2, FKS1, FKS2, GAS2, KNH1, UGP1, XOG1 and MNN2*—were selected for this study. The gene sequences of interest were obtained from *Candida* Genome Database [63] and the primers for quantitative real-time PCR (RT-qPCR) were designed using Primer 3 [64] web-based software and are listed in Table 2. *ACT1* was chosen as a housekeeping gene. In order to verify the specificity of each primer pair for its corresponding target gene, the PCR products were first amplified from *C. glabrata* ATCC2001.

2.5.2. Preparation of Biofilm Cells for RNA Extraction

After biofilm formation, the medium was eliminated, and the wells were washed with sterile water to remove non-adherent cells. The biofilms were scraped from the wells with 1 mL of sterile water and sonicated (Ultrasonic Processor, Cole-Parmer, IL, USA) for 30 s at 30 W to separate the cells from the biofilm matrix. The cells were harvested by centrifugation at 8000× *g* for 5 min at 4 °C [18].

2.5.3. RNA Extraction

RNA extraction was performed using PureLink RNA Mini Kit (Invitrogen, Carlsbad, CA, USA). Prior to RNA extraction, a lysis buffer from PureLink RNA Mini kit was prepared by adding 1% of *ß*-mercaptoethanol to the supplied buffer solution. Then, 500 μL of lysis buffer containing glass beads (0.5 mm diameter) was added to each pellet. The cell suspensions were homogenized twice for 30 s using a Mini-Bead-Beater-8 (Stratech Scientific, Soham, UK). After cell disruption, the PureLink RNA Mini Kit (Invitrogen) was used for total RNA extraction according to the manufacturer's recommended protocol. To avoid potential DNA contamination, the samples were treated with RNase-Free DNase I (Invitrogen) [18].

Table 2. Primers, targets used, and specific function of the genes used for the expression analysis.

Sequence (5′ → 3′)	Primer	Target	Properties and Proposed Function [a]
5′-GGC AAG AAA CTG GAC AGA GC-3′ 5′-GGA AAA CTT GGG TCC TGC TG-3′	F R	*BGL2*	β-1,3-glucanosyltransferase activity; glucan endo-β-1,3-D-glucosidase activity
5′-GTC CTA ACC TTG CAC ACC AG-3′ 5′-CTA CGC CCA AAC ATC AGC-3′	F R	*FKS1*	β-1,3-D-glucan synthase activity
5′-GGG TCA CTG TGA AAT GTT-3 5′-GTA GAC GGG TTC GGA TT-3	F R	*FKS2*	β-1,3-D-glucan synthase activity
5′-ACC AGT CGT ACC ATT ACC GG-3′ 5′-CCT GCC CAA CTT CTA ACA GC-3′	F R	*GAS2*	β-1,3-glucanosyltransferase activity
5′-CGG TGC CAA CGG TTA CTA-3′ 5′-GTG ACA CGG GTT TCA GGA-3′	F R	*KNH1*	β-1,6-D-glucan biosynthetic process
5′-AAT CGC ACA AGG CAG AGA-3′ 5′-ACT TGG GCG ACT TCC AAT-3′	F R	*UGP1*	β-1,6-D-glucan biosynthetic process
5′-GGT GAG TTG CAA CGT GAC AT-3′ 5′-ATT CGG TTA AAG CGG CAC TC-3′	F R	*XOG1*	Glucan endo-β-1,6 and 1,3-glucosidase activity
5′-GAA GCC TGA TGG TGG TGA-3′ 5′-ATT GGG CGA TGA CCT TCT-3′	F R	*MNN2*	α-mannosyltransferase biosynthetic process
5′-GTT GAC CGA GGC TCC AAT GA-3′ 5′-CAC CGT CAC CAG AGT CCA AA-3′	F R	*ACT1*	Housekeeping gene

[a] CGD: Candida Genome Database [63]; F: forward; R: reverse.

2.5.4. Synthesis of Complementary DNA

To synthesize complementary DNA (cDNA), the iScript cDNA Synthesis Kit (Bio-Rad, Hercules, CA, USA) was used according to the manufacturer's instructions. For each sample, 10 μL of the extracted RNA was used in a final reaction volume of 50 μL. cDNA synthesis was performed firstly at 70 °C for 5 min and then at 42 °C for 1 h. The reaction was stopped by heating for 5 min at 95 °C [18].

2.5.5. Quantitative Real-Time PCR

RT-qPCR (CFX96 Real-Time PCR System, Bio-Rad) was performed to determine the relative levels of all genes mRNA transcripts in the RNA samples, with *ACT1* used as a reference *Candida* housekeeping gene. Each reaction mixture consisted of a working concentration of SoFast EvaGreen Supermix (Bio-Rad), 50 μM forward and reverse primers, and 4 μL cDNA, in a final reaction volume of 20 μL. Negative controls (water) as well as non-transcriptase reverse controls (NRT) were included in each run [18]. The relative quantification of gene expression was performed by the $2^{-\Delta C_T}$ method [65]. Each reaction was performed in triplicate, and mean values of relative expression were determined for each gene. The results are presented after calculation of $2^{-\Delta C_T}$.

2.6. Statistical Analysis

All experiments were repeated three times in independent assays. The results were compared using one-way analysis of variance (ANOVA), Dunnett's multiple comparisons tests, using GraphPad™ Prism™ 7 software (GraphPad Software, San Diego, CA, USA). All tests were performed with a confidence level of 95%. In order to determine the similarity of the strains' gene profiles, the Pearson Correlation Coefficient (r) was also applied.

3. Results and Discussion

Candidaemia related to *C. glabrata* has been increasing in the last years in parallel with its high drug resistance, particularly to the azole antifungal class [1,20,66]. Biofilms of *C. glabrata* are highly recalcitrant to treatments with antifungal agents as a consequence of multiple resistance mechanisms, such as those linked to the presence of a strong net of exopolysaccharydes and other biopolymers that protect the cells and hinder the diffusion of the drugs [1,15,67–69]. In order to stress *C. glabrata* biofilm cells, four antifungals were applied (at concentrations based on MBECs values, Table 1) in pre-formed biofilms, and then an evaluation of biofilms' matrix gene expression was performed and compared with the expression of a housekeeping gene.

Figure 1 shows the heatmap with the results of the RT-qPCR expression profiling of biofilm cells of *C. glabrata* ATCC2001 (A), *C. glabrata* 562123 (B), and *C. glabrata* 534784 (C) in the presence of antifungal drugs. The final data are presented in fold-change (FC) in comparison to the expression of the housekeeping gene ($2^{-\Delta C_T}$) [70].

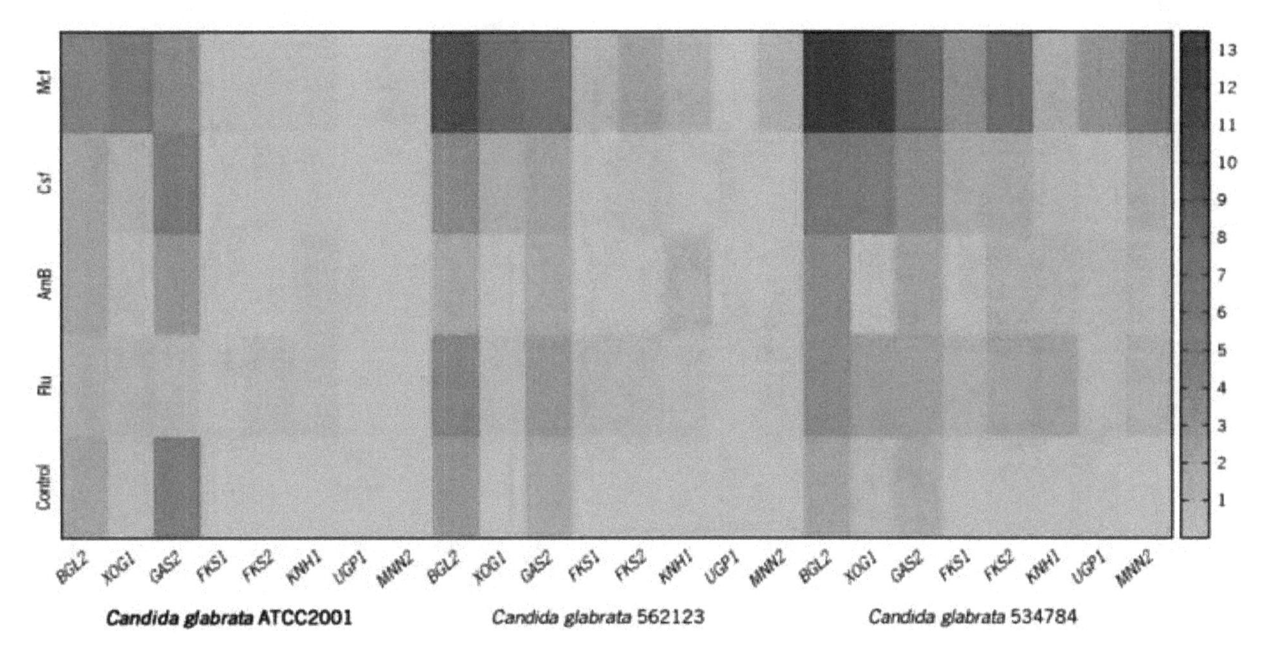

Figure 1. Real-time PCR expression profiling of *BGL2, XOG1, GAS2, FKS1, FKS2, KNH1, UGP1,* and *MNN2* genes in biofilm cells of *Candida glabrata* ATCC2001 (reference strain, in bold) (a), *C. glabrata* 562123 (b), and *C. glabrata* 534784 (c) in the presence of antifungal drugs. The heatmap was generated by a log transformation of the RT-qPCR data and the fold-change (FC) expression determined through $2^{-\Delta C_T}$. The numerical scale on the right represents the FC. (Control: non-treated cells).

Generally, *BGL2, FKS1, FKS2, GAS2,* and *XOG1* displayed higher expression levels in biofilm cells in response to the drugs and, by contrast, *KNH1, UPG1,* and *MNN2* displayed minor expression changes (Figure 1 and Table 3).

BGL2 showed similar expression in the control groups, and its FC expression decreased in the reference strain when Flu and Csf were present (FC: 1.40 and 2.00 respectively) and in the urinary strain when AmB was present (FC: 1.71). In all the other cases, *BGL2* FC expression increased, particularly when the biofilms were treated with Mcf (FC: *C. glabrata* ATCC2001: 5.13; *C. glabrata* 562123: 10.58; *C. glabrata* 534784: 13.49). All changes in *BGL2* expression were statistically significant compared to the untreated cells ($p < 0.0001$). Compared to the controls, *XOG1* gene revealed a statistically significant downregulation after contact with AmB (FC: *C. glabrata* ATCC2001 0.37; *C. glabrata* 562123 0.10; *C. glabrata* 534784: 0.27) and overexpression in the presence of all the other antifungals, in all strains. For this gene, the most noteworthy overexpression was observed in Flu-treated *C. glabrata* 534784 (FC: 2.35, $p < 0.0001$), Csf-treated *C. glabrata* 562123 and *C. glabrata* 534784 (FC: 1.42 and 5.45, $p < 0.0001$, respectively), and in all strains after contact with Mcf (FC: *C. glabrata* ATCC2001: 6.54, *C. glabrata* 562123: 7.89, *C. glabrata* 534784: 12.38, all $p < 0.0001$).

Table 3. Real-time PCR expression profiling of *BGL2, FKS1, FKS2, GAS2, KNH1, UGP1, XOG1,* and *MNN2* genes in biofilm cells of *C. glabrata* ATCC2001, *C. glabrata* 562123, *C. glabrata* 534784 with and without antifungal treatment (FC: $2^{-\Delta C_T}$). The significance of the FC results was determined by comparing the treated groups with the non-treated ones (controls) (* $p < 0.05$; ** $p < 0.001$; *** $p < 0.0005$; **** $p < 0.0001$).

Gene		*Candida glabrata* ATCC2001	*Candida glabrata* 562123	*Candida glabrata* 534784
			Fold-Change	
BGL2	Non treated	2.23	2.77	1.90
	Flu	1.41 ****	4.01 ****	3.66 ****
	AmB	2.04 ****	1.71 ****	3.85 ****
	Csf	2.00 ****	3.82 ****	5.98 ****
	Mcf	5.13 ****	10.58 ****	13.49 ****
XOG1	Non treated	0.57	0.15	0.46
	Flu	0.96 ****	0.54 ****	2.35 ****
	AmB	0.37 ****	0.10 ***	0.27 ****
	Csf	1.08 ****	1.42 ****	5.45 ****
	Mcf	6.54 ****	7.89 ****	12.38 ****
GAS2	Non treated	5.34	1.67	1.39
	Flu	0.76 ****	2.73 ****	2.26 ****
	AmB	3.02 ****	1.56 ****	1.95 ****
	Csf	5.72 ****	2.08 ****	3.39 ****
	Mcf	3.99 ****	7.43 ****	8.18 ****
FKS1	Non treated	0.11	0.11	0.17
	Flu	0.22 ****	0.65 ****	1.07 ****
	AmB	0.16 ****	0.08 ****	0.19 (ns)
	Csf	0.07 ***	0.20 ****	1.27 ****
	Mcf	0.49 ****	0.94 ****	3.55 ****
FKS2	Non treated	0.14	0.20	0.27
	Flu	0.61 ****	0.64 ****	1.77 ****
	AmB	0.06 ***	0.19 *	1.05 ****
	Csf	0.28 ****	0.55 ****	1.66 ****
	Mcf	0.43 ****	2.29 ****	7.50 ****
KNH1	Non treated	0.06	0.20	0.22
	Flu	0.50 ****	0.41 ****	2.08 ****
	AmB	0.51 ****	0.87 ****	0.94 ***
	Csf	0.24 ****	0.21 (ns)	0.72 ****
	Mcf	0.45 ****	1.33 ****	1.43 ****
UGP1	Non treated	0.002	0.07	0.20
	Flu	0.10 ****	0.15 ****	0.55 ****
	AmB	0.17 ****	0.05 ****	0.46 ****
	Csf	0.01 (ns)	0.06 **	0.33 ****
	Mcf	0.04 **	0.21 ****	3.17 ****
MNN2	Non treated	0.02	0.18	0.13
	Flu	0.19 ****	0.34 ****	1.17 ****
	AmB	0.13 ****	0.24 ****	0.71 ****
	Csf	0.18 ****	0.13 ***	1.40 ****
	Mcf	0.31 ****	1.21 ****	7.03 ****

(ns, non-significant; Non-treated, controls).

In an important report, Taff et al. [19] concluded that mutants of *C. albicans* unable to produce Bgl2 and Xog1 enzymes did not show perturbations in the cell wall glucan composition of biofilm cells, and that these enzymes were not necessary for filamentation or biofilm formation. However, the biofilms had a reduced matrix glucan content, reduced total matrix biomass accumulation, and improved susceptibility to antifungal drug therapy [19]. Similarly, Li et al. [71] showed that, in *C. albicans'* persister cells (frequent in biofilms [72,73]), there was an increased expression of cell wall integrity proteins such as Xog1 and Bgl2. These studies recognized a biofilm-specific pathway involving Bgl2 and Xog1 (and Phr1) enzymes and affecting matrix delivery, by which these enzymes release and modify cell wall glucan for deposition in the extracellular space; however, an alternative explanation is that these enzymes act in the extracellular space, being crucial for mature matrix organization and function [19]. These enzymes have been localized in the cell wall, supporting the hypothesis of their activity in the cell wall, but have also secretion sequences that support an extracellular function. As seen earlier, *BGL2* is one of the glucan modifying genes for glucan delivery, and *XOG1* is a glucanase [19], necessary for modification and delivery of carbohydrates to the mature biofilm matrix. Without delivery and accumulation of matrix glucan, the biofilms exhibit enhanced susceptibility to antifungal drugs [19]. The change in the regulation of *BGL2* and *XOG1* in the biofilm cells of *C. glabrata* after drug treatment that we observed is interpreted as a response of the biofilm cells to the reduction of biofilm matrix, specifically of β-1,3-glucans, and it has been described before [7,13,18].

The *GAS* gene family is also a regulator in the production of β-1,3-glucan, and Gas2 is a glycosylphosphatidylinositol (GPI)-anchored cell surface protein [31] involved in the production of β-1,3-glucan in *C. glabrata* [29,30]. Gas2 is a documented putative carbohydrate-active enzyme and consequently it can alter the cell wall polysaccharides in order to build and remodel the cell wall glycan network during growth in *C. glabrata* [29]. In *C. glabrata* ATCC2001, *GAS2* was highly expressed in the non-treated group and after Csf contact (FC: 5.72), while Flu, AmB, and Mcf led to its downregulation. The clinical isolates upregulated the gene in all conditions, except for AmB treatment of *C. glabrata* 562123 (Figure 1 and Table 3, all $p < 0.0001$). Hence, when analyzing the results of *C. glabrata* 562123 and *C. glabrata* 534784, the *GAS2* network seems to be activated also after glycan's loss following drug treatment, in order to replace the lack of 1,3-β-glucans and re-establish biofilm cells' homeostasis. All results compared with those from untreated cells (controls) were statistically significant ($p < 0.0001$).

The resistance to echinocandins increased from 4.9% to 12.3% between 2001 and 2010 [74] with a rapid development of *FKS* mutations in *Candida* spp., especially in *C. glabrata* [75,76]. The amino acid substitutions occurring in *FKS1* [30,77–79] and *FKS2* [30,80] are directly related to the resistance to this class of drugs: acquired *FKS* mutations [81] are reported to confer low β-(1,3)-D-glucan synthase sensitivity and to increase the minimum inhibitory concentration (MIC) values, which are related to clinical failure [82]; intrinsic *FKS* mutations, also lead to elevated MIC levels but have a weaker effect on the reduction of β-(1,3)-D-glucan synthase sensitivity [82–85]. Generally, in the *C. glabrata* strains, a 24 h contact with both echinocandins upregulated *FKS1* and *FKS2* genes and, in the reference strain, the presence of Mcf upregulated *FKS1*. More specifically, the results showed that all strains upregulated the expression of *FKS1* after drug exposure (statistically significant), with the exception of Csf in *C. glabrata* ATCC2001 (FC: 0.07 $p < 0.001$) and Csf in *C. glabrata* 562123 (FC: 0.08; $p < 0.0001$). For *FKS2*, its overexpression was observed for almost all treatments in the three strains, excluding following AmB treatment in *C. glabrata* ATCC2001 (FC: 0.28; $p < 0.001$) and *C. glabrata* 562123 (FC: 0.55; $p < 0.05$). *C. glabrata* 534784 revealed, again, to have the highest capacity to overexpress both genes in response to drug stress. These differences among the strains may be related to the described *Candida* spp. intra-strains variations [62]. Bizerra et al. [76] reported the occurrence of a mutation associated with the resistance phenotype against echinocandins in *C. glabrata* isolated from a single cancer patient with candidemia exposed to antifungal prophylaxis with Mcf. Arendrup et al. [86] revealed that Mcf MICs of *C. glabrata FKS* hot spot mutant isolates were less raised than those obtained for the other echinocandins, showing that the efficacy of Mcf could be differentially dependent on specific *FKS* genes mutations. These reports mention singularities regarding the *FKS* gene and Mcf, which can also be observed in our results (Figure 1 and Table 3). Interestingly, up and downregulations of *FKS1* and *FKS2* were similar in the clinical isolates and parallel to those observed for *BGL2*, which makes sense, since this gene has shown to perform, with *XOG1* (and *PHR1*), in a complementary manner in order to distribute the matrix downstream of the primary β-1,3 glucan synthase encoded by *FKS1* [19]. Previous investigations also found elevated transcript levels of *FKS1*, *BGL2*, and *XOG1* during in vivo *C. albicans* biofilm growth when compared to planktonic growth, which is consistent with our results and with a role in a biofilm-specific function, such as matrix formation [87,88].

The overexpressed values obtained for *BGL1*, *XOG1*, *FKS1*, *FKS2*, and *GAS2* after the stress conditions induced by most antifungals endorse the impact of β-1,3-glucans in the maintenance of the cell and biofilm matrix structure.

Sequencing studies have shown that *C. glabrata* is more closely related to *S. cerevisiae* than to *C. albicans* [89], with some genes functionally interchangeable among the two species [90,91]. An important component of the cell wall and the biofilm matrix is β-1,6-glucan, which is regulated by several genes, such as *KNH1*. Preceding studies have demonstrated that the *KNH1* homologs are essential components of β-1,6-glucan synthesis in *C. glabrata* [35,36,38]. In *S. cerevisiae*, many genes involved in β-1,6-glucan synthesis were isolated through mutations (*kre* [killer resistant] mutations) that are responsible for the resistance to the K1 killer toxin, which kills sensitive yeast cells after binding to β-1,6-glucan [35,38,92]. Dijkgraaf and colleagues [35] reported that the disruption of both *KRE9* and *KNH1* was synthetically lethal for *C. glabrata*, demonstrating the importance of these genes in the maintenance of cell structure. In the present study, after a drug stress, all *C. glabrata* strains upregulated this gene (Figure 1 and Table 3), indicating an effort to replace these β-1,6-glucans after losses due to the aggression of the antifungals, confirming also a certain degree of relevance of these elements in the cell wall and biofilm matrix of *C. glabrata* [35,36,38]. *C. glabrata* ATCC2001 showed to upregulate the *KNH1* gene in the presence of antifungal drugs, and *C. glabrata* 562123 indicated an identical pattern by marginally increasing *KNH1* gene expression in these conditions. Compared to the other two strains, with the exception of Flu treatment (FC: 2.08; $p < 0.0001$), *KNH1* showed a different regulation in the vaginal tract strain (Figure 1 and Table 3). *C. glabrata* 534784 demonstrated to have the highest upregulation capacity, presenting overexpression almost for all genes (Figure 1 and Table 3).

During glucose starvation, a set of genes orthologous to *S. cerevisiae* is induced in *C. glabrata*, including *UGP1*, related to the β-1,6-D-glucan biosynthetic process [39,40], which shows that the environmental stress response is conserved between *S. cerevisiae* and *C. glabrata* [39]. *UPG1* showed to have the lowest expression, compared with other genes and controls. Nonetheless, except for one condition and strain, in the presence of antifungal drugs, several overexpression states were observed (Figure 1 and Table 3). The reference strain displayed overexpression in all conditions, with the highest gene upregulation occurring in the presence of AmB (FC: 0.17; $p < 0.0001$) and the lowest in the presence of Csf (FC: 0.01; non-significant); the urinary tract strain also revealed limited gene upregulation in the presence of Csf and AmB and the highest expression in the presence of Mcf (0.15; $p < 0.0001$). Generally, *C. glabrata* 534784 demonstrated the highest FC expression in all conditions. The lowest gene upregulation was observed in biofilm cells stressed by Csf (FC: 0.33; $p < 0.0001$). Srikantha and colleagues [91] identified a set of genes that are upregulated by the transcription factor Bcr1, involved in impermeability, impenetrability, and drug resistance of *C. albicans'* biofilms. The authors concluded that the induction of Bcr1 overexpression in weak biofilms of *C. albicans* conferred those three characteristics and, in these cases, *UGP1* gene was downregulated [91]. This result supports the FC expression we obtained: since *C. glabrata* biofilms were weakened by the drugs, *UGP1* expression was increased in order to balance this defect (as seen with *KNH1*). The overexpression values we obtained for both *KNH1* and *UPG1* point to the relevance of β-1,6-glucans in the maintenance of a good cell and matrix structure.

Regarding mannans regulation, all strains showed a low or moderate expression of *MNN2* in the controls (non-treated cells), but relevant expression changes arose in the presence of all drugs, particularly when Mcf was added (Figure 1 and Table 3). *C. glabrata* ATCC2001 demonstrated the lowest expression in the control group, among all strains (Figure 1 and Table 3). The urinary strain presented the lowest gene expression, when compared to the untreated group (Figure 1 and Table 3). Flu and Mcf induced the highest *MNN2* values (FC: 0.34 and 1.21, respectively, both $p < 0.0001$), while AmB and Csf were associated with the lowest expressions (FC: 0.24, $p < 0.0001$ and 0.13, $p < 0.0005$). For *C. glabrata* ATCC2001 and *C. glabrata* 534784, the weakest effects were associated with the biofilm cells that were stressed by AmB (FC: 0.13 and 0.71 respectively, both $p < 0.0001$). When Mcf was applied, the biofilm cells of the vaginal strain showed a strong response to the stress, compared to the other two strains (FC: 7.03; $p < 0.0001$). Our team has also found that *C. glabrata* ATCC2001 increased the amounts of mannans on its cell walls in the presence of these drugs (data not shown), revealing a possible adaptation of the cells to the stress caused by the antifungal drugs. Other studies reported analogous adjustments of the cell walls after environmental drug stress, which has been related to high antifungal resistance events [1,2,19,92–94], supporting these results.

Interestingly, and when compared to the rest of the genes, the present results demonstrate that *KNH1*, *UGP1*, and *MNN2* had the lowest values of FC expression. This seems to indicate that, although β-1,6-glucans are an important part of the cell and biofilm matrix, the cells appear to invest more in replacing the lost β-1,3-glucans, leading to consider that these components have a greater significance in the maintenance of the homeostasis of the biofilm matrix and the biofilm cells. In fact, in studies developed in our group [13,18], the total polysaccharides and β-1,3-glucans concentrations increased significantly in *C. glabrata* biofilm matrices after Flu, AmB, and Mcf contact. These higher concentrations in β-1,3 glucans content might explain part of the main biofilm resistance to the drugs that was formerly described [7,95–97].

Finally, downregulation of most genes and strains happened in the presence of AmB whereas, in opposition, Mcf induced the main overexpression alterations (Figure 1 and Table 3). AmB is a fungicidal drug and the most important antifungal polyene used for the treatment of systemic fungal infections [98,99]. This drug binds to the ergosterol of the cell membrane but also induces oxidative stress. This explains the existence of a still low reported rate of resistance and the good effectiveness of AmB [1,13,100–104]. Also, this low resistance may be associated with the lower gene expression effects that we detected after AmB exposure in *C. glabrata* (Figure 1 and Table 3). In opposition, the most acute upregulations occurred in the presence of echinocandins and, particularly, when Mcf was applied. This class of antifungals act by inhibiting β-1,3-glucan synthesis [1,100,105], which affects cell wall and matrix composition. By overexpressing the genes related to β-1,3-glucan synthesis (*BGL2*, *FKS1*, *FKS2*, *GAS2*, *XOG1*), the cells were attempting to compensate and replace the β-1,3-glucan losses in their matrices induced by the drugs and, thus, protect and decrease their susceptibility to the antifungals [19]. This general increase in total carbohydrates and specifically in β-1,3-glucans in *Candida* spp. biofilm matrices has already been described [7,13,18].

Regarding the correlation between the gene expression profiles in *C. glabrata*, the results based on the r are displayed in Table 4.

Table 4. Pearson Correlation Coefficient (r) determined for the expression profiles of *BGL2*, *FKS1*, *FKS2*, *GAS2*, *KNH1*, *UGP1*, *XOG1*, and *MNN2* genes in biofilm cells of *C. glabrata* ATCC2001, *C. glabrata* 562123, *C. glabrata* 534784, in the presence or absence of antifungal drugs.

Gene	ATCC2001 vs. 562123	ATCC2001 vs. 534784	562123 vs. 534784
BGL2	0.9100	0.9145	0.946
XOG1	0.9965	0.9459	0.9646
FKS1	0.8947	0.8723	0.8778
FKS2	0.5074	0.4481	0.9937
GAS2	−0.0514	0.1091	0.9663
KNH1	0.6427	0.8107	0.3697
UGP1	−0.1091	-0.1122	0.8519
MNN2	0.7924	0.8618	0.9728

The results showed a strong positive correlation (r near 1) between the response profiles of *BGL2*, *XOG1*, *FKS1*, and *MNN2* gene expression in the three strains, which means that up and downregulation had a high tendency to occur similarly in all strains. The scores of the r for the profile of the *FKS2* gene revealed a moderate positive correlation between the reference strain (*C. glabrata* ATCC2001) and the isolates (*C. glabrata* 562123 and *C. glabrata* 534784). This indicates that, although the correlation was positive, it was weak, and the profiles of the gene response were variable in the three strains. On the other hand, the clinical isolates showed strong positive correlation between the expression profiles of this gene. *C. glabrata* ATCC2001 demonstrated a moderate positive correlation between the expression profiles of the *GAS2* gene. *C. glabrata* 562123 and *C. glabrata* 534784 had a strong positive correlation between the expression profiles of the *GAS2* gene. *KNH1* gene was the most variable and difficult gene to correlate between the strains. The reference and the 562123 strain showed a moderate correlation, whereas the reference and the 534784 strain showed a strong correlation, and the clinical isolates showed the only weak correlation detected in this study. As for the *UPG1* gene, although there was a negative correlation, the association between its expression in ATCC2001 and in the clinical isolates can be considered weak. Between the isolates, it was determined that *UPG1* up and downregulation had a high tendency to occur similarly in all strains (thus showing strong correlation). In summary, *BGL2*, *XOG1*, *FKS1*, and *MNN2* appeared to be the genes presenting the most similar responses to antifungal drugs within the transcriptome of the three strains; also, the clinical isolates appeared to be nearer each other than to the reference strain. Once more, β-1,3-glucan synthesis was identified as important in *C. glabrata* (three of the four genes affected are responsible for β-1,3-glucan production). These similarities among the two clinical strains may be due to the fact that both were derived from a hospital environment, and it is probable that they had already been challenged by several drugs, so their responses were prompter compared to the reference strain that is a wild type strain.

4. Conclusions

The in vitro high-dose paradox associated with *Candida* spp. isolates is being increasingly reported and connected to slightly elevated MICs, potentially contributing to clinical resistance and failure of antifungal treatments. These drug tolerance and adaptive mechanisms are highly related to *Candida* spp. biofilm forms. *C. glabrata* extracellular matrix is crucial for mature biofilm formation, not only contributing to the adhesive nature of the biofilm cells, but also protecting the cells from antifungal agents and from the host immune system. Understanding the production of the biofilm matrix components and the associated delivery processes is important for the development of effective biofilm therapies. All stakeholders in this process represent potentially attractive targets for detection of and therapeutic interventions against candidiasis.

Acknowledgments: This study was supported by the Portuguese Foundation for Science and Technology (FCT) under the scope of the strategic funding of UID/BIO/04469/2013 unit and COMPETE 2020 (POCI-01-0145-FEDER-006684) and BioTecNorte operation (NORTE-01-0145-FEDER-000004) funded by the European Regional Development Fund under the scope of Norte2020—Programa Operacional Regional do Norte and Célia F. Rodrigues' [SFRH/BD/93078/2013] Ph.D. grant. We also would like to acknowledge MSD® and Astellas® for the kind donation of Caspofungin and Micafungin, respectively.

Author Contributions: Célia F. Rodrigues conceived, designed, performed the experiments, analyzed the data, and wrote the paper; Mariana Henriques conceived, designed the experiments, analyzed the data, and wrote the paper.

References

1. Rodrigues, C.F.; Rodrigues, M.E.; Silva, S.; Henriques, M. *Candida glabrata* biofilms: How far have we come? *J. Fungi* **2017**, *3*, 11. [CrossRef] [PubMed]
2. Silva, S.; Rodrigues, C.F.; Araújo, D.; Rodrigues, M.E.; Henriques, M. *Candida* species biofilms' antifungal resistance. *J. Fungi* **2017**, *3*, 8. [CrossRef] [PubMed]
3. Costa-Orlandi, C.; Sardi, J.; Pitangui, N.; de Oliveira, H.; Scorzoni, L.; Galeane, M.; Medina-Alarcón, K.; Melo, W.; Marcelino, M.; Braz, J.; et al. Fungal biofilms and polymicrobial diseases. *J. Fungi* **2017**, *3*, 22. [CrossRef] [PubMed]
4. McCall, A.; Edgerton, M. Real-time approach to flow cell imaging of *Candida albicans* biofilm development. *J. Fungi* **2017**, *3*, 13. [CrossRef] [PubMed]
5. Nobile, C.J.; Johnson, A.D. *Candida albicans* biofilms and human disease. *Annu. Rev. Microbiol.* **2015**, *69*, 71–92. [CrossRef] [PubMed]
6. Silva, S.; Henriques, M.; Martins, A.; Oliveira, R.; Williams, D.; Azeredo, J. Biofilms of non-*Candida albicans* Candida species: Quantification, structure and matrix composition. *Med. Mycol.* **2009**, *47*, 681–689. [CrossRef] [PubMed]

7. Fonseca, E.; Silva, S.; Rodrigues, C.F.; Alves, C.T.; Azeredo, J.; Henriques, M. Effects of fluconazole on *Candida glabrata* biofilms and its relationship with ABC transporter gene expression. *Biofouling* **2014**, *30*, 447–457. [CrossRef] [PubMed]
8. Rodrigues, C.F.; Silva, S.; Henriques, M. *Candida glabrata*: A review of its features and resistance. *Eur. J. Clin. Microbiol. Infect. Dis.* **2014**, *33*, 673–688. [CrossRef] [PubMed]
9. Negri, M.; Silva, S.; Henriques, M.; Oliveira, R. Insights into *Candida tropicalis* nosocomial infections and virulence factors. *Eur. J. Clin. Microbiol. Infect. Dis.* **2012**, *31*, 1399–1412. [CrossRef] [PubMed]
10. Costerton, J.W.; Lewandowski, Z.; Caldwell, D.E.; Korber, D.R.; Lappin-Scott, H.M. Microbial Biofilms. *Annu. Rev. Microbiol.* **1995**, *49*, 711–745. [CrossRef] [PubMed]
11. Donlan, R.; Costerton, J. Biofilms: Survival mechanisms of clinically relevant microorganisms. *Clin. Microbiol. Rev.* **2002**, *15*, 167–193. [CrossRef] [PubMed]
12. Silva, S.; Negri, M.; Henriques, M.; Oliveira, R.; Williams, D.W.; Azeredo, J. *Candida glabrata*, *Candida parapsilosis* and *Candida tropicalis*: Biology, epidemiology, pathogenicity and antifungal resistance. *FEMS Microbiol. Rev.* **2012**, *36*, 288–305. [CrossRef] [PubMed]
13. Rodrigues, C.F.; Silva, S.; Azeredo, J.; Henriques, M. *Candida glabrata*'s recurrent infections: Biofilm formation during Amphotericin B treatment. *Lett. Appl. Microbiol.* **2016**, *63*, 77–81. [CrossRef] [PubMed]
14. Nett, J.; Lincoln, L.; Marchillo, K.; Massey, R.; Holoyda, K.; Hoff, B.; VanHandel, M.; Andes, D. Putative role of β-1,3 glucans in *Candida albicans* biofilm resistance. *Antimicrob. Agents Chemother.* **2007**, *51*, 510–520. [CrossRef] [PubMed]
15. Nett, J.E.; Sanchez, H.; Cain, M.T.; Andes, D.R. Genetic basis of *Candida* biofilm resistance due to drug-sequestering matrix glucan. *J. Infect. Dis.* **2010**, *202*, 171–175. [CrossRef] [PubMed]
16. Nett, J.E.; Crawford, K.; Marchillo, K.; Andes, D.R.; Nett, J.E.; Crawford, K.; Marchillo, K.; Andes, D.R. Role of Fks1p and matrix glucan in *Candida albicans* biofilm resistance to an echinocandin, pyrimidine, and polyene. *Antimicrob. Agents Chemother.* **2010**, *54*, 3505–3508. [CrossRef] [PubMed]
17. Nett, J.E.; Sanchez, H.; Cain, M.T.; Ross, K.M.; Andes, D.R. Interface of *Candida albicans* biofilm matrix-associated drug resistance and cell wall integrity regulation. *Eukaryot. Cell* **2011**, *10*, 1660–1669. [CrossRef] [PubMed]
18. Rodrigues, C.F.; Gonçalves, B.; Rodrigues, M.E.; Silva, S.; Azeredo, J.; Henriques, M. The effectiveness of voriconazole in therapy of *Candida glabrata*'s biofilms oral infections and its influence on the matrix composition and gene expression. *Mycopathologia* **2017**, *182*, 653–664. [CrossRef] [PubMed]
19. Taff, H.T.; Nett, J.E.; Zarnowski, R.; Ross, K.M.; Sanchez, H.; Cain, M.T.; Hamaker, J.; Mitchell, A.P.; Andes, D.R. A *Candida* biofilm-induced pathway for matrix glucan delivery: Implications for drug resistance. *PLoS Pathog.* **2012**, *8*. [CrossRef] [PubMed]
20. Rodrigues, C.F.; Henriques, M. Oral mucositis caused by *Candida glabrata* biofilms: Failure of the concomitant use of fluconazole and ascorbic acid. *Ther. Adv. Infect. Dis.* **2017**, *1*. [CrossRef] [PubMed]
21. Lopez-Ribot, J.L. Large-scale biochemical profiling of the *Candida albicans* biofilm matrix: New compositional, structural, and functional insights. *MBio* **2014**, *5*, e01314–e01333. [CrossRef] [PubMed]
22. Ramage, G.; Rajendran, R.; Sherry, L.; Williams, C. Fungal biofilm resistance. *Int. J. Microbiol.* **2012**, *2012*. [CrossRef] [PubMed]
23. Mitchell, K.F.; Zarnowski, R.; Andes, D.R. Fungal super glue: The biofilm matrix and its composition, assembly, and functions. *PLoS Pathog.* **2016**, *12*. [CrossRef] [PubMed]
24. Douglas, C. Fungal beta (1,3)-D-glucan synthesis. *Med. Mycol.* **2001**, *39*, 55–66. [CrossRef] [PubMed]
25. Del Mar Gonzalez, M.; Diez-Orejas, R.; Molero, G.; Alvarez, A.M.; Pla, J.; Nombela, C.; Sanchez-Perez, M. Phenotypic characterization of a *Candida albicans* strain deficient in its major exoglucanase. *Microbiology* **1997**, *143*, 3023–3032. [CrossRef] [PubMed]
26. Sarthy, A.V.; McGonigal, T.; Coen, M.; Frost, D.J.; Meulbroek, J.A.; Goldman, R.C. Phenotype in *Candida albicans* of a disruption of the *BGL2* gene encoding a 1,3-β-glucosyltransferase. *Microbiology* **1997**, *143*, 367–376. [CrossRef] [PubMed]
27. Mouyna, I.; Fontaine, T.; Vai, M.; Monod, M.; Fonzi, W.A.; Diaquin, M.; Popolo, L.; Hartland, R.P.; Latgé, J.P. Glycosylphosphatidylinositol-anchored glucanosyltransferases play an active role in the biosynthesis of the fungal cell wall. *J. Biol. Chem.* **2000**, *275*, 14882–14889. [CrossRef] [PubMed]

28. Goldman, R.C.; Sullivan, P.A.; Zakula, D.; Capobianco, J.O. Kinetics of β-1,3 glucan interaction at the donor and acceptor sites of the fungal glucosyltransferase encoded by the *BGL2* gene. *Eur. J. Biochem.* **1995**, *227*, 372–378. [CrossRef] [PubMed]
29. De Groot, P.W.J.; Kraneveld, E.A.; Yin, Q.Y.; Dekker, H.L.; Gross, U.; Crielaard, W.; de Koster, C.G.; Bader, O.; Klis, F.M.; Weig, M. The cell wall of the human pathogen *Candida glabrata*: Differential incorporation of novel adhesin-like wall proteins. *Eukaryot. Cell* **2008**, *7*, 1951–1964. [CrossRef] [PubMed]
30. Garcia-Effron, G.; Lee, S.; Park, S.; Cleary, J.D.; Perlin, D.S. Effect of *Candida glabrata FKS1* and *FKS2* mutations on echinocandin sensitivity and kinetics of 1,3-β-D-glucan synthase: Implication for the existing susceptibility breakpoint. *Antimicrob. Agents Chemother.* **2009**, *53*, 3690–3699. [CrossRef] [PubMed]
31. Vai, M.; Orlandi, I.; Cavadini, P.; Alberghina, L.; Popolo, L. *Candida albicans* homologue of *GGP1/GAS1* gene is functional in *Saccharomyces cerevisiae* and contains the determinants for glycosylphosphatidylinositol attachment. *Yeast* **1996**, *12*, 361–368. [CrossRef]
32. Miyazaki, T.; Nakayama, H.; Nagayoshi, Y.; Kakeya, H.; Kohno, S. Dissection of Ire1 functions reveals stress response mechanisms uniquely evolved in *Candida glabrata*. *PLoS Pathog.* **2013**, *9*. [CrossRef] [PubMed]
33. Montijn, R.C.; van Rinsum, J.; van Schagen, F.A.; Klis, F.M. Glucomannoproteins in the cell wall of *Saccharomyces cerevisiae* contain a novel type of carbohydrate side chain. *J. Biol. Chem.* **1994**, *269*, 19338–19342. [PubMed]
34. Kapteyn, J.C.; Montijn, R.C.; Vink, E.; de la Cruz, J.; Llobell, A.; Douwes, J.E.; Shimoi, H.; Lipke, P.N.; Klis, F.M. Retention of *Saccharomyces cerevisiae* cell wall proteins through a phosphodiester-linked β-1,3-/β-1,6-glucan heteropolymer. *Glycobiology* **1996**, *6*, 337–345. [CrossRef] [PubMed]
35. Dijkgraaf, G.J.; Brown, J.L.; Bussey, H. The *KNH1* gene of *Saccharomyces cerevisiae* is a functional homolog of *KRE9*. *Yeast* **1996**, *12*, 683–692. [CrossRef]
36. Nagahashi, S.; Lussier, M.; Bussey, H. Isolation of *Candida glabrata* homologs of the *Saccharomyces cerevisiae KRE9* and *KNH1* genes and their involvement in cell wall β-1,6-glucan synthesis. *J. Bacteriol.* **1998**, *180*, 5020–5029. [PubMed]
37. Cid, V.J.; Durán, A.; del Rey, F.; Snyder, M.P.; Nombela, C.; Sánchez, M. Molecular basis of cell integrity and morphogenesis in *Saccharomyces cerevisiae*. *Microbiol. Rev.* **1995**, *59*, 345–386. [PubMed]
38. Brown, J.L.; Kossaczka, Z.; Jiang, B.; Bussey, H. A mutational analysis of killer toxin resistance in *Saccharomyces cerevisiae* identifies new genes involved in cell wall (1–6)-β-glucan synthesis. *Genetics* **1993**, *133*, 837–849. [PubMed]
39. Roetzer, A.; Gregori, C.; Jennings, A.M.; Quintin, J.; Ferrandon, D.; Butler, G.; Kuchler, K.; Ammerer, G.; Schüller, C. *Candida glabrata* environmental stress response involves *Saccharomyces cerevisiae* Msn2/4 orthologous transcription factors. *Mol. Microbiol.* **2008**, *69*, 603–620. [CrossRef] [PubMed]
40. Dodgson, A.R.; Pujol, C.; Denning, D.W.; Soll, D.R.; Fox, A.J. Multilocus sequence typing of *Candida glabrata* reveals geographically enriched clades. *J. Clin. Microbiol.* **2003**, *41*, 5709–5717. [CrossRef] [PubMed]
41. Ruiz-Herrera, J.; Victoria Elorza, M.; Valentin, E.; Sentandreu, R. Molecular organization of the cell wall of *Candida albicans* and its relation to pathogenicity. *FEMS Yeast Res.* **2006**, *6*, 14–29. [CrossRef] [PubMed]
42. Latgé, J.-P. The cell wall: A carbohydrate armour for the fungal cell. *Mol. Microbiol.* **2007**, *66*, 279–290. [CrossRef] [PubMed]
43. Netea, M.G.; Brown, G.D.; Kullberg, B.J.; Gow, N.A.R. An integrated model of the recognition of *Candida albicans* by the innate immune system. *Nat. Rev. Microbiol.* **2008**, *6*, 67–78. [CrossRef] [PubMed]
44. Netea, M.G.; Mardi, L. Innate immune mechanisms for recognition and uptake of *Candida* species. *Trends Immunol.* **2010**, *31*, 346–353. [CrossRef] [PubMed]
45. Gow, N.A.R.; Netea, M.G.; Munro, C.A.; Ferwerda, G.; Bates, S.; Mora-Montes, H.M.; Walker, L.; Jansen, T.; Jacobs, L.; Tsoni, V.; et al. Immune recognition of *Candida albicans* β-glucan by Dectin-1. *J. Infect. Dis.* **2007**, *196*, 1565–1571. [CrossRef] [PubMed]
46. Van de Veerdonk, F.L.; Kullberg, B.J.; van der Meer, J.W.; Gow, N.A.; Netea, M.G. Host-microbe interactions: Innate pattern recognition of fungal pathogens. *Curr. Opin. Microbiol.* **2008**, *11*, 305–312. [CrossRef] [PubMed]
47. Reid, D.M.; Gow, N.A.; Brown, G.D. Pattern recognition: Recent insights from Dectin-1. *Curr. Opin. Immunol.* **2009**, *21*, 30–37. [CrossRef] [PubMed]

48. Mora-Montes, H.M.; Bates, S.; Netea, M.G.; Castillo, L.; Brand, A.; Buurman, E.T.; Diaz-Jimenez, D.F.; Jan Kullberg, B.; Brown, A.J.P.; Odds, F.C.; et al. A multifunctional mannosyltransferase family in *Candida albicans* determines cell wall mannan structure and host-fungus interactions. *J. Biol. Chem.* **2010**, *285*, 12087–12095. [CrossRef] [PubMed]
49. Taylor, P.R.; Tsoni, S.V.; Willment, J.A.; Dennehy, K.M.; Rosas, M.; Findon, H.; Haynes, K.; Steele, C.; Botto, M.; Gordon, S.; et al. Dectin-1 is required for β-glucan recognition and control of fungal infection. *Nat. Immunol.* **2007**, *8*, 31–38. [CrossRef] [PubMed]
50. Murciano, C.; Moyes, D.L.; Runglall, M.; Islam, A.; Mille, C.; Fradin, C.; Poulain, D.; Gow, N.A.R.; Naglik, J.R. *Candida albicans* cell wall glycosylation may be indirectly required for activation of epithelial cell proinflammatory responses. *Infect. Immun.* **2011**, *79*, 4902–4911. [CrossRef] [PubMed]
51. Gow, N.A.; Hube, B. Importance of the *Candida albicans* cell wall during commensalism and infection. *Curr. Opin. Microbiol.* **2012**, *15*, 406–412. [CrossRef] [PubMed]
52. Bates, S.; Hughes, H.B.; Munro, C.A.; Thomas, W.P.H.; MacCallum, D.M.; Bertram, G.; Atrih, A.; Ferguson, M.A.J.; Brown, A.J.P.; Odds, F.C.; et al. Outer chain *N*-glycans are required for cell wall integrity and virulence of *Candida albicans*. *J. Biol. Chem.* **2006**, *281*, 90–98. [CrossRef] [PubMed]
53. Bates, S.; MacCallum, D.M.; Bertram, G.; Munro, C.A.; Hughes, H.B.; Buurman, E.T.; Brown, A.J.P.; Odds, F.C.; Gow, N.A.R. *Candida albicans* Pmr1p, a secretory pathway P-type Ca^{2+}/Mn^{2+}-ATPase, is required for glycosylation and virulence. *J. Biol. Chem.* **2005**, *280*, 23408–23415. [CrossRef] [PubMed]
54. Mora-Montes, H.M.; Bates, S.; Netea, M.G.; Diaz-Jimenez, D.F.; Lopez-Romero, E.; Zinker, S.; Ponce-Noyola, P.; Kullberg, B.J.; Brown, A.J.P.; Odds, F.C.; et al. Endoplasmic reticulum -glycosidases of *Candida albicans* are required for *N* glycosylation, cell wall integrity, and normal host-fungus interaction. *Eukaryot. Cell* **2007**, *6*, 2184–2193. [CrossRef] [PubMed]
55. Munro, C.A.; Bates, S.; Buurman, E.T.; Hughes, H.B.; MacCallum, D.M.; Bertram, G.; Atrih, A.; Ferguson, M.A.J.; Bain, J.M.; Brand, A.; et al. Mnt1p and Mnt2p of *Candida albicans* are partially redundant α-1,2-mannosyltransferases that participate in *O*-linked mannosylation and are required for adhesion and virulence. *J. Biol. Chem.* **2005**, *280*, 1051–1060. [CrossRef] [PubMed]
56. Saijo, S.; Ikeda, S.; Yamabe, K.; Kakuta, S.; Ishigame, H.; Akitsu, A.; Fujikado, N.; Kusaka, T.; Kubo, S.; Chung, S.; et al. Dectin-2 recognition of α-mannans and induction of Th17 cell differentiation is essential for host defense against *Candida albicans*. *Immunity* **2010**, *32*, 681–691. [CrossRef] [PubMed]
57. Lal, P.; Sharma, D.; Pruthi, P.; Pruthi, V. Exopolysaccharide analysis of biofilm-forming *Candida albicans*. *J. Appl. Microbiol.* **2010**, *109*, 128–136. [CrossRef] [PubMed]
58. Correia, I.A. *Role of Secreted Aspartyl Proteases in Candida Albicans Virulence, Host Immune Response and Immunoprotection in Murine Disseminated Candidiasis*; Universidade do Minho: Braga, Portugal, 2012.
59. Johnson, C.J.; Cabezas-Olcoz, J.; Kernien, J.F.; Wang, S.X.; Beebe, D.J.; Huttenlocher, A.; Ansari, H.; Nett, J.E. The extracellular matrix of *Candida albicans* biofilms impairs formation of neutrophil extracellular traps. *PLoS Pathog.* **2016**, *12*. [CrossRef] [PubMed]
60. Williams, D.W.; Wilson, M.J.; Lewis, M.A.O.; Potts, A.J.C. Identification of *Candida* species by PCR and restriction fragment length polymorphism analysis of intergenic spacer regions of ribosomal DNA. *J. Clin. Microbiol.* **1995**, *33*, 2476–2479. [PubMed]
61. Arendrup, M.C.; Arikan, S.; Barchiesi, F.; Bille, J.; Dannaoui, E.; Denning, D.W.; Donnelly, J.P.; Fegeler, W.; Moore, C.; Richardson, M.; et al. EUCAST Technical note on the method for the determination of broth dilution minimum inhibitory concentrations of antifungal agents for conidia—Forming moulds. *ESCMID Tech. Notes* **2008**, *14*, 982–984.
62. *European Committee on Antimicrobial Susceptibility Testing, EUCAST Breakpoint Tables for Interpretation of MICs*, Version 8.1; Available online: http://www.eucast.org.
63. Skrzypek, M.S.; Binkley, J.; Binkley, G.; Miyasato, S.R.; Simison, M.; Sherlock, G. The Candida Genome Database (CGD): Incorporation of Assembly 22, systematic identifiers and visualization of high throughput sequencing data. *Nucleic Acids Res.* **2017**, *45*, D592–D596. [CrossRef] [PubMed]
64. Untergasser, A.; Nijveen, H.; Rao, X.; Bisseling, T.; Geurts, R.; Leunissen, J.A.M. Primer3Plus, an enhanced web interface to Primer3. *Nucleic Acids Res.* **2007**, *35*, W71–W74. [CrossRef] [PubMed]
65. Schmittgen, T.D.; Livak, K.J. Analyzing real-time PCR data by the comparative C_T method. *Nat. Protoc.* **2008**, *3*, 1101–1108. [CrossRef] [PubMed]

66. Cho, E.-J.; Shin, J.H.; Kim, S.H.; Kim, H.-K.; Park, J.S.; Sung, H.; Kim, M.N.; Im, H.J. Emergence of multiple resistance profiles involving azoles, echinocandins and Amphotericin B in *Candida glabrata* isolates from a neutropenia patient with prolonged fungaemia. *J. Antimicrob. Chemother.* **2015**, *70*, 1268–1270. [CrossRef] [PubMed]
67. Douglas, L.J. *Candida* biofilms and their role in infection. *Trends Microbiol.* **2003**, *11*, 30–36. [CrossRef]
68. Zarnowski, R.; Westler, W.M.; Lacmbouh, G.A.; Marita, J.M.; Bothe, J.R.; Bernhardt, J.; Sahraoui, A.L.H.; Fontainei, J.; Sanchez, H.; Hatfeld, R.D.; et al. Novel entries in a fungal biofilm matrix encyclopedia. *MBio* **2014**, *5*, 1–13. [CrossRef] [PubMed]
69. Mukherjee, P.K.; Chandra, J. *Candida* biofilm resistance. *Drug. Resist. Updat.* **2004**, *7*, 301–309. [CrossRef] [PubMed]
70. Livak, K.J. *User Bulletin #2 ABI P RISM 7700 Sequence Detection System SUBJECT: Relative Quantitation of Gene Expression—Updated 2001*; Applied Biosystems: Foster City, CA, USA, 1997.
71. Li, P.; Seneviratne, C.; Alpi, E.; Vizcaino, J.; Jin, L. Delicate metabolic control and coordinated stress response critically determine antifungal tolerance of *Candida albicans* biofilm persisters. *Antimicrob. Agents Chemother.* **2015**, *59*, 6101–6112. [CrossRef] [PubMed]
72. Al-Dhaheri, R.S.; Douglas, L.J. Absence of Amphotericin B-tolerant persister cells in biofilms of some *Candida* species. *Antimicrob. Agents Chemother.* **2008**, *52*, 1884–1887. [CrossRef] [PubMed]
73. Sun, J.; Li, Z.; Chu, H.; Guo, J.; Jiang, G.; Qi, Q. *Candida albicans* Amphotericin B-tolerant persister formation is closely related to surface adhesion. *Mycopathologia* **2016**, *181*, 41–49. [CrossRef] [PubMed]
74. Alexander, B.D.; Johnson, M.D.; Pfeiffer, C.D.; Jimenez-Ortigosa, C.; Catania, J.; Booker, R.; Castanheira, M.; Messer, S.A.; Perlin, D.S.; Pfaller, M.A. Increasing echinocandin resistance in *Candida glabrata*: Clinical failure correlates with presence of *FKS* mutations and elevated minimum inhibitory concentrations. *Clin. Infect. Dis.* **2013**, *56*, 1724–1732. [CrossRef] [PubMed]
75. Pinhati, H.M.S.; Casulari, L.A.; Souza, A.C.R.; Siqueira, R.A.; Damasceno, C.M.G.; Colombo, A.L. Outbreak of candidemia caused by fluconazole resistant *Candida parapsilosis* strains in an intensive care unit. *BMC Infect. Dis.* **2016**, *16*. [CrossRef] [PubMed]
76. Bizerra, F.C.; Jimenez-Ortigosa, C.; Souza, A.C.R.; Breda, G.L.; Queiroz-Telles, F.; Perlin, D.S.; Colombo, A.L. Breakthrough candidemia due to multidrug-resistant *Candida glabrata* during prophylaxis with a low dose of micafungin. *Antimicrob. Agents Chemother.* **2014**, *58*, 2438–2440. [CrossRef] [PubMed]
77. Park, S.; Kelly, R.; Kahn, J.N.N.; Robles, J.; Hsu, M.-J.; Register, E.; Li, W.; Vyas, V.; Fan, H.; Abruzzo, G.; et al. Specific substitutions in the echinocandin target Fks1p account for reduced susceptibility of rare laboratory and clinical *Candida* sp. isolates. *Antimicrob. Agents Chemother.* **2005**, *49*, 3264–3273. [CrossRef] [PubMed]
78. Desnos-Ollivier, M.; Moquet, O.; Chouaki, T.; Guerin, A.M.; Dromer, F. Development of echinocandin resistance in *Clavispora lusitaniae* during caspofungin treatment. *J. Clin. Microbiol.* **2011**, *49*, 2304–2306. [CrossRef] [PubMed]
79. Jensen, R.H.; Johansen, H.K.; Arendrup, M.C. Stepwise development of a homozygous S80P substitution in Fks1p, conferring echinocandin resistance in *Candida tropicalis*. *Antimicrob. Agents Chemother.* **2013**, *57*, 614–617. [CrossRef] [PubMed]
80. Lewis, J.S.; Wiederhold, N.P.; Wickes, B.L.; Patterson, T.F.; Jorgensen, J.H. Rapid emergence of echinocandin resistance in *Candida glabrata* resulting in clinical and microbiologic failure. *Antimicrob. Agents Chemother.* **2013**, *57*, 4559–4561. [CrossRef] [PubMed]
81. Shields, R.K.; Nguyen, M.H.; Press, E.G.; Kwa, A.L.; Cheng, S.; Du, C.; Clancy, C.J. The presence of an *FKS* mutation rather than MIC is an independent risk factor for failure of echinocandin therapy among patients with invasive candidiasis due to *Candida glabrata*. *Antimicrob. Agents Chemother.* **2012**, *56*, 4862–4869. [CrossRef] [PubMed]
82. Beyda, N.D.; Lewis, R.E.; Garey, K.W. Echinocandin resistance in *Candida* species: Mechanisms of reduced susceptibility and therapeutic approaches. *Ann. Pharmacother.* **2012**, *46*, 1086–1096. [CrossRef] [PubMed]
83. Barchiesi, F.; Spreghini, E.; Tomassetti, S.; Della Vittoria, A.; Arzeni, D.; Manso, E.; Scalise, G. Effects of caspofungin against *Candida guilliermondii* and *Candida parapsilosis*. *Antimicrob. Agents Chemother.* **2006**, *50*, 2719–2727. [CrossRef] [PubMed]
84. Garcia-Effron, G.; Katiyar, S.K.; Park, S.; Edlind, T.D.; Perlin, D.S. A naturally occurring proline-to-alanine amino acid change in Fks1p in *Candida parapsilosis*, *Candida orthopsilosis*, and *Candida metapsilosis* accounts for reduced echinocandin susceptibility. *Antimicrob. Agents Chemother.* **2008**, *52*, 2305–2312. [CrossRef] [PubMed]

85. Forastiero, A.; Garcia-Gil, V.; Rivero-Menendez, O.; Garcia-Rubio, R.; Monteiro, M.C.; Alastruey-Izquierdo, A.; Jordan, R.; Agorio, I.; Mellado, E. Rapid development of *Candida krusei* echinocandin resistance during caspofungin therapy. *Antimicrob. Agents Chemother.* **2015**, *59*, 6975–6982. [CrossRef] [PubMed]
86. Arendrup, M.; Perlin, D.; Jensen, R.; Howard, S.; Goodwin, J.; Hopec, W. Differential in vivo activities of anidulafungin, caspofungin, and micafungin against *Candida glabrata* isolates with and without *FSK* resistance mutations. *Antim Agents Chemoter.* **2012**, *56*, 2435–2442. [CrossRef] [PubMed]
87. Fanning, S.; Mitchell, A.P. Fungal Biofilms. *PLoS Pathog.* **2012**, *8*. [CrossRef] [PubMed]
88. Nett, J.; Lepak, A.; Marchillo, K.; Andes, D. Time course global gene expression analysis of an in vivo *Candida* biofilm. *J. Infect. Dis.* **2009**, *200*, 307–313. [CrossRef] [PubMed]
89. Barns, S.M.; Lane, D.J.; Sogin, M.L.; Bibeau, C.; Weisburg, W.G. Evolutionary relationships among pathogenic *Candida* species and relatives. *J. Bacteriol.* **1991**, *173*, 2250–2255. [CrossRef] [PubMed]
90. Kitada, K.; Yamaguchi, E.; Arisawa, M. Cloning of the *Candida glabrata TRP1* and HIS3 genes, and construction of their disruptant strains by sequential integrative transformation. *Gene* **1995**, *165*, 203–206. [CrossRef]
91. Nakayama, H.; Ueno, K.; Uno, J.; Nagi, M.; Tanabe, K.; Aoyama, T.; Chibana, H.; Bard, M. Growth defects resulting from inhibiting *ERG20* and *RAM2* in *Candida glabrata*. *FEMS Microbiol. Lett.* **2011**, *317*, 27–33. [CrossRef] [PubMed]
92. Boone, C.; Sommer, S.S.; Hensel, A.; Bussey, H. Yeast KRE genes provide evidence for a pathway of cell wall beta-glucan assembly. *J. Cell Biol.* **1990**, *110*, 1833–1843. [CrossRef] [PubMed]
93. Srikantha, T.; Daniels, K.J.; Pujol, C.; Kim, E.; Soll, D.R. Identification of genes upregulated by the transcription factor Bcr1 that are involved in impermeability, impenetrability, and drug resistance of *Candida albicans* a/α biofilms. *Eukaryot. Cell* **2013**, *12*, 875–888. [CrossRef] [PubMed]
94. Chen, K.-H.; Miyazaki, T.; Tsai, H.-F.; Bennett, J.E. The bZip transcription factor Cgap1p is involved in multidrug resistance and required for activation of multidrug transporter gene *CgFLR1* in *Candida glabrata*. *Gene* **2007**, *386*, 63–72. [CrossRef] [PubMed]
95. Ferrari, S.; Sanguinetti, M.; De Bernardis, F.; Torelli, R.; Posteraro, B.; Vandeputte, P.; Sanglard, D. Loss of mitochondrial functions associated with azole resistance in *Candida glabrata* results in enhanced virulence in mice. *Antimicrob. Agents Chemother.* **2011**, *55*, 1852–1860. [CrossRef] [PubMed]
96. Mathé, L.; Van Dijck, P. Recent insights into *Candida albicans* biofilm resistance mechanisms. *Curr. Genet.* **2013**, *59*, 251–264. [CrossRef] [PubMed]
97. Marco, F.; Pfaller, M.A.; Messer, S.A.; Jones, R.N. Activity of MK-0991 (L-743,872), a new echinocandin, compared with those of LY303366 and four other antifungal agents tested against blood stream isolates of *Candida* spp. *Diagn. Microbiol. Infect. Dis.* **1998**, *32*, 33–37. [CrossRef]
98. Kuhn, D.M.; George, T.; Chandra, J.; Mukherjee, P.K.; Ghannoum, M.A. Antifungal susceptibility of *Candida* biofilms: Unique efficacy of amphotericin B lipid formulations and echinocandins. *Antimicrob. Agents Chemother.* **2002**, *46*, 1773–1780. [CrossRef] [PubMed]
99. Scorzoni, L.; de Paula e Silva, A.C.A.; Marcos, C.M.; Assato, P.A.; de Melo, W.C.; de Oliveira, H.C.; Costa-Orlandi, C.B.; Mendes-Giannini, M.J.; Fusco-Almeida, A.M. Antifungal therapy: New advances in the understanding and treatment of mycosis. *Front. Microbiol.* **2017**, *8*, 1–23. [CrossRef] [PubMed]
100. Pierce, C.G.; Srinivasan, A.; Uppuluri, P.; Ramasubramanian, A.K.; López-Ribot, J.L. Antifungal therapy with an emphasis on biofilms. *Curr. Opin. Pharmacol.* **2013**, *13*. [CrossRef] [PubMed]
101. Pappas, P.G.; Kauffman, C.A.; Andes, D.R.; Clancy, C.J.; Marr, K.A.; Ostrosky-Zeichner, L.; Reboli, A.C.; Schuster, M.G.; Vazquez, J.A.; Walsh, T.J.; et al. Clinical practice guideline for the management of Candidiasis: 2016 update by the infectious diseases society of America. *Clin. Infect. Dis.* **2015**, *62*, e1–e50. [CrossRef] [PubMed]
102. Canuto, M.M.; Rodero, F.G. Antifungal drug resistance to azoles and polyenes. *Lancet Infect. Dis.* **2002**, *2*, 550–563. [CrossRef]
103. Rex, J.H.J.; Walsh, T.J.; Sobel, J.D.J.; Filler, S.G.; Pappas, P.G.; Dismukes, W.E.; Edwards, J.E. Practice guidelines for the treatment of candidiasis. *Clin. Infect. Dis.* **2000**, *30*, 662–678. [CrossRef] [PubMed]
104. Schmalreck, A.F.; Willinger, B.; Haase, G.; Blum, G.; Lass-Flörl, C.; Fegeler, W.; Becker, K.; Antifungal susceptibility testing-AFST study group. Species and susceptibility distribution of 1062 clinical yeast isolates to azoles, echinocandins, flucytosine and amphotericin B from a multi-centre study. *Mycoses* **2012**, *55*, e124–e137. [CrossRef] [PubMed]
105. Perlin, D.S. Mechanisms of echinocandin antifungal drug resistance. *Ann. N. Y. Acad. Sci.* **2015**, *1354*, 1–11. [CrossRef] [PubMed]

11

Invasive Aspergillosis in Pediatric Leukemia Patients: Prevention and Treatment

Savvas Papachristou, Elias Iosifidis and Emmanuel Roilides *

Infectious Diseases Unit, 3rd Department of Pediatrics, Faculty of Medicine, Aristotle University School of Health Sciences, Konstantinoupoleos 49, 54642 Thessaloniki, Greece; savvas.gpap@gmail.com (S.P.); iosifidish@gmail.com (E.I.)

* Correspondence: roilides@med.auth.gr

Abstract: The purpose of this article is to review and update the strategies for prevention and treatment of invasive aspergillosis (IA) in pediatric patients with leukemia and in patients with hematopoietic stem cell transplantation. The major risk factors associated with IA will be described since their recognition constitutes the first step of prevention. The latter is further analyzed into chemoprophylaxis and non-pharmacologic approaches. Triazoles are the mainstay of anti-fungal prophylaxis while the other measures revolve around reducing exposure to mold spores. Three levels of treatment have been identified: (a) empiric, (b) pre-emptive, and (c) targeted treatment. Empiric is initiated in febrile neutropenic patients and uses mainly caspofungin and liposomal amphotericin B (LAMB). Pre-emptive is a diagnostic driven approach attempting to reduce unnecessary use of anti-fungals. Treatment targeted at proven or probable IA is age-dependent, with voriconazole and LAMB being the cornerstones in >2yrs and <2yrs age groups, respectively.

Keywords: *Aspergillus*; anti-fungal agents; hematological malignancies

---

1. Introduction

Aspergillosis can be present in an acute or chronic form [1]. Syndromes of clinical significance include invasive aspergillosis (IA), chronic and saprophytic aspergillosis, and allergic aspergillosis [2]. We focus on IA because it occurs in immuno-compromised hosts. It is associated with notable morbidity and mortality in pediatric patients suffering from immuno-compromising conditions [3–5], with one multi-center retrospective study recording at a 52.5% mortality rate [6]. IA in children has also been related to increased financial costs [7]. While immuno-compromised pediatric patients also display susceptibility to invasive fungal disease (IFD), like IA, differences from adults have been highlighted to pertain to several aspects of these infections [8–12]. These differences are summarized in Table 1.

During recent years, clinicians have been extrapolating evidence from adult studies of IA, due to the lack of respective pediatric data [6]. According to the 2017 European Society for Clinical Microbiology and Infectious Diseases (ESCMID), the European Confederation of Medical Mycology (ECMM) and the European Respiratory Society (ERS) Joint Clinical Guidelines, according to recent guidelines by the Fourth European Conference on Infections in Leukemia (ECIL-4), pediatric recommendations about intervention are based on efficacy data from phase 2 and 3 trials in adults, on pediatric pharmacokinetic (PK), dosing, safety, supportive efficacy data, and on regulatory approvals [8,13].

Table 1. Differences between pediatric and adult Invasive Aspergillosis.

Field of Difference
A) Comorbidities -Biology -Management -Prognosis
B) High-risk populations
C) Epidemiology
D) Diagnostic techniques -Performance -Utility
E) Anti-fungal drugs -Pharmacology -Dosing scheme
F) Phase 3 clinical trials

In pediatric patients, risk factors for IA include primary immunodeficiencies and especially chronic granulomatous disease (CGD), secondary immunodeficiencies (associated with cancer chemotherapy and failure syndromes of the bone marrow), critical illness, chronic diseases of the airways, low birth-weight, and prematurity (the last two are related to neonatal patients) [3–5,14–16]. Additionally, immunosuppressive treatments—including corticosteroids in high doses and biologic agents interacting with immune pathways (like monoclonal antibodies targeting tumor necrosis factor alpha)—are also regarded as risk factors for IA [5,6,17]. Lastly, solid organ transplantation (SOT) is related to IA, which becomes more evident in the case of heart and/or lung recipients [1,18]. Nevertheless, the most significant risk factors are considered to be hematological malignancies and *hematopoietic stem cell transplantation* (HSCT) [1,6,19–21]. These two conditions constitute the two major risk factors that are commonly encountered in patients with IA [1,6,19–21]. Furthermore, the detection of IA in leukemia patients affects the decisions regarding the administration of chemotherapy [5,7]. More specifically, delayed delivery of chemotherapy decreases the risk for IA progression, on one hand, but, conversely, it renders the progression of the malignancy more likely [5]. This delicate balance makes it more urgent to address the management of this group of patients.

This article intends to review the current strategies for prevention and treatment of IA in pediatric leukemia patients. In the section of prevention, the following topics will be covered: (a) epidemiology and risk factors for IA in pediatric patients with leukemia, (b) anti-fungal prophylaxis, and (c) other preventive measures. Treatment will be subdivided into three main areas: (a) empiric treatment, (b) pre-emptive treatment, and (c) treatment for proven/probable IA. The latter will also include an analysis of the therapeutic approaches to invasive pulmonary aspergillosis (IPA) and the central nervous system (CNS) aspergillosis.

2. Prevention

2.1. *Epidemiology and Risk Factors for Invasive Aspergillosis*

The incidence of IA in pediatric patients with hematological malignancies has been estimated by several studies between 4.57% and 9.5% [7,20,22,23]. Identified routes of infection include the respiratory tract, the gastrointestinal tract, and the skin [24]. A retrospective multi-center study incorporating a diverse population [6] found lungs, skin, and paranasal sinuses as the most frequently affected foci of infection. Regarding microbiology, *Aspergillus fumigatus*, *Aspergillus flavus*, *Aspergillus terreus*, and *Aspergillus niger* were the predominant isolates (in order of frequency) in the previous study [6].

Recognizing pediatric patients with leukemia at risk for developing IA is the cornerstone of prevention. This will enable physicians to timely implement the appropriate strategies to reduce modifiable risk factors and initiate anti-fungal prophylaxis in pediatric leukemia and HSCT patients at high risk for invasive *Aspergillus* spp. [8]. Risk factors for IA in the previously mentioned pediatric patients are summarized in Table 2.

Table 2. Risk factors for Invasive Aspergillosis in pediatric patients.

Leukemia Patients	HSCT Recipients
Severe and persistent neutropenia	Severe and persistent neutropenia
Corticosteroids in high-doses	Corticosteroids in high-doses
Mucosal damage	Mucosal damage
Increasing age	Increasing age
AML	Allogeneic transplant
ALL: relapse	GVHD
ALL: de novo	HLA discordance
ALL: high-risk	CMV coinfection
Refractoriness of acute leukemia	Respiratory virus coinfection
	Colonization by *Aspergillus* spp.
	T-cell depletion
	CD 34 selection
Ward-associated factors (local epidemiology, environmental conditions, contamination of hospital water supply systems, construction works)	Ward-associated factors (local epidemiology, environmental conditions, contamination of hospital water supply systems, construction works)

AML, acute myelogenous leukemia. ALL, acute lymphoblastic leukemia. HSCT, hematopoietic stem cell transplantation. GVHD, graft-versus-host disease. HLA, human leukocyte antigen. CMV, cytomegalovirus. References are provided in the text.

Generally, an IFD incidence >10% is considered high-risk [8]. Severe and persistent neutropenia, high-dose corticosteroid regimens, and damage to mucosal surfaces render these two groups of patients susceptible to IA [8,25,26]. A recent systematic review of publications since 1980, that addressed pediatric-specific factors for invasive fungal diseases (IFDs), indicated that increasing age is a risk factor in both groups [27]. In leukemia patients, the type of malignancy determines the risk, with acute myelogenous leukemia (AML) ranking first (3.7–28% risk), while relapse and de novo acute lymphoblastic leukemia (ALL) are associated with a 4–9% and a 0.6–2% risk for IA, respectively [1,20,21,28]. It should be noted, that according to other studies, the risk was nearly equal between AML and ALL patients [6], or even greater in ALL patients [7]. However, these observations could be attributed to the specific characteristics or limitations of the studies. Refractoriness among acute leukemia patients is also a significant risk factor for IA [2]. High-risk ALL is recognized as a risk factor, but the heterogeneity characterizing this group of patients was underlined by the International Pediatric Fever and Neutropenia Guideline Panel [27,29]. In HSCT recipients, an allogeneic transplant is associated with a greater risk for IA than an autologous one [2,30]. Specific risk factors in allogeneic HSCT include the development of graft-versus-host disease (GVHD), the extension of human leukocyte antigen (HLA) discordance, the presence of cytomegalovirus (CMV) or respiratory virus coinfection, and the colonization by *Aspergillus* spp. [1,28,31–33]. In addition, two strategies for reducing GVHD—T-cell depletion and CD34 selection—are also related to IA infection [2,32,34]. Despite the absence of a risk stratification model for IFDs in pediatrics, a differentiation between high-risk and low-risk patients has been attempted [27,29]. More specifically, AML, high-risk ALL, acute leukemia relapse, allogeneic HSCT, protracted granulocytopenia, and administration of corticosteroids in high doses are considered high-risk conditions [29]. All other conditions are low-risk [29]. Lastly, topics in the field of risk factors for further research include the role of lymphopenia in IFDs and IA and the development of a prediction model for IFDs [27].

Certain risk factors for IA in children with leukemia and HSCT are ward-associated. These include local epidemiology, environmental conditions, contamination of hospital water supply systems, and construction work [2,7,8,35–39].

2.2. Anti-Fungal Prophylaxis

Anti-fungal prophylaxis is divided into primary and secondary entities [8]. Primary is defined as the administration of antifungal agents to high-risk patients without infection, whereas secondary lacks a robust definition and occasionally coincides conceptually with treatment for proven/probable IFDs [8].

Initiation of primary prophylaxis for IA is justified due to the lack of efficacy of diagnostic tests and the dismal outcomes of this infection [5,13]. Anti-fungal agents used for primary prophylaxis include the triazoles itraconazole, voriconazole, and posaconazole, liposomal amphotericin B (LAMB) (in the systemic and aerosolized form) and the echinocandins micafungin and caspofungin. Fluconazole has no activity against molds and, thus, it is not used in prophylaxis against IA [8].

Itraconazole is active against both yeasts and molds and is administered at a per os (PO) dose of 2.5 mg/kg/12 h, provided that the patient's age is ≥2 years [8], while therapeutic drug monitoring (TDM) is necessary to achieve the dosing target of ≥0.5 mg/L [40]. Itraconazole has been studied in pediatric cancer and HSCT patients and is considered a reliable option [41,42] even though prospective studies of larger scale are required to reach further conclusions [43]. The use of this azole is restricted by adverse reactions, according to a meta-analysis [44]. It is not approved in EU for patients <18 years of age [8].

Voriconazole has been found to be superior to itraconazole, in terms of tolerability, for allogeneic HSCT in patients ≥12 years. However, the two agents were equally effective in preventing IFD [45]. According to the ECIL-4, the recommended voriconazole dose in pediatrics for ages 2–<12 years, or 12–14 years with a body weight <50 kg, is 9 mg/kg/12 h for the PO forms and 9 mg/kg/12 h the first day, which is followed by 8 mg/kg/12 h on subsequent days for the intravenous (IV) forms [8]. For ages 12–14 years with a body weight ≥50 kg, or ages ≥15 years, the recommended dose for the PO form is 200 mg/12 h and, for the IV form, 6 mg/kg/12 h the first day, which is followed by 4 mg/kg/12 h on subsequent days [8]. In pediatric patients, voriconazole exposure displays substantial variability [46] and, consequently, TDM is required to maintain the plasma concentration of 1–5 mg/L [8,47,48]. In the "Voriproph" study—the largest cohort study of voriconazole chemoprophylaxis in children—the use of this agent has been well tolerated [49]. An age of <2 years is a contraindication to the use of voriconazole [8].

Posaconazole is approved for use as a chemoprophylactic agent with both anti-mold and anti-yeast activity in pediatric patients with AML, GVHD after HSCT and HSCT with a long neutropenic period [10,50–52]. Posaconazole is appropriate only for children ≥13 years, based on scant PK data from two adult clinical trials, which also recruited a few patients older than 13, but younger than 18 years [1,53,54]. This anti-fungal agent is available in two PO formulations including an oral suspension and gastro-resistant tablets [55]. For tablets, the established dose is 300 mg/24 h, whereas, for suspension, it is 200 mg/8 h [50,55]. A recent, non-randomized, single-center study in pediatric patients with HSCT found that the tablets were more reliable than the suspension in terms of plasma trough levels [55]. TDM is, however, still required, when the oral suspension is used, to maintain plasma levels ≥0.5 mg/L [56]. Use of posaconazole in patients <13 years of age is contraindicated due to the lack of PK data, unstable plasma concentrations, lack of an IV preparation, and undependable PO absorption [50].

Liposomal amphotericin B (LAMB), in various dosing regimens, has been assessed in several pediatric studies with positive results regarding its safety, efficacy, and feasibility [57–59]. The IV form is reserved for patients intolerant or with contraindications to the use of triazoles [13]. The recommended dosing scheme is either 1 mg/kg/48 h, or 2.5 mg/kg two times per week [8]. The aerosolized form of LAMB is described in ECIL-4 guidelines as prophylaxis against pulmonary infections. However, the route is not approved and doses in patients younger than 18 years have not been established [8].

Micafungin has been compared to fluconazole in a phase III randomized, double-blind clinical trial including both adults and children with HSCT-associated neutropenia [60]. The results were in favor of micafungin in terms of efficacy as a prophylactic agent [60]. However, the scarcity of pediatric data, the lack of PO forms, and the high cost restrict the use of micafungin in anti-fungal prophylaxis [50]. According to ECIL-4 guidelines, the recommended dose is 1 mg/kg/24 h or 50 mg if the patient's weight ≥50 kg [8]. A dose of 2 mg/kg/24 h has been evaluated and found to be safe in children with allogeneic HSCT [61]. Micafungin is used in cases of intolerance or contra-indication to triazoles [13].

Caspofungin has also been evaluated in the setting of anti-fungal prophylaxis [50]. A randomized study including both adult and pediatric patients with AML and myelodysplastic syndrome (MDS) compared caspofungin to itraconazole and found similar efficiency and tolerability [62]. Studies including only pediatric patients are limited to two retrospective cohort studies [50,63,64]. The first compared caspofungin to LAMB in HSCT recipients and reported comparable efficiency [63]. The second study used micafungin as a comparator and recommended that caspofungin should not be preferred over micafungin [64]. Both studies recognized the need for the conduction of randomized clinical trials [63,64].

In summary, in high-risk, acute leukemia patients, the recommended agents for primary prophylaxis against IA include itraconazole, posaconazole, IV LAMB, aerosolized LAMB, micafungin, and voriconazole [8]. In allogeneic HSCT recipients, the options for IA prophylaxis are itraconazole or voriconazole, micafungin, LAMB, aerosolized LAMB, and posaconazole [8]. When GVHD develops, the recommended anti-fungals are posaconazole, voriconazole, itraconazole, IV LAMB, and micafungin [8]. All the above agents are ranked according to the strength of recommendation and quality of evidence and are summarized in Table 3. The grading system used is the one developed by the Infectious Diseases Society of America (IDSA) [8,65]. According to the previously mentioned system, each recommendation receives a letter (A, B, C, D, or E) reflecting its strength, followed by a Latin number (I, II or III) pertaining to the quality of evidence [65].

Table 3. Primary prophylaxis against Invasive Aspergillosis.

Drug	Route	Dosage	Indications for Use (Recommendation Ranking)	Refs
		AMB formulations		
LAMB	IV	1 mg/kg/48 h, or 2.5 mg/kg two times per week	• High-risk acute leukemia patients (B-II) • Allogeneic HSCT recipients (C-III) • GVHD (no grading)	[8]
LAMB	Aerosolized	Not established	• High-risk acute leukemia patients (no grading) • Allogeneic HSCT recipients (no grading)	[8]
		Azoles		
ITC	PO	2.5 mg/kg/24 h	• High-risk acute leukemia patients (B-I) • Allogeneic HSCT recipients (B-I) • GVHD (C-II)	[8]
VRC	PO	• 9 mg/kg/12 h (ages 2-<12, or 12–14 weighing <50 kg) • 200 mg/12 h (ages ≥15 years, or 12–14 weighing ≥50 kg)	• High-risk acute leukemia patients (no grading) • Allogeneic HSCT recipients (B-I) • GVHD (B-I)	[8]
	IV	• 9 mg/kg/12 h the first day, followed by 8 mg/kg/12 h on next days (ages 2-<12, or 12–14 weighing <50 kg) • 6 mg/kg/12 h the first day, followed by 4 mg/kg/12 h on next days (ages ≥15 years, or 12–14 weighing ≥50 kg)		

Table 3. *Cont.*

Drug	Route	Dosage	Indications for Use (Recommendation Ranking)	Refs
PSC	PO	• 200 mg/8 h (oral susp.) • 300 mg/24 h (tabl.)	• High-risk acute leukemia patients (B-I) • Allogeneic HSCT recipients (no grading) • GVHD (B-I)	[8]
		Echinocandins		
MFG	IV	• 1 mg/kg/24 h (max 50 mg if weight ≥50 kg) • 2 mg/kg/24 h	• High-risk acute leukemia patients (no grading) • Allogeneic HSCT recipients (C-I) • GVHD (no grading) • Allogeneic HSCT recipients	[8,61]

AMB, amphotericin B. LAMB, liposomal amphotericin B. IV, intravenous. HSCT, hematopoietic stem cell transplantation. GVHD, graft-versus-host disease. ITC, itraconazole. PO, per os. VRC, voriconazole. PSC, posaconazole. susp., suspension. tabl., tablets. MFG, micafungin.

Secondary chemoprophylaxis, as mentioned above, is a vague term, which cannot be easily differentiated from continued treatment against previous IA [8]. It should be administered for the entire period during which the risk factors for IA persist (such as granulocytopenia and immunosuppression) and it should be targeted against the previous isolates of *Aspergillus* spp. [8,13]. The "VOSIFI Study" has evaluated voriconazole in the setting of allogeneic HSCT in adult patients and has found that this agent is an efficient option when considering secondary prophylaxis [66]. The regimen consisting of LAMB followed by voriconazole has been evaluated in pediatric patients with acute leukemia and IPA [67]. Other options include itraconazole and caspofungin. However, supporting evidence is derived from adult studies [8].

2.3. Other Preventive Measures

Apart from the administration of drug prophylaxis, several infection control strategies can be applied to prevent IA among pediatric leukemia and HSCT patients. These strategies aim to decrease the exposure to sources of mold spores, which would, otherwise, increase the risk of IA [1].

The cornerstone is the construction of a "protective environment" for inpatients, which regulates room ventilation and involves a specific number of air exchanges/hour, the application of high-efficiency particulate air (HEPA) filters (with or without laminar airflow), appropriate room sealing, automatic-closing doors, pressure monitoring (to maintain a positive pressure differential between the ward and the outside), and directed airflow [2,68]. Furthermore, plants and flowers are not permitted in these rooms, but the installation of shower filters is recommended [13]. In hospitals with a limited number of "protective environments," strict criteria should be implemented regarding which patients will be accommodated in these wards [2,68]. Another option is the admission to a private ward with restricted connections [2].

In outpatients, the previously mentioned measures are not applicable and, thus, different recommendations have been developed for this patient group [69]. These include, among others, avoidance of construction areas, stagnant waters, and areas with increased moisture, avoidance of gardening and lawn mowing, appropriate checking of foods, and hand hygiene [69]. The efficacy against IA of surgical masks and N95 respirators has not been established [2].

In the case of construction or renovation works in the hospital, or in any adjacent sites, infection control strategies should be escalated and interdisciplinary committees should be established to ensure compliance with these strategies [68].

Environmental sampling for microbiological analysis is useful in cases of outbreaks even though its application as part of routine clinical practice has been questioned and there is a lack of data to support it [2,68]. Nevertheless, it is a useful tool to evaluate the function of the filters [13].

3. Treatment

3.1. Empiric Treatment

Empiric treatment should be started in individuals at high-risk for IA presenting with fever and neutropenia, which persist for a minimum of four days after the initiation of broad-spectrum anti-bacteria [1,8,29]. Another group for which empiric therapy is indicated includes neutropenic patients presenting with recurrent febrile episodes after defervescence following the administration of antibiotics [8]. This indication, however, has not been graded in the latest guidelines by ECIL-4. Moreover, ECIL-4 guidelines recommend initiation of empiric therapy in low-risk children with persistent fever accompanied by severe neutropenia and mucosal damage [8]. Nonetheless, according to the clinical practice guideline (CPG) developed by the International Pediatric Fever and Neutropenia Guideline Panel, empiric treatment should not be administered to low-risk pediatric patients with persistent fever and neutropenia [29]. This recommendation is supported by the results of a randomized, prospective study comparing either caspofungin or LAMB to no treatment in low-risk children [70]. In resource-poor settings with inadequate laboratory capabilities, empiric treatment has been associated with better results in individuals at high risk for IA [7].

The anti-fungal agents recommended for this approach are caspofungin and LAMB [8,29]. The first is administered at a loading dose of 70 mg/m^2, which is followed by 50 mg/m^2/24 h (maximum dose 70 mg/24 h) and the second at a dose of 1–3 mg/kg/24 h [8]. Use of these agents in pediatrics has been established by three randomized prospective trials [1]. The results of the first trial underlined the superiority of liposomal to deoxycholate amphotericin B (AMB) [1,71]. According to the second study, amphotericin B colloidal dispersion (ABCD) was superior to deoxycholate AMB in terms of adverse events, but similar in efficacy [1,72]. Lastly, the third study found that caspofungin performed similarly to LAMB in terms of efficiency and adverse effect rates [1,73]. Caspofungin at a dose of 50 mg/m^2/24 h in pediatric patients results in similar exposure levels with adult patients [74]. The use of voriconazole in empiric treatment has also been studied in a randomized, multi-center, open-label trial that compared this second-generation azole with LAMB [2,75]. The trial included both adult and pediatric patients and the results were indicative of a comparable response rate between voriconazole and LAMB groups in high-risk patients [2,75]. In another study, though, oral voriconazole was not preferred over deoxycholate AMB in patients presenting with gastrointestinal symptoms or in those receiving vincristine [7]. Identification of new anti-fungals to be used in empiric treatment remains a "research gap" [29].

Lastly, the duration of empiric anti-fungal schemes is a field that requires further investigation [29]. The 2017 ESCMID-ECMM-ERS Joint Clinical Guidelines recommend that the administration of caspofungin or LAMB should be carried on until defervescence and recovery of the neutrophil count [13,71,73,76].

3.2. Pre-Emptive Treatment

Empiric treatment has the disadvantage of exposing most patients to unnecessary use of anti-fungal drugs, due to the low specificity of fever as a symptom of IA [2]. Furthermore, the low sensitivity of fever as a criterion in the diagnosis of IA might delay the initiation of treatment. On the other hand, establishing a definite diagnosis in these patients is challenging and the risk of a dismal outcome is high [5,77]. Thus, the need for the adoption of a pre-emptive approach is highlighted [5,77].

Pre-emptive treatment is an alternative to empiric treatment that uses clinical and noninvasive, non-culture diagnostic methods to further assess the risk of IFDs and IA in febrile patients with neutropenia or in asymptomatic patients [2,8]. The diagnostic methods utilized in this approach include imaging techniques—mainly computed tomography—and microbiological markers, such as galactomannan (GM) antigen, (1,3)-β-D glucan (BDG), and *Aspergillus* polymerase chain reaction (PCR) [2,8].

Data regarding the application of pre-emptive therapy and the use of the associated biomarkers in children are limited [2,8]. Furthermore, these biomarkers do not perform optimally and consistently in pediatrics [78]. More specifically, the GM antigen displays variable sensitivity and specificity, low positive prognostic value (PPV), and a high false-positive rate [78,79]. The high negative prognostic value (NPV) of this biomarker applies only for *Aspergillus* spp. and not for other pathogens [78,79]. The sensitivity and specificity of BDG ranged from 50% to 82% and from 46% to 82%, respectively [78]. According to the International Pediatric Fever and Neutropenia Guideline Panel, its use is not recommended in the context of empiric anti-fungal treatment [29]. Lastly, the lack of standardization and the false-positive rates limit the use of PCR as well [78]. According to ECIL-4 guidelines, however, there is consensus in the favor of the use of the pre-emptive approach in pediatric patients, which has not received any grading [8]. A recent randomized, multi-center study compared pre-emptive with empiric treatment in pediatric cancer patients with fever and neutropenia with the exception of HSCT recipients [80]. The two methods were found to be comparable in terms of efficiency [80].

3.3. *Targeted Treatment for Proven/Probable Invasive Aspergillosis*

The European Organization for Research and Treatment of Cancer/Invasive Fungal Infections Cooperative Group (EORTC) and the National Institute of Allergy and Infectious Diseases Mycoses Study Group (MSG) provided the definitions for proven and probable IA. Proven IA requires evidence of *Aspergillus* to be identified in tissue by microscopic examination or culture, whereas probable entails a combination of patient risk factors, clinical manifestations, and mycological criteria [77]. Nevertheless, establishing a diagnosis is a challenging and time-consuming task, which may postpone treatment initiation and, thus, the above definitions should be used only for studies and not for routine clinical practice [13,77]. In the current review, the definitions by EORTC/MSG are used.

Targeted treatment for IA includes anti-fungal medications and adjunctive measures [2]. Similarities between pediatric and adult patients do exist, but there are several critical differences, such as the dosing scheme [2]. Targeted anti-fungal drugs (summarized in Table 4) can be further divided into the first and second-line, with the latter being reserved for unresponsive patients or for cases of intolerance to the adverse events [1,8]. Strength of recommendation and quality of evidence are in line with the IDSA grading system [8,65].

Table 4. Anti-fungal drugs for proven/probable Invasive Aspergillosis.

Drug	Route	Dosage	Indications for Use (Recommendation Ranking)	Refs
			AMB formulations	
LAMB	IV	• 3 mg/kg/24 h	• First-line treatment (B-I) (especially for ages <2 years) • Second-line treatment (B-I) (for cases with VRC intolerance, or for settings with azole resistance)	[1,8,13]
		• ≥5 mg/kg/24 h	• Second-line treatment for CNS Aspergillosis	[5,81]
ABLC	IV	5 mg/kg/24 h	• First-line treatment (B-II) • Second-line treatment (B-II)	[8]
			Azoles	
ITC	PO	2.5 mg/kg/12 h (for ages ≥2 years)	Second-line treatment (no grading) (not approved for ages <18 years)	[8]

Table 4. *Cont.*

Drug	Route	Dosage	Indications for Use (Recommendation Ranking)	Refs
VRC	PO	• 9 mg/kg/12 h (ages 2-<12, or 12–14 weighing <50 kg) • 200 mg/12 h (ages ≥15 years, or 12–14 weighing ≥50 kg)	• First-line treatment (A-I) (not approved for ages <2 years) • Second-line treatment (A-I)	[8]
	IV	• 9 mg/kg/12 h the first day, followed by 8 mg/kg/12 h on next days (ages 2-<12, or 12–14 weighing <50 kg) • 6 mg/kg/12 h the first day, followed by 4 mg/kg/12 h on next days (ages ≥15 years, or 12–14 weighing ≥50 kg)		
PSC	PO	800 mg/24 h divided in 2–4 doses (oral susp.)	Second-line treatment (no grading) (not approved for ages <18 years by the EU, approved by the FDA for ages ≥13 years)	[2,8]
ISA	IV PO	Not established	• Evaluation of PK for ages 1–18 years (ClinicalTrials.gov NCT03241550) • Evaluation of PK for ages 6–18 years (ClinicalTrials.gov NCT03241550)	[50]
			Echinocandins	
CAS	IV	70 mg/m^2 loading dose the first day, followed by 50 mg/m^2/24 h the next days (maximum 70 mg)	Second-line treatment (A-II)	[8]
MFG	IV	2–4 mg/kg/24 h (100–200 mg/24 h if patient's weight ≥50 kg)	Second-line treatment (no grading) (non-approved indication by the EU, approved by the FDA for ages ≥4 months)	[1,2,8]
AFG	IV	3 mg/kg loading dose, followed by 1.5 mg/kg/24 h	Not approved by the FDA for ages <18 years	[2,50,82]

AMB, amphotericin B. LAMB, liposomal amphotericin B. IV, intravenous. VRC, voriconazole. CNS, central nervous system. ABLC, amphotericin B lipid complex. ITC, itraconazole. PO, per os. PSC, posaconazole. susp., suspension. EU, European Union. FDA, Food and Drug Administration. ISA, isavuconazole. PK, pharmacokinetics. CAS, caspofungin. MFG, micafungin. AFG, anidulafungin.

3.3.1. First-Line Anti-Fungal Drugs

First-line agents are voriconazole, LAMB, and amphotericin B lipid complex (ABLC) [1,8].

Adequate data and experience have rendered voriconazole the cornerstone of IA treatment in children of all ages, apart from neonates and children <2 years [2,13]. A randomized clinical trial underlying its superiority against deoxycholate AMB in patients ≥12 years of age has played a significant role in establishing the use of this azole [83–85]. The PK of voriconazole is linear in pediatric patients, in contrast to the nonlinear pattern observed in adults and further population analyses of this parameter have facilitated the development of the dosing scheme [2,86,87]. For ages 2–<12 years, or 12–14 years with a body weight <50 kg, the dose is 9 mg/kg/12 h for the PO forms and 9 mg/kg/12 h the first day, followed by 8 mg/kg/12 h on subsequent days for the intravenous (IV) forms [8]. For ages 12–14 years with a body weight ≥50 kg, or ages ≥15 years, it is 200 mg/12 h for the PO form and for the IV form 6 mg/kg/12 h the first day, followed by 4 mg/kg/12 h on subsequent days [8]. According to a meta-analysis that highlighted the value of voriconazole TDM, therapeutic plasma levels increased the probability of a successful result, whereas greater concentrations were likely to cause toxicity [85,88]. TDM is, therefore, recommended, with an optimal range of 1–5 mg/L [8,47,48].

Liposomal amphotericin B (LAMB) is recommended at a dose of 3 mg/kg/24 h, IV [8]. The latter dose has been compared to a regimen of 10 mg/kg/24 h in the "AmBiLoad trial." However, the higher dose has not been associated with greater efficacy [89]. Moreover, in the previous randomized trial, no direct comparison to voriconazole has been attempted [13,89]. LAMB is indicated predominantly for children <2 years and neonates, for whom voriconazole has not been approved, and for settings with increased prevalence of azole-resistance [1,13]. When compared to deoxycholate AMB, LAMB is less nephrotoxic and has fewer infusion-related toxic reactions, but its use is limited due to its high cost [90–92].

Lastly, ABLC is another option for the first-line targeted treatment of IA, which is administered at 5 mg/kg/24 h in one IV dose [8]. However, recommendations for the use of ABLC stem from experience in naïve patients who received the agent in terms of second-line therapy [8].

3.3.2. Second-Line Anti-Fungal Drugs

Second-line agents include caspofungin, micafungin, itraconazole, and posaconazole [8]. The anti-fungal drugs described in the section of first-line treatment may also be used [8]. Voriconazole is reserved for patients naive to this agent and LAMB is used in cases of unresponsiveness or intolerance to the former, as well as in patients naive to AMB [1,8].

Caspofungin is the most preferred echinocandin for pediatric IFDs, based on results from the Antibiotic Resistance and Prescribing in European Children ("ARPEC") study [50,93]. The recommended IV dose is calculated, according to the body surface area (BSA), and is defined as a 70 mg/m^2 loading dose during the first day, which is followed by 50 mg/m^2/24 h on subsequent days (maximum 70 mg) [2,8,74]. Caspofungin is approved for pediatric use both in the United States of America (USA), by the Food and Drug Administration (FDA), and in Europe [2,8]. Two prospective studies verified the efficacy and safety of caspofungin [50]. The first evaluated the drug in the setting of salvage therapy for IA with encouraging results that showed 45% of patients demonstrated complete or partial clinical response [94]. The second study showed that the use of caspofungin in children from 6 months to 17 years of age was efficient against IA in a consistent manner with adult studies [95]. Lastly, caspofungin has been confirmed to be a feasible alternative choice for the treatment of children with IFDs, based on results from a systematic review and meta-analysis [96]. The authors, though, acknowledge the need for further research on this topic [96].

Micafungin—for therapeutic purposes—is administered at a dose of 2 to 4 mg/kg/24 h, IV, which reached a maximum of 100 to 200 mg/24 h if the patient weights $\geq$50 kg [1,8]. ECIL-4 stated that its use as a second-line agent for IA is a non-approved indication, due to a lack of robust evidence [8]. The FDA has approved the use of micafungin in pediatric patients $\geq$4 months of age [2]. Dose adjustments need to be considered in children $\leq$8 years, due to the fact that micafungin clearance increases with decreasing age, which necessitates higher doses in young age groups [2,97]. When compared to triazoles in a meta-analysis evaluating the treatment of IFDs in hematologic patients with neutropenia, micafungin has been associated with higher efficacy, fewer severe adverse events, but displayed similar all-cause mortality [50,98].

Itraconazole, despite being the first in its class exhibiting anti-Aspergillus activity, has fallen into disuse, due to several limitations including unpredictable bioavailability, interactions with chemotherapeutic drugs such as cyclophosphamide, and interactions with drugs that cause QTc prolongation [2,82,85,99]. Using this azole as second-line therapy is not approved in individuals <18 years of age [8] and it is reserved for less critical cases of IA [2]. When used in patients $\geq$2 years old, the dose of the PO form of itraconazole is 2.5 mg/kg/12 h with subsequent TDM, to aim for plasma concentrations $\geq$0.5mg/L [1,8,40].

Posaconazole exists in three different formulations: oral suspension, gastro-resistant tablet, and IV solution [85]. It is not approved in the European Union (EU) for children <18 years of age [8], but has received approval by the FDA for patients $\geq$13 years for both PO formulations and for patients $\geq$18 years for the IV form [2]. Based on limited PK data in pediatric patients $\geq$13 years of age, the dose of oral suspension of posaconazole is 800 mg/24 h divided in 2 to 4 doses [1,8]. TDM is required to maintain plasma concentrations $\geq$0.7–1.5 mg/L [8,56,100]. No need for TDM exists for the PO gastro-resistant tablets [50]. Posaconazole may be used for salvage therapy in cases of refractoriness or intolerance to previous agents among pediatric patients aged 13 years or older [8,84,85,100].

It should be noted that treatment for proven/probable IA in pediatrics is age-dependent. More specifically, the options for children $\geq$2 years of age are voriconazole, LAMB, ABLC, ABCD, caspofungin, itraconazole, and posaconazole (for ages $\geq$13 years), whereas, for children <2 years old, LAMB, ABLC, and caspofungin may be used [5,13]. For neonates, the only option is LAMB [13].

3.3.3. Novel Anti-Fungal Drugs

A brief review has to be made regarding the role that the newer agents of echinocandins and triazoles play in treating IA. These drugs are anidulafungin and isavuconazole, respectively.

Anidulafungin has not received FDA approval yet for use in patients aged <18 years [2,50]. The safety and PK parameters of anidulafungin in neutropenic pediatric patients at risk for IA have been assessed in a multi-center, dose-escalation study [101]. Anidulafungin displayed good tolerability and the regimen of 3 mg/kg loading dose, which was followed by 1.5 mg/kg/24 h that resulted in concentration profiles consistent with the adult dose of 100 mg/24 hours. This is the preferred one for IA [101]. The same regimen has been recently evaluated in a single-center study from Argentina, which underlined the safety and efficiency of this agent [102]. Lastly, the results of an open-label, non-comparative, pediatric study for the use of anidulafungin in invasive candidiasis (IC) have also been in favor of the safety of this drug in children, when administered in the previously mentioned doses [50,103].

Isavuconazole is a novel triazole with an extended spectrum of activity [104]. Its use in adult patients has been established by the "SECURE" trial—a phase 3, multi-center, randomized, double-blind, trial—which highlighted the safety and the non-inferiority of isavuconazole compared to voriconazole for the treatment of IA [85,105]. In pediatric patients, the PK and safety of both oral and IV forms of isavuconazole are currently being evaluated by a phase 1, open-label, multi-center study against no comparators (ClinicalTrials.gov NCT03241550) [50]. Additionally, limited experience has also been reported regarding the successful use of isavuconazole in pediatric hematology-oncology patients with invasive mucormycosis [50,106].

3.3.4. Combination Therapies and Duration of Treatment

The role of combination therapies in pediatric cancer and HSCT patients with IA has not been completely elucidated and requires further assessment [5,85]. However, use of a combination of anti-fungals in such patients has been reported by two multi-center cohort studies [85] including one retrospective [6] and one prospective [107]. Moreover, ECIL-4 provides a recommendation for the use of such therapy—both in the setting of first and second-line treatment for IA—but this is based on poor evidence [8]. Safety and efficacy data have arisen from a study evaluating combination therapy with caspofungin in 40 hematologic pediatric patients with IA [8,108]. In both ECIL-4 and IDSA 2016 Guidelines, the recommended combination is that of a polyene or triazole with an echinocandin [2,8].

There is no consensus regarding the optimal duration for which patients with proven/probable IA need to be treated and, thus, this parameter should be individualized [5,8]. Duration of treatment has been documented to range from 3 to more than 50 weeks [13,83,89,105,109]. Treatment may be discontinued after clinical improvement, microbiological response, and recovery from GVHD [5,8,13].

3.3.5. Breakthrough Infection and *Aspergillus* Resistance

Treatment of breakthrough infection occurring in patients who have received anti-mold prophylaxis includes salvage therapy, changing the antifungal drug class, awareness regarding local epidemiology patterns, and verification of serum triazole levels [2]. Salvage therapy is also used for the refractory or progressive aspergillosis and involves a switch in the class of antifungals, the addition of a second agent, the correction of underlying immunosuppression, and surgery [2,5].

Another essential topic that needs to be addressed is *Aspergillus* resistance. The previously mentioned anti-fungal agents are not active against all *Aspergillus* spp. [13]. Some species may display intrinsic resistance to azoles and polyenes [13,110], while others have acquired resistance to azoles [13,111]. For instance, *Aspergillus calidoustus* and *Aspergillus terreus* exhibit intrinsic resistance to azoles and AMB, respectively [13,110], whereas *Aspergillus fumigatus* may develop acquired resistance to azoles [13,112]. However, anti-fungal susceptibility testing is not recommended for the routine management of the initial infection [2]. More specifically, in patients naïve to azoles or in regions with no documented resistance, anti-fungal susceptibility testing should not be performed [13]. However, such testing is indicated in cases that do not respond to initial treatment, or if there is clinical suspicion of azole resistance [13]. In patients with infections due to *Aspergillus fumigatus* with documented azole-resistance, a group of experts recommended a switch to LAMB or a combination of voriconazole with an echinocandin [112,113]. In cases of environmental resistance to azoles, the latter regimen switch is recommended only if the respective rates are >10% [13]. Azole resistance of *Aspergillus fumigatus* has also been reported in pediatric patients and should be taken into consideration in cases that are unresponsive to azole treatment [114].

3.3.6. Adjunctive Measures

Apart from anti-fungal drugs, treatment for proven/probable IA also includes several adjunctive measures. Colony-stimulating factors (CSFs) may be administered either as prophylaxis, to reduce the duration of granulocytopenia, or as therapy for patients with IA and neutropenia [2]. Granulocyte transfusions may be an option in cases of severe and prolonged neutropenia [8]. An updated Cochrane review, which evaluates the efficacy and safety of this method in the setting of prophylaxis, concluded that the risk of fungemia decreased (low-grade evidence) due to the transfusions, but data was inadequate to support any differences in infection-associated mortality or adverse events [115]. Recombinant interferon gamma (IFN-γ) may be used in cases of severe or refractory IA [2]. Lastly, adoptive transfer of pathogen-specific T lymphocytes, which have derived from donors, is under study for use in IA [85,116,117].

Surgical treatment is reserved for cases of accessible localized lesions [2]. Indications include sinus aspergillosis, localized cutaneous aspergillosis, CNS aspergillosis, pulmonary disease that is localized or adjacent to great vessels/pericardium, or has invaded the pleural space and chest wall, or has caused uncontrolled bleeding [2]. Surgery is contraindicated for unstable patients, or for patients with disseminated disease [6]. In each case, decisions should be individualized [2].

3.3.7. Management of Selected Localized Infections

The following part of this review will be devoted to the management of IPA and CNS aspergillosis in pediatric patients with leukemia and HSCT. These clinical manifestations represent the most common site of infection [6] and one of the most serious complications [118], respectively.

Invasive pulmonary aspergillosis, when suspected, should prompt early initiation of treatment, due to the fact that this practice restricts the development of the disease and due to the unreliability of diagnostic testing [2,89,119]. When the diagnosis is established, the patient should be further evaluated for other foci of IA, such as the CNS [13]. Optimal duration of treatment has not been identified, but a minimum of six to 12 weeks should be applied [2]. Lastly, surgical intervention is reserved for cases with lesions adjacent to great vessels, or vital organs, lesions associated with unmanageable hemoptysis, or lesion causing bone erosions [2].

Aspergillosis of the CNS is most frequently associated with hematologic cancer since the underlying condition in patients >1 year of age [120]. Infection of the CNS has been observed in

6% of children with leukemia with IA [8,121]. The advent of AMB lipid formulations and anti-mold azoles and the progress achieved in early diagnosis have reduced mortality rates from 82.8% to 39.5%, before and after 1990, respectively [121]. Treatment principles of CNS aspergillosis are early diagnosis, initiation of appropriate anti-mold drugs, evaluation of the indications for surgery and decreasing immunosuppression [2,122]. The cornerstone of pharmacotherapy in children has been AMB, with the deoxycholate form having been tolerated in ages <3 months [15,123] and the lipid formulations of AMB having been reserved for older children [15]. Currently, voriconazole is the established first-line treatment of CNS aspergillosis [5,8,81,83,121,124] and the next choice is LAMB at high doses (≥5mg/kg/24 h) [5,125]. TDM for voriconazole is necessary and younger children may require higher doses to reach therapeutic levels [121,126]. However, there is limited evidence regarding the levels of voriconazole in the CSF [121]. Data is also scarce about the use of adjunctive immunotherapy (CSFs, cytokines) in pediatric patients with CNS aspergillosis [120]. Surgical intervention may be indicated in patients with localized lesions [120]. Lastly, the use of corticosteroids and the intrathecal administration of anti-fungals are not recommended [2,127].

4. Conclusions

It has become evident that IA is a major issue in immuno-compromised pediatric patients, especially in those with leukemia and in HSCT recipients. Management of this infection consists of two main components, which includes prevention and treatment. The role of primary anti-fungal prophylaxis is highlighted particularly due to the insufficiency of diagnostic tests. Several agents have been evaluated in this setting including triazoles, polyenes, and echinocandins. Empiric and pre-emptive treatments are two approaches that can be initiated before establishing a definitive diagnosis. The mainstay of targeted treatment is voriconazole for children older than two years of age and LAMB in the younger age group. Further research is required in the field of pediatric IA management in order to reach the evidence quantity and quality of the respective field in adults.

References

1. Frange, P.; Bougnoux, M.-E.; Lanternier, F.; Neven, B.; Moshous, D.; Angebault, C.; Lortholary, O.; Blanche, S. An update on pediatric invasive aspergillosis. *Med. Mal. Infect.* **2015**, *45*, 189–198. [CrossRef] [PubMed]
2. Patterson, T.F.; Thompson, G.R., 3rd; Denning, D.W.; Fishman, J.A.; Hadley, S.; Herbrecht, R.; Kontoyiannis, D.P.; Marr, K.A.; Morrison, V.A.; Nguyen, M.H.; et al. Practice Guidelines for the Diagnosis and Management of Aspergillosis: 2016 Update by the Infectious Diseases Society of America. *Clin. Infect. Dis.* **2016**, *63*, e1–e60. [CrossRef] [PubMed]
3. Steinbach, W.J. Pediatric aspergillosis: Disease and treatment differences in children. *Pediatr. Infect. Dis. J.* **2005**, *24*, 358–364. [CrossRef] [PubMed]
4. Walsh, T.J.; Gonzalez, C.; Lyman, C.A.; Chanock, S.J.; Pizzo, P.A. Invasive fungal infections in children: Recent advances in diagnosis and treatment. *Adv. Pediatr. Infect. Dis.* **1996**, *11*, 187–290. [PubMed]
5. Tragiannidis, A.; Roilides, E.; Walsh, T.J.; Groll, A.H. Invasive aspergillosis in children with acquired immunodeficiencies. *Clin. Infect. Dis.* **2012**, *54*, 258–267. [CrossRef] [PubMed]
6. Burgos, A.; Zaoutis, T.E.; Dvorak, C.C.; Hoffman, J.A.; Knapp, K.M.; Nania, J.J.; Prasad, P.; Steinbach, W.J. Pediatric invasive aspergillosis: A multicenter retrospective analysis of 139 contemporary cases. *Pediatrics* **2008**, *121*, e1286-94. [CrossRef]
7. Jain, S.; Kapoor, G. Invasive aspergillosis in children with acute leukemia at a resource-limited oncology center. *J. Pediatr. Hematol. Oncol.* **2015**, *37*, e1–e5. [CrossRef]
8. Groll, A.H.; Castagnola, E.; Cesaro, S.; Dalle, J.-H.; Engelhard, D.; Hope, W.; Roilides, E.; Styczynski, J.; Warris, A.; Lehrnbecher, T. Fourth European Conference on Infections in Leukaemia (ECIL-4): Guidelines for diagnosis, prevention, and treatment of invasive fungal diseases in paediatric patients with cancer or allogeneic haemopoietic stem-cell transplantation. *Lancet. Oncol.* **2014**, *15*, e327-40. [CrossRef]

9. Sung, L.; Phillips, R.; Lehrnbecher, T. Time for paediatric febrile neutropenia guidelines—children are not little adults. *Eur. J. Cancer* **2011**, *47*, 811–813. [CrossRef]
10. Groll, A.H.; Tragiannidis, A. Update on antifungal agents for paediatric patients. *Clin. Microbiol. Infect.* **2010**, *16*, 1343–1353. [CrossRef]
11. Dornbusch, H.J.; Manzoni, P.; Roilides, E.; Walsh, T.J.; Groll, A.H. Invasive fungal infections in children. *Pediatr. Infect. Dis. J.* **2009**, *28*, 734–737. [CrossRef] [PubMed]
12. Lestner, J.M.; Smith, P.B.; Cohen-Wolkowiez, M.; Benjamin, D.K.J.; Hope, W.W. Antifungal agents and therapy for infants and children with invasive fungal infections: A pharmacological perspective. *Br. J. Clin. Pharmacol.* **2013**, *75*, 1381–1395. [CrossRef] [PubMed]
13. Ullmann, A.J.; Aguado, J.M.; Arikan-Akdagli, S.; Denning, D.W.; Groll, A.H.; Lagrou, K.; Lass-Florl, C.; Lewis, R.E.; Munoz, P.; Verweij, P.E.; et al. Diagnosis and management of Aspergillus diseases: Executive summary of the 2017 ESCMID-ECMM-ERS guideline. *Clin. Microbiol. Infect.* **2018**, *24*, e1–e38. [CrossRef] [PubMed]
14. Antachopoulos, C.; Walsh, T.J.; Roilides, E. Fungal infections in primary immunodeficiencies. *Eur. J. Pediatr.* **2007**, *166*, 1099–1117. [CrossRef] [PubMed]
15. Groll, A.H.; Jaeger, G.; Allendorf, A.; Herrmann, G.; Schloesser, R.; von Loewenich, V. Invasive pulmonary aspergillosis in a critically ill neonate: Case report and review of invasive aspergillosis during the first 3 months of life. *Clin. Infect. Dis.* **1998**, *27*, 437–452. [CrossRef] [PubMed]
16. Groll, A.H.; Shah, P.M.; Mentzel, C.; Schneider, M.; Just-Nuebling, G.; Huebner, K. Trends in the postmortem epidemiology of invasive fungal infections at a university hospital. *J. Infect.* **1996**, *33*, 23–32. [CrossRef]
17. Nedel, W.L.; Kontoyiannis, D.P.; Pasqualotto, A.C. Aspergillosis in patients treated with monoclonal antibodies. *Rev. Iberoam. Micol.* **2009**, *26*, 175–183. [CrossRef] [PubMed]
18. Pappas, P.G.; Alexander, B.D.; Andes, D.R.; Hadley, S.; Kauffman, C.A.; Freifeld, A.; Anaissie, E.J.; Brumble, L.M.; Herwaldt, L.; Ito, J.; et al. Invasive fungal infections among organ transplant recipients: Results of the Transplant-Associated Infection Surveillance Network (TRANSNET). *Clin. Infect. Dis.* **2010**, *50*, 1101–1111. [CrossRef] [PubMed]
19. Abbasi, S.; Shenep, J.L.; Hughes, W.T.; Flynn, P.M. Aspergillosis in children with cancer: A 34-year experience. *Clin. Infect. Dis.* **1999**, *29*, 1210–1219. [CrossRef] [PubMed]
20. Groll, A.H.; Kurz, M.; Schneider, W.; Witt, V.; Schmidt, H.; Schneider, M.; Schwabe, D. Five-year-survey of invasive aspergillosis in a paediatric cancer centre. Epidemiology, management and long-term survival. *Mycoses* **1999**, *42*, 431–442. [CrossRef] [PubMed]
21. Zaoutis, T.E.; Heydon, K.; Chu, J.H.; Walsh, T.J.; Steinbach, W.J. Epidemiology, outcomes, and costs of invasive aspergillosis in immunocompromised children in the United States, 2000. *Pediatrics* **2006**, *117*, e711-6. [CrossRef] [PubMed]
22. Rubio, P.M.; Sevilla, J.; Gonzalez-Vicent, M.; Lassaletta, A.; Cuenca-Estrella, M.; Diaz, M.A.; Riesco, S.; Madero, L. Increasing incidence of invasive aspergillosis in pediatric hematology oncology patients over the last decade: A retrospective single centre study. *J. Pediatr. Hematol. Oncol.* **2009**, *31*, 642–646. [CrossRef] [PubMed]
23. Kaya, Z.; Gursel, T.; Kocak, U.; Aral, Y.Z.; Kalkanci, A.; Albayrak, M. Invasive fungal infections in pediatric leukemia patients receiving fluconazole prophylaxis. *Pediatr. Blood Cancer* **2009**, *52*, 470–475. [CrossRef] [PubMed]
24. Thomas, K.E.; Owens, C.M.; Veys, P.A.; Novelli, V.; Costoli, V. The radiological spectrum of invasive aspergillosis in children: A 10-year review. *Pediatr. Radiol.* **2003**, *33*, 453–460. [CrossRef] [PubMed]
25. Dvorak, C.C.; Fisher, B.T.; Sung, L.; Steinbach, W.J.; Nieder, M.; Alexander, S.; Zaoutis, T.E. Antifungal prophylaxis in pediatric hematology/oncology: New choices & new data. *Pediatr. Blood Cancer* **2012**, *59*, 21–26. [CrossRef] [PubMed]
26. Tragiannidis, A.; Dokos, C.; Lehrnbecher, T.; Groll, A.H. Antifungal chemoprophylaxis in children and adolescents with haematological malignancies and following allogeneic haematopoietic stem cell transplantation: Review of the literature and options for clinical practice. *Drugs* **2012**, *72*, 685–704. [CrossRef] [PubMed]
27. Fisher, B.T.; Robinson, P.D.; Lehrnbecher, T.; Steinbach, W.J.; Zaoutis, T.E.; Phillips, B.; Sung, L. Risk Factors for Invasive Fungal Disease in Pediatric Cancer and Hematopoietic Stem Cell Transplantation: A Systematic Review. *J. Pediatric Infect. Dis. Soc.* **2018**, *7*, 191–198. [CrossRef]

28. Crassard, N.; Hadden, H.; Pondarre, C.; Hadden, R.; Galambrun, C.; Piens, M.A.; Pracros, J.P.; Souillet, G.; Basset, T.; Berthier, J.C.; et al. Invasive aspergillosis and allogeneic hematopoietic stem cell transplantation in children: A 15-year experience. *Transpl. Infect. Dis.* **2008**, *10*, 177–183. [CrossRef]
29. Lehrnbecher, T.; Robinson, P.; Fisher, B.; Alexander, S.; Ammann, R.A.; Beauchemin, M.; Carlesse, F.; Groll, A.H.; Haeusler, G.M.; Santolaya, M.; et al. Guideline for the Management of Fever and Neutropenia in Children With Cancer and Hematopoietic Stem-Cell Transplantation Recipients: 2017 Update. *J. Clin. Oncol.* **2017**, *35*, 2082–2094. [CrossRef]
30. Kontoyiannis, D.P.; Marr, K.A.; Park, B.J.; Alexander, B.D.; Anaissie, E.J.; Walsh, T.J.; Ito, J.; Andes, D.R.; Baddley, J.W.; Brown, J.M.; et al. Prospective surveillance for invasive fungal infections in hematopoietic stem cell transplant recipients, 2001-2006: Overview of the Transplant-Associated Infection Surveillance Network (TRANSNET) Database. *Clin. Infect. Dis.* **2010**, *50*, 1091–1100. [CrossRef]
31. Lass-Florl, C. The changing face of epidemiology of invasive fungal disease in Europe. *Mycoses* **2009**, *52*, 197–205. [CrossRef] [PubMed]
32. Marr, K.A.; Carter, R.A.; Boeckh, M.; Martin, P.; Corey, L. Invasive aspergillosis in allogeneic stem cell transplant recipients: Changes in epidemiology and risk factors. *Blood* **2002**, *100*, 4358–4366. [CrossRef] [PubMed]
33. Segal, B.H. Aspergillosis. *N. Engl. J. Med.* **2009**, *360*, 1870–1884. [CrossRef] [PubMed]
34. van Burik, J.-A.H.; Carter, S.L.; Freifeld, A.G.; High, K.P.; Godder, K.T.; Papanicolaou, G.A.; Mendizabal, A.M.; Wagner, J.E.; Yanovich, S.; Kernan, N.A. Higher risk of cytomegalovirus and aspergillus infections in recipients of T cell-depleted unrelated bone marrow: Analysis of infectious complications in patients treated with T cell depletion versus immunosuppressive therapy to prevent graft-versus-host. *Biol. Blood Marrow Transplant.* **2007**, *13*, 1487–1498. [CrossRef] [PubMed]
35. Hope, W.W.; Castagnola, E.; Groll, A.H.; Roilides, E.; Akova, M.; Arendrup, M.C.; Arikan-Akdagli, S.; Bassetti, M.; Bille, J.; Cornely, O.A.; et al. ESCMID* guideline for the diagnosis and management of Candida diseases 2012: Prevention and management of invasive infections in neonates and children caused by Candida spp. *Clin. Microbiol. Infect.* **2012**, *18*, 38–52. [CrossRef] [PubMed]
36. Panackal, A.A.; Li, H.; Kontoyiannis, D.P.; Mori, M.; Perego, C.A.; Boeckh, M.; Marr, K.A. Geoclimatic influences on invasive aspergillosis after hematopoietic stem cell transplantation. *Clin. Infect. Dis. Off. Publ. Infect. Dis. Soc. Am.* **2010**, *50*, 1588–1597. [CrossRef] [PubMed]
37. Anaissie, E.J.; Stratton, S.L.; Dignani, M.C.; Lee, C.-K.; Mahfouz, T.H.; Rex, J.H.; Summerbell, R.C.; Walsh, T.J. Cleaning patient shower facilities: A novel approach to reducing patient exposure to aerosolized Aspergillus species and other opportunistic molds. *Clin. Infect. Dis.* **2002**, *35*, E86–E88. [CrossRef]
38. Anaissie, E.J.; Stratton, S.L.; Dignani, M.C.; Summerbell, R.C.; Rex, J.H.; Monson, T.P.; Spencer, T.; Kasai, M.; Francesconi, A.; Walsh, T.J. Pathogenic Aspergillus species recovered from a hospital water system: A 3-year prospective study. *Clin. Infect. Dis.* **2002**, *34*, 780–789. [CrossRef]
39. Warris, A.; Klaassen, C.H.W.; Meis, J.F.G.M.; De Ruiter, M.T.; De Valk, H.A.; Abrahamsen, T.G.; Gaustad, P.; Verweij, P.E. Molecular epidemiology of *Aspergillus fumigatus* isolates recovered from water, air, and patients shows two clusters of genetically distinct strains. *J. Clin. Microbiol.* **2003**, *41*, 4101–4106. [CrossRef]
40. Glasmacher, A.; Hahn, C.; Molitor, E.; Marklein, G.; Sauerbruch, T.; Schmidt-Wolf, I.G. Itraconazole trough concentrations in antifungal prophylaxis with six different dosing regimens using hydroxypropyl-beta-cyclodextrin oral solution or coated-pellet capsules. *Mycoses* **1999**, *42*, 591–600. [CrossRef]
41. Simon, A.; Besuden, M.; Vezmar, S.; Hasan, C.; Lampe, D.; Kreutzberg, S.; Glasmacher, A.; Bode, U.; Fleischhack, G. Itraconazole prophylaxis in pediatric cancer patients receiving conventional chemotherapy or autologous stem cell transplants. *Support. Care cancer Off. J. Multinatl. Assoc. Support. Care Cancer* **2007**, *15*, 213–220. [CrossRef] [PubMed]
42. Doring, M.; Blume, O.; Haufe, S.; Hartmann, U.; Kimmig, A.; Schwarze, C.-P.; Lang, P.; Handgretinger, R.; Muller, I. Comparison of itraconazole, voriconazole, and posaconazole as oral antifungal prophylaxis in pediatric patients following allogeneic hematopoietic stem cell transplantation. *Eur. J. Clin. Microbiol. Infect. Dis.* **2014**, *33*, 629–638. [CrossRef] [PubMed]

43. Grigull, L.; Kuehlke, O.; Beilken, A.; Sander, A.; Linderkamp, C.; Schmid, H.; Seidemann, K.; Sykora, K.W.; Schuster, F.R.; Welte, K. Intravenous and oral sequential itraconazole antifungal prophylaxis in paediatric stem cell transplantation recipients: A pilot study for evaluation of safety and efficacy. *Pediatr. Transplant.* **2007**, *11*, 261–266. [CrossRef] [PubMed]

44. Vardakas, K.Z.; Michalopoulos, A.; Falagas, M.E. Fluconazole versus itraconazole for antifungal prophylaxis in neutropenic patients with haematological malignancies: A meta-analysis of randomised-controlled trials. *Br. J. Haematol.* **2005**, *131*, 22–28. [CrossRef] [PubMed]

45. Marks, D.I.; Pagliuca, A.; Kibbler, C.C.; Glasmacher, A.; Heussel, C.-P.; Kantecki, M.; Miller, P.J.S.; Ribaud, P.; Schlamm, H.T.; Solano, C.; et al. Voriconazole versus itraconazole for antifungal prophylaxis following allogeneic haematopoietic stem-cell transplantation. *Br. J. Haematol.* **2011**, *155*, 318–327. [CrossRef] [PubMed]

46. Gastine, S.; Lehrnbecher, T.; Muller, C.; Farowski, F.; Bader, P.; Ullmann-Moskovits, J.; Cornely, O.A.; Groll, A.H.; Hempel, G. Pharmacokinetic Modeling of Voriconazole To Develop an Alternative Dosing Regimen in Children. *Antimicrob. Agents Chemother.* **2018**, *62*. [CrossRef] [PubMed]

47. Park, W.B.; Kim, N.H.; Kim, K.H.; Lee, S.H.; Nam, W.-S.; Yoon, S.H.; Song, K.H.; Choe, P.G.; Kim, N.J.; Jang, I.J.; et al. The effect of therapeutic drug monitoring on safety and efficacy of voriconazole in invasive fungal infections: A randomized controlled trial. *Clin. Infect. Dis.* **2012**, *55*, 1080–1087. [CrossRef]

48. Troke, P.F.; Hockey, H.P.; Hope, W.W. Observational study of the clinical efficacy of voriconazole and its relationship to plasma concentrations in patients. *Antimicrob. Agents Chemother.* **2011**, *55*, 4782–4788. [CrossRef]

49. Pana, Z.D.; Kourti, M.; Vikelouda, K.; Vlahou, A.; Katzilakis, N.; Papageorgiou, M.; Doganis, D.; Petrikkos, L.; Paisiou, A.; Koliouskas, D.; et al. Voriconazole Antifungal Prophylaxis in Children With Malignancies: A Nationwide Study. *J. Pediatr. Hematol. Oncol.* **2018**, *40*, 22–26. [CrossRef]

50. Iosifidis, E.; Papachristou, S.; Roilides, E. Advances in the Treatment of Mycoses in Pediatric Patients. *J. Fungi (Basel, Switzerland)* **2018**, *4*. [CrossRef]

51. Cecinati, V.; Guastadisegni, C.; Russo, F.G.; Brescia, L.P. Antifungal therapy in children: An update. *Eur. J. Pediatr.* **2013**, *172*, 437–446. [CrossRef] [PubMed]

52. Cesaro, S.; Milano, G.M.; Aversa, F. Retrospective survey on the off-label use of posaconazole in pediatric hematology patients. *Eur. J. Clin. Microbiol. Infect. Dis.* **2011**, *30*, 595–596. [CrossRef] [PubMed]

53. Cornely, O.A.; Maertens, J.; Winston, D.J.; Perfect, J.; Ullmann, A.J.; Walsh, T.J.; Helfgott, D.; Holowiecki, J.; Stockelberg, D.; Goh, Y.T.; et al. Posaconazole vs. fluconazole or itraconazole prophylaxis in patients with neutropenia. *N. Engl. J. Med.* **2007**, *356*, 348–359. [CrossRef] [PubMed]

54. Ullmann, A.J.; Lipton, J.H.; Vesole, D.H.; Chandrasekar, P.; Langston, A.; Tarantolo, S.R.; Greinix, H.; Morais de Azevedo, W.; Reddy, V.; Boparai, N.; et al. Posaconazole or fluconazole for prophylaxis in severe graft-versus-host disease. *N. Engl. J. Med.* **2007**, *356*, 335–347. [CrossRef] [PubMed]

55. Doring, M.; Cabanillas Stanchi, K.M.; Queudeville, M.; Feucht, J.; Blaeschke, F.; Schlegel, P.; Feuchtinger, T.; Lang, P.; Muller, I.; Handgretinger, R.; et al. Efficacy, safety and feasibility of antifungal prophylaxis with posaconazole tablet in paediatric patients after haematopoietic stem cell transplantation. *J. Cancer Res. Clin. Oncol.* **2017**, *143*, 1281–1292. [CrossRef]

56. Jang, S.H.; Colangelo, P.M.; Gobburu, J.V.S. Exposure-response of posaconazole used for prophylaxis against invasive fungal infections: Evaluating the need to adjust doses based on drug concentrations in plasma. *Clin. Pharmacol. Ther.* **2010**, *88*, 115–119. [CrossRef]

57. Bochennek, K.; Tramsen, L.; Schedler, N.; Becker, M.; Klingebiel, T.; Groll, A.H.; Lehrnbecher, T. Liposomal amphotericin B twice weekly as antifungal prophylaxis in paediatric haematological malignancy patients. *Clin. Microbiol. Infect.* **2011**, *17*, 1868–1874. [CrossRef]

58. Mehta, P.; Vinks, A.; Filipovich, A.; Vaughn, G.; Fearing, D.; Sper, C.; Davies, S. High-dose weekly AmBisome antifungal prophylaxis in pediatric patients undergoing hematopoietic stem cell transplantation: A pharmacokinetic study. *Biol. Blood Marrow Transplant.* **2006**, *12*, 235–240. [CrossRef]

59. Kolve, H.; Ahlke, E.; Fegeler, W.; Ritter, J.; Jurgens, H.; Groll, A.H. Safety, tolerance and outcome of treatment with liposomal amphotericin B in paediatric patients with cancer or undergoing haematopoietic stem cell transplantation. *J. Antimicrob. Chemother.* **2009**, *64*, 383–387. [CrossRef]

60. van Burik, J.-A.H.; Ratanatharathorn, V.; Stepan, D.E.; Miller, C.B.; Lipton, J.H.; Vesole, D.H.; Bunin, N.; Wall, D.A.; Hiemenz, J.W.; Satoi, Y.; et al. Micafungin versus fluconazole for prophylaxis against invasive fungal infections during neutropenia in patients undergoing hematopoietic stem cell transplantation. *Clin. Infect. Dis.* **2004**, *39*, 1407–1416. [CrossRef]
61. Yoshikawa, K.; Nakazawa, Y.; Katsuyama, Y.; Hirabayashi, K.; Saito, S.; Shigemura, T.; Tanaka, M.; Yanagisawa, R.; Sakashita, K.; Koike, K. Safety, tolerability, and feasibility of antifungal prophylaxis with micafungin at 2 mg/kg daily in pediatric patients undergoing allogeneic hematopoietic stem cell transplantation. *Infection* **2014**, *42*, 639–647. [CrossRef] [PubMed]
62. Mattiuzzi, G.N.; Alvarado, G.; Giles, F.J.; Ostrosky-Zeichner, L.; Cortes, J.; O'brien, S.; Verstovsek, S.; Faderl, S.; Zhou, X.; Raad, I.I.; et al. Open-label, randomized comparison of itraconazole versus caspofungin for prophylaxis in patients with hematologic malignancies. *Antimicrob. Agents Chemother.* **2006**, *50*, 143–147. [CrossRef] [PubMed]
63. Doring, M.; Hartmann, U.; Erbacher, A.; Lang, P.; Handgretinger, R.; Muller, I. Caspofungin as antifungal prophylaxis in pediatric patients undergoing allogeneic hematopoietic stem cell transplantation: A retrospective analysis. *BMC Infect. Dis.* **2012**, *12*, 151. [CrossRef] [PubMed]
64. Maximova, N.; Schillani, G.; Simeone, R.; Maestro, A.; Zanon, D. Comparison of Efficacy and Safety of Caspofungin Versus Micafungin in Pediatric Allogeneic Stem Cell Transplant Recipients: A Retrospective Analysis. *Adv. Ther.* **2017**, *34*, 1184–1199. [CrossRef] [PubMed]
65. Kish, M.A. Guide to development of practice guidelines. *Clin. Infect. Dis.* **2001**, *32*, 851–854. [CrossRef]
66. Cordonnier, C.; Rovira, M.; Maertens, J.; Olavarria, E.; Faucher, C.; Bilger, K.; Pigneux, A.; Cornely, O.A.; Ullmann, A.J.; Bofarull, R.M.; et al. Voriconazole for secondary prophylaxis of invasive fungal infections in allogeneic stem cell transplant recipients: Results of the VOSIFI study. *Haematologica* **2010**, *95*, 1762–1768. [CrossRef]
67. Allinson, K.; Kolve, H.; Gumbinger, H.G.; Vormoor, H.J.; Ehlert, K.; Groll, A.H. Secondary antifungal prophylaxis in paediatric allogeneic haematopoietic stem cell recipients. *J. Antimicrob. Chemother.* **2008**, *61*, 734–742. [CrossRef]
68. Tomblyn, M.; Chiller, T.; Einsele, H.; Gress, R.; Sepkowitz, K.; Storek, J.; Wingard, J.R.; Young, J.-A.H.; Boeckh, M.J. Guidelines for preventing infectious complications among hematopoietic cell transplantation recipients: A global perspective. *Biol. Blood Marrow Transplant.* **2009**, *15*, 1143–1238. [CrossRef]
69. Partridge-Hinckley, K.; Liddell, G.M.; Almyroudis, N.G.; Segal, B.H. Infection control measures to prevent invasive mould diseases in hematopoietic stem cell transplant recipients. *Mycopathologia* **2009**, *168*, 329–337. [CrossRef]
70. Caselli, D.; Cesaro, S.; Ziino, O.; Ragusa, P.; Pontillo, A.; Pegoraro, A.; Santoro, N.; Zanazzo, G.; Poggi, V.; Giacchino, M.; et al. A prospective, randomized study of empirical antifungal therapy for the treatment of chemotherapy-induced febrile neutropenia in children. *Br. J. Haematol.* **2012**, *158*, 249–255. [CrossRef]
71. Prentice, H.G.; Hann, I.M.; Herbrecht, R.; Aoun, M.; Kvaloy, S.; Catovsky, D.; Pinkerton, C.R.; Schey, S.A.; Jacobs, F.; Oakhill, A.; et al. A randomized comparison of liposomal versus conventional amphotericin B for the treatment of pyrexia of unknown origin in neutropenic patients. *Br. J. Haematol.* **1997**, *98*, 711–718. [CrossRef] [PubMed]
72. Sandler, E.S.; Mustafa, M.M.; Tkaczewski, I.; Graham, M.L.; Morrison, V.A.; Green, M.; Trigg, M.; Abboud, M.; Aquino, V.M.; Gurwith, M.; et al. Use of amphotericin B colloidal dispersion in children. *J. Pediatr. Hematol. Oncol.* **2000**, *22*, 242–246. [CrossRef] [PubMed]
73. Maertens, J.A.; Madero, L.; Reilly, A.F.; Lehrnbecher, T.; Groll, A.H.; Jafri, H.S.; Green, M.; Nania, J.J.; Bourque, M.R.; Wise, B.A.; et al. A randomized, double-blind, multicenter study of caspofungin versus liposomal amphotericin B for empiric antifungal therapy in pediatric patients with persistent fever and neutropenia. *Pediatr. Infect. Dis. J.* **2010**, *29*, 415–420. [CrossRef] [PubMed]
74. Walsh, T.J.; Adamson, P.C.; Seibel, N.L.; Flynn, P.M.; Neely, M.N.; Schwartz, C.; Shad, A.; Kaplan, S.L.; Roden, M.M.; Stone, J.A.; et al. Pharmacokinetics, safety, and tolerability of caspofungin in children and adolescents. *Antimicrob. Agents Chemother.* **2005**, *49*, 4536–4545. [CrossRef] [PubMed]
75. Walsh, T.J.; Pappas, P.; Winston, D.J.; Lazarus, H.M.; Petersen, F.; Raffalli, J.; Yanovich, S.; Stiff, P.; Greenberg, R.; Donowitz, G.; et al. Voriconazole compared with liposomal amphotericin B for empirical antifungal therapy in patients with neutropenia and persistent fever. *N. Engl. J. Med.* **2002**, *346*, 225–234. [CrossRef] [PubMed]

76. Caselli, D.; Paolicchi, O. Empiric antibiotic therapy in a child with cancer and suspected septicemia. *Pediatr. Rep.* **2012**, *4*, e2. [CrossRef]
77. De Pauw, B.; Walsh, T.J.; Donnelly, J.P.; Stevens, D.A.; Edwards, J.E.; Calandra, T.; Pappas, P.G.; Maertens, J.; Lortholary, O.; Kauffman, C.A.; et al. Revised definitions of invasive fungal disease from the European Organization for Research and Treatment of Cancer/Invasive Fungal Infections Cooperative Group and the National Institute of Allergy and Infectious Diseases Mycoses Study Group (EORTC/MSG) C. *Clin. Infect. Dis.* **2008**, *46*, 1813–1821. [CrossRef]
78. Huppler, A.R.; Fisher, B.T.; Lehrnbecher, T.; Walsh, T.J.; Steinbach, W.J. Role of Molecular Biomarkers in the Diagnosis of Invasive Fungal Diseases in Children. *J. Pediatric Infect. Dis. Soc.* **2017**, *6*, S32–S44. [CrossRef]
79. Lehrnbecher, T.; Robinson, P.D.; Fisher, B.T.; Castagnola, E.; Groll, A.H.; Steinbach, W.J.; Zaoutis, T.E.; Negeri, Z.F.; Beyene, J.; Phillips, B.; et al. Galactomannan, beta-D-Glucan, and Polymerase Chain Reaction-Based Assays for the Diagnosis of Invasive Fungal Disease in Pediatric Cancer and Hematopoietic Stem Cell Transplantation: A Systematic Review and Meta-Analysis. *Clin. Infect. Dis.* **2016**, *63*, 1340–1348. [CrossRef]
80. Santolaya, M.E.; Alvarez, A.M.; Acuna, M.; Aviles, C.L.; Salgado, C.; Tordecilla, J.; Varas, M.; Venegas, M.; Villarroel, M.; Zubieta, M.; et al. Efficacy of pre-emptive versus empirical antifungal therapy in children with cancer and high-risk febrile neutropenia: A randomized clinical trial. *J. Antimicrob. Chemother.* **2018**, *73*, 2860–2866. [CrossRef]
81. Felton, T.; Troke, P.F.; Hope, W.W. Tissue penetration of antifungal agents. *Clin. Microbiol. Rev.* **2014**, *27*, 68–88. [CrossRef] [PubMed]
82. Marr, K.A.; Leisenring, W.; Crippa, F.; Slattery, J.T.; Corey, L.; Boeckh, M.; McDonald, G.B. Cyclophosphamide metabolism is affected by azole antifungals. *Blood* **2004**, *103*, 1557–1559. [CrossRef] [PubMed]
83. Herbrecht, R.; Denning, D.W.; Patterson, T.F.; Bennett, J.E.; Greene, R.E.; Oestmann, J.-W.; Kern, W.V.; Marr, K.A.; Ribaud, P.; Lortholary, O.; et al. Voriconazole versus amphotericin B for primary therapy of invasive aspergillosis. *N. Engl. J. Med.* **2002**, *347*, 408–415. [CrossRef] [PubMed]
84. Walsh, T.J.; Anaissie, E.J.; Denning, D.W.; Herbrecht, R.; Kontoyiannis, D.P.; Marr, K.A.; Morrison, V.A.; Segal, B.H.; Steinbach, W.J.; Stevens, D.A.; et al. Treatment of aspergillosis: Clinical practice guidelines of the Infectious Diseases Society of America. *Clin. Infect. Dis.* **2008**, *46*, 327–360. [CrossRef] [PubMed]
85. Wattier, R.L.; Ramirez-Avila, L. Pediatric Invasive Aspergillosis. *J. fungi (Basel, Switzerland)* **2016**, *2*. [CrossRef] [PubMed]
86. Walsh, T.J.; Karlsson, M.O.; Driscoll, T.; Arguedas, A.G.; Adamson, P.; Saez-Llorens, X.; Vora, A.J.; Arrieta, A.C.; Blumer, J.; Lutsar, I.; et al. Pharmacokinetics and safety of intravenous voriconazole in children after single- or multiple-dose administration. *Antimicrob. Agents Chemother.* **2004**, *48*, 2166–2172. [CrossRef] [PubMed]
87. Friberg, L.E.; Ravva, P.; Karlsson, M.O.; Liu, P. Integrated population pharmacokinetic analysis of voriconazole in children, adolescents, and adults. *Antimicrob. Agents Chemother.* **2012**, *56*, 3032–3042. [CrossRef]
88. Luong, M.-L.; Al-Dabbagh, M.; Groll, A.H.; Racil, Z.; Nannya, Y.; Mitsani, D.; Husain, S. Utility of voriconazole therapeutic drug monitoring: A meta-analysis. *J. Antimicrob. Chemother.* **2016**, *71*, 1786–1799. [CrossRef]
89. Cornely, O.A.; Maertens, J.; Bresnik, M.; Ebrahimi, R.; Ullmann, A.J.; Bouza, E.; Heussel, C.P.; Lortholary, O.; Rieger, C.; Boehme, A.; et al. Liposomal amphotericin B as initial therapy for invasive mold infection: A randomized trial comparing a high-loading dose regimen with standard dosing (AmBiLoad trial). *Clin. Infect. Dis.* **2007**, *44*, 1289–1297. [CrossRef]
90. Girois, S.B.; Chapuis, F.; Decullier, E.; Revol, B.G.P. Adverse effects of antifungal therapies in invasive fungal infections: Review and meta-analysis. *Eur. J. Clin. Microbiol. Infect. Dis.* **2006**, *25*, 138–149. [CrossRef]
91. Pappas, P.G.; Rex, J.H.; Sobel, J.D.; Filler, S.G.; Dismukes, W.E.; Walsh, T.J.; Edwards, J.E. Guidelines for treatment of candidiasis. *Clin. Infect. Dis.* **2004**, *38*, 161–189. [CrossRef] [PubMed]
92. Filioti, I.; Iosifidis, E.; Roilides, E. Therapeutic strategies for invasive fungal infections in neonatal and pediatric patients. *Expert Opin. Pharmacother.* **2008**, *9*, 3179–3196. [CrossRef] [PubMed]

93. Lestner, J.M.; Versporten, A.; Doerholt, K.; Warris, A.; Roilides, E.; Sharland, M.; Bielicki, J.; Goossens, H. Systemic antifungal prescribing in neonates and children: Outcomes from the Antibiotic Resistance and Prescribing in European Children (ARPEC) Study. *Antimicrob. Agents Chemother.* **2015**, *59*, 782–789. [CrossRef] [PubMed]
94. Maertens, J.; Raad, I.; Petrikkos, G.; Boogaerts, M.; Selleslag, D.; Petersen, F.B.; Sable, C.A.; Kartsonis, N.A.; Ngai, A.; Taylor, A.; et al. Efficacy and safety of caspofungin for treatment of invasive aspergillosis in patients refractory to or intolerant of conventional antifungal therapy. *Clin. Infect. Dis.* **2004**, *39*, 1563–1571. [CrossRef] [PubMed]
95. Zaoutis, T.E.; Jafri, H.S.; Huang, L.-M.; Locatelli, F.; Barzilai, A.; Ebell, W.; Steinbach, W.J.; Bradley, J.; Lieberman, J.M.; Hsiao, C.-C.; et al. A prospective, multicenter study of caspofungin for the treatment of documented Candida or Aspergillus infections in pediatric patients. *Pediatrics* **2009**, *123*, 877–884. [CrossRef] [PubMed]
96. Rosanova, M.T.; Bes, D.; Serrano Aguilar, P.; Cuellar Pompa, L.; Sberna, N.; Lede, R. Efficacy and safety of caspofungin in children: Systematic review and meta-analysis. *Arch. Argent. Pediatr.* **2016**, *114*, 305–312. [CrossRef] [PubMed]
97. Seibel, N.L.; Schwartz, C.; Arrieta, A.; Flynn, P.; Shad, A.; Albano, E.; Keirns, J.; Lau, W.M.; Facklam, D.P.; Buell, D.N.; et al. Safety, tolerability, and pharmacokinetics of Micafungin (FK463) in febrile neutropenic pediatric patients. *Antimicrob. Agents Chemother.* **2005**, *49*, 3317–3324. [CrossRef]
98. Lee, C.-H.; Lin, J.-C.; Ho, C.-L.; Sun, M.; Yen, W.-T.; Lin, C. Efficacy and safety of micafungin versus extensive azoles in the prevention and treatment of invasive fungal infections for neutropenia patients with hematological malignancies: A meta-analysis of randomized controlled trials. *PLoS ONE* **2017**, *12*, e0180050. [CrossRef]
99. de Repentigny, L.; Ratelle, J.; Leclerc, J.M.; Cornu, G.; Sokal, E.M.; Jacqmin, P.; De Beule, K. Repeated-dose pharmacokinetics of an oral solution of itraconazole in infants and children. *Antimicrob. Agents Chemother.* **1998**, *42*, 404–408. [PubMed]
100. Walsh, T.J.; Raad, I.; Patterson, T.F.; Chandrasekar, P.; Donowitz, G.R.; Graybill, R.; Greene, R.E.; Hachem, R.; Hadley, S.; Herbrecht, R.; et al. Treatment of invasive aspergillosis with posaconazole in patients who are refractory to or intolerant of conventional therapy: An externally controlled trial. *Clin. Infect. Dis.* **2007**, *44*, 2–12. [CrossRef] [PubMed]
101. Benjamin, D.K.J.; Driscoll, T.; Seibel, N.L.; Gonzalez, C.E.; Roden, M.M.; Kilaru, R.; Clark, K.; Dowell, J.A.; Schranz, J.; Walsh, T.J. Safety and pharmacokinetics of intravenous anidulafungin in children with neutropenia at high risk for invasive fungal infections. *Antimicrob. Agents Chemother.* **2006**, *50*, 632–638. [CrossRef] [PubMed]
102. Rosanova, M.T.; Sarkis, C.; Escarra, F.; Epelbaum, C.; Sberna, N.; Carnovale, S.; Figueroa, C.; Bologna, R.; Lede, R. Anidulafungin in children: Experience in a tertiary care children's hospital in Argentina. *Arch. Argent. Pediatr.* **2017**, *115*, 374–376. [CrossRef] [PubMed]
103. Roilides, E.; Carlesse, F.; Leister-Tebbe, H.; Conte, U.; Yan, J.L.; Liu, P.; Tawadrous, M.; Aram, J.A.; Queiroz-Telles, F. A Prospective, Open-label Study to Assess the Safety, Tolerability, and Efficacy of Anidulafungin in the Treatment of Invasive Candidiasis in Children 2 to <18 Years of Age. *Pediatr. Infect. Dis. J.* **2018**. [CrossRef]
104. Miceli, M.H.; Kauffman, C.A. Isavuconazole: A New Broad-Spectrum Triazole Antifungal Agent. *Clin. Infect. Dis.* **2015**, *61*, 1558–1565. [CrossRef] [PubMed]
105. Maertens, J.A.; Raad, I.I.; Marr, K.A.; Patterson, T.F.; Kontoyiannis, D.P.; Cornely, O.A.; Bow, E.J.; Rahav, G.; Neofytos, D.; Aoun, M.; et al. Isavuconazole versus voriconazole for primary treatment of invasive mould disease caused by Aspergillus and other filamentous fungi (SECURE): A phase 3, randomised-controlled, non-inferiority trial. *Lancet (London, England)* **2016**, *387*, 760–769. [CrossRef]
106. Barg, A.A.; Malkiel, S.; Bartuv, M.; Greenberg, G.; Toren, A.; Keller, N. Successful treatment of invasive mucormycosis with isavuconazole in pediatric patients. *Pediatr. Blood Cancer* **2018**, *65*, e27281. [CrossRef] [PubMed]
107. Wattier, R.L.; Dvorak, C.C.; Hoffman, J.A.; Brozovich, A.A.; Bin-Hussain, I.; Groll, A.H.; Castagnola, E.; Knapp, K.M.; Zaoutis, T.E.; Gustafsson, B.; et al. A Prospective, International Cohort Study of Invasive Mold Infections in Children. *J. Pediatric Infect. Dis. Soc.* **2015**, *4*, 313–322. [CrossRef] [PubMed]

108. Cesaro, S.; Giacchino, M.; Locatelli, F.; Spiller, M.; Buldini, B.; Castellini, C.; Caselli, D.; Giraldi, E.; Tucci, F.; Tridello, G.; et al. Safety and efficacy of a caspofungin-based combination therapy for treatment of proven or probable aspergillosis in pediatric hematological patients. *BMC Infect. Dis.* **2007**, *7*, 28. [CrossRef] [PubMed]
109. Marr, K.A.; Schlamm, H.T.; Herbrecht, R.; Rottinghaus, S.T.; Bow, E.J.; Cornely, O.A.; Heinz, W.J.; Jagannatha, S.; Koh, L.P.; Kontoyiannis, D.P.; et al. Combination antifungal therapy for invasive aspergillosis: A randomized trial. *Ann. Intern. Med.* **2015**, *162*, 81–89. [CrossRef] [PubMed]
110. Van Der Linden, J.W.M.; Warris, A.; Verweij, P.E. Aspergillus species intrinsically resistant to antifungal agents. *Med. Mycol.* **2011**, *49*, S82–S89. [CrossRef] [PubMed]
111. Anderson, J.B. Evolution of antifungal-drug resistance: Mechanisms and pathogen fitness. *Nat. Rev. Microbiol.* **2005**, *3*, 547–556. [CrossRef] [PubMed]
112. Verweij, P.E.; Chowdhary, A.; Melchers, W.J.G.; Meis, J.F. Azole Resistance in *Aspergillus fumigatus*: Can We Retain the Clinical Use of Mold-Active Antifungal Azoles? *Clin. Infect. Dis.* **2016**, *62*, 362–368. [CrossRef] [PubMed]
113. Verweij, P.E.; Ananda-Rajah, M.; Andes, D.; Arendrup, M.C.; Bruggemann, R.J.; Chowdhary, A.; Cornely, O.A.; Denning, D.W.; Groll, A.H.; Izumikawa, K.; et al. International expert opinion on the management of infection caused by azole-resistant *Aspergillus fumigatus*. *Drug Resist. Updat.* **2015**, *21–22*, 30–40. [CrossRef] [PubMed]
114. Thors, V.S.; Bierings, M.B.; Melchers, W.J.G.; Verweij, P.E.; Wolfs, T.F.W. Pulmonary aspergillosis caused by a pan-azole-resistant *Aspergillus fumigatus* in a 10-year-old boy. *Pediatr. Infect. Dis. J.* **2011**, *30*, 268–270. [CrossRef] [PubMed]
115. Estcourt, L.J.; Stanworth, S.; Doree, C.; Blanco, P.; Hopewell, S.; Trivella, M.; Massey, E. Granulocyte transfusions for preventing infections in people with neutropenia or neutrophil dysfunction. *Cochrane database Syst. Rev.* **2015**, CD005341. [CrossRef] [PubMed]
116. Perruccio, K.; Tosti, A.; Burchielli, E.; Topini, F.; Ruggeri, L.; Carotti, A.; Capanni, M.; Urbani, E.; Mancusi, A.; Aversa, F.; et al. Transferring functional immune responses to pathogens after haploidentical hematopoietic transplantation. *Blood* **2005**, *106*, 4397–4406. [CrossRef]
117. Papadopoulou, A.; Kaloyannidis, P.; Yannaki, E.; Cruz, C.R. Adoptive transfer of Aspergillus-specific T cells as a novel anti-fungal therapy for hematopoietic stem cell transplant recipients: Progress and challenges. *Crit. Rev. Oncol. Hematol.* **2016**, *98*, 62–72. [CrossRef]
118. Lin, S.J.; Schranz, J.; Teutsch, S.M. Aspergillosis case-fatality rate: Systematic review of the literature. *Clin. Infect. Dis.* **2001**, *32*, 358–366. [CrossRef]
119. Greene, R.E.; Schlamm, H.T.; Oestmann, J.-W.; Stark, P.; Durand, C.; Lortholary, O.; Wingard, J.R.; Herbrecht, R.; Ribaud, P.; Patterson, T.F.; et al. Imaging findings in acute invasive pulmonary aspergillosis: Clinical significance of the halo sign. *Clin. Infect. Dis.* **2007**, *44*, 373–379. [CrossRef]
120. Dotis, J.; Iosifidis, E.; Roilides, E. Central nervous system aspergillosis in children: A systematic review of reported cases. *Int. J. Infect. Dis.* **2007**, *11*, 381–393. [CrossRef]
121. Palmisani, E.; Barco, S.; Cangemi, G.; Moroni, C.; Dufour, C.; Castagnola, E. Need of voriconazole high dosages, with documented cerebrospinal fluid penetration, for treatment of cerebral aspergillosis in a 6-month-old leukaemic girl. *J. Chemother.* **2017**, *29*, 42–44. [CrossRef] [PubMed]
122. McCarthy, M.; Rosengart, A.; Schuetz, A.N.; Kontoyiannis, D.P.; Walsh, T.J. Mold infections of the central nervous system. *N. Engl. J. Med.* **2014**, *371*, 150–160. [CrossRef] [PubMed]
123. Starke, J.R.; Mason, E.O.J.; Kramer, W.G.; Kaplan, S.L. Pharmacokinetics of amphotericin B in infants and children. *J. Infect. Dis.* **1987**, *155*, 766–774. [CrossRef] [PubMed]
124. Schwartz, S.; Ruhnke, M.; Ribaud, P.; Corey, L.; Driscoll, T.; Cornely, O.A.; Schuler, U.; Lutsar, I.; Troke, P.; Thiel, E. Improved outcome in central nervous system aspergillosis, using voriconazole treatment. *Blood* **2005**, *106*, 2641–2645. [CrossRef] [PubMed]
125. Groll, A.H.; Giri, N.; Petraitis, V.; Petraitiene, R.; Candelario, M.; Bacher, J.S.; Piscitelli, S.C.; Walsh, T.J. Comparative efficacy and distribution of lipid formulations of amphotericin B in experimental *Candida albicans* infection of the central nervous system. *J. Infect. Dis.* **2000**, *182*, 274–282. [CrossRef] [PubMed]
126. Bartelink, I.H.; Wolfs, T.; Jonker, M.; de Waal, M.; Egberts, T.C.G.; Ververs, T.T.; Boelens, J.J.; Bierings, M. Highly variable plasma concentrations of voriconazole in pediatric hematopoietic stem cell transplantation patients. *Antimicrob. Agents Chemother.* **2013**, *57*, 235–240. [CrossRef] [PubMed]

12

Pathogenesis of the *Candida parapsilosis* Complex in the Model Host *Caenorhabditis elegans*

Ana Carolina Remondi Souza [1,2], Beth Burgwyn Fuchs [2,*], Viviane de Souza Alves [3], Elamparithi Jayamani [2], Arnaldo Lopes Colombo [1] and Eleftherios Mylonakis [2,*]

1 Special Mycology Laboratory, Division of Infectious Diseases, Federal University of São Paulo-UNIFESP, 04039-032 São Paulo, SP, Brazil; carolina.remondi@yahoo.com.br (A.C.R.S.); arnaldolcolombo@gmail.com (A.L.C.)
2 Division of Infectious Diseases, Rhode Island Hospital, Alpert Medical School of Brown University, Providence, RI 02903, USA; ejayamani@partners.org
3 Microorganisms Cell Biology Laboratory, Microbiology Department, Biological Sciences Institute, Federal University of Minas Gerais, Belo Horizonte 31270-901, MG, Brazil; gouveiava@ufmg.br
* Correspondence: helen_fuchs@brown.edu (B.B.F.); emylonakis@lifespan.org (E.M.)

Abstract: *Caenorhabditis elegans* is a valuable tool as an infection model toward the study of *Candida* species. In this work, we endeavored to develop a *C. elegans-Candida parapsilosis* infection model by using the fungi as a food source. Three species of the *C. parapsilosis* complex (*C. parapsilosis* (*sensu stricto*), *Candida orthopsilosis* and *Candida metapsilosis*) caused infection resulting in *C. elegans* killing. All three strains that comprised the complex significantly diminished the nematode lifespan, indicating the virulence of the pathogens against the host. The infection process included invasion of the intestine and vulva which resulted in organ protrusion and hyphae formation. Importantly, hyphae formation at the vulva opening was not previously reported in *C. elegans-Candida* infections. Fungal infected worms in the liquid assay were susceptible to fluconazole and caspofungin and could be found to mount an immune response mediated through increased expression of *cnc-4*, *cnc-7*, and *fipr-22/23*. Overall, the *C. elegans-C. parapsilosis* infection model can be used to model *C. parapsilosis* host-pathogen interactions.

Keywords: *Candida parapsilosis*; *Caenorhabditis elegans*; hyphae; invertebrate infection model; host-pathogen interaction

---

1. Introduction

Candida parapsilosis is a common human opportunistic pathogen able to cause superficial and invasive diseases. Most notably, it causes bloodstream infections (BSIs) in very low birth weight neonates and in patients with catheter-associated candidemia and/or intravenous hyperalimentation [1–3]. In 2005, the genetically heterogeneous taxon *C. parapsilosis* was reclassified into three species: *C. parapsilosis* (*sensu stricto*), *Candida orthopsilosis*, and *Candida metapsilosis* [4]. As of yet, it is unclear if there are putative differences between virulence traits among species within the *C. parapsilosis* complex [5–7]. *C. parapsilosis* (*sensu lato*) is the most common non-*albicans Candida* species (NAC) isolated from BSIs in Spain, Italy, many countries in Latin America, while being described as prevalent in U.S. medical centers [8–11].

Over the past decade, invertebrate models have become increasingly valuable to facilitate the study of fungal pathogenesis [12]. Several factors triggered the development of these models, including ethical issues, costs, and physiological simplicity. Moreover, the innate immune mechanisms between invertebrates and mammals share evolutionary conservation, which provides insight into common virulence factors involved in fungal pathogenesis of different types of hosts [13–17].

In particular, *Caenorhabditis elegans* has been used successfully as a candidiasis infection model [18], and its utility has been demonstrated in the assessment of fungal virulence traits and identification of new anti-fungal compounds [18–20]. Nematodes consume fungal pathogens, substituting for the normal laboratory diet, *Escherichia coli*. The ingested fungi establish an infection within the worm gut that can be characterized by the accumulation of yeast and distention of the intestine. The infected nematodes can be followed with either solid or liquid media assay conditions. In liquid medium assays, yeast form hyphae that protrude through the worm cuticle [18,21]. Both *Candida albicans* and non-*albicans* species have been found to cause lethal infections in *C. elegans* [18].

Although *C. elegans* has proven to be a valuable host to study *C. albicans* and a limited number of non-*albicans* species, there are still limited evaluations applied to the study of *C. parapsilosis* species complex infections [22]. In this study, we developed a *C. parapsilosis* (*sensu lato*)-*C. elegans* infection model and demonstrated the utility of this model to study virulence traits of this pathogenic yeast. Furthermore, our endeavors provide insight into the host's defense mechanisms involved against *C. parapsilosis* infection. We describe the reduced lifespan of worms that ingest *C. parapsilosis* and the host symptoms that follow, which differ from those involved in *C. albicans*-*C. elegans* infection.

2. Materials and Methods

2.1. Strains and Media

The *Candida* strains used in these experiments were obtained from the American Type Culture Collection (ATCC) and included: *C. parapsilosis* (*sensu stricto*) ATCC 22019, *C. orthopsilosis* ATCC 96141, and *C. metapsilosis* ATCC 96143. The *C. elegans* strains described in this study were: *N2* bristol [23], *glp-4; sek-1* [24] and *pmk-1* [24] (Table 1). The *C. elegans* strains were maintained at 25 °C (*N2* and *pmk-1*) and 15 °C (*glp-4; sek-1*) and propagated on Nematode Growth Medium (NGM) agar plates seeded with the *E. coli* strain HB101 using established procedures [23]. Yeast cultures were grown in yeast extract, peptone, dextrose (YPD) medium at 30 °C. *E. coli* (HB101) was grown in Luria Broth (LB, Sigma Aldrich, Saint Louis, MO, USA) at 37 °C.

Table 1. *Candida parapslosis* complex and *Caenorhabditis elegans* strains used in this study.

Strain	Description	Purpose	Reference
***Candida* species**			
C. parapsilosis (*sensu stricto*) ATCC 22019	WT [a]	All experiments	ATCC
Candida orthopsilosis ATCC 96141	WT	All experiments	ATCC
Candida metapsilosis ATCC 93143	WT	All experiments	ATCC
C. elegans			
N2	WT	Immunity response	[23]
glp-4; sek-1	*glp-4(bn2) I; sek-1(km4)*	Killing assay, treatment with antifungal drugs, microscopic studies	[24]
pmk-1	*pmk-1(km25)*	Immunity response	[24]

[a] Wild-type.

2.2. Caenorhabditis elegans Liquid Medium Killing Assays

The infection assay was performed as previously described [14]. In brief, worms grown on NGM plates were washed with M9 buffer and placed on 24 h old *C. parapsilosis* species complex lawns (on brain heart infusion (BHI) agar plates) for 4 h. After this, the worms were washed off the plates and transferred to wells (n = 30 per well) in a twelve-well plate that contained 2 mL of liquid medium

(20% BHI, 80% M9, 45 μg/mL Kan). The plates were incubated at 25 °C and nematode survival was examined at 24 h intervals for the subsequent 144 h.

2.3. *Antifungal Drug Treatments*

To study the efficacy of antifungal agents against the *C. parapsilosis* complex in this model, fluconazole (Sigma Aldrich) and caspofungin (Merck, Kenilworth, NJ, USA) were dissolved in dimethyl sulphoxide (DMSO) and added to the liquid assay. The exposure was relative to the minimum inhibitory concentration (MIC) (Table 2): 1×MIC, 2×MIC, and 0.5×MIC. Therefore, for fluconazole, the following concentrations were tested: 1.0 μg/mL (1×), 2.0 μg/mL (2×) and 0.5 μg/mL (0.5×). For caspofungin, the concentrations tested were: 0.5 μg/mL (1×), 1 μg/mL (2×) and 0.25 μg/mL (0.5×) (Table 2). Worms were incubated at 25 °C and survival was monitored daily [18]. Worms were considered dead when they failed to respond to the touch of a platinum wire pick [14].

Table 2. In vitro activity against *C. parapsilosis* species complex reference strains.

	MIC (μg/mL)	
Strain	**Fluconazole**	**Caspofungin**
ATCC 22019	1.0	0.5
ATCC 96141	1.0	0.5
ATCC 96143	1.0	0.5

MIC: minimum inhibitory concentration.

2.4. *Microscopic Studies*

To study *C. parapsilosis* colonization in *C. elegans*, *glp-4*, *sek-1* nematodes were pre-infected with *C. parapsilosis* reference strains for 4 h at 25 °C [14]. Then, the worms were washed three times in M9 buffer and transferred to fresh BHI:M9 medium and incubated at 25 °C for 20 and 48 h. The worms were paralyzed with 1 mM sodium azide solution and placed on 2% agarose pads to capture images at 20 and 48 h post infection [25]. A confocal laser microscope was used for observation (Carl Zeiss M1, Oberkochen, Germany).

2.5. *Quantitative RT-PCR Analyses of Candida parapsilosis Infected Nematodes*

Following infection, N2 worms were treated and RNA was extracted as previously described [26]. The sample quality was assessed through RNA concentration and the 260/280 or 260/230 ratios using a Nanovue spectrophotometer (GE LifeSciences, Piscataway, NJ, USA).

RNA was reverse transcribed to cDNA using the Verso cDNA Synthesis Kit (Thermo Scientific, Waltham, MA, USA). cDNA was analyzed by quantitative real-time (qRT-PCR) using iTaq Universal SYBR Green Supermix® (Bio-Rad, Hercules, CA, USA) at CFX1000 machine (Bio-Rad) and specific primers to the following targets: *Fipr22/23*, *abf-1*, *abf-2*, *cnc-4*, and *cnc-7* (Table 3). All values were normalized against the reference gene *act-1* [19,27–30]. The thermal cycling conditions were comprised of an initial step at 95 °C for 30 s, followed by thirty-five cycles involving denaturation at 95 °C for 5 s, annealing at 58 °C for 15 s and extension at 72 °C for 1 min. The $2^{-\Delta\Delta Ct}$ was calculated for relative quantification of gene expression.

Table 3. Oligonucleotide sequences used in this study.

Oligonucleotide [a]	Sequence 5′ to 3′	Reference
ABF-1/Fw	CTGCCTTCTCCTTGTTCTCCTACT	[19]
ABF-1/Rv	CCTCTGCATTACCGGAACATC	[19]
ABF-2/Fw	TTTCCTTGCACTTCTCCTGG	This study [b]
ABF-2/Rv	CGGTTCCACAGTTTTGCATAC	This study
CNC-4/Fw	ACAATGGGGCTACGGTCCATAT	This study
CNC-4/Rv	ACTTTCCAATGAGCATTCCGAGGA	This study
CNC-7/Fw	CAGGTTCAATGCAGTATGGCTATGG	This study
CNC-7/Rv	GGACGGTACATTCCCATACC	This study
FIPR-22/23 Fw	GCTGAAGCTCCACACATCC	[19]
FIPR-22/23 Rv	TATCCCATTCCTCCGTATCC	[19]

[a] The letters Fw and Rv in the primers names describe the orientation of the primers 5′ to 3′: F for forward (sense) and R for reverse (antisense); [b] the efficiency of primers was evaluated based on the slope of the standard curve constructed by a 10 fold-dilution series using the cDNA.

2.6. Statistics

Killing curves were plotted and the estimation of differences in survival (log-rank and Wilcoxon tests) was performed by the Kaplan-Meier method using GraphPad Prism 5 (GraphPad Software, La Jolla, CA, USA). A *p*-value of <0.05 was considered significant. Relative gene expression was compared using Bonferroni's Method with GraphPad Prism 5 (GraphPad Software). Each experiment was repeated at least three times, and each independent experiment gave similar results.

3. Results

3.1. Killing *Caenorhabditis elegans* by *Candida parapsilosis* Species Complex

First, we assessed the ability of different species within the *C. parapsilosis* complex to cause infection. The results showed that all three species (*C. parapsilosis* (*sensu stricto*), *C. orthopsilosis* and *C. metapsilosis*) were able to kill *C. elegans*. More specifically, in triplicate experiments, we found that *C. parapsilosis* (*sensu stricto*) 22019, *C. orthopsilosis* 96141 and *C. metapsilosis* 96143 killed *C. elegans* with the time to 50% mortality ranging from four to six days for *C. parapsilosis* (*sensu stricto*), *C. orthopsilosis* and *C. metapsilosis*. In all cases, mortality was higher than the *E. coli* (HB101) control group (*p*-value = 0.003 for *C. parapsilosis* (*sensu stricto*); *p*-value = 0.009 for *C. orthopsilosis* and m for *C. metapsilosis*). Interestingly, there was no significant difference between the *C. parapsilosis* species complex infection groups.

3.2. Hyphal Formation of *Candida parapsilosis* Species Complex within *Caenorhabditis elegans*

Candida albicans directed killing of *C. elegans* host is characterized by the formation of filaments that pierce the worm cuticle in a liquid media assay [18]. We investigated whether filaments could be observed in the *C. parapsilosis-C. elegans* infection model. Nematode morphology of the infected worms was observed at 20 h and 48 h post-infection with the reference strains ATCC22019, ATCC96141, and ATCC96143. Worms that consumed *E. coli* appeared in good health (Figure 1A). As shown in Figure 1, the progress of infection and death of *C. elegans* was similar when they were infected by *C. parapsilosis* (*sensu stricto*) (Figure 1B), *C. orthopsilosis* (Figure 1C) or *C. metapsilosis* (Figure 1D). The intestine was distended after ingesting fungal cells (Figure 1B–D), and hyphae start to accumulate within the intestines of live animals 20 h after infection. Filaments were observed breaching the worm at the vulva by 48 h and were observed fully protruding only in lethally infected nematodes.

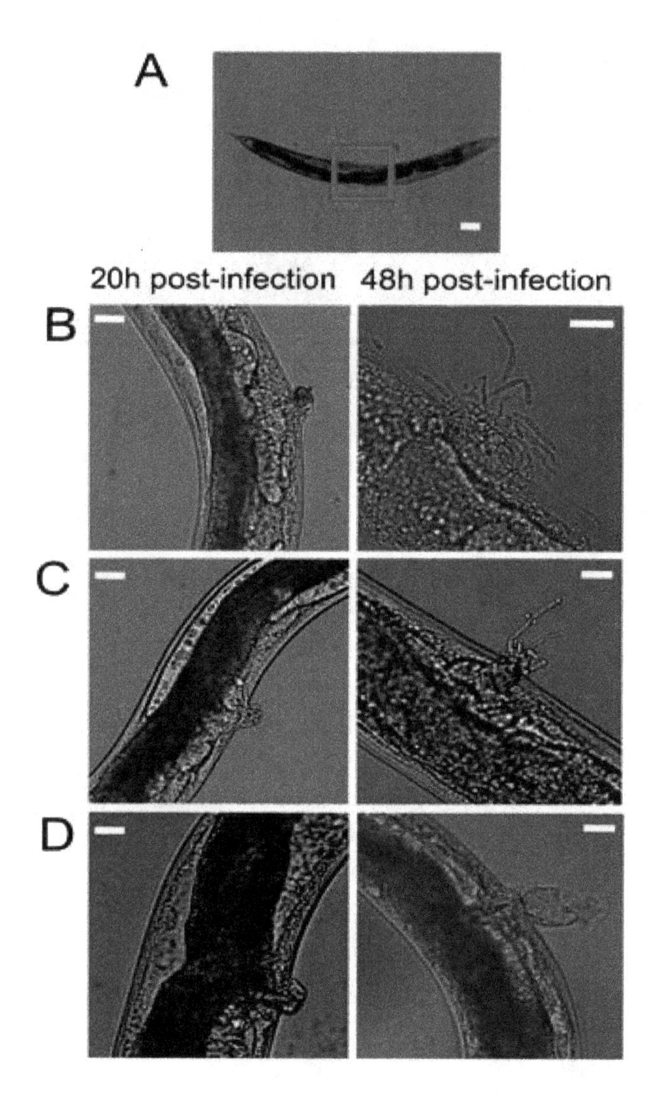

Figure 1. *C. elegans* physiological effects after *C. parapsilosis* exposure. (**A**) *Escherichia coli* exposure shows no adverse effects. The vulva region is highlighted in red; (**B**) *C. parapsilosis* (*sensu stricto*); (**C**) *C. orthopsilosis*; (**D**) *C. metapsilosis*. Scale Bar-20 μm.

3.3. Treatment with Antifungal Drugs

To investigate if compounds could inhibit the fungal infection, *C. elegans* were challenged with *C. parapsilosis* (*sensu stricto*), *C. orthopsilosis*, or *C. metapsilosis* by ingesting the three investigational strains individually on solid media and were then transferred to liquid media, where they were treated with fluconazole and caspofungin. As demonstrated in Figure 2, we observed a dose dependent prolonged survival in response to caspofungin or fluconazole, so that the wells containing antifungal at a concentration below the effective dose (0.5×MIC) resulted in dead worms, similar to those observed in wells that contained no antifungal. On the other hand, a statistically significant increase in survival ($p < 0.001$) was detected when the worms were treated with caspofungin and fluconazole in concentrations at 1×MIC and 2×MIC. In fact, on average, after administration for 144 days, 1×MIC and 2×MIC of fluconazole allowed for, at least, 57% of the nematodes to survive, while 0.5×MIC of this azole resulted in only 38% of the worms being alive. Similarly, caspofungin treatment at 0.5×MIC resulted in 31% nematode survival, whereas at a dose of 1×MIC and 2×MIC, this increased to 69% and 74%, respectively.

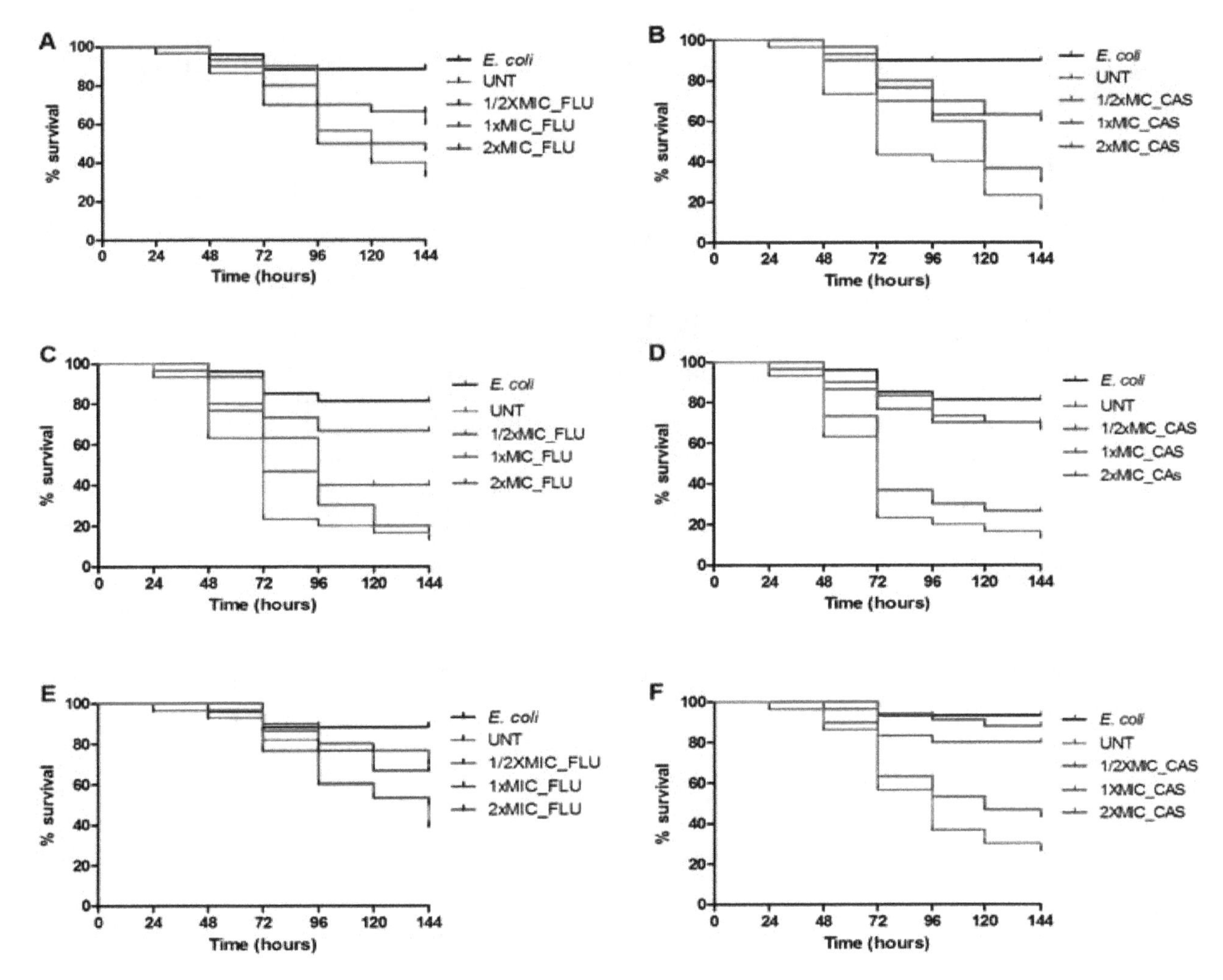

Figure 2. Efficacy of fluconazole (FLU) and caspofungin (CAS) during *C. elegans* infection with *C. parapsilosis* (*sensu stricto*), *C. orthopsilosis* and *C. metapsilosis* reference strains. (**A**) *C. parapsilosis* ATCC 22019, FLU treatment. $p < 0.05$ to FLU_1×MIC. (**B**) *C. parapsilosis* ATCC 22019, CAS treatment. $p = 0.0004$ to CAS_1×MIC. (**C**) *C. orthopsilosis* ATCC 96141, FLU treatment. $p < 0.007$ to FLU_1×MIC. (**D**) *C. orthopsilosis* ATCC 96141, CAS treatment. $p = 0.0001$ to CAS_1×MIC. (**E**) *C. metapsilosis* ATCC 96143, FLU treatment. $p < 0.05$ to FLU_1×MIC. (**D**) *C. metapsilosis* ATCC 96143, CAS treatment. $p < 0.0001$ to CAS_1×MIC.

3.4. Caenorhabditis elegans Immune Response to Candida parapsilosis Complex Infection

In order to understand the defense mechanisms involved against *C. parapsilosis* infection, we focused our attention on five antimicrobial peptides (AMP), which are postulated to have antifungal activity in vivo [19,31]. As shown in Figure 3, 4 h after exposure to *C. parapsilosis* (*sensu stricto*), *C. orthopsilosis*, and *C. metapsilosis*, the expression of the AMP's *cnc-4*, *cnc-7*, and *fipr22/23* increased significantly in response to the presence of the fungal pathogens compared to an *E. coli* control group. The expression of *abf-1* increased only when *C. elegans* was challenged with *C. orthopsilosis* and the *abf-2* expression was unchanged for any of the species involved. Corroborating these findings, we found that *C. elegans pmk-1* (*km25*) mutants were hyper-susceptible to infection with *C. parapsilosis* complex ($p < 0.05$). The PMK-1 mitogen-activated protein (MAP) kinase, orthologous to the mammalian p38 MAPK, in *C. elegans* immunity, is a central regulator of nematode defenses and is required for the basal and pathogen-induced expression of three antifungal immune effectors (*cnc-4*, *cnc-7* and *fipr22/23*), but not *abf-2* (Figure 4).

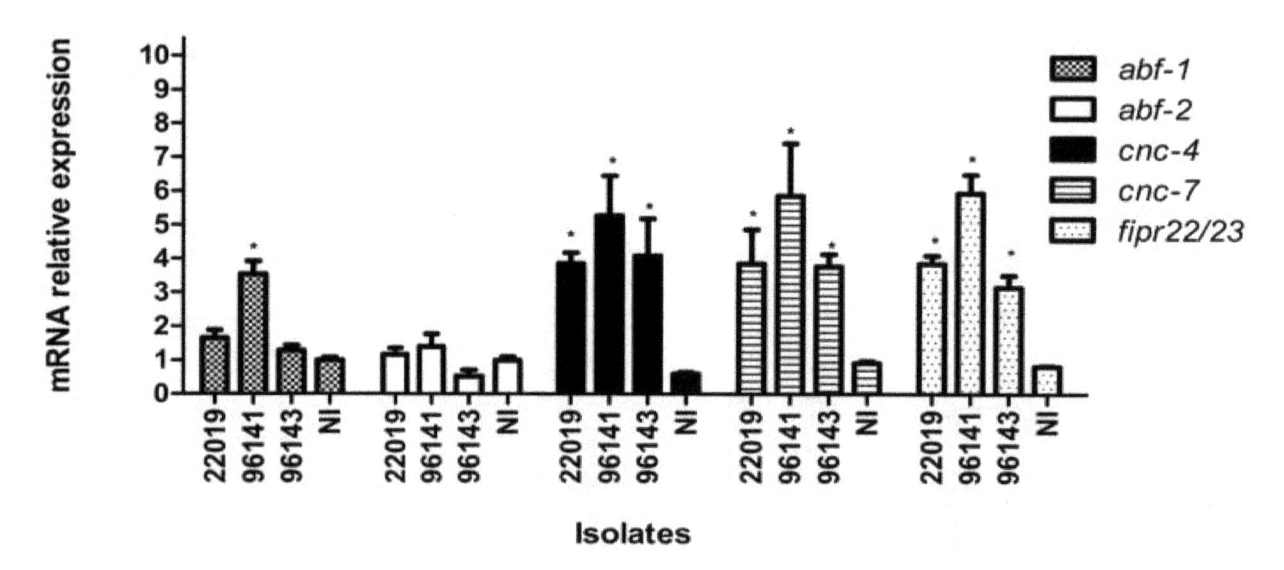

Figure 3. Relative expression of antimicrobial peptides after infection with *C. parapsilosis* (*sensu stricto*) (ATCC 22019), *C. orthopsilosis* (ATCC 96141), *C. metapsilosis* (ATCC 96143), and in non-infected worms (NI). Data are presented as the average of three biological replicates each normalized to a control gene. The error bars represent the standard errors of the mean for three independent biological replicates. * $p < 0.005$.

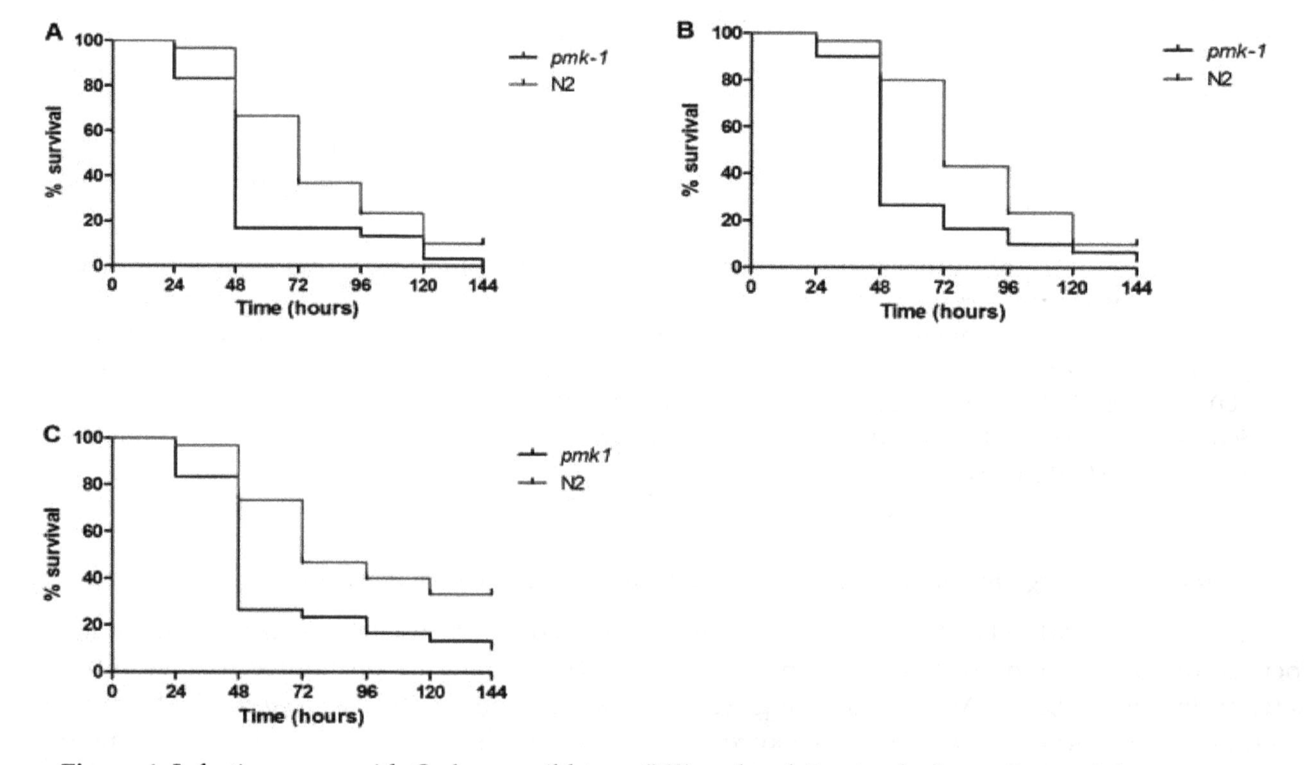

Figure 4. Infection assay with *C. elegans* wild-type (N2) and *pmk-1* animals shows that *pmk-1* was more susceptible to (**A**) *C. parapsilosis* (*sensu stricto*) (ATCC 22019), (**B**) *C. orthopsilosis* (ATCC 96141) and (**C**) *C. metapsilosis* (ATCC 96143) infection ($p < 0.05$).

4. Discussion

In this study, the *C. elegans* invertebrate infection model emerges as a valuable tool for the study of *C. parapsilosis*. Using reference strains, we revealed that *C. parapsilosis* species complex cells, ingested by *C. elegans*, infect and kill the nematode. These data corroborate previous studies showing that

C. orthopsilosis and *C. metapsilosis* are capable of causing invasive infection in mouse models and in humans [8,32,33].

Although the pathogenic potential of the three *C. parapsilosis* complex species is characterized, little is known about the putative differences of their virulence. Previous studies have reported that *C. metapsilosis* seems to be the least virulent member of the *C. parapsilosis* complex in both in vitro and in vivo assays [17,34–36]. However, there is still some controversy within the topic. In a study by Treviño-Rangel et al. (2014), the authors suggest that the three species of the *C. parapsilosis* group possess a similar pathogenic potential in disseminated candidiasis [32]. Corroborating these data, we found that, in *C. elegans* infection model, *C. metapsilosis* had similar mortality rates to that of *C. parapsilosis* (*sensu stricto*) and *C. orthopsilosis*.

Candida hyphal formation is a key virulence determinant that allows cells to invade host tissues and escape phagocytic destruction [37,38]. Pukkila-Worley et al. (2009) demonstrated that the switching from budding yeast to a filamentous (hyphal) form results in aggressive tissue destruction and death of the nematode [21]. In this context, we investigated if this change also took place during infection by *C. parapsilosis*. As expected, all three *C. parapsilosis* strains produced filaments and this phenomenon seems to be associated with *C. elegans* killing. The *C. elegans*-*C. albicans* infection model revealed that *C. albicans* infections within this host begin to accumulate hyphae at the upper part of the gut at the initial stages of the infection process that then spread to consume the entire worm [21]. By contrast, in *C. elegans* infected with *C. parapsilosis* species complex filaments initiated at the vulva rather than the gut, indicating differences in the *C. elegans*-*C. albicans* versus *C. elegans*-*C. parapsilosis* infection processes and host-pathogen interactions.

Although the pathogenicity process may be altered between the two host-pathogen infection models, a response to drug treatment remains conserved. Different studies have demonstrated the utility of invertebrate infection models, including *C. elegans*, as a screening method for potential antifungal compounds [18,25]. In this study, we found a correlation between the in vivo efficacy of antifungals during *C. parapsilosis* (*sensu stricto*), *C. orthopsilosis*, and *C. metapsilosis* infection and their in vitro susceptibility profiles for the standard care therapeutic agents, fluconazole and caspofungin, in a dose dependent manner. By demonstrating that both therapeutic agents have a protective effect during infection in the *C. elegans* model, we gave evidence that drug discovery assays applied to other *Candida* spp.-*C. elegans* models are potentially applicable.

As a whole organism, the model also yields the ability to investigate host immune responses to the pathogen. The nematode *C. elegans* is able to specifically recognize and defend itself against bacterial and fungal pathogens due the presence of complex, inducible, antimicrobial, innate immune responses, which involve the activation of antifungal effectors and core immune genes [19,31,39–42]. Pukkila-Worley et al. (2011) showed that exposure to *C. albicans* stimulated a rapid host response involving approximately 1.6% of the genome, with the majority of the genes encoding antimicrobial, secreted, or detoxification proteins [19]. We therefore used qPCR to check the expression of five AMP genes (*abf-1*, *abf-2*, *cnc-4*, *cnc-7*, and *fipr22/23*) and two transcriptional factors (*zip-2* and *atf-7*). The *abf-1* and *abf*-2 genes belong to a family of six genes encoding antibacterial factors (ABFs) in *C. elegans* and their antimicrobial action were previously described [19,31,43,44]. Expression of these two ABFs seems to be species-specific [41,45]. We found induction of *abf-1* only after *C. orthopsilosis* infection. Regarding *abf-2*, no expression was observed for any of the species.

The second family of AMP we evaluated was the caenacins (CNCs), which are expressed in the *C. elegans* epidermis and, therefore, play a direct role against pathogens that infect worms via the intestinal lumen or cuticle [41,46–48]. Induction of *cnc-4* and *cnc-7* after *C. albicans* has been previously described [41]. Accordingly, we found that the expression of both genes increases significantly in worms fed with any of the three *C. parapsilosis* species, when compared to an *E. coli* control.

In 2008, Pujol et al. described a group of uncharacterized genes that seem to be specifically induced upon fungal infection and could potentially encode AMPs [47]. These genes were called fungus-induced peptide related (*fipr*). During infection with *C. albicans*, the expression of *fipr22/23* is up-regulated [19]. In our study, we also observed induction of the *fipr22/23* gene by the three species belonging to the *C. parapsilosis* complex, suggesting that the FIPRs are involved in the defense mechanisms against *C. parapsilosis* infection. Taken together, our data demonstrated that *C. elegans* mounts a specific defense response against the three different species of *C. parapsilosis* complex.

5. Conclusions

In summary, we demonstrated that *C. elegans* can be used as an appropriate infection model to study the pathogenicity of *C. parapsilosis* (*sensu stricto*), *C. orthopsilosis* and *C. metapsilosis*, not only for evaluating the virulence traits of these species, but also to screen antifungal agents, and study mechanisms of innate immune response against these yeasts. Future studies should expand the model described here to yield more insights about the pathogenicity of these species, especially *C. orthopsilosis* and *C. metapsilosis*.

Author Contributions: B.B.F., A.C.R.S., A.L.C., E.J. and E.M. conceived and designed the experiments. A.C.R.S., V.d.S.A., and E.J. performed the experiments and contributed to analysis. A.L.C. contributed strains and materials. A.C.R.S. and B.B.F wrote the first draft of the article and all other authors contributed to the final version of the text.

References

1. Pammi, M.; Holland, L.; Butler, G.; Gacser, A.; Bliss, J.M. *Candida parapsilosis* is a significant neonatal pathogen: A systematic review and meta-analysis. *Pediatr. Infect. Dis. J.* **2013**, *32*, e206–e216. [CrossRef] [PubMed]
2. Trofa, D.; Gacser, A.; Nosanchuk, J.D. *Candida parapsilosis*, an emerging fungal pathogen. *Clin. Microbiol. Rev.* **2008**, *21*, 606–625. [CrossRef] [PubMed]
3. Quindos, G. Epidemiology of candidaemia and invasive candidiasis. A changing face. *Rev. Iberoam. Micol.* **2014**, *31*, 42–48. [CrossRef] [PubMed]
4. Tavanti, A.; Davidson, A.D.; Gow, N.A.; Maiden, M.C.; Odds, F.C. *Candida orthopsilosis* and *Candida metapsilosis* spp. Nov. to replace *Candida parapsilosis* groups II and III. *J. Clin. Microbiol.* **2005**, *43*, 284–292. [CrossRef] [PubMed]
5. Gomez-Lopez, A.; Alastruey-Izquierdo, A.; Rodriguez, D.; Almirante, B.; Pahissa, A.; Rodriguez-Tudela, J.L.; Cuenca-Estrella, M.; Barcelona Candidemia Project Study Group. Prevalence and susceptibility profile of *Candida metapsilosis* and *Candida orthopsilosis*: Results from population-based surveillance of candidemia in Spain. *Antimicrob. Agents Chemother.* **2008**, *52*, 1506–1509. [CrossRef] [PubMed]
6. Lockhart, S.R.; Messer, S.A.; Pfaller, M.A.; Diekema, D.J. Geographic distribution and antifungal susceptibility of the newly described species *Candida orthopsilosis* and *Candida metapsilosis* in comparison to the closely related species *Candida parapsilosis*. *J. Clin. Microbiol.* **2008**, *46*, 2659–2664. [CrossRef] [PubMed]
7. Goncalves, S.S.; Amorim, C.S.; Nucci, M.; Padovan, A.C.; Briones, M.R.; Melo, A.S.; Colombo, A.L. Prevalence rates and antifungal susceptibility profiles of the *Candida parapsilosis* species complex: Results from a nationwide surveillance of candidaemia in Brazil. *Clin. Microbiol. Infect.* **2010**, *16*, 885–887. [CrossRef] [PubMed]
8. Nucci, M.; Queiroz-Telles, F.; Alvarado-Matute, T.; Tiraboschi, I.N.; Cortes, J.; Zurita, J.; Guzman-Blanco, M.; Santolaya, M.E.; Thompson, L.; Sifuentes-Osornio, J.; et al. Epidemiology of candidemia in Latin America: A laboratory-based survey. *PLoS ONE* **2013**, *8*, e59373. [CrossRef] [PubMed]
9. Marcos-Zambrano, L.J.; Escribano, P.; Sanchez, C.; Munoz, P.; Bouza, E.; Guinea, J. Antifungal resistance to fluconazole and echinocandins is not emerging in yeast isolates causing fungemia in a Spanish tertiary care center. *Antimicrob. Agents Chemother.* **2014**, *58*, 4565–4572. [CrossRef] [PubMed]
10. Colombo, A.L.; Guimaraes, T.; Sukienik, T.; Pasqualotto, A.C.; Andreotti, R.; Queiroz-Telles, F.; Nouer, S.A.; Nucci, M. Prognostic factors and historical trends in the epidemiology of candidemia in critically ill patients:

An analysis of five multicenter studies sequentially conducted over a 9-year period. *Intens. Care Med.* **2014**, *40*, 1489–1498. [CrossRef] [PubMed]

11. Pfaller, M.A.; Jones, R.N.; Castanheira, M. Regional data analysis of *Candida non-albicans* strains collected in United States medical sites over a 6-year period, 2006–2011. *Mycoses* **2014**, *57*, 602–611. [CrossRef] [PubMed]
12. Arvanitis, M.; Glavis-Bloom, J.; Mylonakis, E. Invertebrate models of fungal infection. *Biochim. Biophys. Acta* **2013**, *1832*, 1378–1383. [CrossRef] [PubMed]
13. Desalermos, A.; Fuchs, B.B.; Mylonakis, E. Selecting an invertebrate model host for the study of fungal pathogenesis. *PLoS Pathog.* **2012**, *8*, e1002451. [CrossRef] [PubMed]
14. Muhammed, M.; Coleman, J.J.; Mylonakis, E. *Caenorhabditis elegans*: A nematode infection model for pathogenic fungi. *Methods Mol. Biol.* **2012**, *845*, 447–454. [PubMed]
15. Junqueira, J.C. *Galleria mellonella* as a model host for human pathogens: Recent studies and new perspectives. *Virulence* **2012**, *3*, 474–476. [CrossRef] [PubMed]
16. Fallon, J.P.; Reeves, E.P.; Kavanagh, K. The *Aspergillus fumigatus* toxin fumagillin suppresses the immune response of *Galleria mellonella* larvae by inhibiting the action of haemocytes. *Microbiology* **2011**, *157*, 1481–1488. [CrossRef] [PubMed]
17. Gago, S.; Garcia-Rodas, R.; Cuesta, I.; Mellado, E.; Alastruey-Izquierdo, A. *Candida parapsilosis, Candida orthopsilosis,* and *Candida metapsilosis* virulence in the non-conventional host *Galleria mellonella. Virulence* **2014**, *5*, 278–285. [CrossRef] [PubMed]
18. Breger, J.; Fuchs, B.B.; Aperis, G.; Moy, T.I.; Ausubel, F.M.; Mylonakis, E. Antifungal chemical compounds identified using a *C. elegans* pathogenicity assay. *PLoS Pathog.* **2007**, *3*, e18. [CrossRef] [PubMed]
19. Pukkila-Worley, R.; Ausubel, F.M.; Mylonakis, E. *Candida albicans* infection of *Caenorhabditis elegans* induces antifungal immune defenses. *PLoS Pathog.* **2011**, *7*, e1002074. [CrossRef] [PubMed]
20. Mylonakis, E.; Moreno, R.; El Khoury, J.B.; Idnurm, A.; Heitman, J.; Calderwood, S.B.; Ausubel, F.M.; Diener, A. *Galleria mellonella* as a model system to study *Cryptococcus neoformans* pathogenesis. *Infect. Immun.* **2005**, *73*, 3842–3850. [CrossRef] [PubMed]
21. Pukkila-Worley, R.; Peleg, A.Y.; Tampakakis, E.; Mylonakis, E. *Candida albicans* hyphal formation and virulence assessed using a *Caenorhabditis elegans* infection model. *Eukaryot. Cell* **2009**, *8*, 1750–1758. [CrossRef] [PubMed]
22. Ortega-Riveros, M.; De-la-Pinta, I.; Marcos-Arias, C.; Ezpeleta, G.; Quindos, G.; Eraso, E. Usefulness of the non-conventional *Caenorhabditis elegans* model to assess *Candida* virulence. *Mycopathologia* **2017**, *182*, 785–795. [CrossRef] [PubMed]
23. Brenner, S. The genetics of *Caenorhabditis elegans*. *Genetics* **1974**, *77*, 71–94. [PubMed]
24. Kim, D.H.; Feinbaum, R.; Alloing, G.; Emerson, F.E.; Garsin, D.A.; Inoue, H.; Tanaka-Hino, M.; Hisamoto, N.; Matsumoto, K.; Tan, M.W.; et al. A conserved p38 map kinase pathway in *Caenorhabditis elegans* innate immunity. *Science* **2002**, *297*, 623–626. [CrossRef] [PubMed]
25. Huang, X.; Li, D.; Xi, L.; Mylonakis, E. *Caenorhabditis elegans*: A simple nematode infection model for *Penicillium marneffei*. *PLoS ONE* **2014**, *9*, e108764. [CrossRef] [PubMed]
26. Rohlfing, A.K.; Miteva, Y.; Hannenhalli, S.; Lamitina, T. Genetic and physiological activation of osmosensitive gene expression mimics transcriptional signatures of pathogen infection in *C. elegans*. *PLoS ONE* **2010**, *5*, e9010. [CrossRef] [PubMed]
27. Hoogewijs, D.; Houthoofd, K.; Matthijssens, F.; Vandesompele, J.; Vanfleteren, J.R. Selection and validation of a set of reliable reference genes for quantitative *sod* gene expression analysis in *C. elegans*. *BMC Mol. Biol.* **2008**, *9*, 9. [CrossRef] [PubMed]
28. Li, J.; Ebata, A.; Dong, Y.; Rizki, G.; Iwata, T.; Lee, S.S. *Caenorhabditis elegans* HCF-1 functions in longevity maintenance as a DAF-16 regulator. *PLoS Biol.* **2008**, *6*, e233. [CrossRef] [PubMed]
29. Pujol, N.; Cypowyj, S.; Ziegler, K.; Millet, A.; Astrain, A.; Goncharov, A.; Jin, Y.; Chisholm, A.D.; Ewbank, J.J. Distinct innate immune responses to infection and wounding in the *C. elegans* epidermis. *Curr. Biol.* **2008**, *18*, 481–489. [CrossRef] [PubMed]
30. Zhang, Y.; Chen, D.; Smith, M.A.; Zhang, B.; Pan, X. Selection of reliable reference genes in *Caenorhabditis elegans* for analysis of nanotoxicity. *PLoS ONE* **2012**, *7*, e31849. [CrossRef] [PubMed]

31. Kato, Y.; Aizawa, T.; Hoshino, H.; Kawano, K.; Nitta, K.; Zhang, H. abf-1 and abf-2, ASABF-type antimicrobial peptide genes in *Caenorhabditis elegans*. *Biochem. J.* **2002**, *361*, 221–230. [CrossRef] [PubMed]
32. Treviño-Rangel Rde, J.; Rodriguez-Sanchez, I.P.; Elizondo-Zertuche, M.; Martinez-Fierro, M.L.; Garza-Veloz, I.; Romero-Diaz, V.J.; Gonzalez, J.G.; Gonzalez, G.M. Evaluation of in vivo pathogenicity of *Candida parapsilosis*, *Candida orthopsilosis*, and *Candida metapsilosis* with different enzymatic profiles in a murine model of disseminated candidiasis. *Med. Mycol.* **2014**, *52*, 240–245. [CrossRef] [PubMed]
33. Blanco-Blanco, M.T.; Gomez-Garcia, A.C.; Hurtado, C.; Galan-Ladero, M.A.; Lozano Mdel, C.; Garcia-Tapias, A.; Blanco, M.T. *Candida orthopsilosis* fungemias in a Spanish tertiary care hospital: Incidence, epidemiology and antifungal susceptibility. *Rev. Iberoam. Micol.* **2014**, *31*, 145–148. [CrossRef] [PubMed]
34. Gacser, A.; Trofa, D.; Schafer, W.; Nosanchuk, J.D. Targeted gene deletion in *Candida parapsilosis* demonstrates the role of secreted lipase in virulence. *J. Clin. Investig.* **2007**, *117*, 3049–3058. [CrossRef] [PubMed]
35. Nemeth, T.; Toth, A.; Szenzenstein, J.; Horvath, P.; Nosanchuk, J.D.; Grozer, Z.; Toth, R.; Papp, C.; Hamari, Z.; Vagvolgyi, C.; et al. Characterization of virulence properties in the *C. parapsilosis sensu lato* species. *PLoS ONE* **2013**, *8*, e68704. [CrossRef] [PubMed]
36. Bertini, A.; De Bernardis, F.; Hensgens, L.A.; Sandini, S.; Senesi, S.; Tavanti, A. Comparison of *Candida parapsilosis*, *Candida orthopsilosis*, and *Candida metapsilosis* adhesive properties and pathogenicity. *Int. J. Med. Microbiol.* **2013**, *303*, 98–103. [CrossRef] [PubMed]
37. Liu, H. Transcriptional control of dimorphism in *Candida albicans*. *Curr. Opin. Microbiol.* **2001**, *4*, 728–735. [CrossRef]
38. Navarro-Garcia, F.; Sanchez, M.; Nombela, C.; Pla, J. Virulence genes in the pathogenic yeast *Candida albicans*. *FEMS Microbiol. Rev.* **2001**, *25*, 245–268. [CrossRef] [PubMed]
39. Engelmann, I.; Pujol, N. Innate immunity in *C. elegans*. *Adv. Exp. Med. Biol.* **2010**, *708*, 105–121. [PubMed]
40. Ermolaeva, M.A.; Schumacher, B. Insights from the worm: The *C. elegans* model for innate immunity. *Semin. Immunol.* **2014**, *26*, 303–309. [CrossRef] [PubMed]
41. Engelmann, I.; Griffon, A.; Tichit, L.; Montanana-Sanchis, F.; Wang, G.; Reinke, V.; Waterston, R.H.; Hillier, L.W.; Ewbank, J.J. A comprehensive analysis of gene expression changes provoked by bacterial and fungal infection in *C. elegans*. *PLoS ONE* **2011**, *6*, e19055. [CrossRef] [PubMed]
42. Zugasti, O.; Bose, N.; Squiban, B.; Belougne, J.; Kurz, C.L.; Schroeder, F.C.; Pujol, N.; Ewbank, J.J. Activation of a G protein-coupled receptor by its endogenous ligand triggers the innate immune response of *Caenorhabditis elegans*. *Nat. Immunol.* **2014**, *15*, 833–838. [CrossRef] [PubMed]
43. Kato, Y.; Komatsu, S. ASABF, a novel cysteine-rich antibacterial peptide isolated from the nematode *Ascaris suum*. Purification, primary structure, and molecular cloning of cDNA. *J. Biol. Chem.* **1996**, *271*, 30493–30498. [CrossRef] [PubMed]
44. Zhang, H.; Kato, Y. Common structural properties specifically found in the CSαβ-type antimicrobial peptides in nematodes and mollusks: Evidence for the same evolutionary origin? *Dev. Comp. Immunol.* **2003**, *27*, 499–503. [CrossRef]
45. Means, T.K.; Mylonakis, E.; Tampakakis, E.; Colvin, R.A.; Seung, E.; Puckett, L.; Tai, M.F.; Stewart, C.R.; Pukkila-Worley, R.; Hickman, S.E.; et al. Evolutionarily conserved recognition and innate immunity to fungal pathogens by the scavenger receptors SCARF1 and CD36. *J. Exp. Med.* **2009**, *206*, 637–653. [CrossRef] [PubMed]
46. Couillault, C.; Pujol, N.; Reboul, J.; Sabatier, L.; Guichou, J.F.; Kohara, Y.; Ewbank, J.J. TIR-independent control of innate immunity in *Caenorhabditis elegans* by the TIR domain adaptor protein TIR-1, an ortholog of human SARM. *Nat. Immunol.* **2004**, *5*, 488–494. [CrossRef] [PubMed]
47. Pujol, N.; Zugasti, O.; Wong, D.; Couillault, C.; Kurz, C.L.; Schulenburg, H.; Ewbank, J.J. Anti-fungal innate immunity in *C. elegans* is enhanced by evolutionary diversification of antimicrobial peptides. *PLoS Pathog.* **2008**, *4*, e1000105. [CrossRef] [PubMed]
48. Zugasti, O.; Ewbank, J.J. Neuroimmune regulation of antimicrobial peptide expression by a noncanonical TGF-β signaling pathway in *Caenorhabditis elegans* epidermis. *Nat. Immunol.* **2009**, *10*, 249–256. [CrossRef] [PubMed]

13

Pre-Existing Liver Disease and Toxicity of Antifungals

Nikolaos Spernovasilis [1] and Diamantis P. Kofteridis [1,2,*]

[1] Department of Internal Medicine and Infectious Diseases, University Hospital of Heraklion, P.O. Box 1352, 71110 Heraklion, Crete, Greece; nikspe@hotmail.com
[2] School of Medicine, University of Crete, 71110 Heraklion, Crete, Greece
* Correspondence: kofterid@med.uoc.gr

Abstract: Pre-existing liver disease in patients with invasive fungal infections further complicates their management. Altered pharmacokinetics and tolerance issues of antifungal drugs are important concerns. Adjustment of the dosage of antifungal agents in these cases can be challenging given that current evidence to guide decision-making is limited. This comprehensive review aims to evaluate the existing evidence related to antifungal treatment in individuals with liver dysfunction. This article also provides suggestions for dosage adjustment of antifungal drugs in patients with varying degrees of hepatic impairment, after accounting for established or emerging pharmacokinetic–pharmacodynamic relationships with regard to antifungal drug efficacy in vivo.

Keywords: liver disease; hepatic impairment; invasive fungal infection; antifungal agent; antifungal drug; toxicity

---

1. Introduction

Invasive fungal infection (IFI) is a leading cause of morbidity and mortality among immunocompromised and critically ill patients [1,2]. Although antifungal drug options have increased in recent years, effective management of IFI depends mainly on early and appropriate individualized treatment that optimizes efficacy and safety based on local epidemiology, drug spectrum of activity, pharmacokinetic (PK) and pharmacodynamic (PD) properties of the antifungal agent, and patient related factors [3].

Pre-existing liver disease in patients with IFIs raises significant concern about the safety of antifungal agent administration. The liver is the primary site of drug metabolism, and hepatic disease can significantly alter the PKs of antifungal drugs, mainly through impaired clearance [4]. Moreover, other variables that affect PKs such as liver blood flow, biliary excretion and plasma protein binding may be altered in patients with pre-existing hepatic dysfunction [4]. These patients may also tolerate drug-induced liver injury (DILI) more poorly than healthy individuals [5]. Furthermore, in the cirrhotic patients, drug-related extrahepatic effects, such as renal failure, gastrointestinal bleeding and hepatic encephalopathy, are more likely to occur [6]. Hepatic functional status is also an important determinant of the drug–drug interaction (DDI) magnitude due to enzyme inhibition or induction in the liver [7].

It is important to distinguish isolated biochemical injury from hepatic dysfunction [8]. In general, DILI is characterized by elevations in hepatic enzymes, resulting from the effect of an active drug or its metabolites to the liver [9]. This biochemical abnormality is not necessarily accompanied by clinically significant liver dysfunction, since liver has a notable healing capacity [8]. However, DILI can be the cause of hepatic dysfunction, manifested by hyperbilirubinemia and coagulopathy [10], or even acute liver failure, presented with jaundice and hepatic encephalopathy [11].

Liver injury induced by a drug is generally classified as either intrinsic, which is predictable, dose-dependent and reproducible in preclinical models, or idiosyncratic, which is unpredictable and dose-independent [12–14]. An international expert group of clinicians and scientists comprehensibly proposed the clinical chemistry criteria for the diagnosis of DILI, taking also into account the possibility of pre-existing liver enzymes abnormalities (Table 1) [15]. Furthermore, the ratio of serum alanine aminotransferase (ALT) to alkaline phosphatase (ALP), expressed as multiples of upper limit of normal (ULN), is called R ratio or value, and is used to classify DILI in individuals with previous normal liver tests into three categories: hepatocellular ($R > 5$), cholestatic ($R < 2$) and mixed (R of 2–5) [16]. Bilirubin, although not incorporated into the R ratio, remains an essential marker in calculating the Model for End-Stage Liver Disease (MELD) score and the Child–Pugh score [17,18]. Both these prognostic models are also used to assess hepatic function, with the Child–Pugh score being the most commonly used method in cirrhotic patients among studies submitted to the US Food and Drug Administration (FDA) although it is not associated directly with PK changes [19] and does not represent a reliable estimator of liver function [20].

Table 1. Clinical chemistry criteria for DILI.

Anyone of the Following *:
ALT elevation $\geq 5 \times$ ULN ¶
ALP elevation $\geq 2 \times$ ULN ¶, especially with accompanying elevations in concentrations of 5′-NT or GGT
ALT elevation $\geq 3 \times$ ULN ¶ and simultaneous TB elevation $\geq 2 \times$ ULN ¶

DILI: drug-induced liver injury; ALT: alanine transaminase; ULN: upper limit of normal; AST: aspartate transaminase; ALP: alkaline phosphatase; 5′-NT: 5′-nucleotidase; GGT: γ-glutamyl transpeptidase; TB: total bilirubin. * After other causes have been ruled-out [15]. ¶ In cases of pre-existing abnormal biochemistry before the administration of the implicated drug, ULN is replaced by the mean baseline values obtained prior to drug exposure [15].

The risk of developing liver injury and possible hepatic dysfunction by an antifungal agent depends on several factors. The chemical properties of the agent, demographics, genetic predisposition, comorbidities including underlying hepatic disease, concomitant hepatotoxic drugs and DDIs, severity of the illness, and liver involvement by the fungal infection, all affect the possibility for hepatotoxicity [21]. Under these circumstances, it can be difficult to attribute DILI due to antifungals to only one factor.

In general, published literature regarding the use of antifungal agents in patients with pre-existing liver disease is somewhat inconclusive. A clear understanding of antifungal-caused liver injury in patients with underlying hepatic impairment is lacking, and recommendations for dosage adjustments in these cases are not straightforward [3,22]. Most of the information about antifungal dosing regimens is derived from clinical trials and PK studies, in which only few patients with a varying level of liver impairment were included [20]. For some antifungals, a dose reduction is recommended in the manufacturers' product characteristics in cases of pre-existing hepatic dysfunction, while for other antifungal agents no dosage adjustment is required or recommended [22].

The aim of the present review is to provide an overview of the safety profile of the various antifungal agents in patients with underlying liver disease. The intention is to summarize current data on the PKs of antifungals in these individuals and to increase clinical awareness of how various antifungal compounds should be used under these circumstances.

2. Antifungal Agents

The current antifungal armory for IFIs includes polyenes (amphotericin B-based preparations), flucytosine, triazoles (fluconazole, itraconazole, voriconazole, posaconazole, and isavuconazole), and echinocandins (caspofungin, micafungin, and anidulafungin) [23]. These compounds differ from each other in their spectrum of activity, pharmacokinetics/pharmacodynamics (PK/PD) properties, indications, dosing, safety profile, cost, and ease of use [3,24,25].

2.1. Polyenes

Amphotericin was introduced in therapy in 1958 as amphotericin B deoxycholate (AmBD), but its clinical usefulness is limited because of nephrotoxicity and infusion-related reactions [24,26]. Three lipid formulations of amphotericin B (AmB), liposomal amphotericin B (LAmB), amphotericin B lipid complex (ABLC), and amphotericin colloidal dispersion (ABCD; discontinued in most countries) were developed in the 1990s to reduce the toxicity observed with AmBD [24]. AmB interacts with ergosterol in the fungal membranes leading to the formation of membrane-spanning pores, ion leakage, and ultimately fungal cell death [27]. Additional cytotoxic mechanisms of AmB are inhibition of the fungal proton-ATPase and lipid peroxidation [28]. It is eliminated unchanged mainly via urine and feces [29]. Because of its broad antimycotic spectrum, AmB is a cornerstone in the treatment of serious and life-threatening fungal infections. The daily dose for AmBD ranges from 0.3 to 1.5 mg/kg, while the recommended standard doses for the lipid formulations of AmB are much higher [29,30]. Specifically, for LAmB the usual daily dose ranges from 3 to 5 mg/kg, but doses up to 10 mg/kg/d can be administered in cases of rhino-orbital-cerebral mucormycosis [29]. For ABLC the usual dose is 5mg/kg/d, while for ABCD the daily dose ranges from 3 to 4 mg/kg [30].

Generally, lipid-based formulations of AmB present at least the same efficacy as AmBD and are even superior in the treatment of certain fungal infections, such as mild to moderate disseminated histoplasmosis in patients with acquired immunodeficiency syndrome (AIDS), while they are associated with a safer profile [30–32]. Notably, in some studies, the administration of LAmB was associated with lower toxicity rates, namely infusional and kidney toxicity, compared to other lipid formulations [33–35]. However, differences in drug-induced nephrotoxicity between lipid-based formulations of AmB continue to be a subject of debate [36,37]. Other commonly encountered adverse effects of AmB preparations, apart from nephrotoxicity and infusion reactions, include hypokalemia, hypomagnesemia, and anemia [27,38]. Liver injury due to AmB therapy is relatively subtle and reversible, with its incidence reaching 32% for LAmB and 41% for ABLC in some clinical studies [21,39,40]. Interestingly, lipid formulations of AmB, mainly LAmB, seem to have a stronger association with DILI than AmBD, probably due to the carriers of these formulations [24,33,40,41]. In any case, clinically evident liver injury and treatment discontinuation due to AmB preparations are rare [21,27].

No specific recommendations are available for AmB preparations in the case of pre-existing hepatic impairment, but considering their limited hepatic metabolism, dosage adjustment is unlikely to be necessary [22]. Data on the PKs of AmB in pre-existing liver disease are sparse and clinical studies are lacking so far. In a retrospective single-center non-randomized autopsy-controlled study, Chamilos et al. compared hepatic enzymes elevations and histopathological findings in the livers of 64 patients with hematologic malignancies who had received LAmB or ABLC for at least 7 days, as a treatment for IFIs [42]. Among these patients, there were 22 patients with elevated liver enzymes at baseline, more than five times the ULN. None of the patients with acute liver injury, including those with abnormal baseline hepatic biochemical parameters, showed the histopathological changes induced by liposomal formulations of AmB that have been reported in animal studies [42]. Another study assessed the PK properties of ABCD in 11 patients with cholestatic liver disease compared to 9 subjects with normal liver enzymes [43]. Pre-existing cholestatic liver disease had no significant influence on steady state PKs of liberated AmB, and the authors concluded that the standard dosage of ABCD is probably appropriate for these patients [43].

2.2. Flucytosine

Flucytosine became available in 1968 [44]. It is taken up by fungal cells by cytosine permease and converted intracellularly into fluorouracil, which is further metabolized into 5-fluorouridine triphosphate and 5-fluorodeoxyuridine monophosphate, resulting in inhibition of fungal protein and DNA synthesis [45]. It is mainly eliminated by the kidneys, while it is minimally metabolized in the liver [46]. The high occurrence of resistance precludes its use as a single agent. Nowadays, flucytosine is used in combination therapy with AmB as first-line therapy in cryptococcal meningoencephalitis [47]. Furthermore, it may be added to other regimens for the treatment of severe pulmonary cryptococcosis, central nervous system candidiasis, *Candida* endocarditis, and *Candida* urinary tract infections [47–49]. Flucytosine's recommended dosage in individuals with normal renal function ranges from 50 to 150 mg/kg/d divided in four doses for both oral and intravenous formulation, while dosages up to 200 mg/kg/d can be administered [29,50].

Flucytosine's most significant adverse effects is myelotoxicity, mainly neutropenia and thrombocytopenia, and hepatotoxicity, and both are thought to be due to the effects of fluorouracil [46,51]. Because human intestinal flora is capable of converting flucytosine into fluorouracil in vitro, oral administration of the drug might be associated with more side effects than intravenous administration [51]. Liver injury is frequently encountered during treatment with flucytosine and the incidence varies from 0% to 41%, probably due to the different definition of liver injury in different studies [24]. The elevation in liver enzymes is usually mild to moderate and reversible on discontinuation, while two cases of severe liver necrosis have been reported in patients who received flucytosine for candidal endocarditis [46,52]. Both myelotoxicity and liver toxicity have been associated with high flucytosine concentrations in the blood. Therapeutic drug monitoring (TDM) is advisable 3–5 days after initiating therapy and after any changes in the glomerular filtration rate (GFR) to keep the 2 h flucytosine post-dose levels between 30 to 80 mg/L [53]. DDIs involving the cytochrome P450 (CYP450) pose a minor concern for flucytosine administration [29].

For patients with pre-existing hepatic impairment, limited data are available regarding the PK properties and the safety of flucytosine. In 1973, Block studied for the first time the effect of hepatic insufficiency on flucytosine concentrations in the serum of rabbits with chemically induced acute hepatitis [54]. No influence of the hepatic function on serum concentration of the drug was observed. In the same paper, a single patient with biopsy-proven cirrhosis was described as treated with flucytosine for cryptococcal meningitis. Drug concentrations in serum were measured at 1, 2, and 6 h after a dose and did not differ from concentrations determined simultaneously in 10 patients with cryptococcal infection and normal liver function being treated with the same dose of flucytosine [54]. However, given the fact that liver injury due to flucytosine treatment is a common adverse effect in many studies, this antifungal agent should be used with extreme caution or even be avoided in this patient population, although there are no dosage adjustments provided in the manufacturer's labeling [29,50]. Combined treatment with AmB may lead to the accumulation of flucytosine because of AmB-induced nephrotoxicity, further complicating the matter [55]. In addition, a recent study examining the hepatotoxicity induced by combined therapy of flucytosine and AmB in animal models showed a synergistic inflammatory activation in a dose-dependent manner, through the NF-κB pathway, which promoted an inflammatory cascade in the liver. The authors suggested that the combination of flucytosine and AmB for the treatment of IFIs in patients with hepatic dysfunction requires careful clinical, biochemical, and drug monitoring [56].

2.3. *Azoles*

The azole antifungals are synthetic compounds that can be divided into two subclasses, the imidazoles and the triazoles, according to the number of nitrogen atoms in the five-membered azole ring [29]. The imidazoles include ketoconazole, miconazole, and clotrimazole [21]. Miconazole was at one time administered intravenously for the treatment of certain IFIs, but soon this formulation was withdrawn due to toxicity associated with drug solvent [57]. Ketoconazole was frequently applied for systemic mycoses in the past, but it is now avoided due to its liver and hormonal toxicity [23]. The triazoles consist of fluconazole, itraconazole, voriconazole, posaconazole, and isavuconazole [29]. Azole antifungals inhibit the synthesis of ergosterol in the fungal cell membrane [29]. Despite this mechanism of action, azoles are generally fungistatic against yeasts, while the newer members of this subclass possess fungicidal activity against certain molds [23,48]. At present, these agents are considered the backbone of IFI therapy [23,49,58].

The most common adverse events (AEs) with all the triazoles, and especially with oral itraconazole, are nausea, vomiting, diarrhea, and abdominal pain [23,59]. Liver injury has been described also with all triazoles, ranging from mild elevations in transaminases to fatal hepatic failure [60–62]. Generally, in most cases of hepatic injury due to triazoles, normalization of the liver enzymes and resolution of the clinical symptoms occurred gradually after the discontinuation of the drug [21,63]. Additionally, triazoles are involved in numerous DDIs because they are substrates and inhibitors of CYP450 isoenzymes [63,64].

2.3.1. Fluconazole

Fluconazole, unlike the other triazoles, is characterized by high water solubility and approximately 60–80% of the drug is eliminated by the kidneys, while hepatic metabolism does not play an important role in the elimination of the drug [29]. The fluconazole dosage regimen for IFIs is guided by the indication, and the daily dose recommended by the manufacturer is up to 400 mg, but in clinical practice it usually ranges from 400 to 800 mg [49,65]. It is well tolerated, even in cases requiring long-term administration of the drug [21]. Nevertheless, up to 10% of patients treated with fluconazole developed asymptomatic liver injury, with those with AIDS or bone marrow transplantation being at greater risk [40,66–69]. Hepatic injury was typically transient and usually resolved despite drug continuation [21]. Cholestatic and mixed patterns of hepatic injury have been reported, and reinstitution of fluconazole resulted in recurrences in many cases [67,70–72]. Furthermore, there are some limited data to suggest that liver injury is dose-related [67,73]. In a large meta-analysis of antifungals tolerability and hepatotoxicity, the risk of liver injury with standard dose of fluconazole not requiring treatment discontinuation was 9.3%, while the risk of drug discontinuation due to elevated liver enzymes was 0.7% [74]. Despite the fact that the risk of acute liver failure due to fluconazole treatment is minimal [74,75], there are some case reports describing deaths attributable to liver dysfunction [66,76–78].

Few reports exist regarding the use of fluconazole in patients with pre-existing liver disease. Ruhnke et al. evaluated the PKs of a single 100 mg dose of fluconazole in 9 patients with cirrhosis, classified as group B or group C according to Child-Pugh score, compared with 10 healthy subjects [79]. They found that in cirrhotic patients the terminal elimination constant for fluconazole was lower, and that the total plasma clearance was reduced and the mean residence time increased. The authors assumed that this may be due to kidney dysfunction not reflected in creatinine clearance or the DDIs between fluconazole and diuretics that cirrhotic individuals were receiving. Nevertheless, the authors argued that dosage adjustment of fluconazole in patients with liver impairment is unnecessary, because of the wide range of values they found and the known low toxicity of fluconazole [79]. At the clinical level, Gearhart first described a 50-year old woman with hepatitis who received fluconazole for *Candida* infection and experienced worsening of liver function, which returned to baseline after discontinuation of the drug [80].

A population-based study by Lo Re et al. assessed the risk of acute liver injury with oral azole antifungals in the outpatient setting [81]. Liver aminotransferase levels and development of hepatic dysfunction were examined in 195,334 new initiators of these drugs, for a period of 182 days after the last day's supply. Fluconazole initiators were 178,879 and, among them, 7073 individuals had pre-existing liver disease. The authors found that the risk of transaminitis (liver aminotransferases > 200 U/L) and severe liver injury [international normalized ratio (INR) ≥ 1.5 and total bilirubin (TB) $> 2\times$ ULN] in patients without history of chronic liver disease was lower among users of fluconazole, compared to other azoles. Nevertheless, it should be taken into account that, with the exception of itraconazole, patients administered other azoles were probably of worse health status compared to those administered fluconazole. More interestingly, compared to patients without chronic liver disease who received fluconazole, patients with pre-existing liver disease who were treated with the same drug had higher absolute risk and incidence rate of transaminitis (p value interaction < 0.001) and of severe liver injury (p value interaction < 0.001) [81]. Whether this observation was due to fluconazole, the natural history of the disease, or both, is unclear [81]. However, no dosage adjustment is provided by the manufacturer for patients with liver impairment, although prescribing information includes a warning that fluconazole should be administered with caution to patients with hepatic dysfunction [65].

2.3.2. Itraconazole

Itraconazole is highly lipophilic, undergoes extensive hepatic metabolism, and is eliminated mostly via feces and urine [29]. It is available as capsule, oral solution, and intravenous formulation [82]. The oral solution has higher bioavailability than capsule formulation, and thus they should not be used interchangeably [83]. The adults recommended by the manufacturer dosage depends on the drug formulation and the indication, usually ranging from 200 mg to 400 mg per day, and doses above 200 mg should be divided [82,83]. However, for the treatment of certain fungal infections, such as blastomycosis and histoplasmosis, doses of 200 mg t.i.d. for 3 days and then 200 mg q.d. or b.i.d. as long-term therapy are recommended, while for coccidioidal meningitis doses up to 800 mg per day can be administrated [84–86]. Itraconazole-induced liver injury is not uncommon, and the pattern is typically cholestatic, although hepatocellular injury has been described in cases of acute liver failure [21]. In a large meta-analysis, 31.5% of patients treated with itraconazole developed hepatotoxicity, but a great variability of hepatotoxicity definition was noted in the included studies and many patients may have developed liver injury owing to the underlying IFI itself, limiting the validity of these results [87]. Treatment discontinuation due to itraconazole-induced liver injury was observed in 1.6% of patients [87]. In a more recent meta-analysis, Wang et al. estimated the risk of elevation of liver enzymes not requiring discontinuation of therapy at 17.4% among itraconazole recipients, while the respective risk of treatment discontinuation due to liver injury was 1.5% [74].

The use of itraconazole in patients with liver disease is not well studied. In a PK study, a single 100 mg dose of itraconazole was administered in 12 cirrhotic and 6 healthy individuals [88]. Compared with healthy volunteers, a statistically significant reduction in C_{max} and an increase in the elimination half-time of the drug were observed in patients with cirrhosis. Nevertheless, based on the area under the curve (AUC), cirrhotic and healthy individuals had comparable overall exposure to the drug [88]. In the already mentioned observational study of Lo Re et al., 55 patients with chronic liver disease received itraconazole, and onychomycosis was the most common indication for treatment initiation [81]. Interestingly, none of them developed transaminitis or severe acute liver injury [81]. The fact that, in this study, itraconazole was prescribed mainly for a less severe condition such as onychomycosis and probably in lower doses than those recommended for severe IFIs treatment, may be the reasons for its decreased hepatotoxic potential, compared with what has been observed in other studies which included patients with severe fungal infections and multiple comorbidities. No dose adjustment is available for patients with hepatic impairment, but it is recommended that these patients should be carefully monitored when treated with itraconazole [83]. Apart from the periodic assessment of a patient's liver enzymes levels while on itraconazole, TDM is generally recommended, in order to assure adequate exposure and to minimize potential toxicities [55,58,82,89,90].

2.3.3. Voriconazole

Voriconazole's chemical structure is similar to fluconazole, but its spectrum of activity is much broader [48]. It is metabolized by CYP450, mainly CYP2C19, which exhibit significant genetic polymorphism, and it is involved in many DDIs. In addition, recent data suggest that voriconazole metabolism can be inhibited in cases of severe inflammation [91]. It is available as tablet, oral suspension, and intravenous solution [92]. The manufacturer's recommended dose of intravenous formulation for most IFIs is 6 mg/kg b.i.d. on day 1 as a loading dose, followed by 4 mg/kg b.i.d. as a maintenance dose [92,93]. The oral dose for adult patients is 400 mg b.i.d. on the first day followed by 200 mg b.i.d., while if patient response is inadequate, the maintenance dose may be increased from 200 mg b.i.d. to 300 mg b.i.d. [92,93]. A 50% reduction of both loading and maintenance oral doses is recommended for adult patients with a body weight less than 40 kg [92,93]. The incidence of liver injury in patients treated with voriconazole varies significantly among studies, depending mostly on the characteristics of the study population, while the pattern of liver enzyme abnormality is not uniform [94–97]. Wang et al. found in their meta-analysis that 19.7% of 881 patients who received voriconazole developed elevation of liver enzymes without the need for treatment discontinuation [74]. A more recent meta-analysis of the utility of voriconazole's TDM included 11 studies and reported a pooled incidence rate of liver injury among voriconazole recipients at 5.7% [98].

Compared with other triazoles, more data exist regarding the use of voriconazole in patients with underlying hepatic impairment. After a single oral dose of 200 mg of voriconazole in 12 patients with mild to moderate hepatic impairment (Child–Pugh Classes A and B), AUC was 3.2-fold higher than in age and weight matched controls with normal liver function [92]. In an oral multiple-dose PK study, AUC at steady state ($AUC\tau$) was similar in individuals with Child–Pugh Class B cirrhosis given a maintenance dose of 100 mg twice daily and individuals with normal liver function given 200 mg twice daily [99]. Based on the aforementioned data, the medication label of voriconazole recommends that individuals with mild to moderate cirrhosis (Child–Pugh Class A and B) receive the same loading dose as individuals with hepatic function, but half the maintenance dose, while no recommendation is given for individuals with Child–Pugh Class C cirrhosis [92].

In a cohort study of 29 patients with severe liver dysfunction, defined as MELD score > 9, who received at least four doses of voriconazole, a deterioration of hepatic biochemistry was observed in 69% of them [100]. The pattern of the liver injury was mixed; hepatocellular and cholestatic in 45%, 35% and 15% of patients, respectively. None of them developed clinical or laboratory signs of worsening hepatic function. The biochemical parameters returned to baseline levels in all patients after the cessation of voriconazole treatment [100]. Lo Re et al included in their study 97 patients with pre-existing liver disease who received oral voriconazole. Among them, 4 developed ALT or AST > 200 U/L and 2 developed severe liver injury (INR > 1.5 and TB > 2 × ULN), but none of them experienced acute liver failure. Individuals with pre-existing liver disease treated with voriconazole had higher rates of severe liver injury than recipients of voriconazole without underlying hepatic disease [81]. A recent single-center retrospective study compared 6 patients with severe liver cirrhosis (Child–Pugh Class C) who were treated with oral voriconazole based on TDM, with 56 individuals without severe liver cirrhosis who received voriconazole in the recommended dosage for IFIs, also under TDM [101]. The daily maintenance doses of voriconazole of the severe cirrhotic patients were in the range of 50 to 200 mg, with a median daily dose at one-third of the median daily dose of the individuals without severe cirrhosis. The median trough serum concentration of the drug was within recommended levels in both groups of patients. Thus, the authors argued that a dose reduction to about one-third that of the standard maintenance dose is required in patients with Child–Pugh Class C cirrhosis [101].

A multicenter retrospective study aimed to investigate the voriconazole trough concentrations and safety in cirrhotic patients receiving the drug [102]. Seventy-eight patients with Child–Pugh Class B or C cirrhosis who had been treated with voriconazole under TDM were allocated to two groups, according to the dosage regimen they had received. Patients in the first group had received the recommended dosage by the manufacturer or a fixed dose of 200 mg twice daily. Patients in the second group had received a loading dose of 200 mg twice daily on day 1, followed by 100 mg twice daily, or a fixed dose of 100 mg twice daily. The steady-state trough concentration of voriconazole was measured in all patients and its relationship with AEs was analyzed. Voriconazole C_{min} values were significantly different between the two groups, and the proportion of C_{min} higher than the super-therapeutic concentration (defined as 5 mg/L) was 63% in the first group and 28% in the second group of patients. While no statistically significant differences were observed in the incidence of AEs between the two groups, these incidences were considered excessively high (26.5% of patients in the first group and 15.9% of patients in the second group). Interestingly, voriconazole C_{min} between patients with an AE and those without AEs in both groups was similar. However, based on the high C_{min} and incidence of AEs in these patients, both the recommended maintenance dose and halved maintenance dose were considered as inappropriately high [102].

The same authors conducted another study including solely patients with Child–Pugh Class C cirrhosis [103]. Patients were allocated to two groups, according to the dosage schedule of voriconazole's maintenance dose. The first group included those who received 100 mg of voriconazole twice daily, while the second group included those who received 200 mg of voriconazole once daily. There was no significant difference in voriconazole C_{min} between the two groups. However, the proportion of voriconazole C_{min} higher than the upper limit of therapeutic level (defined again as 5 mg/L) in the first and second groups was 34% and 48%, respectively. The incidence of AEs was 21% in the first group and 27% in the second group, with no statistically significant difference. Further analysis revealed that the increasing C_{min} of voriconazole was associated with increasing incidence of AEs, although no statistical significance was found. It was suggested that in patients with Child–Pugh Class C cirrhosis the halved maintenance dose is probably inappropriate, and that lower dosage should be considered in conjunction with early TDM [103].

Voriconazole TDM is generally recommended because of its highly variable PKs, in order to enhance efficacy, to evaluate therapeutic failure due to possible suboptimal drug exposure, and to avoid associated toxicity due to increased serum drug levels [55,58,104]. It is well established in the literature that an elevated drug's level in the serum is correlated with increased risk of toxicity [104–106]. Thus, voriconazole TDM is of paramount importance in patients with pre-existing liver disease, since the drug is extensively metabolized by the liver and this population is more difficult to tolerate a deterioration of hepatic function due to voriconazole-induced liver injury [101–103,107]. Various target trough concentrations associated with efficacy and safety have been reported, and most experts aim for voriconazole trough serum concentration of more than 1–1.5 μg/mL for efficacy but less than 5–6 μg/mL for avoiding toxicity [58,89,98,104].

2.3.4. Posaconazole

Posaconazole's chemical structure resembles that of itraconazole, but it has a wider antimycotic spectrum [29]. Initially, posaconazole was available only as an oral suspension which displays poor and highly variable absorption [108]. Recently, tablet and intravenous formulations with improved bioavailability were approved [109–111]. Posaconazole is metabolized in the liver by UDP-glucuronic-transferase, usually without previous oxidation by CYP450, and is eliminated mainly in the feces and, secondarily, in the urine [112]. Noticeably, posaconazole is a potent inhibitor of CYP3A4, thus clinically relevant DDIs may occur [29]. Regarding IFIs, the adult recommended therapeutic dose for oral suspension is 200 mg q.i.d., while the prophylactic dose is 200 mg t.i.d. [113,114]. In addition, for both tablet and intravenous formulation a loading dose of 300 mg b.i.d. on day 1, followed by a maintenance dose of 300 mg once daily, is recommended as prophylactic as well as therapeutic dosage regimen for several IFIs [113,114]. Liver injury occurs in up to 25% of patients receiving posaconazole regardless of the formulation, but this may be multifactorial and not only attributable to the drug [81,115–118]. The dominant pattern of hepatic injury varies among studies, partly depending on the studied population [21,115–117]. In addition, hepatic failure due to posaconazole treatment is generally uncommon [81,110,111,115–117].

Regarding the use of posaconazole in individuals with pre-existing hepatic impairment, Moton et al. conducted a PK study to evaluate the need of posaconazole dose adjustment in this population [119]. In their single-center study, the researchers aimed to compare the PKs of a single dose 400 mg of posaconazole oral suspension in 19 patients with varying degrees of hepatic dysfunction with 18 matched healthy individuals who received the same regimen. No clear trend was observed of an increase or decrease in posaconazole exposure linked with increasing degrees of hepatic dysfunction. The detected differences of PKs between healthy individuals and those with hepatic dysfunction were not clinically significant, and the authors suggested that posaconazole dosage adjustment may not be required in individuals with hepatic impairment [119]. A case-report also described a patient with Child–Pugh Class B cirrhosis suffering from maxillary mucormycosis who, after surgical debridement and initial treatment with AmB followed by itraconazole, was successfully treated with oral posaconazole suspension 400 mg twice daily for nine months without hepatic decompensation [120]. In addition, Lo Re et al included in their observational study 9 patients with chronic liver disease who received posaconazole, and only one of them developed severe acute liver injury (INR > 1.5 and TB > 2× ULN) [81].

In a recent single-center retrospective cohort study, Tverdek et al. assessed the real-life safety and effectiveness of primary antifungal prophylaxis with new tablet and intravenous posaconazole formulations in high-risk patients with leukemia and/or hematopoietic stem cell transplantation (HSCT) [116]. A total of 343 patients were included, 62% of whom received 300 mg of posaconazole twice daily on day 1, while 99% received the maintenance dose of 300 mg per day. Among them, 316 patients had baseline liver assessment, including 144 patients with baseline elevations of ALT, ALP, or/and TB, of which 23 had grade 3 or 4 liver injury [121]. Concerning the 121 patients with baseline liver injury but no grade 3 or 4 abnormalities, 34 (28%) of them developed grade 3 or 4 liver injury. Liver abnormalities were developed in nearly 20% of all patients, primarily manifested as hyperbilirubinemia. These abnormalities were more frequent in individuals with pre-existing liver injury, but this may not be solely due to DILI, as the underlying disease and concomitant drugs may also have contributed [116].

Noticeably, in patients with new-onset hepatotoxicity due to voriconazole administration for IFIs, sequential use of posaconazole seems to be safe and effective, with favorable outcomes and improvement of liver biochemistry in most of the cases [122–124]. Independently of the acute or chronic nature of pre-existing liver injury, no dosage adjustments are recommended for individuals with hepatic impairment treated with posaconazole [113]. In addition, while many guidelines recommend TDM in patients receiving posaconazole oral suspension for IFI prophylaxis or treatment to confirm adequate absorption and ensure efficacy [58,89], PK/PD analyses conducted with oral posaconazole suspension do not support a relationship between plasma concentrations and toxicity [125,126]. On the contrary, Tverdek et al identified a potential association between elevated serum posaconazole levels and hepatotoxicity in patients treated with the new tablet and intravenous formulations of the drug, but further evaluation is needed [116].

2.3.5. Isavuconazole

Isavuconazole is the newest member of triazoles antifungals. In both oral and intravenous formulations, it is administered as a water-soluble prodrug, isavuconazonium sulfate [127]. After intravenous administration, the prodrug is rapidly hydrolyzed to isavuconazole by plasma esterases, while oral formulation of isavuconazonium sulfate sustains chemical hydrolysis in the gastrointestinal lumen [112]. Metabolism of isavuconazole takes place in the liver by CYP450 isoenzymes, with subsequent glucuronidation by uridine diphosphate-glucuronosyl transferase (UGT) [127]. Isavuconazole is generally well tolerated and safe, and has fewer DDIs compared with voriconazole and posaconazole, but clinical experience is still limited [60,61]. It is approved by the FDA and the European Medicine Agency (EMA) for the treatment of adult patients with invasive aspergillosis or invasive mucormycosis, with a loading dose of 200 mg t.i.d. for the first two days, followed by a maintenance dose of 200 mg q.d., via oral or intravenous administration [127,128]. Elevations in liver enzymes have been reported in clinical trials but they are generally reversible and rarely only require treatment discontinuation [129–131]. However, cases of severe liver injury have occurred during treatment with this antifungal agent [127,129]. In a phase 3 comparative study evaluating isavuconazole versus voriconazole for the treatment of invasive aspergillosis, there were significantly higher liver disorders in the voriconazole arm (p value = 0.016), but the protocol of the study did not allow TDM [131]. Since voriconazole displays highly variable non-linear pharmacokinetics in adults and, thus, TDM is recommended, these results should be interpreted with caution, and further research is needed.

An initial single-dose PK study aimed to assess the effect of mild to moderate hepatic impairment due to alcoholic cirrhosis on the disposition of isavuconazole [132]. Clearance values of isavuconazole were significantly decreased and half-life values were significantly increased in cirrhotic patients compared with healthy individuals, leading the authors to recommend a 50% decrease in the maintenance dose of the drug for patients with mild or moderate liver disease [132]. However, a subsequent population PK analysis used data from the aforementioned study and from another study and reported different results [133]. The PK and safety results showed that dose adjustment appears to be unnecessary for patients with Child–Pugh Class A or Class B cirrhosis treated with isavuconazole, since there was a less than twofold increase in trough concentrations for those compared with healthy subjects, while the AEs profile was similar between cirrhotic and healthy individuals [133].

Notwithstanding, both these PK studies did not take PD into consideration, which may affect the dose of isavuconazole against different fungi in this population of patients. In a recently published PK/PD study, Zheng et al. examined the efficacy of various isavuconazole dosing regimens for healthy individuals and patients with renal and hepatic impairment, namely Child-Pugh Class A or B cirrhosis, against *Aspergillus* spp. and other fungi [134]. The Monte Carlo simulation was used in each scenario to calculate target attainment and cumulative fractions of response probabilities. The clinically recommended dose of 200 mg isavuconazole per day was effective for all individuals against *A. fumigatus*, *A. flavus*, *A. nidulans*, *A. terreus*, and *A. versicolor*. [134].

In the manufacturer's labeling, the standard dose of isavuconazole is recommended for patients with mild or moderate liver dysfunction, while the drug has not been studied in patients with Child–Pugh Class C hepatic impairment, and should be used in these individuals only when the benefits outweigh the risks [127]. Although TDM of isavuconazole may be considered in selected patients, such as those with severe hepatic impairment, routine TDM for isavuconazole is not recommended [135].

2.4. Echinocandins

Echinocandins inhibit the synthesis of 1,3-β-D-glucan, a fungal cell wall component, resulting in instability of the cell wall, cell lysis, and death [136]. The fact that this class of antifungals agents targets the fungal cell wall and not the cell membrane explains the absence of cross-reactivity with mammalian cells and the excellent tolerability of this class of compounds in humans [48]. They are fungicidal to *Candida*, including several non-albicans strains, and fungistatic to *Aspergilli*, thus they are considered the first-line treatment for *Candida* spp. infections [29,49]. At present, the available agents of this class include caspofungin, micafungin, and anidulafungin [23]. Common AEs related with echinocandins treatment include phlebitis, nausea, diarrhea, headache and pruritus, but also other drug reactions such as leukopenia, anemia, hypokalemia, and liver injury have been reported [29]. Noticeably, the echinocandins have less than half the likelihood of discontinuation of therapy due to AEs, compared with triazoles [137].

2.4.1. Caspofungin

Caspofungin bounds to plasma proteins at 95%; it is transformed in the liver but only minimally undergoes degradation by CYP450 isoenzymes, and the metabolites are eliminated via urine [138,139]. The recommended dosage for adults is 70 mg as a single loading dose on day 1, followed by a maintenance dose of 50 mg once daily [140,141]. The EMA recommends an increase of maintenance dose to 70 mg daily when patient's body weight exceeds 80 kg [140]. Generally, hepatic abnormalities related to caspofungin treatment are uncommon and severe hepatic AEs are rare [21]. In most studies, elevated hepatic enzymes were observed in up to 9% of patients, and they were often clinically irrelevant [24].

Regarding patients with pre-existing liver disease treated with caspofungin, Mistry et al. conducted single- and multiple-dose open-label studies to assess dosage and safety of caspofungin in hepatic impairment [142]. Patients with Child–Pugh score 5–6 or 7–9 hepatic impairment were matched with healthy individuals. Patients with Child–Pugh score 5–6 hepatic impairment had a mild elevation in caspofungin serum concentration, which was considered as clinically irrelevant. Patients with Child–Pugh score 7–9 hepatic impairment needed a reduced maintenance dose of caspofungin in order to achieve drug concentrations similar with the healthy individuals in the control group [142]. Based mainly on these data, a reduction of caspofungin maintenance dose from 50 mg to 35 mg per day is recommended for patients with Child–Pugh Class 7–9 hepatic impairment, while no recommendation is given for patients with Child–Pugh score 10–15 hepatic impairment [141].

However, Spriet et al. initially described a patient with Child-Pugh Score 9 cirrhosis diagnosed with acute myeloid leukemia, who was treated for a severe IFI with a full dose of caspofungin 70 mg per day, since his body weight was over 80 kg [143]. The PK data of this case-report indicated that if the reduced dose of caspofungin had been used, it would probably have resulted in a low caspofungin systemic exposure and a possible therapeutic failure [143]. A subsequent population PK analysis concluded that a reduction of caspofungin maintenance dose in non-cirrhotic intensive-care unit (ICU) patients, who are misclassified due to hypoalbuminemia as with Child–Pugh Class B hepatic impairment, is not recommended, because it may result in significantly lower drug exposure and possible therapeutic failure [144]. On the contrary, authors suggested that, depending on pathogens MIC, a caspofungin maintenance dose of 70–100 mg daily may be reasonable in many cases [144].

Furthermore, data from the aforementioned population PK analysis in non-cirrhotic ICU patients were used in another PK study of a single-dose of 70 mg of caspofungin in patients with decompensated Child–Pugh Class B or C cirrhosis to evaluate the impact of cirrhosis and hepatic impairment severity on the PK of the drug [145]. Remarkably, their data showed that cirrhosis had a limited impact on clearance of caspofungin. Also, it was the first study providing PK data of caspofungin for patients with Child–Pugh Class C cirrhosis and compared with patients with Child–Pugh Class B cirrhosis, no further decrease of caspofungin clearance was observed in the former group of individuals. Thus, the researchers concluded that reducing the dose of caspofungin in patients with Child–Pugh Class B or C cirrhosis leads to a decrease in exposure and this may result in a suboptimal clinical outcome [145]. In another recent PK study for general patients, ICU patients, and patients with hepatic impairment receiving caspofungin, a whole-body physiology-based PK model was developed and was combined with Monte Carlo stimulation to optimize dosage regimen of the drug in patients with different characteristics [146]. The results of this study indicated that the caspofungin maintenance dose should not be reduced to 35 mg per day for ICU patients classified as Child–Pugh Class B when this classification is driven by hypoalbuminemia, as lower drug exposure occurs. On the contrary, authors argued that, in any other case, a reduction of caspofungin maintenance dose to 35 mg per day for patients with moderate hepatic impairment classified as Child–Pugh Class B, may be reasonable [146].

2.4.2. Micafungin

Micafungin is highly bound to proteins, it is metabolized in the liver by enzymes unrelated to CYP450, and the metabolites are excreted primarily via feces [147]. The recommended dosage for patients weighing greater than 40 kg is 100 once daily for the treatment of invasive candidiasis, and 150 mg once daily for the treatment of *Candida* esophagitis [148,149]. It is a well-tolerated antifungal agent with few AEs requiring cessation of the drug [21]. Mild elevations of hepatic enzymes may occur, but clinically overt liver toxicity is rare [23,150]. Nevertheless, rat models demonstrated an association between micafungin and foci of altered hepatocytes and hepatocellular tumors when this was given for more than 3 months, but this finding has not been replicated in humans [23,29].

Micafungin has a low hepatic extraction ratio with high protein binding in plasma, and while its total plasma concentration may decrease in some clinical cases, the unbound fraction of the drug is likely to remain stable [151,152]. A phase I parallel group open-label PK study of a single-dose of micafungin included 8 patients with Child–Pugh Score 7–9 hepatic dysfunction and did not find significant difference in unbound plasma concentration of the drug compared with healthy controls, while a lower AUC was found in the patients with hepatic impairment [153]. The latter was attributed to the differences in body weight among patients, and no dose adjustment was recommended [153]. In an another open-label single-dose PK study, 8 patients with Child–Pugh score 10-12 hepatic impairment and 8 healthy individuals received 100 mg of micafungin [154]. Compared with healthy subjects, patients with hepatic dysfunction had lower C_{max} and AUC values, but the magnitude of differences was considered as clinically meaningless and no dose reduction was recommended in patients with severe hepatic impairment [154]. In addition, Luque et al. conducted a prospective observational study to assess the possibility of DILI due to micafungin use in daily practice including 12 patients, 8 of whom had elevated liver enzymes at the beginning of the treatment [155]. The daily dose of micafungin was 100 mg for 10 patients and 150 mg for the remaining two. There was no correlation between the degree of the pre-existing liver injury and micafungin levels. In steady state, C_{max} and C_{min} were similar in subjects with and without initial liver abnormalities. Hepatic enzymes levels remained stable or even improved in all but one patient. These results further support the safety of micafungin in patients with pre-existing liver injury and IFIs [155]. Based on most of the aforementioned studies, the summary of manufacturers' product characteristics approved by the FDA recommends that no dosage adjustment is required in patients with hepatic impairment [149]. Contrarily, EMA recommends avoidance of micafungin use in patients with severe hepatic impairment, while it has issued a black-box warning for hepatotoxicity and potential for liver tumors [148].

2.4.3. Anidulafungin

Anidulafungin has a very high protein binding of 99%; it is degraded non-hepatically in the blood, and the metabolites are eliminated via feces [156]. The recommended adult dosage for invasive candidiasis is a single loading dose of 200 mg on day 1, followed by a maintenance dose of 100 mg once daily [157,158]. Anidulafungin AEs, including DILI, are generally infrequent [159,160]. With regard to patients with pre-existing hepatic disease treated with this antifungal agent, Dowel et al. conducted a phase I, open-label, single-dose study including 20 patients with varying degrees of hepatic impairment and 7 healthy controls [161]. No statistically significant differences in PK parameters were observed between healthy controls and patients with mild or moderate hepatic impairment. However, compared with healthy controls, subjects with severe hepatic impairment (Child–Pugh Class C) showed statistically significant decreases in C_{max} and AUC values, most likely secondary to ascites and edema, but anidulafungin exposure remained significantly above MIC_{90} of many common fungal pathogens. Additionally, the values of all PK parameters still remained within the range that had been previously reported in healthy subjects. No evidence of dose-depended toxicity or serious AEs was observed. Thus, the authors suggested that anidulafungin can be safely used in patients with hepatic dysfunction without dosage adjustment [161].

In a retrospective cohort study, Verma et al. assessed the safety and efficacy of anidulafungin in the treatment of IFIs in patients with hepatic impairment or multiorgan failure [162]. Fifty patients were included, among them 30 with a calculated baseline MELD score, of whom 13 had a score $\geq$ 30. A dose of 200 mg was given to all patients on day 1, followed by 100 mg per day onwards. Before initiation of treatment with anidulafungin, at least one abnormal liver function test (LFT) was observed in 49 of 50 patients (98%). During treatment, LFTs worsened in many patients, but fewer patients had elevated LFTs at the completion of treatment than at the beginning. A favorable outcome was seen in more than 75% of patients. The latter further supports indications that anidulafungin is efficacious and safe in patients with decompensated hepatic disease and, in agreement with package insert recommendations, no dosage reduction is needed in patients with any degree of hepatic impairment [157,162].

3. Clinical Implications and Future Directions

Patients treated with antifungal agents for IFIs may have underlying hepatic impairment of varying degrees and origin. Clinicians should be aware of that, since it further complicates management with regard to efficacy and safety of the antifungal therapy. Firstly, metabolism and elimination of many antifungals are significantly altered by hepatic dysfunction, while DDIs are somewhat unpredictable compared to individuals with intact liver function. Moreover, it may be difficult to attribute further deterioration of liver biochemistry or function only to antifungals in patients with severe comorbidities and concomitant administration of other hepatotoxic drugs. In addition, precise estimates of hepatic function are currently unavailable. The Child–Pugh system, on which most dosage modifications in hepatic impairment are based, was initially developed to assess the prognosis of chronic liver disease and not the degree of hepatic dysfunction [20]. For all the above reasons, the optimal use of antifungals in patients with pre-existing liver disease with IFIs is still unfolding. Data discussed in the present review give rise to useful clinical suggestions for the optimization of treatment. Table 2 summarizes the dosage adjustments of antifungal agents that are approved and recommended by FDA and/or EMA for patients with hepatic impairment treated for IFIs, and also presents the recommendations included in many guidelines regarding TDM of certain antifungal drugs for optimizing efficacy and safety.

Table 2. Antifungal agent dosage adjustment for patients with hepatic impairment.

Antifungal Agent	Severity of Hepatic Impairment by Child–Pugh Score		
	Score 5–6 (Class A)	**Score 7–9 (Class B)**	**Score 10–15 (Class C)**
AmB preparations	No recommendations available		
Flucytosine	No recommendations available, use with caution, TDM recommended Authors' comment: extra caution when combined with AmB preparations		
Fluconazole	No recommendations available, use with caution		
Itraconazole	No recommendations available, strongly discouraged unless benefit exceeds risk, use with caution and under close monitoring, TDM is recommended		
Voriconazole	50% reduction of maintenance dosage and TDM are recommended		No recommendations available, use only if benefit outweighs risk, close monitoring and TDM are recommended Authors' comment: reduction of maintenance dosage to about one-third may be considered
Posaconazole	No dosage adjustment is recommended, TDM when oral suspension is used Authors' comment: TDM may also be considered when tablet or intravenous drug formulation is used		
Isavuconazole	No dosage adjustment is recommended		No recommendations available, use only if benefit outweighs risk
Caspofungin	No dosage adjustment is recommended	Reduced maintenance dose from 50 mg to 35 mg daily Authors' comment: in critically ill patients, reduced dosage may lead to decreased drug exposure	No recommendations available
Micafungin	No dosage adjustment is recommended		US FDA recommends no dosage adjustment, EMA recommends avoidance of its use
Anidulafungin	No dosage adjustment is recommended		

AmB: amphotericin B; TDM: therapeutic drug monitoring; US FDA: United States Food and Drug Administration; EMA: European Medicines Agency.

With regard to AmB, to date few data exist on the necessity for dosage adjustment of any AmB formulations in patients with hepatic impairment. However, the lipid formulations of the drug seem to have a higher potential for hepatotoxicity compared to AmBD. In addition, AmB formulations combined with flucytosine for the treatment of certain fungal infections may lead to increased flucytosine serum levels due to kidney injury and accumulation of the renally eliminated drug. Flucytosine TDM is of clinical importance generally, in order to assure efficacy and to prevent AEs, including hepatotoxicity.

Fluconazole dosage modification for hepatic impairment per se is not required. Nevertheless, it should be used cautiously in this subset of patients due to the increased risk of further deterioration of hepatic enzymes levels and/or hepatic function compared to subjects with normal liver function. For itraconazole there are no dosage adjustment recommendations available for patients with hepatic dysfunction, however its use is discouraged in this subset of patients unless benefit exceeds risk. In the latter case, close monitoring, including TDM, is recommended, but further work is necessary for establishing clear drug target levels.

Use of voriconazole has also an increased risk for severe live injury in patients with chronic liver disease. While reduction of voriconazole's maintenance dose by 50% is recommended in patients with Child–Pugh Class A or B cirrhosis, data for patients with more severe hepatic impairment were lacking until recently. New evidence suggests that dose should be lowered more than 50% in patients with Child–Pugh Class C hepatic dysfunction, and always under TDM for safety and efficacy enhancement [101–103,163]. However, optimal dosage in this setting has not formally been defined and this is a noteworthy area of active research. Likewise, posaconazole and isavuconazole have not been studied sufficiently in patients with severe hepatic impairment and more research on that topic is of paramount importance. Furthermore, only recently a possible relationship between increased posaconazole serum levels and liver toxicity was identified in patients receiving the new intravenous and tablet drug formulations, thus more PK studies are needed, especially in patients with underlying liver disease [116]. Regarding isavuconazole, generally it demonstrates a favorable safety profile in relation to DDIs and hepatotoxicity. Nevertheless, compared with other triazoles, published clinical experience and post-marketing data, including its use in special patient populations, are still limited.

Compared with triazoles, echinocandin use in patients with underlying hepatic impairment is considered relatively safe. A reduction to caspofungin maintenance dose is recommended for patients classified with Child–Pugh Score 7–9 hepatic dysfunction, yet this has been challenged recently and clinicians should be aware of that, since it may result in suboptimal exposure in critically ill patients [144–146]. With regard to micafungin, no dosage modification is recommended in mild and moderate hepatic insufficiency, but additional research seems necessary for patients with severe hepatic impairment. Among this class of antifungal agents, anidulafungin may have an advantage for use in cirrhotic patients due to its non-hepatic metabolism, more predictable PK, and favorable tolerability. However, this remains to be further evaluated with future comparative studies in this subset of patients.

4. Conclusions

Treatment of IFIs in patients with pre-existing liver disease poses a significant challenge for clinicians. These patients are often more vulnerable to the hepatotoxic potential of many antifungal agents, while possible alterations of the PKs of these drugs may trigger adverse effects not localized only to the liver. Current evidence from PK studies and safety data from the existing clinical trials and post-marketing studies can help physicians optimize IFIs treatment in this special group of patients. However, most of the existing evidence is limited to subjects with mild to moderate hepatic disease, and clear recommendations for dosage adjustments in cases of severe hepatic impairment are not yet available for the majority of antifungal agents. This raises the need for more PK and clinical studies in this subset of patients. Furthermore, additional attention should be paid to future pharmacovigilance monitoring of antifungal agent use in patients with liver disease of any degree. In any case, close clinical and laboratory monitoring, including TDM for specific antifungal drugs, is essential in the majority of these patients in order to prevent or promptly recognize further deterioration of the hepatic function, thus avoiding unfavorable outcomes.

References

1. Limper, A.H.; Adenis, A.; Le, T.; Harrison, T.S. Fungal infections in HIV/AIDS. *Lancet Infect. Dis.* **2017**, *17*, e334–e343. [CrossRef]
2. Colombo, A.L.; de Almeida Júnior, J.N.; Slavin, M.A.; Chen, S.C.A.; Sorrell, T.C. Candida and invasive mould diseases in non-neutropenic critically ill patients and patients with haematological cancer. *Lancet Infect. Dis.* **2017**, *17*, e344–e356. [CrossRef]

3. Kontoyiannis, D.P. Invasive mycoses: Strategies for effective management. *Am. J. Med.* **2012**, *125*, S25–S38. [CrossRef] [PubMed]
4. Rodighiero, V. Effects of liver disease on pharmacokinetics. An update. *Clin. Pharmacokinet.* **1999**, *37*, 399–431. [CrossRef] [PubMed]
5. Gupta, N.K.; Lewis, J.H. Review article: The use of potentially hepatotoxic drugs in patients with liver disease. *Aliment. Pharmacol. Ther.* **2008**, *28*, 1021–1041. [CrossRef] [PubMed]
6. Lewis, J.H.; Stine, J.G. Review article: Prescribing medications in patients with cirrhosis—A practical guide. *Aliment. Pharmacol. Ther.* **2013**, *37*, 1132–1156. [CrossRef] [PubMed]
7. Palatini, P.; De Martin, S. Pharmacokinetic drug interactions in liver disease: An update. *World J. Gastroenterol.* **2016**, *22*, 1260–1278. [CrossRef]
8. Navarro, V.J.; Senior, J.R. Drug-related hepatotoxicity. *New Eng. J. Med.* **2006**, *354*, 731–739. [CrossRef]
9. Lee, W.M. Drug-induced hepatotoxicity. *New Eng. J. Med.* **2003**, *349*, 474–485. [CrossRef]
10. Lo Re, V.; Haynes, K.; Goldberg, D.; Forde, K.A.; Carbonari, D.M.; Leidl, K.B.F.; Hennessy, S.; Reddy, K.R.; Pawloski, P.A.; Daniel, G.W.; et al. Validity of diagnostic codes to identify cases of severe acute liver injury in the U.S. Food and Drug Administration's Mini-Sentinel Distributed Database. *Pharmacoepidemiol. Drug Saf.* **2013**, *22*, 861–872. [CrossRef]
11. Bernal, W.; Wendon, J. Acute Liver Failure. *New Eng. J. Med.* **2013**, *369*, 2525–2534. [CrossRef] [PubMed]
12. Ortega-Alonso, A.; Stephens, C.; Lucena, M.I.; Andrade, R.J. Case Characterization, Clinical Features and Risk Factors in Drug-Induced Liver Injury. *Int. J. Mol. Sci.* **2016**, *17*, 714. [CrossRef] [PubMed]
13. Kullak-Ublick, G.A.; Andrade, R.J.; Merz, M.; End, P.; Benesic, A.; Gerbes, A.L.; Aithal, G.P. Drug-induced liver injury: Recent advances in diagnosis and risk assessment. *Gut* **2017**, *66*, 1154–1164. [CrossRef] [PubMed]
14. Alempijevic, T.; Zec, S.; Milosavljevic, T. Drug-induced liver injury: Do we know everything? *World J. Hepatol.* **2017**, *9*, 491–502. [CrossRef] [PubMed]
15. Aithal, G.P.; Watkins, P.B.; Andrade, R.J.; Larrey, D.; Molokhia, M.; Takikawa, H.; Hunt, C.M.; Wilke, R.A.; Avigan, M.; Kaplowitz, N.; et al. Case definition and phenotype standardization in drug-induced liver injury. *Clin. Pharmacol. Ther.* **2011**, *89*, 806–815. [CrossRef]
16. Ahmad, J.; Odin, J.A. Epidemiology and Genetic Risk Factors of Drug Hepatotoxicity. *Clin. Liver Dis.* **2017**, *21*, 55–72. [CrossRef]
17. Temple, R. Hy's law: Predicting serious hepatotoxicity. *Pharmacoepidemiol. Drug Saf.* **2006**, *15*, 241–243. [CrossRef]
18. Lewis, J.H. The Art and Science of Diagnosing and Managing Drug-induced Liver Injury in 2015 and Beyond. *Clin. Gastroenterol. Hepatol.* **2015**, *13*, 2173–2189. [CrossRef]
19. Pena, M.A.; Horga, J.F.; Zapater, P. Variations of pharmacokinetics of drugs in patients with cirrhosis. *Expert Rev. Clin. Pharmacol.* **2016**, *9*, 441–458. [CrossRef]
20. Cota, J.M.; Burgess, D.S. Antifungal Dose Adjustment in Renal and Hepatic Dysfunction: Pharmacokinetic and Pharmacodynamic Considerations. *Curr. Fungal Infect. Rep.* **2010**, *4*, 120–128. [CrossRef]
21. Tverdek, F.P.; Kofteridis, D.; Kontoyiannis, D.P. Antifungal agents and liver toxicity: A complex interaction. *Expert Rev. Anti Infect. Ther.* **2016**, *14*, 765–776. [CrossRef] [PubMed]
22. Pea, F.; Lewis, R.E. Overview of antifungal dosing in invasive candidiasis. *J. Antimicrob. Chemother.* **2018**, *73*, i33–i43. [PubMed]
23. Mourad, A.; Perfect, J.R. Tolerability profile of the current antifungal armoury. *J. Antimicrob. Chemother.* **2018**, *73*, i26–i32. [CrossRef] [PubMed]
24. Kyriakidis, I.; Tragiannidis, A.; Munchen, S.; Groll, A.H. Clinical hepatotoxicity associated with antifungal agents. *Expert Opin. Drug Saf.* **2017**, *16*, 149–165. [CrossRef]
25. Bader, J.C.; Bhavnani, S.M.; Andes, D.R.; Ambrose, P.G. We can do better: A fresh look at echinocandin dosing. *J. Antimicrob. Chemother.* **2018**, *73*, i44–i50. [CrossRef] [PubMed]
26. Utz, J.P.; Treger, A.; Mc, C.N.; Emmons, C.W. Amphotericin B: Intravenous use in 21 patients with systemic fungal diseases. *Antibiot. Annu.* **1958**, *6*, 628–634.
27. Loo, A.S.; Muhsin, S.A.; Walsh, T.J. Toxicokinetic and mechanistic basis for the safety and tolerability of liposomal amphotericin B. *Expert Opin. Drug Saf.* **2013**, *12*, 881–895. [CrossRef]
28. Brajtburg, J.; Bolard, J. Carrier effects on biological activity of amphotericin B. *Clin. Microbiol. Rev.* **1996**, *9*, 512–531. [CrossRef]

29. Bellmann, R.; Smuszkiewicz, P. Pharmacokinetics of antifungal drugs: Practical implications for optimized treatment of patients. *Infection* **2017**, *45*, 737–779. [CrossRef]
30. Steimbach, L.M.; Tonin, F.S.; Virtuoso, S.; Borba, H.H.; Sanches, A.C.; Wiens, A.; Fernandez-Llimos, F.; Pontarolo, R. Efficacy and safety of amphotericin B lipid-based formulations—A systematic review and meta-analysis. *Mycoses* **2017**, *60*, 146–154. [CrossRef]
31. Hamill, R.J. Amphotericin B formulations: A comparative review of efficacy and toxicity. *Drugs* **2013**, *73*, 919–934. [CrossRef] [PubMed]
32. Johnson, P.C.; Wheat, L.J.; Cloud, G.A.; Goldman, M.; Lancaster, D.; Bamberger, D.M.; Powderly, W.G.; Hafner, R.; Kauffman, C.A.; Dismukes, W.E. Safety and efficacy of liposomal amphotericin B compared with conventional amphotericin B for induction therapy of histoplasmosis in patients with AIDS. *Ann. Intern. Med.* **2002**, *137*, 105–109. [CrossRef] [PubMed]
33. Fleming, R.V.; Kantarjian, H.M.; Husni, R.; Rolston, K.; Lim, J.; Raad, I.; Pierce, S.; Cortes, J.; Estey, E. Comparison of amphotericin B lipid complex (ABLC) vs. ambisome in the treatment of suspected or documented fungal infections in patients with leukemia. *Leuk. Lymphoma* **2001**, *40*, 511–520. [CrossRef] [PubMed]
34. Wade, R.L.; Chaudhari, P.; Natoli, J.L.; Taylor, R.J.; Nathanson, B.H.; Horn, D.L. Nephrotoxicity and other adverse events among inpatients receiving liposomal amphotericin B or amphotericin B lipid complex. *Diagn. Microbiol. Infect. Dis.* **2013**, *76*, 361–367. [CrossRef] [PubMed]
35. Wingard, J.R.; White, M.H.; Anaissie, E.; Raffalli, J.; Goodman, J.; Arrieta, A. A randomized, double-blind comparative trial evaluating the safety of liposomal amphotericin B versus amphotericin B lipid complex in the empirical treatment of febrile neutropenia. L Amph/ABLC Collaborative Study Group. *Clin. Infect. Dis.* **2000**, *31*, 1155–1163. [CrossRef]
36. Safdar, A.; Ma, J.; Saliba, F.; Dupont, B.; Wingard, J.R.; Hachem, R.Y.; Mattiuzzi, G.N.; Chandrasekar, P.H.; Kontoyiannis, D.P.; Rolston, K.V.; et al. Drug-induced nephrotoxicity caused by amphotericin B lipid complex and liposomal amphotericin B: A review and meta-analysis. *Medicine* **2010**, *89*, 236–244. [CrossRef]
37. Stone, N.R.; Bicanic, T.; Salim, R.; Hope, W. Liposomal Amphotericin B (AmBisome((R))): A Review of the Pharmacokinetics, Pharmacodynamics, Clinical Experience and Future Directions. *Drugs* **2016**, *76*, 485–500. [CrossRef]
38. Shigemi, A.; Matsumoto, K.; Ikawa, K.; Yaji, K.; Shimodozono, Y.; Morikawa, N.; Takeda, Y.; Yamada, K. Safety analysis of liposomal amphotericin B in adult patients: Anaemia, thrombocytopenia, nephrotoxicity, hepatotoxicity and hypokalaemia. *Int. J. Antimicrob. Agents* **2011**, *38*, 417–420. [CrossRef]
39. Inselmann, G.; Inselmann, U.; Heidemann, H.T. Amphotericin B and liver function. *Eur. J. Int. Med.* **2002**, *13*, 288–292. [CrossRef]
40. Fischer, M.A.; Winkelmayer, W.C.; Rubin, R.H.; Avorn, J. The hepatotoxicity of antifungal medications in bone marrow transplant recipients. *Clin. Infect. Dis.* **2005**, *41*, 301–307. [CrossRef]
41. Patel, G.P.; Crank, C.W.; Leikin, J.B. An evaluation of hepatotoxicity and nephrotoxicity of liposomal amphotericin B (L-AMB). *J. Med. Toxicol.* **2011**, *7*, 12–15. [CrossRef] [PubMed]
42. Chamilos, G.; Luna, M.; Lewis, R.E.; Chemaly, R.; Raad, I.I.; Kontoyiannis, D.P. Effects of liposomal amphotericin B versus an amphotericin B lipid complex on liver histopathology in patients with hematologic malignancies and invasive fungal infections: A retrospective, nonrandomized autopsy study. *Clin. Ther.* **2007**, *29*, 1980–1986. [CrossRef] [PubMed]
43. Weiler, S.; Überlacher, E.; Schöfmann, J.; Stienecke, E.; Dunzendorfer, S.; Joannidis, M.; Bellmann, R. Pharmacokinetics of Amphotericin B Colloidal Dispersion in Critically Ill Patients with Cholestatic Liver Disease. *Antimicrob. Agents Chemother.* **2012**, *56*, 5414–5418. [CrossRef]
44. Tassel, D.; Madoff, M.A. Treatment of Candida sepsis and Cryptococcus meningitis with 5-fluorocytosine. A new antifungal agent. *JAMA* **1968**, *206*, 830–832. [CrossRef]
45. Waldorf, A.R.; Polak, A. Mechanisms of action of 5-fluorocytosine. *Antimicrob. Agents Chemother.* **1983**, *23*, 79–85. [CrossRef] [PubMed]
46. Vermes, A.; Guchelaar, H.J.; Dankert, J. Flucytosine: A review of its pharmacology, clinical indications, pharmacokinetics, toxicity and drug interactions. *J. Antimicrob. Chemother.* **2000**, *46*, 171–179. [CrossRef]
47. Maziarz, E.K.; Perfect, J.R. Cryptococcosis. *Infect. Dis. Clin. N. Am.* **2016**, *30*, 179–206. [CrossRef] [PubMed]
48. Ashley, E.S.D.; Lewis, R.; Lewis, J.S.; Martin, C.; Andes, D. Pharmacology of Systemic Antifungal Agents. *Clin. Infect. Dis.* **2006**, *43*, S28–S39. [CrossRef]

49. Pappas, P.G.; Kauffman, C.A.; Andes, D.R.; Clancy, C.J.; Marr, K.A.; Ostrosky-Zeichner, L.; Reboli, A.C.; Schuster, M.G.; Vazquez, J.A.; Walsh, T.J.; et al. Clinical Practice Guideline for the Management of Candidiasis: 2016 Update by the Infectious Diseases Society of America. *Clin. Infect. Dis.* **2016**, *62*, e1–e50. [CrossRef] [PubMed]
50. *Ancobon*; Valeant Pharmaceuticals: Bridgewater, NJ, USA, 2017.
51. Brouwer, A.E.; van Kan, H.J.; Johnson, E.; Rajanuwong, A.; Teparrukkul, P.; Wuthiekanun, V.; Chierakul, W.; Day, N.; Harrison, T.S. Oral versus intravenous flucytosine in patients with human immunodeficiency virus-associated cryptococcal meningitis. *Antimicrob. Agents Chemother.* **2007**, *51*, 1038–1042. [CrossRef]
52. Record, C.O.; Skinner, J.M.; Sleight, P.; Speller, D.C. Candida endocarditis treated with 5-fluorocytosine. *Br. Med. J.* **1971**, *1*, 262–264. [CrossRef] [PubMed]
53. Pasqualotto, A.C.; Howard, S.J.; Moore, C.B.; Denning, D.W. Flucytosine therapeutic monitoring: 15 years experience from the UK. *J. Antimicrob. Chemother.* **2007**, *59*, 791–793. [CrossRef] [PubMed]
54. Block, E.R. Effect of hepatic insufficiency on 5-fluorocytosine concentrations in serum. *Antimicrob. Agents Chemother.* **1973**, *3*, 141–142. [CrossRef] [PubMed]
55. Ashbee, H.R.; Barnes, R.A.; Johnson, E.M.; Richardson, M.D.; Gorton, R.; Hope, W.W. Therapeutic drug monitoring (TDM) of antifungal agents: Guidelines from the British Society for Medical Mycology. *J. Antimicrob. Chemother.* **2014**, *69*, 1162–1176. [CrossRef] [PubMed]
56. Folk, A.; Cotoraci, C.; Balta, C.; Suciu, M.; Herman, H.; Boldura, O.M.; Dinescu, S.; Paiusan, L.; Ardelean, A.; Hermenean, A. Evaluation of Hepatotoxicity with Treatment Doses of Flucytosine and Amphotericin B for Invasive Fungal Infections. *BioMed Res. Int.* **2016**, *2016*, 9. [CrossRef] [PubMed]
57. Fothergill, A.W. Miconazole: A historical perspective. *Expert Rev. Anti Infect. Ther* **2006**, *4*, 171–175. [CrossRef] [PubMed]
58. Patterson, T.F.; Thompson, G.R., III; Denning, D.W.; Fishman, J.A.; Hadley, S.; Herbrecht, R.; Kontoyiannis, D.P.; Marr, K.A.; Morrison, V.A.; Nguyen, M.H.; et al. Practice Guidelines for the Diagnosis and Management of Aspergillosis: 2016 Update by the Infectious Diseases Society of America. *Clin. Infect. Dis.* **2016**, *63*, e1–e60. [CrossRef]
59. Tucker, R.M.; Haq, Y.; Denning, D.W.; Stevens, D.A. Adverse events associated with itraconazole in 189 patients on chronic therapy. *J. Antimicrob. Chemother.* **1990**, *26*, 561–566. [CrossRef]
60. Natesan, S.K.; Chandrasekar, P.H. Isavuconazole for the treatment of invasive aspergillosis and mucormycosis: Current evidence, safety, efficacy, and clinical recommendations. *Infect. Drug Resist.* **2016**, *9*, 291–300. [CrossRef]
61. Wilson, D.T.; Dimondi, V.P.; Johnson, S.W.; Jones, T.M.; Drew, R.H. Role of isavuconazole in the treatment of invasive fungal infections. *Ther. Clin. Risk Manag.* **2016**, *12*, 1197–1206. [CrossRef]
62. Raschi, E.; Poluzzi, E.; Koci, A.; Caraceni, P.; Ponti, F.D. Assessing liver injury associated with antimycotics: Concise literature review and clues from data mining of the FAERS database. *World J. Hepatol.* **2014**, *6*, 601–612. [CrossRef] [PubMed]
63. Song, J.C.; Deresinski, S. Hepatotoxicity of antifungal agents. *Curr. Opin. Investig. Drugs* **2005**, *6*, 170–177. [PubMed]
64. Bruggemann, R.J.; Alffenaar, J.W.; Blijlevens, N.M.; Billaud, E.M.; Kosterink, J.G.; Verweij, P.E.; Burger, D.M. Clinical relevance of the pharmacokinetic interactions of azole antifungal drugs with other coadministered agents. *Clin. Infect. Dis.* **2009**, *48*, 1441–1458. [CrossRef] [PubMed]
65. *Diflucan*; Pfizer Inc.: New York, NY, USA, 2018.
66. Muñoz, P.P.; Moreno, S.S.; Berenguer, J.J.; de Quirós, J.; Bouza, E.E. Fluconazole-related hepatotoxicity in patients with acquired immunodeficiency syndrome. *Arch. Intern. Med.* **1991**, *151*, 1020–1021. [CrossRef] [PubMed]
67. Wells, C.; Lever, A.M. Dose-dependent fluconazole hepatotoxicity proven on biopsy and rechallenge. *J. Infect.* **1992**, *24*, 111–112. [CrossRef]
68. Hay, R.J. Risk/benefit ratio of modern antifungal therapy: Focus on hepatic reactions. *J. Am. Acad. Dermatol.* **1993**, *29*, S50–S54. [CrossRef]
69. Como, J.A.; Dismukes, W.E. Oral azole drugs as systemic antifungal therapy. *N. Engl. J. Med.* **1994**, *330*, 263–272. [PubMed]
70. Franklin, I.M.; Elias, E.; Hirsch, C. Fluconazole-induced jaundice. *Lancet* **1990**, *336*, 565. [CrossRef]

71. Trujillo, M.A.; Galgiani, J.N.; Sampliner, R.E. Evaluation of hepatic injury arising during fluconazole therapy. *Arch. Intern. Med.* **1994**, *154*, 102–104. [CrossRef]
72. Guillaume, M.P.; De Prez, C.; Cogan, E. Subacute mitochondrial liver disease in a patient with AIDS: Possible relationship to prolonged fluconazole administration. *Am. J. Gastroenterol.* **1996**, *91*, 165–168.
73. Anaissie, E.J.; Kontoyiannis, D.P.; Huls, C.; Vartivarian, S.E.; Karl, C.; Prince, R.A.; Bosso, J.; Bodey, G.P. Safety, plasma concentrations, and efficacy of high-dose fluconazole in invasive mold infections. *J. Infect. Dis.* **1995**, *172*, 599–602. [CrossRef] [PubMed]
74. Wang, J.L.; Chang, C.H.; Young-Xu, Y.; Chan, K.A. Systematic review and meta-analysis of the tolerability and hepatotoxicity of antifungals in empirical and definitive therapy for invasive fungal infection. *Antimicrob. Agents Chemother.* **2010**, *54*, 2409–2419. [CrossRef] [PubMed]
75. Garcia Rodriguez, L.A.; Duque, A.; Castellsague, J.; Perez-Gutthann, S.; Stricker, B.H. A cohort study on the risk of acute liver injury among users of ketoconazole and other antifungal drugs. *Br. J. Clin. Pharmacol.* **1999**, *48*, 847–852. [CrossRef] [PubMed]
76. Jacobson, M.A.; Hanks, D.K.; Ferrell, L.D. Fatal acute hepatic necrosis due to fluconazole. *Am. J. Med.* **1994**, *96*, 188–190. [CrossRef]
77. Chmel, H. Fatal acute hepatic necrosis due to fluconazole. *Am. J. Med.* **1995**, *99*, 224–225. [CrossRef]
78. Bronstein, J.A.; Gros, P.; Hernandez, E.; Larroque, P.; Molinie, C. Fatal acute hepatic necrosis due to dose-dependent fluconazole hepatotoxicity. *Clin. Infect. Dis.* **1997**, *25*, 1266–1267. [CrossRef]
79. Ruhnke, M.; Yeates, R.A.; Pfaff, G.; Sarnow, E.; Hartmann, A.; Trautmann, M. Single-dose pharmacokinetics of fluconazole in patients with liver cirrhosis. *J. Antimicrob. Chemother.* **1995**, *35*, 641–647. [CrossRef]
80. Gearhart, M.O. Worsening of Liver Function with Fluconazole and Review of Azole Antifungal Hepatotoxicity. *Ann. Pharmacother.* **1994**, *28*, 1177–1181. [CrossRef]
81. Lo Re, V., 3rd; Carbonari, D.M.; Lewis, J.D.; Forde, K.A.; Goldberg, D.S.; Reddy, K.R.; Haynes, K.; Roy, J.A.; Sha, D.; Marks, A.R.; et al. Oral Azole Antifungal Medications and Risk of Acute Liver Injury, Overall and by Chronic Liver Disease Status. *Am. J. Med.* **2016**, *129*, 283–291. [CrossRef]
82. Lestner, J.; Hope, W.W. Itraconazole: An update on pharmacology and clinical use for treatment of invasive and allergic fungal infections. *Expert Opin. Drug Metab. Toxicol.* **2013**, *9*, 911–926. [CrossRef]
83. *Sporanox*; Janssen Pharmaceuticals: Titusville, FL, USA, 2017.
84. Chapman, S.W.; Dismukes, W.E.; Proia, L.A.; Bradsher, R.W.; Pappas, P.G.; Threlkeld, M.G.; Kauffman, C.A. Clinical Practice Guidelines for the Management of Blastomycosis: 2008 Update by the Infectious Diseases Society of America. *Clin. Infect. Dis.* **2008**, *46*, 1801–1812. [CrossRef] [PubMed]
85. Galgiani, J.N.; Ampel, N.M.; Blair, J.E.; Catanzaro, A.; Geertsma, F.; Hoover, S.E.; Johnson, R.H.; Kusne, S.; Lisse, J.; MacDonald, J.D.; et al. 2016 Infectious Diseases Society of America (IDSA) Clinical Practice Guideline for the Treatment of Coccidioidomycosis. *Clin. Infect. Dis.* **2016**, *63*, e112–e146. [CrossRef] [PubMed]
86. Wheat, L.J.; Freifeld, A.G.; Kleiman, M.B.; Baddley, J.W.; McKinsey, D.S.; Loyd, J.E.; Kauffman, C.A. Clinical practice guidelines for the management of patients with histoplasmosis: 2007 Update by the Infectious Diseases Society of America. *Clin. Infect. Dis.* **2007**, *45*, 807–825. [CrossRef] [PubMed]
87. Girois, S.B.; Chapuis, F.; Decullier, E.; Revol, B.G. Adverse effects of antifungal therapies in invasive fungal infections: Review and meta-analysis. *Eur. J. Clin. Microbiol. Infect. Dis.* **2006**, *25*, 138–149. [CrossRef]
88. Levron, J.C.; Chwetzoff, E.; Perrichon, P.; Autic, A.; Berthelot, P.; Boboc, D. *Pharmacokinetics of Itraconazole in Cirrhotic Patients*; Clinical Research Report R 51211; Laboratoires Janssen: Val-de-Reuil, France, 1987.
89. Ullmann, A.J.; Aguado, J.M.; Arikan-Akdagli, S.; Denning, D.W.; Groll, A.H.; Lagrou, K.; Lass-Flörl, C.; Lewis, R.E.; Munoz, P.; Verweij, P.E.; et al. Diagnosis and management of Aspergillus diseases: Executive summary of the 2017 ESCMID-ECMM-ERS guideline. *Clin. Microbiol. Infect.* **2018**, *24*, e1–e38. [CrossRef]
90. Lestner, J.M.; Roberts, S.A.; Moore, C.B.; Howard, S.J.; Denning, D.W.; Hope, W.W. Toxicodynamics of itraconazole: Implications for therapeutic drug monitoring. *Clin. Infect. Dis.* **2009**, *49*, 928–930. [CrossRef]
91. Veringa, A.; Ter Avest, M.; Span, L.F.; van den Heuvel, E.R.; Touw, D.J.; Zijlstra, J.G.; Kosterink, J.G.; van der Werf, T.S.; Alffenaar, J.C. Voriconazole metabolism is influenced by severe inflammation: A prospective study. *J. Antimicrob. Chemother.* **2017**, *72*, 261–267. [CrossRef]
92. *Vfend*; Pfizer Inc.: New York, NY, USA, 2018.

93. European Medicines Agency. Summary of Product Characteristics: Vfend. Available online: https://www.ema.europa.eu/documents/product-information/vfend-epar-product-information_en.pdf (accessed on 22 November 2018).
94. Denning, D.W.; Ribaud, P.; Milpied, N.; Caillot, D.; Herbrecht, R.; Thiel, E.; Haas, A.; Ruhnke, M.; Lode, H. Efficacy and safety of voriconazole in the treatment of acute invasive aspergillosis. *Clin. Infect. Dis.* **2002**, *34*, 563–571. [CrossRef]
95. Zonios, D.; Yamazaki, H.; Murayama, N.; Natarajan, V.; Palmore, T.; Childs, R.; Skinner, J.; Bennett, J.E. Voriconazole metabolism, toxicity, and the effect of cytochrome P450 2C19 genotype. *J. Infect. Dis.* **2014**, *209*, 1941–1948. [CrossRef]
96. Amigues, I.; Cohen, N.; Chung, D.; Seo, S.; Plescia, C.; Jakubowski, A.; Barker, J.; Papanicolaou, G.A. Hepatic Safety of Voriconazole after Allogeneic Hematopoietic Stem Cell Transplantation. *Biol. Blood Marrow Transplant.* **2010**, *16*, 46–52. [CrossRef]
97. Saito, T.; Fujiuchi, S.; Tao, Y.; Sasaki, Y.; Ogawa, K.; Suzuki, K.; Tada, A.; Kuba, M.; Kato, T.; Kawabata, M.; et al. Efficacy and safety of voriconazole in the treatment of chronic pulmonary aspergillosis: Experience in Japan. *Infection* **2012**, *40*, 661–667. [CrossRef] [PubMed]
98. Luong, M.L.; Al-Dabbagh, M.; Groll, A.H.; Racil, Z.; Nannya, Y.; Mitsani, D.; Husain, S. Utility of voriconazole therapeutic drug monitoring: A meta-analysis. *J. Antimicrob. Chemother.* **2016**, *71*, 1786–1799. [CrossRef] [PubMed]
99. Tan, K.K.C.; Wood, N.; Weil, A. Multiple-dose pharmacokinetics of voriconazole in chronic hepatic impairment. In Proceedings of the 41st Interscience Conference on Antimicrobial Agents and Chemotherapy, Chicago, IL, USA, 16–19 December 2001.
100. Solis-Munoz, P.; Lopez, J.C.; Bernal, W.; Willars, C.; Verma, A.; Heneghan, M.A.; Wendon, J.; Auzinger, G. Voriconazole hepatotoxicity in severe liver dysfunction. *J. Infect.* **2013**, *66*, 80–86. [CrossRef] [PubMed]
101. Yamada, T.; Imai, S.; Koshizuka, Y.; Tazawa, Y.; Kagami, K.; Tomiyama, N.; Sugawara, R.; Yamagami, A.; Shimamura, T.; Iseki, K. Necessity for a Significant Maintenance Dosage Reduction of Voriconazole in Patients with Severe Liver Cirrhosis (Child-Pugh Class C). *Biol. Pharm. Bull.* **2018**, *41*, 1112–1118. [CrossRef] [PubMed]
102. Wang, T.; Yan, M.; Tang, D.; Xue, L.; Zhang, T.; Dong, Y.; Zhu, L.; Wang, X.; Dong, Y. Therapeutic drug monitoring and safety of voriconazole therapy in patients with Child-Pugh class B and C cirrhosis: A multicenter study. *Int. J. Infect. Dis.* **2018**, *72*, 49–54. [CrossRef]
103. Wang, T.; Yan, M.; Tang, D.; Xue, L.; Zhang, T.; Dong, Y.; Zhu, L.; Wang, X.; Dong, Y. A retrospective, multicenter study of voriconazole trough concentrations and safety in patients with Child-Pugh class C cirrhosis. *J. Clin. Pharm. Ther.* **2018**. [CrossRef]
104. Hashemizadeh, Z.; Badiee, P.; Malekhoseini, S.A.; Raeisi Shahraki, H.; Geramizadeh, B.; Montaseri, H. Observational Study of Associations between Voriconazole Therapeutic Drug Monitoring, Toxicity, and Outcome in Liver Transplant Patients. *Antimicrob. Agents Chemother.* **2017**, *61*. [CrossRef] [PubMed]
105. Pascual, A.; Calandra, T.; Bolay, S.; Buclin, T.; Bille, J.; Marchetti, O. Voriconazole therapeutic drug monitoring in patients with invasive mycoses improves efficacy and safety outcomes. *Clin. Infect. Dis.* **2008**, *46*, 201–211. [CrossRef] [PubMed]
106. Pasqualotto, A.C.; Xavier, M.O.; Andreolla, H.F.; Linden, R. Voriconazole therapeutic drug monitoring: Focus on safety. *Expert Opin. Drug Saf.* **2010**, *9*, 125–137. [CrossRef]
107. Liu, X.; Su, H.; Tong, J.; Chen, J.; Yang, H.; Xiao, L.; Hu, J.; Zhang, L. Significance of monitoring plasma concentration of voriconazole in a patient with liver failure: A case report. *Medicine* **2017**, *96*, e8039. [CrossRef]
108. Courtney, R.; Pai, S.; Laughlin, M.; Lim, J.; Batra, V. Pharmacokinetics, safety, and tolerability of oral posaconazole administered in single and multiple doses in healthy adults. *Antimicrob. Agents Chemother.* **2003**, *47*, 2788–2795. [CrossRef] [PubMed]
109. Sime, F.B.; Stuart, J.; Butler, J.; Starr, T.; Wallis, S.C.; Pandey, S.; Lipman, J.; Roberts, J.A. Pharmacokinetics of Intravenous Posaconazole in Critically Ill Patients. *Antimicrob. Agents Chemother.* **2018**, *62*. [CrossRef] [PubMed]
110. Strommen, A.; Hurst, A.L.; Curtis, D.; Abzug, M.J. Use of Intravenous Posaconazole in Hematopoietic Stem Cell Transplant Patients. *J. Pediatr. Hematol. Oncol.* **2018**, *40*, e203–e206. [CrossRef]

111. Wiederhold, N.P. Pharmacokinetics and safety of posaconazole delayed-release tablets for invasive fungal infections. *Clin. Pharmacol.* **2016**, *8*, 1–8. [CrossRef]
112. Jovic, Z.; Jankovic, S.M.; Ruzic Zecevic, D.; Milovanovic, D.; Stefanovic, S.; Folic, M.; Milovanovic, J.; Kostic, M. Clinical Pharmacokinetics of Second-Generation Triazoles for the Treatment of Invasive Aspergillosis and Candidiasis. *Eur. J. Drug Metab. Pharmacokinet.* **2018**. [CrossRef]
113. *Noxafil*; Merk & Co., Inc.: Whitehouse Station, NJ, USA, 2017.
114. European Medicines Agency. Summary of Product Characteristics: Noxafil. Available online: https://www.ema.europa.eu/documents/product-information/noxafil-epar-product-information_en.pdf (accessed on 22 November 2018).
115. Cornely, O.A.; Duarte, R.F.; Haider, S.; Chandrasekar, P.; Helfgott, D.; Jimenez, J.L.; Candoni, A.; Raad, I.; Laverdiere, M.; Langston, A.; et al. Phase 3 pharmacokinetics and safety study of a posaconazole tablet formulation in patients at risk for invasive fungal disease. *J. Antimicrob. Chemother.* **2016**, *71*, 718–726. [CrossRef]
116. Tverdek, F.P.; Heo, S.T.; Aitken, S.L.; Granwehr, B.; Kontoyiannis, D.P. Real-Life Assessment of the Safety and Effectiveness of the New Tablet and Intravenous Formulations of Posaconazole in the Prophylaxis of Invasive Fungal Infections via Analysis of 343 Courses. *Antimicrob. Agents Chemother.* **2017**, *61*. [CrossRef] [PubMed]
117. Boglione-Kerrien, C.; Picard, S.; Tron, C.; Nimubona, S.; Gangneux, J.P.; Lalanne, S.; Lemaitre, F.; Bellissant, E.; Verdier, M.C.; Petitcollin, A. Safety study and therapeutic drug monitoring of the oral tablet formulation of posaconazole in patients with haematological malignancies. *J. Cancer Res. Clin. Oncol.* **2018**, *144*, 127–134. [CrossRef] [PubMed]
118. Zhang, S.; He, Y.; Jiang, E.; Wei, J.; Yang, D.; Zhang, R.; Zhai, W.; Zhang, G.; Wang, Z.; Zhang, L.; et al. Efficacy and safety of posaconazole in hematopoietic stem cell transplantation patients with invasive fungal disease. *Future Microbiol.* **2017**, *12*, 1371–1379. [CrossRef] [PubMed]
119. Moton, A.; Krishna, G.; Ma, L.; O'Mara, E.; Prasad, P.; McLeod, J.; Preston, R.A. Pharmacokinetics of a single dose of the antifungal posaconazole as oral suspension in subjects with hepatic impairment. *Curr. Med. Res. Opin.* **2010**, *26*, 1–7. [CrossRef]
120. Lin, S.Y.; Lu, P.L.; Tsai, K.B.; Lin, C.Y.; Lin, W.R.; Chen, T.C.; Chang, Y.T.; Huang, C.H.; Chen, C.Y.; Lai, C.C.; et al. A mucormycosis case in a cirrhotic patient successfully treated with posaconazole and review of published literature. *Mycopathologia* **2012**, *174*, 499–504. [CrossRef] [PubMed]
121. National Cancer Institute. *Common Terminology Criteria for Adverse Events*; Version 4.0; NIH, U.S. Department of Health and Human Services: Washington, DC, USA, 2009.
122. Heinz, W.J.; Egerer, G.; Lellek, H.; Boehme, A.; Greiner, J. Posaconazole after previous antifungal therapy with voriconazole for therapy of invasive aspergillus disease, a retrospective analysis. *Mycoses* **2013**, *56*, 304–310. [CrossRef]
123. Foo, H.; Gottlieb, T. Lack of Cross-Hepatotoxicity between Voriconazole and Posaconazole. *Clin. Infect. Dis.* **2007**, *45*, 803–805. [CrossRef] [PubMed]
124. Martinez-Casanova, J.; Carballo, N.; Luque, S.; Sorli, L.; Grau, S. Posaconazole achieves prompt recovery of voriconazole-induced liver injury in a case of invasive aspergillosis. *Infect. Drug Resist.* **2018**, *11*, 317–321. [CrossRef] [PubMed]
125. Jang, S.H.; Colangelo, P.M.; Gobburu, J.V. Exposure-response of posaconazole used for prophylaxis against invasive fungal infections: Evaluating the need to adjust doses based on drug concentrations in plasma. *Clin. Pharmacol. Ther.* **2010**, *88*, 115–119. [CrossRef] [PubMed]
126. Catanzaro, A.; Cloud, G.A.; Stevens, D.A.; Levine, B.E.; Williams, P.L.; Johnson, R.H.; Rendon, A.; Mirels, L.F.; Lutz, J.E.; Holloway, M.; et al. Safety, tolerance, and efficacy of posaconazole therapy in patients with nonmeningeal disseminated or chronic pulmonary coccidioidomycosis. *Clin. Infect. Dis.* **2007**, *45*, 562–568. [CrossRef]
127. *Cresemba*; Astellas Pharma Inc.: Northbrook, IL, USA, 2018.
128. European Medicines Agency. Summary of Product Characteristics: Cresemba. Available online: https://www.ema.europa.eu/documents/product-information/cresemba-epar-product-information_en.pdf (accessed on 22 November 2018).

129. Marty, F.M.; Ostrosky-Zeichner, L.; Cornely, O.A.; Mullane, K.M.; Perfect, J.R.; Thompson, G.R., 3rd; Alangaden, G.J.; Brown, J.M.; Fredricks, D.N.; Heinz, W.J.; et al. Isavuconazole treatment for mucormycosis: A single-arm open-label trial and case-control analysis. *Lancet Infect. Dis.* **2016**, *16*, 828–837. [CrossRef]
130. Jenks, J.D.; Salzer, H.J.; Prattes, J.; Krause, R.; Buchheidt, D.; Hoenigl, M. Spotlight on isavuconazole in the treatment of invasive aspergillosis and mucormycosis: Design, development, and place in therapy. *Drug Des. Devel. Ther.* **2018**, *12*, 1033–1044. [CrossRef]
131. Maertens, J.A.; Raad, I.I.; Marr, K.A.; Patterson, T.F.; Kontoyiannis, D.P.; Cornely, O.A.; Bow, E.J.; Rahav, G.; Neofytos, D.; Aoun, M.; et al. Isavuconazole versus voriconazole for primary treatment of invasive mould disease caused by Aspergillus and other filamentous fungi (SECURE): A phase 3, randomised-controlled, non-inferiority trial. *Lancet* **2016**, *387*, 760–769. [CrossRef]
132. Schmitt-Hoffmann, A.; Roos, B.; Spickermann, J.; Heep, M.; Peterfai, E.; Edwards, D.J.; Stoeckel, K. Effect of mild and moderate liver disease on the pharmacokinetics of isavuconazole after intravenous and oral administration of a single dose of the prodrug BAL8557. *Antimicrob. Agents Chemother.* **2009**, *53*, 4885–4890. [CrossRef]
133. Desai, A.; Schmitt-Hoffmann, A.H.; Mujais, S.; Townsend, R. Population Pharmacokinetics of Isavuconazole in Subjects with Mild or Moderate Hepatic Impairment. *Antimicrob. Agents Chemother.* **2016**, *60*, 3025–3031. [CrossRef] [PubMed]
134. Zheng, X.; Xu, G.; Zhu, L.; Fang, L.; Zhang, Y.; Ding, H.; Tong, Y.; Sun, J.; Huang, P. Pharmacokinetic/ Pharmacodynamic Analysis of Isavuconazole against *Aspergillus* spp. and *Candida* spp. in Healthy Subjects and Patients With Hepatic or Renal Impairment by Monte Carlo Simulation. *J. Clin. Pharmacol.* **2018**, *58*, 1266–1273. [CrossRef] [PubMed]
135. Desai, A.V.; Kovanda, L.L.; Hope, W.W.; Andes, D.; Mouton, J.W.; Kowalski, D.L.; Townsend, R.W.; Mujais, S.; Bonate, P.L. Exposure-Response Relationships for Isavuconazole in Patients with Invasive Aspergillosis and Other Filamentous Fungi. *Antimicrob. Agents Chemother.* **2017**, *61*. [CrossRef] [PubMed]
136. Wiederhold, N.P.; Lewis, R.E. The echinocandin antifungals: An overview of the pharmacology, spectrum and clinical efficacy. *Expert Opin. Investig. Drugs* **2003**, *12*, 1313–1333. [CrossRef] [PubMed]
137. Wang, J.F.; Xue, Y.; Zhu, X.B.; Fan, H. Efficacy and safety of echinocandins versus triazoles for the prophylaxis and treatment of fungal infections: A meta-analysis of RCTs. *Eur. J. Clin. Microbiol. Infect. Dis.* **2015**, *34*, 651–659. [CrossRef]
138. Dekkers, B.G.J.; Veringa, A.; Marriott, D.J.E.; Boonstra, J.M.; van der Elst, K.C.M.; Doukas, F.F.; McLachlan, A.J.; Alffenaar, J.C. Invasive Candidiasis in the Elderly: Considerations for Drug Therapy. *Drugs Aging* **2018**, *35*, 781–789. [CrossRef]
139. Balani, S.K.; Xu, X.; Arison, B.H.; Silva, M.V.; Gries, A.; DeLuna, F.A.; Cui, D.; Kari, P.H.; Ly, T.; Hop, C.E.; et al. Metabolites of caspofungin acetate, a potent antifungal agent, in human plasma and urine. *Drug Metab. Dispos.* **2000**, *28*, 1274–1278.
140. European Medicines Agency. Summary of Product Characteristics: Cancidas. Available online: https://www.ema.europa.eu/documents/product-information/cancidas-epar-product-information_en.pdf (accessed on 22 November 2018).
141. *Cancidas*; Merk & Co., Inc.: Whitehouse Station, NJ, USA, 2018.
142. Mistry, G.C.; Migoya, E.; Deutsch, P.J.; Winchell, G.; Hesney, M.; Li, S.; Bi, S.; Dilzer, S.; Lasseter, K.C.; Stone, J.A. Single- and multiple-dose administration of caspofungin in patients with hepatic insufficiency: Implications for safety and dosing recommendations. *J. Clin. Pharmacol.* **2007**, *47*, 951–961. [CrossRef]
143. Spriet, I.; Meersseman, W.; Annaert, P.; de Hoon, J.; Willems, L. Pharmacokinetics of caspofungin in a critically ill patient with liver cirrhosis. *Eur. J. Clin. Pharmacol.* **2011**, *67*, 753–755. [CrossRef]
144. Martial, L.C.; Bruggemann, R.J.; Schouten, J.A.; van Leeuwen, H.J.; van Zanten, A.R.; de Lange, D.W.; Muilwijk, E.W.; Verweij, P.E.; Burger, D.M.; Aarnoutse, R.E.; et al. Dose Reduction of Caspofungin in Intensive Care Unit Patients with Child Pugh B Will Result in Suboptimal Exposure. *Clin. Pharmacokinet.* **2016**, *55*, 723–733. [CrossRef]
145. Gustot, T.; Ter Heine, R.; Brauns, E.; Cotton, F.; Jacobs, F.; Bruggemann, R.J. Caspofungin dosage adjustments are not required for patients with Child-Pugh B or C cirrhosis. *J. Antimicrob. Chemother.* **2018**, *73*, 2493–2496. [CrossRef]

146. Yang, Q.T.; Zhai, Y.J.; Chen, L.; Zhang, T.; Yan, Y.; Meng, T.; Liu, L.C.; Chen, L.M.; Wang, X.; Dong, Y.L. Whole-body physiology-based pharmacokinetics of caspofungin for general patients, intensive care unit patients and hepatic insufficiency patients. *Acta Pharmacol. Sin.* **2018**, *39*, 1533–1543. [CrossRef] [PubMed]
147. Kofla, G.; Ruhnke, M. Pharmacology and metabolism of anidulafungin, caspofungin and micafungin in the treatment of invasive candidosis: Review of the literature. *Eur. J. Med. Res.* **2011**, *16*, 159–166. [CrossRef] [PubMed]
148. European Medicines Agency. Summary of Product Characteristics: Mycamine. Available online: https://www.ema.europa.eu/documents/product-information/mycamine-epar-product-information_en.pdf (accessed on 15 October 2018).
149. *Mycamine*; Astellas Pharma Inc.: Northbrook, IL, USA, 2018.
150. Lee, C.H.; Lin, J.C.; Ho, C.L.; Sun, M.; Yen, W.T.; Lin, C. Efficacy and safety of micafungin versus extensive azoles in the prevention and treatment of invasive fungal infections for neutropenia patients with hematological malignancies: A meta-analysis of randomized controlled trials. *PLoS ONE* **2017**, *12*, e0180050. [CrossRef] [PubMed]
151. Yeoh, S.F.; Lee, T.J.; Chew, K.L.; Lin, S.; Yeo, D.; Setia, S. Echinocandins for management of invasive candidiasis in patients with liver disease and liver transplantation. *Infect. Drug Resist.* **2018**, *11*, 805–819. [CrossRef] [PubMed]
152. Wasmann, R.E.; Muilwijk, E.W.; Burger, D.M.; Verweij, P.E.; Knibbe, C.A.; Bruggemann, R.J. Clinical Pharmacokinetics and Pharmacodynamics of Micafungin. *Clin. Pharmacokinet.* **2018**, *57*, 267–286. [CrossRef] [PubMed]
153. Hebert, M.F.; Smith, H.E.; Marbury, T.C.; Swan, S.K.; Smith, W.B.; Townsend, R.W.; Buell, D.; Keirns, J.; Bekersky, I. Pharmacokinetics of micafungin in healthy volunteers, volunteers with moderate liver disease, and volunteers with renal dysfunction. *J. Clin. Pharmacol.* **2005**, *45*, 1145–1152. [CrossRef]
154. Undre, N.; Pretorius, B.; Stevenson, P. Pharmacokinetics of micafungin in subjects with severe hepatic dysfunction. *Eur. J. Drug Metab. Pharmacokinet.* **2015**, *40*, 285–293. [CrossRef]
155. Luque, S.; Campillo, N.; Alvarez-Lerma, F.; Ferrandez, O.; Horcajada, J.P.; Grau, S. Pharmacokinetics of micafungin in patients with pre-existing liver dysfunction: A safe option for treating invasive fungal infections. *Enferm. Infecc. Microbiol. Clin.* **2016**, *34*, 652–654. [CrossRef]
156. Damle, B.D.; Dowell, J.A.; Walsky, R.L.; Weber, G.L.; Stogniew, M.; Inskeep, P.B. In vitro and in vivo studies to characterize the clearance mechanism and potential cytochrome P450 interactions of anidulafungin. *Antimicrob. Agents Chemother.* **2009**, *53*, 1149–1156. [CrossRef]
157. *Eraxis*; Pfizer Inc.: New York, NY, USA, 2018.
158. European Medicines Agency. Summary of Product Characteristics: Ecalta. Available online: https://www.ema.europa.eu/documents/product-information/ecalta-epar-product-information_en.pdf (accessed on 22 November 2018).
159. Reboli, A.C.; Rotstein, C.; Pappas, P.G.; Chapman, S.W.; Kett, D.H.; Kumar, D.; Betts, R.; Wible, M.; Goldstein, B.P.; Schranz, J.; et al. Anidulafungin versus fluconazole for invasive candidiasis. *N. Eng. J. Med.* **2007**, *356*, 2472–2482. [CrossRef] [PubMed]
160. Aguado, J.M.; Varo, E.; Usetti, P.; Pozo, J.C.; Moreno, A.; Catalan, M.; Len, O.; Blanes, M.; Sole, A.; Munoz, P.; et al. Safety of anidulafungin in solid organ transplant recipients. *Liver Transplant.* **2012**, *18*, 680–685. [CrossRef]
161. Dowell, J.A.; Stogniew, M.; Krause, D.; Damle, B. Anidulafungin does not require dosage adjustment in subjects with varying degrees of hepatic or renal impairment. *J. Clin. Pharmacol.* **2007**, *47*, 461–470. [CrossRef] [PubMed]
162. Verma, A.; Auzinger, G.; Kantecki, M.; Campling, J.; Spurden, D.; Percival, F.; Heaton, N. Safety and Efficacy of Anidulafungin for Fungal Infection in Patients With Liver Dysfunction or Multiorgan Failure. *Open Forum Infect. Dis.* **2017**, *4*. [CrossRef]
163. Weiler, S.; Zoller, H.; Graziadei, I.; Vogel, W.; Bellmann-Weiler, R.; Joannidis, M.; Bellmann, R. Altered Pharmacokinetics of Voriconazole in a Patient with Liver Cirrhosis. *Antimicrob. Agents Chemother.* **2007**, *51*, 3459–3460. [CrossRef] [PubMed]

14

Dual RNA-Seq Analysis of *Trichophyton rubrum* and HaCat Keratinocyte Co-Culture Highlights Important Genes for Fungal-Host Interaction

Monise Fazolin Petrucelli [1], Kamila Peronni [2], Pablo Rodrigo Sanches [3], Tatiana Takahasi Komoto [1], Josie Budag Matsuda [1], Wilson Araújo da Silva Jr. [2,4], Rene Oliveira Beleboni [1], Nilce Maria Martinez-Rossi [3], Mozart Marins [1] and Ana Lúcia Fachin [1,*]

[1] Biotechnology Unit, University of Ribeirão Preto-UNAERP, São Paulo 2201, Brazil; mofazolin@gmail.com (M.F.P.); tattytk@hotmail.com (T.T.K.); josie@unidavi.edu.br (J.B.M.); rbeleboni@unaerp.br (R.O.B.); mmarins@gmb.bio.br (M.M.)

[2] Laboratory of Molecular Genetics and Bioinformatics, Regional Hemotherapy Center of Ribeirão Preto, Ribeirão Preto 2501, Brazil; kcperoni@gmail.com (K.P.); wilsonjr@usp.br (W.A.d.S.J.)

[3] Department of Genetics, Ribeirão Preto Medical School, University of São Paulo, Ribeirão Preto 14049-900, Brazil; psanches@gmail.com (P.R.S.); nmmrossi@usp.br (N.M.M.-R.)

[4] Center for Medical Genomics at the Clinics Hospital of Ribeirão Preto Medical School, University of São Paulo, Ribeirão Preto 14049-900, Brazil

* Correspondence: afachin@unaerp.br

Abstract: The dermatophyte *Trichophyton rubrum* is the major fungal pathogen of skin, hair, and nails that uses keratinized substrates as the primary nutrients during infection. Few strategies are available that permit a better understanding of the molecular mechanisms involved in the interaction of *T. rubrum* with the host because of the limitations of models mimicking this interaction. Dual RNA-seq is a powerful tool to unravel this complex interaction since it enables simultaneous evaluation of the transcriptome of two organisms. Using this technology in an in vitro model of co-culture, this study evaluated the transcriptional profile of genes involved in fungus-host interactions in 24 h. Our data demonstrated the induction of glyoxylate cycle genes, *ERG6* and *TERG_00916*, which encodes a carboxylic acid transporter that may improve the assimilation of nutrients and fungal survival in the host. Furthermore, genes encoding keratinolytic proteases were also induced. In human keratinocytes (HaCat) cells, the *SLC11A1*, *RNASE7*, and *CSF2* genes were induced and the products of these genes are known to have antimicrobial activity. In addition, the *FLG* and *KRT1* genes involved in the epithelial barrier integrity were inhibited. This analysis showed the modulation of important genes involved in *T. rubrum*–host interaction, which could represent potential antifungal targets for the treatment of dermatophytoses.

Keywords: dermatophytes; *ERG6*; epithelial barrier; glyoxylate cycle; fungal-host interaction

---

1. Introduction

Dermatophytoses are superficial infections of keratinized tissues caused by a group of filamentous fungi called dermatophytes [1]. Although these infections are restricted to the superficial layers of the epidermis, they can become invasive and can lead to severe diseases in immunocompromised [2] and diabetic patients [3]. Data from the World Health Organization estimate that approximately 25% of the world's population have skin infections caused by fungi.

Most human dermatophytoses are caused by anthropophilic dermatophytes. Among these species, *Trichophyton rubrum* is the main cause of dermatophytoses in the world [4,5]. It is estimated that *T. rubrum* is the etiological agent of 69.5% of all cases of dermatophytosis caused by species of the genus *Trichophyton*, followed by *Trichophyton interdigitale*, *Trichophyton verrucosum* and *Trichophyton tonsurans* [6].

Despite the importance of these infections in clinical practice, knowledge of the molecular mechanisms involved in the dermatophyte-host interaction is limited, possibly because of the technical difficulties of the models mimicking this interaction, as well as the lack of genetic tools that allow for a more in-depth study of these organisms [7]. However, this scenario has been changing with the sequencing of mixed transcriptomes, also called dual RNA-seq, an approach widely used for the study of the complex interaction that exists between the host and pathogen [8] including bacteria [9], viruses [10], and fungi [11,12].

With the advent of this technology and the published sequence of the *T. rubrum* genome, the present study evaluated the transcriptional profile of *T. rubrum* co-cultured with human keratinocytes (HaCat) for 24 h by dual RNA-seq to identify important genes involved in the host defense and fungal pathogenicity in order to increase our understanding of the molecular aspects of this interaction. After 24 h of co-culture, we observed the induction of specific genes of the glyoxylate cycle and of a carboxylic acid transporter in *T. rubrum*, which may contribute to metabolic flexibility in nutrient-limited host niches, as well as of the *ERG6* gene involved in plasma membrane permeability, which may favor the assimilation of nutrients and fungal survival in the host. In addition, we found that the modulation of the *LAP2* and *DPPV* genes involved in the production of keratinolytic proteases that are important for the virulence of this dermatophyte. In contrast, in keratinocytes, genes involved in the repair of the epithelial barrier, in the increase of cell migration and the *RNASE7*, *SLC11A1* and *CSF2* genes (whose gene products have potential antimicrobial activity) were induced. Furthermore, the inhibition of *FLG* and *KRT1* genes whose products are directly involved in the maintenance of skin barrier integrity was observed.

2. Materials and Methods

2.1. Strains, Media and Growth Conditions

The *T. rubrum* strain CBS 118892 (CBS-KNAW Fungal Biodiversity Center, Utrech, The Netherlands) sequenced by the Broad Institute (Cambridge, MA, USA) was cultured on Sabouraud dextrose agar (Oxoid, Hampshire, UK) for 15 days at 28 °C.

2.2. Keratinocytes, Media and Growth Conditions

The immortalized human keratinocytes cell line HaCat was purchased from Cell Lines Service GmbH (Eppelheim, Germany). The cells were cultured in an RPMI medium (Sigma Aldrich, St. Louis, MO, USA) supplemented with 10% fetal bovine serum at 37 °C in a humidified atmosphere containing 5% CO_2. Antibiotics (100 U/mL penicillin and 100 µg/mL streptomycin) were added to the medium to prevent bacterial contamination.

2.3. Co-Culture Assay and Conditions

For co-culture assay, a ratio of 2.5×10^5 cells/mL of keratinocytes to 1×10^7 conidia/mL of *T. rubrum* solution was used, and the co-culture was performed as described in [13]. The assays were carried out in three independent experiments performed in triplicate. Cultured keratinocytes and *T. rubrum* conidia were used as controls and were cultured similarly to the co-infection in RPMI Medium (Sigma Aldrich). Scanning electron microscopy was performed with a JEOL JEM 100CXII electron microscope at the Multiuser Electron Microscopy Laboratory of the Department of Cell and Molecular Biology (Ribeirão Preto Medical School, São Paulo, Brazil) to determine whether the penetration of fungal hyphae into keratinocytes occurred within 24 h of co-culture. The cell viability of HaCat keratinocytes prior to *T. rubrum* inoculation and after 24 h of co-culture was determined by measuring the release of the enzyme lactate dehydrogenase (LDH) (TOX7 kit from Sigma-Aldrich) in the RPMI Medium (Sigma Aldrich) according to the manufacturer's instructions and described in [14].

The absorbance was read in a microplate reader (Elx 800 UV Bio-Tek Instruments, Inc., Winooski, VT, USA) at 490 nm.

2.4. RNA Isolation and Integrity Analysis

After 24 h of incubation, fungi and human cells were recovered by scraping and centrifuging at 1730× *g* for 10 min. For the disruption of the fungal cell wall, the samples (co-culture and controls) were treated with lysis solution (20 mg/mL of lysing enzymes from *Trichoderma harzianum* purchased from Sigma-Aldrich; 0.7 M KCl and 1 M $MgSO_4$, pH 6.8) for 1 h at 28 °C under gentle shaking, followed by centrifugation at 1000× *g* for 10 min, as described in [13]. Total RNA was extracted using the Illustra RNAspin Mini RNA Isolation Kit (GE Healthcare, Chicago, IL, USA) according to the manufacturer's instructions. After extraction, the absence of proteins and phenol in the RNA was analyzed in a MidSci Nanophotometer (Midwest Scientific, St. Louis, MO, USA) and the RNA integrity was assessed by microfluidic electrophoresis in an Agilent 2100 Bioanalyzer (Agilent Technologies, Santa Clara, CA, USA). Only RNA with an RNA integrity number (RIN) >7.0 was used. These RNAs were quantified in a Quantus™ Fluorometer (Promega Corporation, Madison, WI, USA) to verify if they had the adequate concentration for library construction.

2.5. Library Construction and Sequencing

The cDNA libraries for RNA sequencing were constructed in triplicate for each condition (cultured keratinocytes and *T. rubrum* conidia as control and co-culture). The libraries were constructed using the TrueSeq® RNA Sample Preparation Kit v2 (Illumina, San Diego, CA, USA) according to manufacturer's instructions and the libraries were validated according to the Library quantitative PCR (qPCR) Quantification Guide (Illumina). A pool of 11 pM of each library was distributed on the flowcell lanes and cluster amplification was performed in a cBot (Illumina) according to the manufacturer's instructions.

Single read and paired-end sequencing were performed in a Genome Analyzer IIx and Hiseq 2000 (Illumina), respectively, according to the manufacturer's instructions. The RNA-seq data are deposited in the GEO (Gene Expression Omnibus) database [15] under the accession number GSE110073

2.6. Sequence Data Analysis

The reads generated for each library were filtered using the FastQC software (https://www.bioinformatics.babraham.ac.uk/index.html) for removal of Illumina adapters and poor-quality reads. Only those with a Phred score > 20 were considered high-quality reads.

The high-quality reads were aligned to the *T. rubrum* reference genome of the Broad Institute's Dermatophyte Comparative Database and to the *Homo sapiens* reference genome HG19 [16].

After alignment, the triplicate of each library was normalized according to each library size and the number of reads was calculated using the summarize Overlaps function in the Genomic Ranges Bioconductor package, obtaining the expression levels of the transcripts in the samples. For statistical evaluation of the gene expression data between the samples, the false discovery rate (FDR) procedure was applied using the DEseq package [17] implemented in the R/Bioconductor software. Genes exhibiting statistical significance <0.05 and a $\log_2$ fold change ratio ≥ 1 or ≤ -1 were defined as differentially expressed genes (DEGs). The functional categorization of *T. rubrum* and keratinocyte DEGs in co-culture was performed according to Gene Ontology [18] using the Blast2GO algorithm [19] for *T. rubrum* and the website http://www.geneontology.org/ for human keratinocyte DEGs. For functional enrichment, the BayGO algorithm [20] and Enrichr enrichment tool [21,22], were used for the *T. rubrum* and keratinocyte DEGs, respectively. A p-value < 0.05 indicated the over-represented categories.

2.7. *qPCR Validation*

A set of 14 genes, including the *T. rubrum* and keratinocyte genes, were selected for validation by qPCR. For the reaction, 1 µg of the total RNA used for sequencing was treated with DNAse 1 Amplification Grade® (Sigma Aldrich) to remove any genomic DNA contamination. The High-Capacity cDNA Reverse Transcription® Kit (Applied Biosystems, Foster city, CA, USA) was used for cDNA conversion according to the manufacturer's instructions. Quantitative Real Time (RT)-PCR experiments were performed in triplicate using the SYBR Taq Ready Mix Kit (Sigma Aldrich) in a Mx3300 qPCR System (Stratagene, San Diego, CA, USA). The cycling conditions were initial denaturation at 94 °C for 10 min, followed by 40 cycles at 94 °C for 2 min, at 60 °C for 60 s and at 72 °C for 1 min. A dissociation curve was constructed at the end of each PCR cycle to verify single product amplification. Gene expression levels were calculated using the $2^{-\Delta\Delta C_T}$ comparative method. GAPDH [23] and β-actin [24] were used as normalizer genes for keratinocytes and 18S [25] and β-tubulin [26] as normalizer genes for *T. rubrum*. The results are reported as the mean ± standard deviation of three experiments. Pearson's correlation test was used to evaluate the correlation between the qPCR and RNA-seq techniques. The primers used for qPCR validation are available in Table S4.

3. Results

3.1. *Electron Microscopy of T. rubrum and HaCat Co-Culture*

Figure 1B shows the penetration of a *T. rubrum* hypha into a HaCat cell after 24 h of co-culture. Thus, the period of co-culture was considered appropriate for the evaluation of the fungal-host interaction.

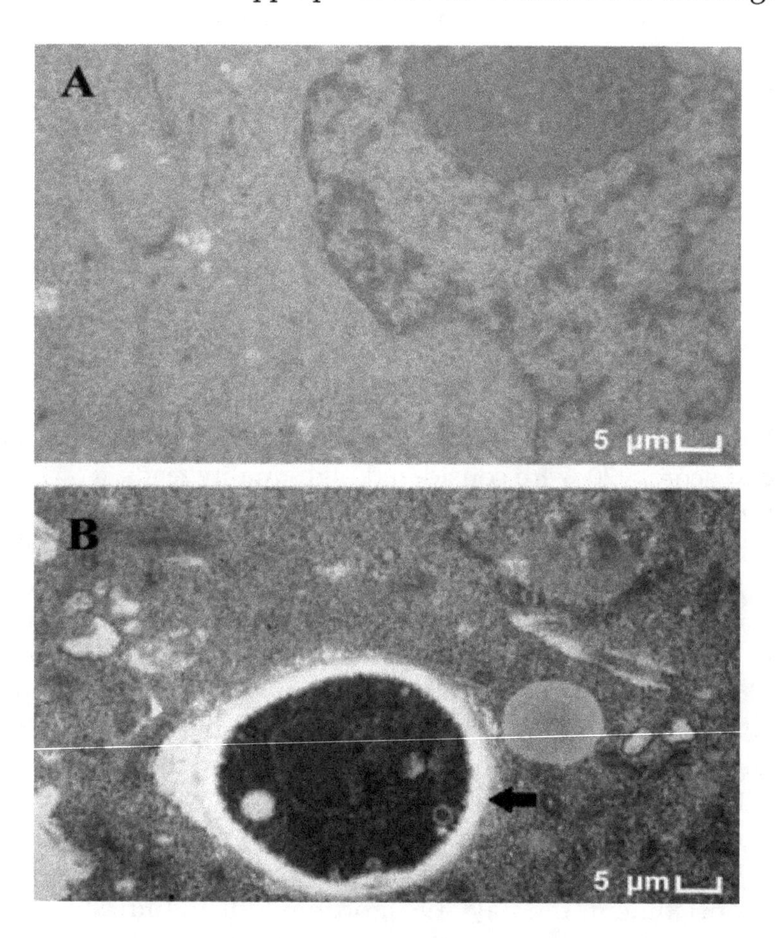

Figure 1. The transmission electron microscopy of the *Trichophyton rubrum*-HaCat co-culture after 24 h. (**A**) Human keratinocytes (HaCat) keratinocyte as the control (14kx); (**B**) Co-culture (14kx). The arrow indicates a fragment of *T. rubrum* hyphae inside the HaCat cells.

We performed the LDH assay with 24 h of co-culture to evaluate the keratinocyte cell viability. The percentage of LDH release was 18%. This LDH release may be due to the penetration of some fungal hyphae into keratinocyte cells (as observed in Figure 1B). LDH release was also evaluated at 0 h to assess cell viability prior to the addition of the fungus. The LDH release rate at 0 h was 1%. As a positive control, Triton X-100 (1%) was used in which 100% of the LDH release was obtained. Considering that we used 2.5×10^5 cells/mL prior to inoculation of the fungus and that the percentage of LDH was 18%, we can estimate that approximately 2×10^5 cells/mL are still viable in 24 h of co-cultivation.

3.2. Dual RNA-Seq Analysis of the Fungal-Host Interaction

Sequencing resulted in an average of 40, 34 and 47 million raw reads corresponding to the libraries of *T. rubrum* conidia, co-culture, and keratinocytes, respectively. Low-quality reads were then removed, and the resulting reads were aligned to the references genomes of *T. rubrum* and *Homo sapiens* HG19 (UCSC Genome Bioinformatics site, Santa Cruz, CA, USA). On average, 85% and 5% of the quality reads of the *T. rubrum* conidia and co-culture libraries, respectively, aligned to the *T. rubrum* reference genome (CBS 118892). These percentages were 84% and 85%, respectively, when the quality reads of the co-culture and keratinocyte cell line were aligned to the HG19 reference genome. The total number of filtered and aligned reads of each library is shown in Table S1.

3.3. Transcriptional Profile Analysis of Differentially Expressed Genes in the T. rubrum-Keratinocyte Co-Culture System

Tables 1 and 2 show the genes that are up-regulated and down-regulated in keratinocytes and *T. rubrum*, respectively. According to the distribution of the genes, those showing a p-value < 0.05 and $\log_2$ fold change ≥ 1 or ≥ -1 in each condition were considered differentially expressed (Figure S1). A total of 353 HaCat genes and 70 *T. rubrum* genes were differentially expressed during 24 h of co-culture (Tables S2 and S3).

Table 1. The major up- and down-regulated genes in HaCat cells after 24 h of co-culture.

ID	Gene Product Name	$\log_2$ Fold Change
SLC9A2	Sodium/hydrogen exchanger 2	5.01
ANGPTL4	Angiopoietin-related protein 4	4.71
DES	Desmin	4.53
C4orf47	UPF0602 protein C4orf47	4.51
KISS1R	KiSS-1 receptor	4.49
NSA2	Ribosome biogenesis protein NSA2 homolog	4.35
HIST1H3C	Histone cluster 1 H3 family member c	4.04
SEC11C	Signal peptidase complex catalytic subunit	3.87
KPNA7	Importin subunit alpha-8	3.83
CASP14	Caspase 14	3.74
SLC2A3	Facilitated glucose transporter member 3	3.73
ALDOC	Fructose-bisphosphate aldolase C	3.70
MT1B	Metallothionein-1B	3.62
SERPINE1	Plasminogen activator inhibitor 1	3.55
MAF	Transcription factor Maf	3.54
CA9	Carbonic anhydrase 9	3.36
TGM2	Transglutaminase 2	3.35
PADI1	Protein-arginine deiminase type-1	3.29
STC1	Stanniocalcin 1	3.14
BNIP3	BCL2 interacting protein 3	3.08
LSS	Lanosterol synthase	3.06
MT1H	Metallothionein 1H	3.05
MT1X	Metallothionein 1X	2.97
PLA2G2F	Group IIF secretory phospholipase A2	2.96
CALB1	Calbindin 1	2.93
POTEM	Putative POTE ankyrin domain family member M	−5.31

Table 1. *Cont.*

ID	Gene Product Name	Log_2 Fold Change
SNORA51	Small nucleolar RNA. H/ACA box	−4.90
ANP32A-IT1	ANP32A intronic transcript 1	−4.64
UCKL1	Uridine-cytidine kinase 1 like 1	−4.50
FNDC3B	Fibronectin type III domain containing	−4.37
KRT1	Keratin 1	−4.02
MMP12	Matrix metallopeptidase 12	−3.22
NSD1	Nuclear receptor binding SET domain	−3.06
CYCSP52	Cytochrome c. somatic pseudogene	−3.02
EME2	Essential meiotic structure-specific endonuclease subunit 2	−3.00
COL12A1	Collagen type XII alpha 1 chain	−2.88
SNORD45A	Small nucleolar RNA. C/D box	−2.82
FBXL19-AS1	FBXL19 antisense RNA 1 (head to head)	−2.80
TRIM26	Tripartite motif containing 26	−2.76
IARS	Isoleucyl-tRNA synthetase	−2.76
KIF14	Kinesin family member 14	−2.74
MEGF8	Multiple EGF like domains 8	−2.67
HNRNPL	Heterogeneous nuclear ribonucleoprotein	−2.66

Table 2. The major up- and down-regulated genes in *T. rubrum* after 24 h of co-culture.

ID	Gene Product Name	Log_2 Fold Change
TERG_12606	Dipeptidyl peptidase V (DPPV)	2.16
TERG_01280	Hypothetical protein	2.06
TERG_03102	Sterol 24-C-methyltransferase- ERG6	2.05
TERG_08104	Potassium/sodium efflux P-type ATPase	1.98
TERG_01281	Malate synthase	1.72
TERG_04399	Phthalate transporter	1.62
TERG_00215	MFS peptide transporter	1.47
TERG_00348	Galactose-proton symporter	1.47
TERG_02811	Hypothetical protein	1.42
TERG_12645	Hypothetical protein	1.40
TERG_07017	Oxidoreductase	1.35
TERG_08333	1-pyrroline-5-carboxylate dehydrogenase	1.34
TERG_02671	Hypothetical protein	1.34
TERG_02023	Extracellular matrix protein	1.32
TERG_08405	Leucine aminopeptidase 2	1.30
TERG_00916	Carboxylic acid transporter	1.29
TERG_11638	Isocitrate lyase	1.28
TERG_04952	ABC transporter	1.26
TERG_01406	Hypothetical protein	−2.91
TERG_07726	Hypothetical protein	−2.25
TERG_03174	MFS siderochrome iron transporter	−1.99
TERG_06355	Hypothetical protein	−1.90
TERG_07035	Hypothetical protein	−1.85
TERG_04156	Hypothetical protein	−1.77
TERG_05655	AN1 zinc finger protein	−1.73
TERG_01622	Hypothetical protein	−1.63
TERG_07477	Hypothetical protein	−1.57
TERG_06186	Protein disulfide-isomerase domain-containing protein	−1.57
TERG_03708	Hypothetical protein	−1.53
TERG_03855	Hypothetical protein	−1.50
TERG_00499	Hypothetical protein	−1.45
TERG_03175	Hypothetical protein	−1.45
TERG_04073	Glutathione synthetase	−1.41
TERG_12563	Hypothetical protein	−1.37
TERG_08139	NAD dependent epimerase/dehydratase	−1.34
TERG_06963	Hsp90-like protein	−1.33
TERG_01731	Hypothetical protein	−1.32
TERG_04006	Rho guanyl nucleotide exchange factor	−1.32

3.4. Functional Categorization of Differentially Expressed Genes

To evaluate the molecular and biological mechanisms involved in the fungal-host interaction, the DEGs were categorized according to biological processes and molecular functions. The most enriched categories considering a $p < 0.05$ are shown in Figure 2.

Most of the up-regulated *T. rubrum* genes (Figure 2A) belong to categories related to metabolic processes, membrane proteins, and substance transport, while the down-regulated genes are mainly involved in ATP binding. However, categories important to the fungus-host relationship, such as those including genes involved in the glyoxylate cycle and pathogenicity, should also be highlighted. Table 3 shows some functional categories that are important for the interaction of *T. rubrum* with HaCat keratinocytes. Within these categories, we selected some genes considered to play a fundamental role in the attack mechanisms and survival of the fungus when in contact with the host for validation and discussion: genes involved in protease secretion (*TERG_12606; TERG_08405*), metabolic flexibility for nutrient assimilation (*TERG_01281; TERG_11638; TERG_11639; TERG_00916*), and plasma membrane permeability (*TERG_03102*). On the other hand, up-regulated genes in keratinocytes (Figure 2B) are mainly found in the categories related to RNA binding, translation, and rRNA processing, while most of the down-regulated genes belong to the RNA binding category. Furthermore, Table 4 shows some functional categories that are important for the cell defense mechanisms of human keratinocytes during co-culture with *T. rubrum*, such as the genes involved in the innate immune response, epidermal cell differentiation, regulation of cell migration, and establishment of the skin barrier.

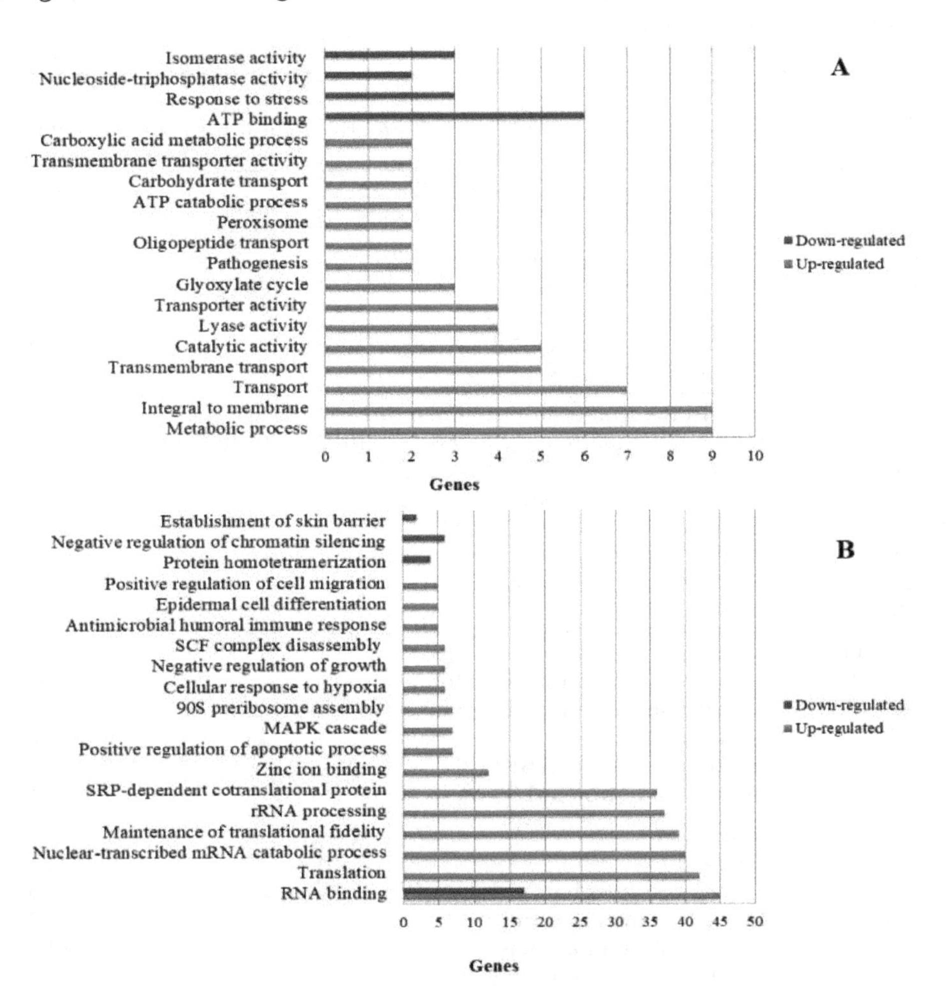

Figure 2. The gene Ontology-based functional categorization of differentially expressed genes. The main representative functional categories ($p < 0.05$) of genes differentially expressed in *T. rubrum* (**A**) and HaCat (**B**).

Table 3. Some functional categories and related genes important for the pathogenesis of *T. rubrum*.

ID	Gene Product Name	Log_2 Fold Change
Metabolic process		
TERG_03102	Sterol 24-C-methyltransferase	2.05
TERG_08104	Sodium transport ATPase	1.98
TERG_02811	Hypothetical protein	1.40
TERG_08333	Delta 1-pyrroline-5-carboxylate dehydrogenase	1.34
TERG_11638	Isocitrate lyase	1.26
TERG_01270	AMP-dependent ligase	1.13
TERG_07691	Nonspecific lipid-transfer protein	1.13
TERG_07222	Carbonic anhydrase	1.05
Transmembrane transport		
TERG_04399	Phthalate transporter (MFS transporter)	1.62
TERG_00348	Galactose-proton symporter (MFS transporter)	1.42
TERG_00916	Carboxylic acid transporter (MFS transporter)	1.28
TERG_04952	ABC transporter	1.25
TERG_04356	Amino acid permease	1.06
Pathogenesis		
TERG_12606	Dipeptidyl peptidase V	2.16
TERG_08405	Leucine Aminopeptidase 2	1.29
Glyoxylate cycle		
TERG_01281	Malate synthase	1.72
TERG_11638	Isocitrate lyase	1.26
TERG_11639	Isocitrate lyase	1.13

Table 4. Some functional categories and related genes important for human host defense.

ID	Gene Product Name	Log_2 Fold Change
Positive regulation of cell migration		
TCAF2	TRPM8 channel-associated factor 2	2.11
MMP9	Matrix metalloproteinase-9	2.06
LAMC2	Laminin subunit gamma-2	1.97
HBEGF	Proheparin-binding EGF-like growth factor	1.84
HAS2	Hyaluronan synthase 2	1.46
MAPK cascade involved in the innate immune response		
CSF2	Granulocyte-macrophage colony-stimulating factor	2.86
HBEGF	Proheparin-binding EGF-like growth factor	1.84
DUSP5	Dual specificity protein phosphatase 5	1.57
PSMB3	Proteasome subunit beta type-3	1.46
PPP5C	Serine/threonine-protein phosphatase 5	1.37
PSMB2	Proteasome subunit beta type-2	1.28
UBB	Polyubiquitin-B	1.24
Antimicrobial humoral immune response		
SERPINE1	Plasminogen activator inhibitor 1	3.55
SLC11A1	Natural resistance-associated macrophage protein 1	2.28
RNASE7	Ribonuclease 7	2.27
RPS19	40S ribosomal protein S19	1.66
RPL30	60S ribosomal protein L30	1.41
Epidermal cell differentiation		
CASP14	Caspase-14	3.74
ALDOC	Fructose-bisphosphate aldolase C	3.70
AKR1C1	Aldo-keto reductase family	2.80
LAMC2	Laminin subunit gamma-2	1.97
PGK1	Phosphoglycerate kinase 1	1.62
Establishment of the skin barrier		
KRT1	Keratin type II cytoskeletal 1	−4.02
FLG	Filaggrin	−1.86

3.5. Validation by qPCR

Pearson's correlation test was used to evaluate the correlation between dual RNA-seq and qPCR. For this purpose, 14 genes were chosen for validation, including 6 *T. rubrum* genes (*TERG_11638; TERG_01281; TERG_08405; TERG_12606; TERG_00916; TERG_03102*) and 8 HaCat genes (*HAS2; CSF2; SLC11A1; RNASE7; CASP14; MMP9; KRT1; FLG*). Figure 3 shows the comparison of the $\log_2$ fold change values obtained with the two techniques. The gene expression results obtained by RNA-seq showed a strong correlation ($r = 0.80$, $p < 0.001$) with the gene modulation values obtained by qPCR. This finding suggests that sequencing provided reliable results, demonstrating the reproducibility and accuracy of the technique.

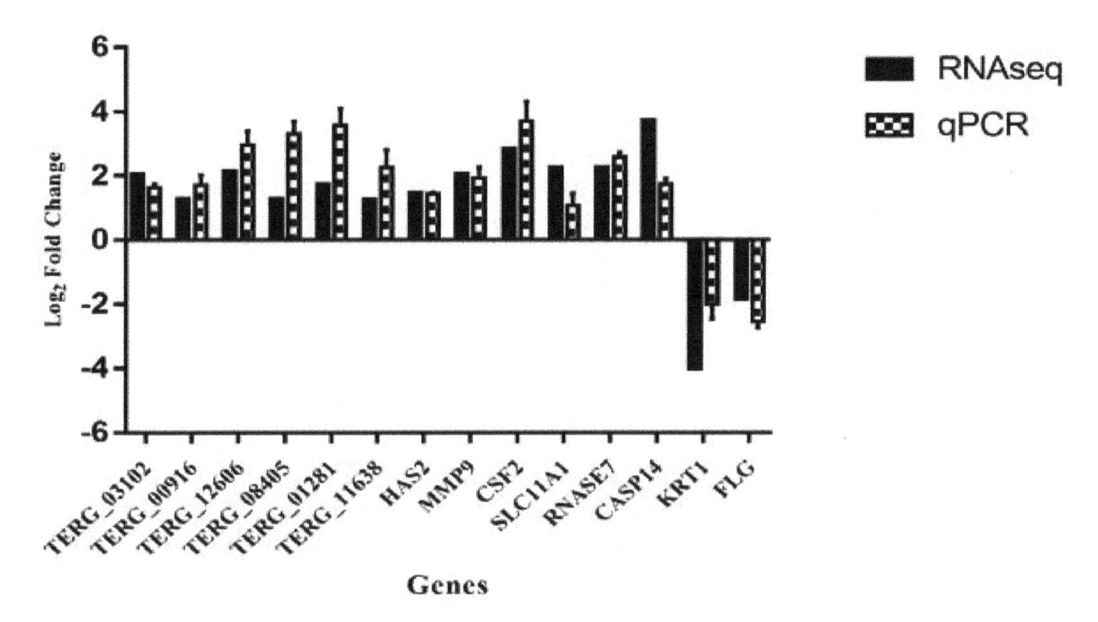

Figure 3. The comparison of gene modulation obtained by RNA-seq and quantitative PCR (qPCR). The error bars represent the standard error of three independent replicates. Pearson's test indicated a strong correlation between the two techniques ($r = 0.80$; $p < 0.01$).

4. Discussion

Through the analysis of mixed transcriptomes, this is the first study to sequence by dual RNA-seq the dermatophyte *T. rubrum* with a HaCat cells in an in vitro model of co-culture for 24 h.

Based on the sequencing data generated, only about 5% of the quality reads of the co-culture could be aligned to the *T. rubrum* reference genome (CBS 118892), indicating a predominance of human reads in this library. Indeed, a major challenge encountered in the sequencing of mixed transcriptomes is the difference in the amount of RNA between different cell types. Whereas a human cell contains about 20–25 pg RNA, a fungal cell contains 0.5–1 pg [8,27], a fact that may explain the smaller number of reads generated for *T. rubrum* compared to human keratinocytes. This obstacle was also observed in dual RNA-seq analysis of a *Magnaporthe oryzae* and *Oryza sativa* co-culture, in which the percentage of alignment of fungal reads to the *M. oryzae* reference genome ranged from 0.1–0.2% [28]. As the latest example, in [12] also obtained a low percentage (~1%) of reads corresponding to the pathogen *Phytophthora cinnamomi* in dual RNAseq with *Eucalyptus nitens*. In that study, the authors obtained 283 genes of *Phytophthora cinnamomi* in a genome comprising approximately 58.38 Mb (National Center for Biotechnology). Comparing these data with our study, we obtained about 5% of read alignment and 70 modulated *T. rubrum* genes considering a fold change ≥ 1 or ≤ -1 within a genome of 22.5 Mb.

However, we reached coverage of 90.7% of the 8.616 annotated genes in the *T. rubrum* genome considering the genes with at least one count read.

Seventy DEGs of *T. rubrum* were identified after 24 h of co-culture, which could be allocated to different categories according to biological function. Categories that were relevant for the understanding of the attack mechanisms of *T. rubrum* against keratinocytes included those containing *TERG_12606* and *TERG_08405* which encode important proteases for tissue invasion by the fungus, *TERG_03102* or the *ERG6* gene which is considered a promising target for the development of new antifungal agents [29], and *TERG_01281*, *TERG_11638* and *TERG_ 00916* which may be involved in the metabolic flexibility of *T. rubrum*, improving the adaptation and development in the host.

Regarding the functional categories containing the 353 HaCat DEGs, we highlight the following genes as important for the host defense mechanisms: *SLC11A1*, *RNASE7* and *CSF2* involved in innate immune response signaling; *MMP9* and *HAS2* involved in the regulation of epithelial cell migration; *KRT1* and *FLG* involved in maintaining skin barrier integrity, and *CASP14* involved in epidermal cells differentiation.

4.1. Genes Involved in Protease Secretion Are Important for the Pathogenicity of T. rubrum

During the course of infection, dermatophytes such as *T. rubrum* secrete endo- and exoproteases that degrade the keratin of the host tissue into oligopeptides and amino acids [30]. These compounds are used as a source of carbon, nitrogen, phosphorus, and sulfur for nutrition of the fungus [31].

The results of dual RNA-seq showed the induction of *TERG_12606* ($\log_2$ fold change: 2.16) and *TERG_08405* ($\log_2$ fold change: 1.29) (functional category: pathogenicity), which encode exoproteases (dipeptidyl peptidase V and leucine aminopeptidase 2, respectively). These findings corroborate the results of Reference [32] which evaluated gene expression by microarray in *T. rubrum* grown in a keratin-containing medium, and in [33] which evaluated the secretion of exoproteases, including dipeptidyl peptidase V, by *T. rubrum* in a keratin-containing medium. The secretion of endo- and exo-proteases by dermatophytes is one of the best-characterized virulence factors [32,33] and is of fundamental importance for invasion and dissemination of the fungus through the stratum corneum of the host [34].

4.2. The ERG6 *Gene Is a Promising Target for Developing a New Antifungal Agent Against T. rubrum*

In addition to the need of effective degradation of skin protein components for penetration of the fungus into tissue, the maintenance of fungal plasma membrane permeability and fluidity is essential for the correct assimilation of nutrients and the consequent growth and survival of *T. rubrum* in the host. In the present study, we observed the induction of *TERG_03102* ($\log_2$ fold change: 2.05) (functional category: metabolic process), which corresponds to the *ERG6* gene. This gene encodes the enzyme 24-C-methyltransferase, which participates in ergosterol biosynthesis [35]. Ergosterol is known to be responsible for fungal plasma membrane fluidity and permeability and it is important for the adequate function of membrane-anchored proteins [36].

The ergosterol biosynthesis pathway, which is absent in mammals, is the target of antifungal agents such as terbinafine. However, new genes of this pathway should be explored as potential targets because of reports of resistance of *T. rubrum* to this commercial antifungal drug [37]. One example of a promising potential target of new antifungals is the *ERG6* gene whose expression was found to be modulated in this study. Altered expression of this gene results in plasma membrane changes, impairing the transport of nutrients into the fungal cell [38]. The importance of this gene as a new therapeutic strategy has also been reported in [29]. In a comparative genomics study, these authors identified this gene in important human fungal pathogens such as *Candida albicans* and *Aspergillus fumigatus*.

4.3. Glyoxylate Cycle Genes and a Carboxylic Acid Transporter May Be Associated with Mechanisms of Metabolic Flexibility in the T. rubrum-Host Relationship

Additionally, regarding the importance of nutrient assimilation by the fungus for its development during infection, the metabolic flexibility of some pathogenic fungi is worth noting. This flexibility enables the fungus to obtain nutrients through the assimilation of alternative carbon sources in nutrient-limited host niches [39,40]. Knowledge of the genes that are induced to favor this metabolic flexibility is still limited. Thus, these genes are interesting targets for the development of more selective antifungals since the induction of alternative metabolic pathways is an exclusive property of pathogenic fungi [41].

In the present study, genes involved in metabolic flexibility were modulated: *TERG_01281* ($\log_2$ fold change: 1.72), *TERG_11638* ($\log_2$ fold change: 1.26) and *TERG_11639* ($\log_2$ fold change: 1.13) (functional category: glyoxylate cycle), which encode malate synthase and isocitrate lyases, respectively, are enzymes that participate in the glyoxylate cycle. In other clinical fungi, the activation of this cycle permits cell survival in low-glucose environments through the synthesis of glucose from lipids and other carbon sources [41]. We suggest this strategy could favor the growth and persistence of *T. rubrum* in the host since the fungus infects tissues rich in keratin and lipids. Furthermore, this cycle provides pathogenicity and virulence to other pathogens such as *C. albicans*, since the alternative assimilation of nutrients in nutrient-limited host niches favors pathogen survival and adaptation to the host [39].

The role of the glyoxylate cycle in the pathogenicity of *T. rubrum* is still not well established considering that this fungus causes superficial infections. However, we also showed the induction of genes encoding isocitrate lyase and malate synthase during the co-culture of HaCat keratinocytes with *T. rubrum*. The same genes were repressed in the presence of antifungal compounds licochalcone and caffeic acid in the co-culture for 24 h [42]. We also highlight the induction of *TERG_00916* ($\log_2$ fold change: 1.28), which encodes a carboxylic acid transporter (functional category: transport), and suggest that the fungus can use this transporter to facilitate the assimilation of carboxylic acids as an alternative carbon source during infection. The expression of two short-chain carboxylic acid transporters has been demonstrated in *C. albicans* when glucose availability in the host is low. These findings indicate the importance of these transporters in the early stages of infection, contributing to the virulence of the pathogen [43].

4.4. The Modulation of Genes Involved in the Maintenance of the Skin Barrier, Cell Migration, and Differentiation May Be Associated with the Defense Strategies of Human Keratinocytes

The degradation of keratin present in the epidermis through the secretion of proteases such as those modulated in this study (*TERG_12606* and *TERG_08405*) causes marked changes in the function and structure of the epithelial barrier [44]. Repression of the *FLG* ($\log_2$ fold change: -1.86) and *KRT1* ($\log_2$ fold change: -4.02) genes that encode filaggrin and keratin 1, respectively, was observed during the 24 h of co-culture of *T. rubrum* with HaCat cells. We suggest the repression of the *FLG* and *KRT1* genes to be related to the loss of skin barrier integrity, favoring the installation and tissue invasion by the fungus since the proteins encoded by these genes act together during the transition of keratinocytes to corneocytes that will compose the epithelial barrier [45,46]. These results corroborate the findings reported in [47] which identified the reduced expression of filaggrin in cases of tinea corporis caused by *T. rubrum*, and in [48] which observed the loss of skin barrier integrity in *KRT1*-deficient mice.

In the case of damage to the skin barrier, creating a portal of entry for exogenous microorganisms, epithelial cells respond rapidly to close the wound by increasing cell proliferation. In addition, the remodeling of affected tissue occurs and the migration of epithelial and immunocompetent cells to the site of infection is facilitated [49,50].

Among the genes allocated to the functional category of epidermal cell differentiation, the most modulated gene was *CASP14* ($\log_2$ fold change: 3.74), which encodes caspase 14 (Table 4). This is the only caspase not involved in apoptotic pathways [51,52] and an increase in its expression is associated with the differentiation of keratinocytes into corneocytes [53,54], demonstrating a low accumulation of filaggrin fragments in the stratum corneum and increased epithelial water loss in caspase 14-deficient mice. Thus, the induction of *CASP14* expression might be related to the increased differentiation of keratinocytes into corneocytes in an attempt to strengthen the epithelial barrier. Another possibility is that the increased expression of the *CASP14* gene is involved in the repair of damage caused by the repression of the *FLG* and *KRT1* genes as a host defense response during infection with *T. rubrum*.

Regarding the functional category containing genes involved in the regulation of cell migration, the induction of the *MMP9* gene (log_2 fold change: 1.46), which encodes matrix metalloproteinase 9, should be highlighted (Table 4). In addition to the role of matrix metalloproteinases in the remodeling of damaged tissues through the degradation of extracellular matrix, studies have shown that matrix metalloproteinase 9 is necessary for the migration of inflammatory cells to the epidermis [55]. Considering the data available so far, the induction of this gene may indicate an important role in the regulation of the flow of immunocompetent cells through the epidermal compartment in infections caused by *T. rubrum*. Since this protein is produced in its inactive form [56], the present results do not permit to establish whether the matrix metalloproteinase 9 becomes active in keratinocytes during dermatophyte infections. Furthermore, the increased expression of this enzyme in its active form may be associated with an increase in inflammation and the occurrence of ulcers in some diseases such as ocular herpes [57] and leishmaniasis [58], in addition to facilitating the dissemination of the pathogen through tissues by excessive cleavage of collagen IV present in the basement membrane [59].

With respect to other genes involved in the regulation of cell migration, the induction of the *HAS2* gene was observed (log_2 fold change: 1.46), which encodes hyaluronan synthase 2, an enzyme that participates in the synthesis of hyaluronic acid. This acid is one of the main components of the extracellular matrix and plays an important role in the repair of damaged tissues, contributing to the activation of inflammatory cells and the stimulation of chemokines and cytokines through its interaction with Toll-like receptors [60]. Studies also indicate a potential antifungal effect of hyaluronic acid, which inhibits the growth of *C. albicans* in vitro [61].

4.5. The Induction of Genes Involved in the Immune Response of Human Keratinocytes that Encode Compounds with Antimicrobial Activity

Among the functional categories studied, the most important to be evaluated during the fungal-host interaction are those containing the set of genes involved in the human cellular defense. These genes participate not only in the signaling and recruitment of immune system cells, but also in the production of compounds by the host that have a potential antimicrobial effect. These include genes allocated to the MAPK cascade involved in the innate immune response and antimicrobial humoral immune response categories (Table 4).

As an innate cellular defense mechanism, keratinocytes produce peptides with antimicrobial activity, such as cathelicidins, defensins, and ribonucleases [62]. We observed the induction of the *RNASE7* gene (log_2 fold change: 2.27) that encodes ribonuclease 7. This ribonuclease is known for its marked antimicrobial activity against Gram-positive and -negative bacteria, *C. albicans* [63] and dermatophytes [64], suggesting its use as a new antifungal agent.

Compounds that can be used as new approaches to the treatment of fungal diseases are increasingly being explored because of the growing resistance of pathogenic fungi to conventional antifungal agents [65]. In addition to the *RNASE7* gene, we highlight the induction of the *CSF2* gene (log_2 fold change: 2.86), which encodes the cytokine granulocyte-macrophage colony-stimulating factor (GM-CSF). Studies indicate the clinical use of this cytokine as an immunological adjuvant for the treatment of fungal diseases. GM-CSF has already been used to treat neutropenic patients undergoing chemotherapy, HIV-infected patients, and bone marrow transplant recipients [66,67]. The effects of this cytokine have been evaluated in species of the genera *Candida* [68] and *Aspergillus* [69], administered alone or in combination with other commercial antifungals.

The induction of the *CSF2* and *RNASE7* genes during co-culture of human keratinocytes with *T. rubrum* may indicate an important cellular defense response of the host when in contact with this fungus since both genes encode compounds with antimicrobial activity. Furthermore, the production of these compounds favors the recruitment of immunocompetent cells to the affected sites that are important for the host's innate immune mechanisms [63,70,71]

Another gene that was found to be induced in this study and that is also known for its antimicrobial activity is *SLC11A1*. This gene encodes an integral membrane protein [72] that mediates the transport of divalent ions, activating macrophages and exerting other pleiotropic effects on the innate immune system [73]. The available data indicate that this protein protects the host against intracellular pathogens such as *Salmonella* by controlling iron homeostasis inside macrophages, limiting the access of the pathogen to this essential element inside the host, and by concomitantly promoting an increase in the production of antimicrobial effector molecules [72].

Although more elucidated in macrophages, the increased expression of this gene was also observed in keratinocytes of patients with severe burns, suggesting that this gene participates in the innate immune response in the presence of tissue injury [74]. Tissue damage also occurs in dermatophytoses as a result of the secretion of keratinolytic proteases by the fungus. We, therefore, suggest that the induction of this gene during co-culture of keratinocytes with *T. rubrum* may be associated with a defense mechanism of the host, since the *SLC11A1* gene can also exert some signaling effects on the immune system such as macrophage activation, the regulation of interleukin 1-β, and the induction of iNOS, major histocompatibility (MHC) class II molecules, and tumor necrosis factor α (TNFα), among others [75,76]. However, more in-depth studies are necessary to elucidate this possible mechanism of defense

In summary, within the complex interaction between the fungus and host, we highlight the importance of the modulation of expression of *T. rubrum* genes that contribute to the acquisition and assimilation of nutrients. In this respect, genes responsible for the secretion of keratinolytic proteases (*TERG_12606*; *TERG_08405*) and metabolic adaptation (*TERG_01281*; *TERG_11638*; *TERG_11639*; *TERG_00916*) were found to be induced, as well as the *ERG6* gene that is responsible for maintaining the integrity and permeability of the plasma membrane. In contrast, in the presence of keratinocytes, genes encoding proteins with antimicrobial activity (*RNASE7*; *SLC11A1*; *CSF2*) and genes involved in the maintenance of the skin barrier (*MMP9*; *HAS2*; *CASP14*) are induced, while two genes essential for the stability and integrity of the skin barrier (*FLG*; *KRT1*) are repressed (Figure 4). Considering the limited knowledge, the use of dual RNA-seq allowed for a better understanding of some of the molecular mechanisms involved in the *T. rubrum*-host relationship.

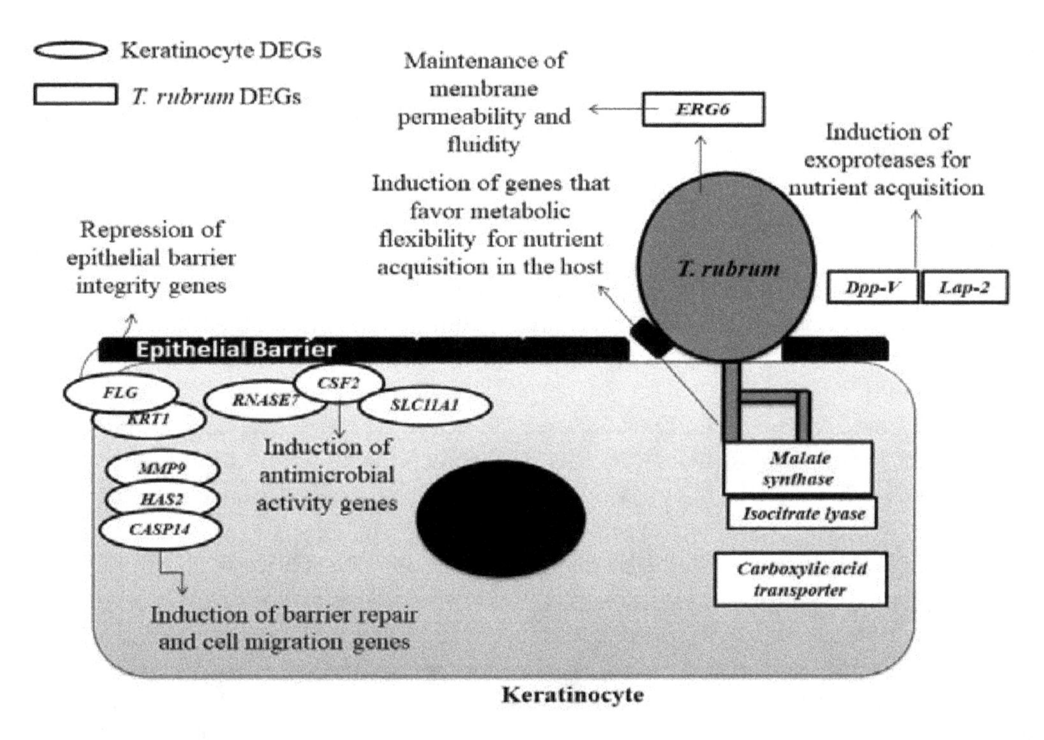

Figure 4. The schematic overview of the *T. rubrum*-keratinocyte interaction. The genes differentially expressed (DEGs) during host-pathogen interaction discussed in this paper are shown.

Supplementary Materials:
Figure S1: The distribution of differentially expressed genes after 24 h of co-culture. The red points indicate differentially expressed genes. Table S1: The general features of dual RNA-seq sequences against reference genomes. CBS I, CBS II, CBS III: *T. rubrum* libraries; CO I, CO II, CO III: co-culture libraries; H I, H II, H III: human keratinocyte libraries. The libraries were constructed in triplicate, with I, II and III corresponding to the sample number of each condition. PE: paired-end sequence; SR: single read sequence. Table S2: The complete list of genes differentially expressed in keratinocytes after 24 h of co-culture. Table S3: The complete list of genes differentially expressed in *T. rubrum* after 24 h of co-culture. Table S4: The primers used for qPCR analysis.

Author Contributions: M.F.P. performed the laboratory experiments and bioinformatics analysis and wrote the manuscript. K.P. and W.A.d.S.J. constructed the libraries and performed the sequencing. P.R.S. performed the bioinformatics analysis. T.T.K. performed the co-culture experiments. J.B.M. helped with the laboratory experiments. R.O.B. discussed the manuscript. N.M.M.-R. supervised the bioinformatics analysis. M.M. supervised the research and contributed reagents and materials. A.L.F. designed the project, supervised the research, contributed reagents/materials/analysis tools, and wrote the manuscript. All authors have read and approved the final version of the manuscript.

Acknowledgments: We thank the Multiuser Electron Microscopy Laboratory of the Department of Cellular and Molecular Biology, Ribeirão Preto Medical School, for the electron microscopy experiment, and Nilce M. Martinez-Rossi for kindly providing the CBS *T. rubrum* strain.

References

1. Bouchara, J.P.; Mignon, B.; Chaturvedi, V. Dermatophytes and dermatophytoses: a thematic overview of state of the art, and the directions for future research and developments. *Mycopathologia* **2017**, *182*, 1–4. [CrossRef] [PubMed]
2. Rodwell, G.E.; Bayles, C.L.; Towersey, L.; Aly, R. The prevalence of dermatophyte infection in patients infected with human immunodeficiency virus. *Int. J. Dermatol.* **2008**, *47*, 339–343. [CrossRef] [PubMed]
3. Romano, C.; Massai, L.; Asta, F.; Signorini, A.M. Prevalence of dermatophytic skin and nail infections in diabetic patients. *Mycoses* **2001**, *44*, 83–86. [CrossRef] [PubMed]
4. Aly, R. Ecology and epidemiology of dermatophyte infections. *J. Am. Acad. Dermatol.* **1994**, *31*, S21–S25. [CrossRef]
5. Havlickova, B.; Czaika, V.A.; Fredrich, M. Epidemiological trends in skin mycosis worldwide. *Mycosis* **2008**, *51*, 2–15. [CrossRef] [PubMed]
6. Hube, B.; Hay, R.; Brasch, J.; Veraldi, S.; Schaller, M. Dermatomycoses and inflammation: The adaptive balance between growth, damage, and survival. *J. Mycol. Med.* **2015**, *25*, e44–e58. [CrossRef] [PubMed]
7. Achterman, R.R.; White, T.C. A foot in the door for dermatophyte research. *PLoS Pathog.* **2012**, *8*, 6–9. [CrossRef] [PubMed]
8. Wolf, T.; Kämmer, P.; Brunke, S.; Linde, J. Two's company: Studying interspecies relationships with dual RNA-seq. *Curr. Opin. Microbiol.* **2018**, *42*, 7–12. [CrossRef] [PubMed]
9. Aprianto, R.; Slager, J.; Holsappel, S.; Veening, J.-W. Time-resolved dual RNA-seq reveals extensive rewiring of lung epithelial and pneumococcal transcriptomes during early infection. *Genome Biol.* **2016**, *17*, 198. [CrossRef] [PubMed]
10. Wesolowska-Andersen, A.; Everman, J.L.; Davidson, R.; Rios, C.; Herrin, R.; Eng, C.; Janssen, W.J.; Liu, A.H.; Oh, S.S.; Kumar, R.; et al. Dual RNA-seq reveals viral infections in asthmatic children without respiratory illness which are associated with changes in the airway transcriptome. *Genome Biol.* **2017**, *18*, 12. [CrossRef] [PubMed]
11. Tierney, L.; Linde, J.; Müller, S.; Brunke, S.; Molina, J.C.; Hube, B.; Schöck, U.; Guthke, R.; Kuchler, K. An interspecies regulatory network inferred from simultaneous RNA-seq of *Candida albicans* invading innate immune cells. *Front. Microbiol.* **2012**, *3*, 1–14. [CrossRef] [PubMed]
12. Meyer, F.E.; Shuey, L.S.; Naidoo, S.; Mamni, T.; Berger, D.K.; Myburg, A.A.; Van den Berg, N.; Naidoo, S. Dual RNA-sequencing of *Eucalyptus nitens* during *Phytophthora cinnamomi* challenge reveals pathogen and host factors influencing compatibility. *Front. Plant Sci.* **2016**, *7*, 1–15. [CrossRef] [PubMed]

13. Komoto, T.T.; Bitencourt, T.A.; Silva, G.; Beleboni, R.O.; Marins, M.; Fachin, A.L. Gene expression response of *Trichophyton rubrum* during coculture on keratinocytes exposed to antifungal agents. *Evid. Based Complement. Altern. Med.* **2015**, *2015*, 1–7. [CrossRef] [PubMed]
14. Santiago, K.; Bomfim, G.F.; Criado, P.R.; Almeida, S.R. Monocyte-derived dendritic cells from patients with dermatophytosis restrict the growth of *Trichophyton rubrum* and induce CD4-T cell activation. *PLoS ONE* **2014**, *9*, e110879. [CrossRef] [PubMed]
15. Edgar, R. Gene Expression Omnibus: NCBI gene expression and hybridization array data repository. *Nucleic Acids Res.* **2002**, *30*, 207–210. [CrossRef] [PubMed]
16. Langmead, B.; Salzberg, S.L. Fast gapped-read alignment with Bowtie 2. *Nat. Methods* **2012**, *9*, 357–359. [CrossRef] [PubMed]
17. Anders, S.; Huber, W. Differential expression analysis for sequence count data. *Genome Biol.* **2010**, *11*, 1–12. [CrossRef] [PubMed]
18. Blake, J.A.; Harris, M.A. The Gene Ontology (GO) Project: Structured vocabularies for molecular biology and their application to genome and expression analysis. In *Current Protocols in Bioinformatics*; John Wiley & Sons, Inc.: Hoboken, NJ, USA, 2008; pp. 1–9. ISBN 0471250953.
19. Gotz, S.; García-Gómez, J.M.; Terol, J.; Williams, T.D.; Nagaraj, S.H.; Nueda, M.J.; Robles, M.; Talón, M.; Dopazo, J.; Conesa, A. High-throughput functional annotation and data mining with the Blast2GO suite. *Nucleic Acids Res.* **2008**, *36*, 3420–3435. [CrossRef] [PubMed]
20. Vêncio, R.Z.N.; Koide, T.; Gomes, S.L.; de Pereira, C.A.B. BayGO: Bayesian analysis of ontology term enrichment in microarray data. *BMC Bioinf.* **2006**, *7*, 86. [CrossRef] [PubMed]
21. Chen, E.Y.; Tan, C.M.; Kou, Y.; Duan, Q.; Wang, Z.; Meirelles, G.; Clark, N.R.; Ma'ayan, A. Enrichr: Interactive and collaborative HTML5 gene list enrichment analysis tool. *BMC Bioinf.* **2013**, *14*, 128. [CrossRef] [PubMed]
22. Kuleshov, M.V.; Jones, M.R.; Rouillard, A.D.; Fernandez, N.F.; Duan, Q.; Wang, Z.; Koplev, S.; Jenkins, S.L.; Jagodnik, K.M.; Lachmann, A.; et al. Enrichr: A comprehensive gene set enrichment analysis web server 2016 update. *Nucleic Acids Res.* **2016**, *44*, W90–W97. [CrossRef] [PubMed]
23. Ma, R.; Zhang, D.; Hu, P.-C.; Li, Q.; Lin, C.-Y. HOXB7-S3 inhibits the proliferation and invasion of MCF-7 human breast cancer cells. *Mol. Med. Rep.* **2015**, 4901–4908. [CrossRef] [PubMed]
24. Dai, Z.; Ma, X.; Kang, H.; Gao, J.; Min, W.; Guan, H.; Diao, Y.; Lu, W.; Wang, X. Antitumor activity of the selective cyclooxygenase-2 inhibitor, celecoxib, on breast cancer in vitro and in vivo. *Cancer Cell Int.* **2012**, *19*, 1–8. [CrossRef] [PubMed]
25. Bitencourt, T.A.; Komoto, T.T.; Massaroto, B.G.; Miranda, C.E.S.; Beleboni, R.O.; Marins, M.; Fachin, A.L. Trans-chalcone and quercetin down-regulate fatty acid synthase gene expression and reduce ergosterol content in the human pathogenic dermatophyte *Trichophyton rubrum*. *BMC Complement. Altern. Med.* **2013**, *13*, 229. [CrossRef] [PubMed]
26. Jacob, T.R.; Peres, N.T.A.; Persinoti, G.F.; Silva, L.G.; Mazucato, M.; Rossi, A.; Martinez-Rossi, N.M. *Rpb2* is a reliable reference gene for quantitative gene expression analysis in the dermatophyte *Trichophyton rubrum*. *Med. Mycol.* **2012**, *50*, 368–377. [CrossRef] [PubMed]
27. Westermann, A.J.; Barquist, L.; Vogel, J. Resolving host–pathogen interactions by dual RNA-seq. *PLoS Pathog.* **2017**, *13*, 1–19. [CrossRef] [PubMed]
28. Kawahara, Y.; Oono, Y.; Kanamori, H.; Matsumoto, T.; Itoh, T.; Minami, E. Simultaneous RNA-seq analysis of a mixed transcriptome of rice and blast fungus interaction. *PLoS ONE* **2012**, *7*, e49423. [CrossRef] [PubMed]
29. Abadio, A.K.R.; Kioshima, E.S.; Teixeira, M.M.; Martins, N.F.; Maigret, B.; Felipe, M.S.S. Comparative genomics allowed the identification of drug targets against human fungal pathogens. *BMC Genom.* **2011**, *12*, 75. [CrossRef] [PubMed]
30. Baldo, A.; Monod, M.; Mathy, A.; Cambier, L.; Bagut, E.T.; Defaweux, V.; Symoens, F.; Antoine, N.; Mignon, B. Mechanisms of skin adherence and invasion by dermatophytes. *Mycoses* **2012**, *55*, 218–223. [CrossRef] [PubMed]
31. Peres, N.T.D.A.; Maranhão, F.C.A.; Rossi, A.; Martinez-Rossi, N.M. Dermatophytes: Host-pathogen interaction and antifungal resistance. *An. Bras. Dermatol.* **2010**, *85*, 657–667. [CrossRef] [PubMed]
32. Bitencourt, T.A.; Macedo, C.; Franco, M.E.; Assis, A.F.; Komoto, T.T.; Stehling, E.G.; Beleboni, R.O.; Malavazi, I.; Marins, M.; Fachin, A.L. Transcription profile of *Trichophyton rubrum* conidia grown on keratin reveals the induction of an adhesin-like protein gene with a tandem repeat pattern. *BMC Genom.* **2016**, *17*, 249. [CrossRef] [PubMed]

33. Monod, M.; Léchenne, B.; Jousson, O.; Grand, D.; Zaugg, C.; Stöcklin, R.; Grouzmann, E. Aminopeptidases and dipeptidyl-peptidases secreted by the dermatophyte *Trichophyton rubrum*. *Microbiology* **2005**, *151*, 145–155. [CrossRef] [PubMed]
34. Leng, W.; Liu, T.; Wang, J.; Li, R.; Jin, Q. Expression dynamics of secreted protease genes in *Trichophyton rubrum* induced by key host's proteinaceous components. *Med. Mycol.* **2009**, *47*, 759–765. [CrossRef] [PubMed]
35. Azam, S.S.; Abro, A.; Raza, S.; Saroosh, A. Structure and dynamics studies of sterol 24-C-methyltransferase with mechanism based inactivators for the disruption of ergosterol biosynthesis. *Mol. Biol. Rep.* **2014**, *41*, 4279–4293. [CrossRef] [PubMed]
36. Iwaki, T.; Iefuji, H.; Hiraga, Y.; Hosomi, A.; Morita, T.; Giga-Hama, Y.; Takegawa, K. Multiple functions of ergosterol in the fission yeast *Schizosaccharomyces pombe*. *Microbiology* **2008**, *154*, 830–841. [CrossRef] [PubMed]
37. Osborne, C.S.; Leitner, I.; Favre, B.; Neil, S.; Osborne, C.S.; Leitner, I.; Favre, B.; Ryder, N.S. Amino acid substitution in *Trichophyton rubrum* squalene epoxidase associated with resistance to terbinafine. **2005**, *49*, 2840–2844. [CrossRef] [PubMed]
38. Gaber, R.F.; Copple, D.M.; Kennedy, B.K.; Vidal, M.; Bard, M. The yeast gene *ERG6* is required for normal membrane function but is not essential for biosynthesis of the cell-cycle-sparking sterol. *Mol. Cell. Biol.* **1989**, *9*, 3447–3456. [CrossRef] [PubMed]
39. Mayer, F.L.; Wilson, D.; Hube, B. *Candida albicans* pathogenicity mechanisms. *Virulence* **2013**, *4*, 119–128. [CrossRef] [PubMed]
40. Cheah, H.L.; Lim, V.; Sandai, D. Inhibitors of the glyoxylate cycle enzyme ICL1 in *Candida albicans* for potential use as antifungal agents. *PLoS ONE* **2014**, *9*. [CrossRef] [PubMed]
41. Fleck, C.B.; Schöbel, F.; Brock, M. Nutrient acquisition by pathogenic fungi: Nutrient availability, pathway regulation, and differences in substrate utilization. *Int. J. Med. Microbiol.* **2011**, *301*, 400–407. [CrossRef] [PubMed]
42. Cantelli, B.A.M.; Bitencourt, T.A.; Komoto, T.T.; Beleboni, R.O.; Marins, M.; Fachin, A.L. Caffeic acid and licochalcone A interfere with the glyoxylate cycle of *Trichophyton rubrum*. *Biomed. Pharmacother.* **2017**. [CrossRef] [PubMed]
43. Vieira, N.; Casal, M.; Johansson, B.; MacCallum, D.M.; Brown, A.J.P.; Paiva, S. Functional specialization and differential regulation of short-chain carboxylic acid transporters in the pathogen *Candida albicans*. *Mol. Microbiol.* **2010**, *75*, 1337–1354. [CrossRef] [PubMed]
44. Lee, W.J.; Kim, J.Y.; Song, C.H.; Jung, H.D.; Lee, S.H.; Lee, S.J.; Kim, D.W. Disruption of barrier function in dermatophytosis and pityriasis versicolor. *J. Dermatol.* **2011**, *38*, 1049–1053. [CrossRef] [PubMed]
45. Brown, S.J.; Irvine, A.D. Atopic eczema and the filaggrin story. *Semin. Cutan. Med. Surg.* **2008**, *27*, 128–137. [CrossRef] [PubMed]
46. McGrath, J.A. Filaggrin and the great epidermal barrier grief. *Australas. J. Dermatol.* **2008**, *49*, 67–74. [CrossRef] [PubMed]
47. Jensen, J.-M.; Pfeiffer, S.; Akaki, T.; Schröder, J.-M.; Kleine, M.; Neumann, C.; Proksch, E.; Brasch, J. Barrier function, epidermal differentiation, and human β-defensin 2 expression in Tinea Corporis. *J. Investig. Dermatol.* **2007**, *127*, 1720–1727. [CrossRef] [PubMed]
48. Roth, W.; Kumar, V.; Beer, H.-D.; Richter, M.; Wohlenberg, C.; Reuter, U.; Thiering, S.; Staratschek-Jox, A.; Hofmann, A.; Kreusch, F.; et al. Keratin 1 maintains skin integrity and participates in an inflammatory network in skin through interleukin-18. *J. Cell Sci.* **2012**, *125*, 5269–5279. [CrossRef] [PubMed]
49. Parks, W.; Wilson, C.; López-Boado, Y. Matrix metalloproteinases as modulators of inflammation and innate immunity. *Nat. Rev. Immunol.* **2004**, *4*, 617–629. [CrossRef] [PubMed]
50. Purwar, R.; Kraus, M.; Werfel, T.; Wittmann, M. Modulation of keratinocyte-derived MMP-9 by IL-13: A possible role for the pathogenesis of epidermal inflammation. *J. Investig. Dermatol.* **2008**, *128*, 59–66. [CrossRef] [PubMed]
51. Hvid, M.; Johansen, C.; Deleuran, B.; Kemp, K.; Deleuran, M.; Vestergaard, C. Regulation of caspase 14 expression in keratinocytes by inflammatory cytokines—A possible link between reduced skin barrier function and inflammation? *Exp. Dermatol.* **2011**, *20*, 633–636. [CrossRef] [PubMed]

52. Gkegkes, I.D.; Aroni, K.; Agrogiannis, G.; Patsouris, E.S.; Konstantinidou, A.E. Expression of caspase-14 and keratin-19 in the human epidermis and appendages during fetal skin development. *Arch. Dermatol. Res.* **2013**, *305*, 379–387. [CrossRef] [PubMed]
53. Lippens, S.; Kockx, M.; Knaapen, M.; Mortier, L.; Polakowska, R.; Verheyen, A.; Garmyn, M.; Zwijsen, A.; Formstecher, P.; Huylebroeck, D.; et al. Epidermal differentiation does not involve the pro-apoptotic executioner caspases, but is associated with caspase-14 induction and processing. *Cell Death Differ.* **2000**, *7*, 1218–1224. [CrossRef] [PubMed]
54. Denecker, G.; Hoste, E.; Gilbert, B.; Hochepied, T.; Ovaere, P.; Lippens, S.; Van den Broecke, C.; Van Damme, P.; D'Herde, K.; Hachem, J.-P.; et al. Caspase-14 protects against epidermal UVB photodamage and water loss. *Nat. Cell Biol.* **2007**, *9*, 666–674. [CrossRef] [PubMed]
55. Ratzinger, G.; Stoitzner, P.; Ebner, S.; Lutz, M.B.; Layton, G.T.; Rainer, C.; Senior, R.M.; Shipley, J.M.; Fritsch, P.; Schuler, G.; et al. Matrix metalloproteinases 9 and 2 are necessary for the migration of langerhans cells and dermal dendritic cells from human and murine skin. *J. Immunol.* **2002**, *168*, 4361–4371. [CrossRef] [PubMed]
56. Visse, R.; Nagase, H. Matrix metalloproteinases and tissue inhibitors of metalloproteinases: Structure, function, and biochemistry. *Circ. Res.* **2003**, *92*, 827–839. [CrossRef] [PubMed]
57. Lee, S.; Zheng, M.; Kim, B.; Rouse, B.T. Role of matrix metalloproteinase-9 in angiogenesis caused by ocular infection with herpes simplex virus. *J. Clin. Investig.* **2002**, *110*, 1105–1111. [CrossRef] [PubMed]
58. Campos, T.M.; Passos, S.T.; Novais, F.O.; Beiting, D.P.; Costa, R.S.; Queiroz, A.; Mosser, D.; Scott, P.; Carvalho, E.M.; Carvalho, L.P. Matrix metalloproteinase 9 production by monocytes is enhanced by TNF and participates in the pathology of human Cutaneous Leishmaniasis. *PLoS Negl. Trop. Dis.* **2014**, *8*. [CrossRef] [PubMed]
59. Murphy, G.; Nagase, H. Progress in matrix metalloproteinase research. *Mol. Aspects Med.* **2009**, *29*, 290–308. [CrossRef] [PubMed]
60. Jiang, D.; Liang, J.; Noble, P.W. Hyaluronan in tissue injury and repair. *Annu. Rev. Cell Dev. Biol.* **2007**, *23*, 435–461. [CrossRef] [PubMed]
61. Sakai, A.; Akifusa, S.; Itano, N.; Kimata, K.; Kawamura, T.; Koseki, T.; Takehara, T.; Nishihara, T. Potential role of high molecular weight hyaluronan in the anti-*Candida* activity of human oral epithelial cells. *Med. Mycol.* **2007**, *45*, 73–79. [CrossRef] [PubMed]
62. Becknell, B.; Spencer, J.D. A Review of ribonuclease 7's structure, regulation, and contributions to host defense. *Int. J. Mol. Sci.* **2016**, *17*, 423. [CrossRef] [PubMed]
63. Harder, J.; Schröder, J.M. RNase 7, a novel innate immune defense antimicrobial protein of healthy human skin. *J. Biol. Chem.* **2002**, *277*, 46779–46784. [CrossRef] [PubMed]
64. Fritz, P.; Beck-Jendroschek, V.; Brasch, J. Inhibition of dermatophytes by the antimicrobial peptides human β-defensin-2, ribonuclease 7 and psoriasin. *Med. Mycol.* **2012**, *50*, 579–584. [CrossRef] [PubMed]
65. Mehra, T.; Köberle, M.; Braunsdorf, C.; Mailänder-Sanchez, D.; Borelli, C.; Schaller, M. Alternative approaches to antifungal therapies. *Exp. Dermatol.* **2012**, *21*, 778–782. [CrossRef] [PubMed]
66. Hubel, K.; Dale, D.C.; Liles, W.C. Therapeutic use of cytokines to modulate phagocyte function for the treatment of infectious diseases: Current status of granulocyte colony-stimulating factor, granulocyte-macrophage colony-stimulating factor, macrophage colony-stimulating factor, and interferon-gamma. *J. Infect. Dis.* **2002**, *185*, 1490–1501. [CrossRef] [PubMed]
67. Shi, Y.; Liu, C.H.; Roberts, A.I.; Das, J.; Xu, G.; Ren, G.; Zhang, Y.; Zhang, L.; Yuan, Z.R.; Tan, H.S.W.; et al. Granulocyte-macrophage colony-stimulating factor (GM-CSF) and T-cell responses: What we do and don't know. *Cell Res.* **2006**, *16*, 126–133. [CrossRef] [PubMed]
68. Liehl, E.; Hildebrandt, J.; Lam, C.; Mayer, P. Prediction of the role of granulocyte-macrophage colony-stimulating factor in animals and man from in vitro results. *Eur. J. Clin. Microbiol. Infect. Dis.* **1994**, *13*, 9–17. [CrossRef]
69. Bodey, G.P.; Anaissie, E.; Gutterman, J.; Vadhan-Raj, S. Role of granulocyte-macrophage colony-stimulating factor as adjuvant treatment in neutropenic patients with bacterial and fungal infection. *Eur. J. Clin. Microbiol. Infect. Dis.* **1994**, *13* (Suppl. 2), S18–S22. [CrossRef] [PubMed]
70. Hamilton, J.A. GM-CSF in inflammation and autoimmunity. *Trends Immunol.* **2002**, *23*, 403–408. [CrossRef]
71. Rademacher, F.; Simanski, M.; Harder, J. RNase 7 in cutaneous defense. *Int. J. Mol. Sci.* **2016**, *17*. [CrossRef] [PubMed]

72. Nairz, M.; Fritsche, G.; Crouch, M.L.V.; Barton, H.C.; Fang, F.C.; Weiss, G. Slc11a1 limits intracellular growth of *Salmonella enterica* sv. Typhimurium by promoting macrophage immune effector functions and impairing bacterial iron acquisition. *Cell. Microbiol.* **2009**, *11*, 1365–1381. [CrossRef] [PubMed]
73. Stober, C.B.; Brode, S.; White, J.K.; Popoff, J.-F.; Blackwell, J.M. Slc11a1, formerly Nramp1, is expressed in dendritic cells and influences major histocompatibility complex class II expression and antigen-presenting cell function. *Infect. Immun.* **2007**, *75*, 5059–5067. [CrossRef] [PubMed]
74. Noronha, S.A.; Noronha, S.M.; Lanziani, L.E.; Ferreira, L.M.; Gragnani, A. Innate and adaptive immunity gene expression of human keratinocytes cultured of severe burn injury. *Acta Cir. Bras.* **2014**, *29* (Suppl. 3), 60–67. [CrossRef] [PubMed]
75. Blackwell, J.M.; Searle, S.; Goswami, T.; Miller, E.N. Understanding the multiple functions of Nrampl. *Microbes Infect.* **2000**, *2*, 317–321. [CrossRef]
76. Blackwell, J.M.; Goswami, T.; Evans, C.A.W.; Sibthorpe, D.; Papo, N.; White, J.K.; Searle, S.; Miller, E.N.; Peacock, C.S.; Mohammed, H.; et al. SLC11A1 (formerly NRAMP1) and disease resistance. *Cell. Microbiol.* **2001**, *3*, 773–784. [CrossRef] [PubMed]

Permissions

All chapters in this book were first published by MDPI; hereby published with permission under the Creative Commons Attribution License or equivalent. Every chapter published in this book has been scrutinized by our experts. Their significance has been extensively debated. The topics covered herein carry significant findings which will fuel the growth of the discipline. They may even be implemented as practical applications or may be referred to as a beginning point for another development.

The contributors of this book come from diverse backgrounds, making this book a truly international effort. This book will bring forth new frontiers with its revolutionizing research information and detailed analysis of the nascent developments around the world.

We would like to thank all the contributing authors for lending their expertise to make the book truly unique. They have played a crucial role in the development of this book. Without their invaluable contributions this book wouldn't have been possible. They have made vital efforts to compile up to date information on the varied aspects of this subject to make this book a valuable addition to the collection of many professionals and students.

This book was conceptualized with the vision of imparting up-to-date information and advanced data in this field. To ensure the same, a matchless editorial board was set up. Every individual on the board went through rigorous rounds of assessment to prove their worth. After which they invested a large part of their time researching and compiling the most relevant data for our readers.

The editorial board has been involved in producing this book since its inception. They have spent rigorous hours researching and exploring the diverse topics which have resulted in the successful publishing of this book. They have passed on their knowledge of decades through this book. To expedite this challenging task, the publisher supported the team at every step. A small team of assistant editors was also appointed to further simplify the editing procedure and attain best results for the readers.

Apart from the editorial board, the designing team has also invested a significant amount of their time in understanding the subject and creating the most relevant covers. They scrutinized every image to scout for the most suitable representation of the subject and create an appropriate cover for the book.

The publishing team has been an ardent support to the editorial, designing and production team. Their endless efforts to recruit the best for this project, has resulted in the accomplishment of this book. They are a veteran in the field of academics and their pool of knowledge is as vast as their experience in printing. Their expertise and guidance has proved useful at every step. Their uncompromising quality standards have made this book an exceptional effort. Their encouragement from time to time has been an inspiration for everyone.

The publisher and the editorial board hope that this book will prove to be a valuable piece of knowledge for researchers, students, practitioners and scholars across the globe.

List of Contributors

Cara M. Dunaiski
Department of Health and Applied Sciences, Namibia University of Science and Technology, 13 Jackson Kaujeua Street, Windhoek 9000, Namibia

David W. Denning
National Aspergillosis Centre, Wythenshawe Hospital and the University of Manchester, Manchester M23 9LT, UK

Aline M. F. Matos, Bianca F. Barczewski, Lucas X. de Matos and Jordane B. V. de Oliveira
Department of Ophthalmology, University Hospital of the Federal University of Juiz de Fora, Juiz de Fora 36038-330, Brazil

Lucas M. Moreira and Rodrigo Almeida-Paes
Laboratório de Micologia, Instituto Nacional de Infectologia Evandro Chagas, Fundação Oswaldo Cruz, Rio de Janeiro 21040-900, Brazil

Maria Ines F. Pimentel
Laboratório de Pesquisa Clínica e Vigilância em Leishmanioses, Instituto Nacional de Infectologia Evandro Chagas, Fundação Oswaldo Cruz, Rio de Janeiro 21040-900, Brazil

Murilo G. Oliveira
Department of Pharmacy, Federal University of Juiz de Fora, Juiz de Fora 36036-900, Brazil

Tatiana C. A. Pinto
Instituto de Microbiologia Paulo de Goes, Universidade Federal do Rio de Janeiro, Rio de Janeiro 21941 -901, Brazil

Nelson Lima
CEB—Biological Engineering Centre, University of Minho, Campus de Gualtar, 4710-057 Braga, Portugal

Magnum de O. Matos
Imaging Department of Instituto Oncológico, Hospital Nove de Julho, Juiz de Fora 36010-510, Brazil

Louise G. de M. e Costa
Department of Pathology, Faculty of Medicine of Federal University of Juiz de Fora, Juiz de Fora 36038 -330, Brazil

Cledir Santos
Department of Chemical Science and Natural Resources, BIOREN-UFRO, Universidad de La Frontera, 4811-230 Temuco, Chile

Manoel Marques Evangelista Oliveira
Laboratório de Pesquisa Clínica em Dermatozoonoses em Animais Domésticos, Instituto Nacional de Infectologia Evandro Chagas, Fundação Oswaldo Cruz, Rio de Janeiro 21040-900, Brazil
Laboratório de Taxonomia, Bioquímica e Bioprospecção de Fungos, Instituto Oswaldo Cruz, Fundação Oswaldo Cruz, Rio de Janeiro 21040-900, Brazil

Rosana Puccia
Departamento de Microbiologia, Imunologia e Parasitologia da Escola Paulista de Medicina-UNIFESP, São Paulo, SP 04023-062, Brazil

Roberta Peres da Silva
Departamento de Microbiologia, Imunologia e Parasitologia da Escola Paulista de Medicina-UNIFESP, São Paulo, SP 04023-062, Brazil
School of Life Sciences, University of Nottingham, Nottingham NG7 2RD, UK

Sharon de Toledo Martins, Flavia C. G. dos Reis, Samuel Goldenberg and Lysangela R. Alves
Instituto Carlos Chagas, Fundação Oswaldo Cruz, Fiocruz-PR, Curitiba, PR 81310-020, Brazil

Juliana Rizzo
Instituto de Microbiologia Professor Paulo de Góes, Universidade Federal do Rio de Janeiro, Rio de Janeiro, RJ 21941-901, Brazil

Luna S. Joffe and Débora L. Oliveira
Centro de Desenvolvimento Tecnológico em Saúde (CDTS), Fundação Oswaldo Cruz, Rio de Janeiro, RJ 21040-900, Brazil

Marilene Vainstein and Livia Kmetzsch
Centro de Biotecnologia e Departamento de Biologia Molecular e Biotecnologia, Universidade Federal do Rio Grande do Sul, Porto Alegre, RS 91501-970, Brazil

Marcio L. Rodrigues
Instituto Carlos Chagas, Fundação Oswaldo Cruz, Fiocruz-PR, Curitiba, PR 81310-020, Brazil
Instituto de Microbiologia Professor Paulo de Góes, Universidade Federal do Rio de Janeiro, Rio de Janeiro, RJ 21941-901, Brazil

Flavio Queiroz-Telles
Department of Public Health, Federal University of Paraná, Curitiba 80060-000, Brazil

Renata Buccheri
Emilio Ribas Institute of Infectious Diseases, São Paulo 05411-000, Brazil

Gil Benard
Laboratory of Medical Mycology, Department of Dermatology, and Tropical Medicine Institute, University of São Paulo, Sao Paulo 05403-000, Brazil

Ulrike Binder, Verena Naschberger and Cornelia Lass-Flörl
Division of Hygiene and Medical Microbiology, Medical University Innsbruck, Schöpfstrasse 41, 6020 Innsbruck, Austria

Maria Isabel Navarro-Mendoza, Francisco E. Nicolas and Victoriano Garre
Departamento de Genética y Microbiología, Facultad de Biología, Universidad de Murcia, 30100 Murcia, Spain

Ingo Bauer
Division of Molecular Biology, Biocenter, Medical University of Innsbruck, Innrain 80-82, 6020 Innsbruck, Austria

Johannes D. Pallua
Institute of Pathology, Neuropathology and Molecular Pathology, Medical University of Innsbruck, Müllerstraße 44, 6020 Innsbruck, Austria

Toni Ciudad, Alberto Bellido, Encarnación Andaluz, Belén Hermosa and Germán Larriba
Departamento de Microbiología, Facultad de Ciencias, Universidad de Extremadura, 06071 Badajoz, Spain

Robin Patel
Department of Laboratory Medicine and Pathology, Division of Clinical Microbiology, Mayo Clinic, Rochester, MN 55905, USA
Department of Medicine, Mayo Clinic, Division of Infectious Diseases, Rochester, MN 55905, USA

Eta E. Ashu
Department of Biology, McMaster University, 1280 Main St. W, Hamilton, Ontario, ON L8S 4K1, Canada

Jianping Xu
Department of Biology, McMaster University, 1280 Main St. W, Hamilton, Ontario, ON L8S 4K1, Canada Public Research Laboratory, Hainan Medical University, Haikou, Hainan 571199, China

Nathan P. Wiederhold and Connie F. C. Gibas
Fungus Testing Laboratory, Department of Pathology and Laboratory Medicine, University of Texas Health Science Center at San Antonio, San Antonio, TX 78229, USA

Célia F. Rodrigues and Mariana Henriques
Laboratório de Investigação em Biofilmes Rosário Oliveira (LIBRO), Centre of Biological Engineering, University of Minho, 4710-057 Braga, Portugal

Savvas Papachristou, Elias Iosifidis and Emmanuel Roilides
Infectious Diseases Unit, 3rd Department of Pediatrics, Faculty of Medicine, Aristotle University School of Health Sciences, Konstantinoupoleos 49, 54642 Thessaloniki, Greece

Arnaldo Lopes Colombo
Special Mycology Laboratory, Division of Infectious Diseases, Federal University of São Paulo-UNIFESP, 04039-032 São Paulo, SP, Brazil

Ana Carolina Remondi Souza
Special Mycology Laboratory, Division of Infectious Diseases, Federal University of São Paulo-UNIFESP, 04039-032 São Paulo, SP, Brazil
Division of Infectious Diseases, Rhode Island Hospital, Alpert Medical School of Brown University, Providence, RI 02903, USA

Beth Burgwyn Fuchs, Elamparithi Jayamani and Eleftherios Mylonakis
Division of Infectious Diseases, Rhode Island Hospital, Alpert Medical School of Brown University, Providence, RI 02903, USA

Viviane de Souza Alves
Microorganisms Cell Biology Laboratory, Microbiology Department, Biological Sciences Institute, Federal University of Minas Gerais, Belo Horizonte 31270-901, MG, Brazil

Nikolaos Spernovasilis
Department of Internal Medicine and Infectious Diseases, University Hospital of Heraklion, P.O. Box 1352, 71110 Heraklion, Crete, Greece

Diamantis P. Kofteridis
Department of Internal Medicine and Infectious Diseases, University Hospital of Heraklion, P.O. Box 1352, 71110 Heraklion, Crete, Greece
School of Medicine, University of Crete, 71110 Heraklion, Crete, Greece

Monise Fazolin Petrucelli, Tatiana Takahasi Komoto, Josie Budag Matsuda, Rene Oliveira Beleboni, Mozart Marins and Ana Lúcia Fachin
Biotechnology Unit, University of Ribeirão Preto-UNAERP, São Paulo 2201, Brazil

Kamila Peronni
Laboratory of Molecular Genetics and Bioinformatics, Regional Hemotherapy Center of Ribeirão Preto, Ribeirão Preto 2501, Brazil

Pablo Rodrigo Sanches and Nilce Maria Martinez-Rossi
Department of Genetics, Ribeirão Preto Medical School, University of São Paulo, Ribeirão Preto 14049 -900, Brazil

Wilson Araújo da Silva Jr.
Laboratory of Molecular Genetics and Bioinformatics, Regional Hemotherapy Center of Ribeirão Preto, Ribeirão Preto 2501, Brazil
Center for Medical Genomics at the Clinics Hospital of Ribeirão Preto Medical School, University of São Paulo, Ribeirão Preto 14049-900, Brazil

Index

Printed in the USA
CPSIA information can be obtained
at www.ICGtesting.com
JSHW051401091023
49903JS00006B/227

9 781639 276738